T0192873

# Fundamentals of Fibre Reinforced Composite Materials

# Fundamentals of Fibre Reinforced Composite Materials

## Second Edition

A.R. Bunsell
S. Joannès
A. Thionnet

CRC Press
Taylor & Francis Group
Boca Raton  London  New York

CRC Press is an imprint of the
Taylor & Francis Group, an **informa** business

A CHAPMAN & HALL BOOK

[Second] edition published [2021]
by CRC Press
6000 Broken Sound Parkway NW, Suite 300, Boca Raton, FL 33487-2742

and by CRC Press
2 Park Square, Milton Park, Abingdon, Oxon, OX14 4RN

© 2021 Taylor & Francis Group, LLC

[First edition published by CRC Press 2005]

CRC Press is an imprint of Taylor & Francis Group, LLC

ISBN: 978-0-367-02373-7 (hbk)
ISBN: 978-0-429-39990-9 (ebk)

Typeset in Computer Modern font
by Cenveo Publisher Services

# Contents

# Preface to the first edition

The contents of this book on composite materials reflects courses that both authors have developed and presented over a period of more than 25 years to undergraduate and postgraduate students in France at the Ecole des Mines de Paris and the University of Paris Sud, and in Belgium at the University of Leuven as well as numerous other courses at universities on three continents. The plan, however, is considerably influenced by workshops and courses both authors have developed separately for the Conservatoire National des Arts et Métiers in Paris which has attracted hundreds of professional people looking to broaden their experience so as to work in this exciting field. The authors have experience both in industrial and in academic research. The approach therefore has been to present composites not just as an academic subject but one which is increasingly entering into the day-to-day experience of all of us and having a growing influence on modern industry. Many of the challenges that society will face in this 21$^{st}$ century will require the use of composite materials. The synergistic combination of two very different materials to form these materials offers the means of overcoming many of the limitations of traditional structural materials. It is hoped that this book will encourage interest in and the development of fibre reinforced composites.

Anthony R. BUNSELL & Jacques RENARD (2005)

# Preface

This new edition has been influenced by the first edition but the contents have been considerably updated. This is to reflect the expanding use of fibre reinforced composite materials and the increasing ability to more fully exploit their potential provided by the increasing power and speed of readily available computers to simulate and predict their behaviour. Markets for composites have grown impressively and an ever expanding range of industries is finding applications for them. Asia has increased both production and use of composites and reinforced its place as a major market alongside North America and Europe. Markets in other areas of the world are also developing, making composites an international phenomenon. New reinforcements, such as natural fibres, are making their mark as applications have broadened whilst advanced fibre reinforcements are continuing to enable remarkable composite structures to be realised. Man is reaching, if not for the stars, at least for the planets with the help of advanced composites. Manufacturing processes have progressed and thermoplastic matrix materials are catching up on thermosets so encouraging the use of composites in large scale manufacturing. Advanced composites are presenting solutions to some of the major issues now confronting mankind enabling environmental problems to be tackled by providing non-polluting hydrogen-based technologies as alternatives to oil-based energy production. The subject of composite materials is multidisciplinary and the various topics in the book have been presented in as accessible manner as possible with the individual chapters being able to be consulted as they meet the reader's needs. This edition is aimed both at engineers using composites and wishing to learn how to exploit them and also undergraduates and graduate students wishing to consider careers in this exciting and fast developing technology.

Anthony R. BUNSELL, Sébastien JOANNÈS & Alain THIONNET (2020)

# 1

# Introduction

Fibre reinforced composite materials are now firmly placed in the vanguard of advanced materials and are used in an ever widening number of applications from planes to fishing rods. They exploit the remarkable properties of fibres which can be natural or man-made. Natural composites can be seen to have evolved to be the basis of the living world as evolution has exploited the ability of a composite structure to best respond to environmental forces. Modern composites have largely been developed based on man-made fibres which have been produced since the 1930s and advanced composites based on fibres produced since the 1960s together with resin matrices which also began to be produced in the 1930s. However, the technical use of fibres in the form of cords has a surprisingly long history with archaeological evidence showing that it even predates the arrival of modern Homo-Sapian man in Europe. Recent discoveries in the Abri du Maras located in a valley near the Ardèche River, a tributary of the Rhône River in France have confirmed earlier suspicions that our ancient cousins, Neanderthal man, created cords for a number of applications. The cordage which was found dated from approximately 50 000 years ago, about 5000 years before modern man arrived there, and consisted of three bundles of fibres with S-twist plied together with a Z-twist to form a 3-ply cord (Hardy et al., 2020).

Natural fibres have found wide use over many centuries for sails, ropes other types of cords and of course textiles. Wood has been widely used for structures and is a very good example of a natural composite. Adobe building material consisting of baked mud reinforced with natural fibres has been used for millennia and is an example of a man-made composite structure. Nevertheless the advantages of fibres were largely overlooked in the development of structural materials in the modern world, until the second half of the twentieth century, in favour of metals, concrete and wood. Now, in the twenty first century, fibre reinforced composites have become ubiquitous due to their low density and remarkable physical properties. This is prompting impressive growth and innovation in composites which this book will explain.

The greatest use of high-performance composite materials, today, is in civil applications but initially it was military and in particular aerospace structures which used composites. The reason is obvious; weight saving. To go higher and further, it is necessary to have sufficient strength and rigidity for the aircraft, for example, but weight will limit performance. That is why there are baggage weight restrictions when boarding a plane and why some low-cost carriers have greater weight restrictions than other more expensive carriers. We can quantify the effect of weight for a given material by dividing its elastic modulus (or any other property) by its density to give what is known as the specific properties of the material. These specific properties values, constituting performance indices, can easily be compared but a graph is often more meaningful. A Young's modulus versus density plot (Ashby chart, named for Michael ASHBY of Cambridge University) (Ashby, 2016), is thus of particular interest to compare the main classes of materials to composites. Figure 1.1 shows that composite materials are very well positioned on the graph, like wood and wood products, also well placed, which are natural composite materials. An equivalent graph can be obtained with the ultimate strength properties but this can be misleading. As it will be seen later, strength is not an intrinsic property of the material, it depends upon geometry as well as the characteristics of the material.

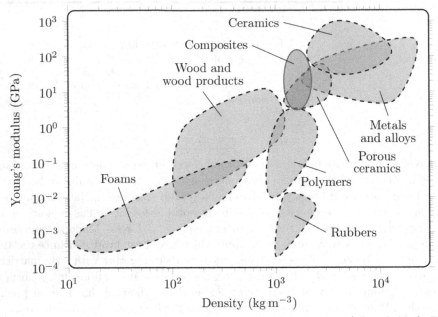

Figure 1.1: Young's modulus versus density chart with log-log scaling (after Ashby). Regarding the specific stiffness, composites materials are very well positioned on the graph compared to other materials.

This book concentrates on resin matrix composite materials as these are by far the most important type of man-made composites (if steel reinforced concrete is not considered) and they are showing impressive rates of market growth.

## 1.1   Book content description

All aspects of composite technology will be explained from the fibres and resin matrices used, composite manufacturing processes, the physical mechanisms controlling composite character-istics, computer modelling of composite behaviour to damage processes in composites due to mechanical loading or environmental effects. The choice of fibre reinforcement and matrix is dic-tated by the end-use to which the composite part will be put and will depend on the properties of the components used as well as the manufacturing route to be adopted. Although the topics covered are broad, this book does not intend to be completely exhaustive but rather to introduce the fundamentals of fibre reinforced composite materials.

**Chapter 2: Fibre reinforcements** . . . . . . . . . . . . . . . . . . . . . . . . . . . . . . . . . . . . **21**

Chapter 2 is dedicated to fibres and their extraordinary mechanical properties of strength and stiffness. The long thin form of fibres is one reason why they are well suited to reinforce composites. With this form, fibres can be very stiff in tension but flexible in bending, as explained in Chapter 2, so that they can be easily formed into complicated shapes. They also very efficiently transfer stresses between broken fibres and neighbouring intact fibres so limiting the effect of fibre breaks. Composite materials, or at least most of them, exploit these features. Chapter 2 also explains the reasons why fibres are so much stronger and often stiffer than the same material in bulk form. For the present, let us just remember that a glass fibre can be hundreds of times stronger than the same glass in bulk form. What is true for glass fibres is also true for other materials in fibre form so that fibres are really an extraordinary form of matter.

**Chapter 3: Organic matrices** . . . . . . . . . . . . . . . . . . . . . . . . . . . . . . . . . . . . . . . . **59**

While fibres are important in any composite part we do not simply knit or weave a wing of a plane, a hull of a boat or a body part for a car. The fibres are held together in a matrix. Organic matrices are the subject of Chapter 3 and details most of the common polymer matrix system from a physico-chemical point of view.

**Chapter 4: Composite manufacturing processes** . . . . . . . . . . . . . . . . . . . . . . . **97**

The assembly of the fibres with the matrix is a key step to obtain the targeted performances for the composite materials. Most composites are made up of layers of fibres, similar to cloth and many of the ways of handling them have their origins in the textile industry. Composite materials are fundamentally two dimensional whereas traditional materials are usually made as a block in three dimensions and then formed into their final shape. This is an advantage in forming composites as, just as cloth can be draped and made to take the shape of complex structures, such as a person, so can composite materials be formed into complex shapes. Chapter 4 describes the most widely used composite manufacturing processes. For composite materials, manufacturing processes are numerous and many variations exist. Moreover, some processes are resin dependent but others, at least in their philosophy, can be applied to both thermoplastics and thermosets. Emphasis is placed on the main processes by highlighting the key steps.

**Chapter 5: Constitutive relations** . . . . . . . . . . . . . . . . . . . . . . . . . . . . . . . . . . . **123**

If we reconsider the assembly of fibres mentioned above, it should be obvious though that the strength we are considering is parallel to the fibres' axes. It is obvious that if we pull the bundle at right angles to the fibres the strength will be negligible. Composites, at the level of layers of fibres, are inherently anisotropic whereas most traditional structural materials are isotropic. This notion of anisotropy is dealt with in detail in Chapter 5 by highlighting the mechanical implications and constitutive relations that will be used in the following chapters.

## Chapter 6: Micromechanical models for composite materials . . . . . . . . . . . 138

Chapter 6 deals with micromechanical modelling and directly considers modelling approaches which are particularly well adapted to fibre reinforced composite materials: the homogenisation process and as a consequence, the multi-scale processes. These models were developed in the 1980s and are now widely used for the study of composites and the calculation of industrial composite structures.

## Chapter 7: Laminated composites . . . . . . . . . . . . . . . . . . . . . . . . . . . . . . . 183

The use of anisotropic "layers" to make up a composite part opens up many possibilities which are not available to the designer using conventional bulk materials. For instance the tensile modulus can be decoupled from the shear modulus in a composite part which allows designs of structures which are not possible with conventional materials. In order to calculate the behaviour of such complex materials a greater number of variables have to be considered. This is explained in detail in Chapter 7 on laminated composites which expands on a particular approach to that treated in Chapter 6. It considers laminated, or stratified, composites. The widely used classical theory of laminates is introduced but also a less well known approach to deal with thick, periodically layered laminates is described.

## Chapter 8: Failure criteria . . . . . . . . . . . . . . . . . . . . . . . . . . . . . . . . . . . . 231

Chapter 8 on failure criteria refines the preceding two chapters and extends them to consider in detail the criteria governing damage development in laminates.

## Chapter 9: Multiscale modelling of the intralaminar cracking phenomenon. From Fracture to Damage Mechanics . . . . . . . . . . . . . . . . . . . . . . . . . . . . . 243

Chapter 9 on intralaminar cracking presents one of the most often encountered types of damage in composite structures which is that of crack growth within plies. This type of damage is usually not very damaging for the structure however it can initiate other types of damage which can be important. It is therefore necessary to understand intralaminar cracking. It is usually called transverse cracking as the cracks are normal to the direction of the fibres making up the plies. Chapter 9 extends what is generally considered to be an all or nothing phenomenon. This original approach presents in detail the concepts of homogenisation and multi-scale modelling so as to construct a detailed model within the framework of Damage Mechanics but also to take into account the concepts of the Thermodynamics of Continua so as to obtain a model which agrees with the Second Law of Thermodynamics.

## Chapter 10: Multiscale modelling of the fibre break phenomenon. Composite pressure vessel design . . . . . . . . . . . . . . . . . . . . . . . . . . . . . . . . . . . . . . . . 268

Going down in scales, Chapter 10 deals with the fibre failure phenomenon and its consequences. This type of damage is first introduced for simple unidirectional composites but leads onto the design of and damage accumulation in filament wound composite pressure vessels which have become very important structures for tackling environmental issues. This chapter draws on studies over many years and allows the numerical predictions of the failure and therefore reliability of such pressure vessels and other composite structures which until now have not been possible with such precision.

Chapter 11: Environmental ageing . . . . . . . . . . . . . . . . . . . . . . . . . . . . . . . . . . . 311

Finally, Chapter 11, which is the last in the book, introduces the reader to the processes involved in environmental ageing and to an even wider approach to the modelling of the behaviour of composite structures. As composites find ever increasing numbers of applications, their long-term resistance to the loads and environments they encounter will become of great importance.

Mechanics, like all fundamental sciences, requires the support of tensorial calculations and therefore to numerous and various forms of notations. It has been chosen in the chapters dealing with these subjects to employ simple notations. The choice could have been to use different notations for the manipulation of tensors and matrices for each tensorial or matricial operation and for each type of tensor of a different order. We have decided not to do that so as to avoid overloading the reader. The notations are therefore necessarily associated to the context in which they are employed so the reader must pay particular attention to their use. However the corresponding list of indices is always given which gives the reader, without any ambiguity, a clear and precise idea of the objects which are manipulated. Many calculations and concepts in the book are given in detail so as to allow the interested reader to understand the development of the model used.

Chapters dealing with the modelling of composites have been constructed in a logical manner so as to give a complete overview of the techniques which allow most important problems for fibres composite structures to be resolved. Continuous advanced fibre composites have been particularly described but all the tools used, concepts and techniques presented can be adapted to other fibre composites, including short or long discontinuous fibre composites, woven reinforcements and others. The chapters can be read as standalone chapters so that the reader can choose chapters treating his or her own interests. For this reason certain concepts, formulae and equations appear more than once so as to avoid repeatedly sending the reader back to their explanations earlier in the book. In this way the authors believe that the subject is more easily assimilated.

## 1.2 The global composites market – Key figures

The economic worldwide down turn at the beginning of the twenty first century and later the corona virus epidemic has introduced uncertainty into world markets but composites, particularly advanced composites, have shown resilience and have high growth rates, which for some are over 10% per year.

Although there are very large regional differences, nearly three quarters of organic matrix composites are reinforced with glass fibres but carbon, aramid and plant fibres are increasing their market share each year. The same is true for thermoplastic matrices which represent in 2020 nearly 40% of the world market compared to around 30% in the 2000s. This rough analysis is detailed in the following paragraphs by providing the key figures of the global composite market, based on regional differences, raw materials (fibre and resin types) and application sectors.

### 1.2.1 By geographical zones

The first applications for advanced composites were for military jets, both in the fuselage and wings as well as in the outer casing of their engines. The markets for these applications were primarily in those countries which had developed industrial military complexes so that the USA and the UK and then France became leading actors with Germany and Italy also becoming involved. Russia was also there but playing largely out of sight. The rise of China as an important manufacturer and consumer of composites makes that country a major player in this area and although clearly the Chinese also invested in military composite development it is also true that they have benefited from the experience of others so as to move quickly and also produce and use a very wide range of all types of composites. Japan has invested heavily in the development of advanced fibres but from the start retained a much broader approach to markets so that from an early date the use of high-performance composites in such areas as civil engineering were investigated. Japan was also very active in developing sports goods and in particular the ubiquitous carbon fibre composite golf club which is a necessity for every Japanese businessman. Very few types of sports goods are made today without a major part being in composite materials. Most tennis and squash racquets are made of composites although light metal alloys are also used. Taiwan became the main and dominant producer of such composite racquets, but this type of market is evolving quickly and manufacture moved to China as there labour costs are lower. Skis are almost inevitably made up of complex layers of composites, often with several types of fibre reinforcement used. India is strong in producing hockey sticks and China is producing advanced competition rowing boats and sculls.

From the first markets for composites others have developed so that the producers and the markets are truly worldwide. Figure 1.2 shows that the distribution of composite markets North America and Europe remain important markets but Asian countries, led by China, now represent 50% of the market in volume.

As nearly three quarters of composites produced are reinforced with glass fibres, Figure 1.2 is based on that market. Growth rates vary from one year to the next and between sources however this distribution has changed little since the last decade of the twentieth century. Rates of growth suggest some saturation of the overall composite market in North America but impressive growth in Asia, principally in China and India.

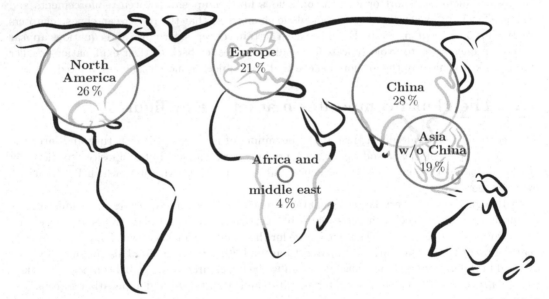

Figure 1.2: Volume breakdown of the global composite market. Asia alone accounts for half of the market (JEC, 2017).

## 1.2.2 By fibre and resin type

### Glass fibre market

Glass fibres are a low value added product so that although they represent such a large part of the industry they represent a lower percentage of the value of the world's composite market. Advanced composites, based on higher performance fibres, account for a disproportionate part of the overall value of the market even though they account for a small percentage of market shares. As already indicated, the rate of growth of advanced composites is considerably greater than that of glass fibre reinforced plastics (GFRP).

Glass is an interesting border line case as it is amorphous and even in the form of a fibre its microstructure does not become aligned as is the case for most other fibres. This results in glass fibres having the same elastic modulus as bulk glass and there is no weight saving for stiffness to be made by replacing steel, aluminium or even titanium metals by glass fibres because their specific moduli are equivalent. In addition the specific modulus of any GFRP will be less than the value obtained from glass as the fibres have to be bound together with the resin matrix material and this has low stiffness. Some natural fibres can compete with glass fibres when their specific properties are considered and they do have other advantages as well as some disadvantages which are discussed in Chapter 2.

Other fibres however show much higher specific moduli than the conventional materials and even allowing for the need to combine them with a resin, with say a fibre volume fraction of 60%, it is clear that there are considerable weight savings to be made by using composites. Even though glass fibres offer no weight savings they are the most widely used reinforcement for composites as their low cost and ease of use, which facilitates manufacturing, are often determining points in their favour. For high-performance structures however it was the development of high-performance fibres, such as carbon and Kevlar®fibres, in the 1960s and 70s which allowed composites to become the material of choice for an increasing number of applications.

The little use of GFRP in aerospace and military applications clearly reflects the lack of weight advantage which is often at a premium for these sectors. Nevertheless, the ease of manufacture coupled with the lower cost of GFRP means that the interior cladding including baggage racks in civil planes are made in this composite. For the same reasons the bodies of civil helicopters are also produced in GFRP. Glass fibres are also used in the GLARE materials which is a sandwich of GFRP and aluminium foil (Glass Laminate Aluminium Reinforced Epoxy) used in the construction of aircraft such as the A380.

### Carbon fibre market

Carbon fibre reinforced plastic (CFRP) composites have become a standard choice for the aerospace industry. The Airbus A350 and the Boeing 787 airliners make enormous use of advanced composites reinforced by both carbon and aramid fibres. The aeronautical industry together with other fast growing industries such as gas pressure vessels, wind turbines, offshore oil applications as well as sports goods and a slowly emerging utility and sports car sector are stretching the worldwide carbon fibre production capacity which is around 240 000 metric tons in 2020 (JEC, 2017). Note that the global demand is only about half as much, i.e. 120 000 metric tons but in the past, some markets experienced a shortage of carbon fibres and huge investments have been made by companies.

There is an increasing number of companies which make carbon fibres. They all have an international outlook which reflects the growing market. Some have evolved due to acquisitions which have meant that some names have disappeared or been subsumed by another name. Cytec based in the USA was bought by the Belgium company Solvay in 2015 but retains its brand name. Hexcel® also from the USA is a large producer, however carbon fibre manufacture, with its conception in Europe, has been for many years overwhelmingly dominated by Asiatic, with a dominant role for Japanese, producers who have installed production plants in both the USA and Europe. The Japanese company, Toray, is the leading manufacturer in the world of polyacrylonitrile (PAN) carbon fibres, which is the most important type of carbon fibre and has been making them from about 1970. The production of carbon fibres by the Toray company represented about a third of the total global production capacity in 2020.

Earlier producers, who manufactured the fibres from around 1967, have, in some cases got out of the market or have had their carbon fibre technology bought by other companies. The acquisition in 2014 by Toray of Zoltek, a company with large tow carbon fibre manufacturing plants in Hungary, the USA and Mexico revealed the changing nature of the industry. Until the beginning of the twenty first century the carbon fibre market was dominated first by the aerospace industry but other ever expanding industrial markets requiring cheaper fibres, such as for wind turbines and pressure vessels, created a business for large tow manufacture. Other Japanese companies are important such as Teijin which integrated Toho Tenax, the second biggest PAN based carbon fibre producer in Japan, fully into its organisation in 2018. Mitsubishi Rayon was incorporated into Mitsubishi Chemical Corporation together with Mitsubishi Plastics in 2010 which, through earlier acquisitions, became the third biggest Japanese carbon fibre producer making both PAN and pitch based carbon fibres. There are now carbon fibre producers in South Korea and China.

At the end of the twentieth century Taiwan developed an impressive carbon fibre production capability mainly for sports goods and at one time it was claimed that they produced 80% of composite tennis rackets. However this is a fast moving industry and particularly labour intensive manufacturing processes have moved to cheaper sources. Now Taiwan imports sports goods made in Dongguan in China.

Worldwide growth rate of carbon fibre reinforced plastics (CFRP) use is difficult to evaluate from one year to the next but is several times that of GFRP, although for a smaller market and the market is far from saturated. Indeed CFRP has matured, during the first part of the twenty first century, into a large industry that is finding applications in many other sectors other than aerospace and sports goods.

The broadening of the market offers opportunities to other producers to follow Zoltek's example in producing fibres in very large tows, with up to twenty five to thirty times the number of fibres per tow generally produced for the aerospace industry. In this way production costs are being dramatically reduced. These large tows can be separated into more manageable tows containing smaller numbers of fibres which then can be processed.

Alternatives to PAN based carbon fibres exist of which the most important are fibres made from pitch which is a natural by-product of the oil or coking industries. They are discussed in Chapter 2. The advantage with pitch is that it is made up of more than 80% carbon but being a natural product its refinement is costly so countering the initial low cost of the precursor material. These fibres can be made into very high-moduli carbon fibres which make them attractive for some niche markets requiring very stiff structures. Low stiffness carbon fibres are also made from pitch with the leading manufacturer being Kureha in Japan. These fibres which are of low price are very well adapted for reinforcing cement in which they are inert. Finally there is some production of carbon fibres made from cellulose. These fibres have low Young's moduli due to their poor atomic organisation but this does confer on them lower thermal reactivity so that they are of interest for carbon-carbon applications used for their resistance to high temperatures.

With increasing volumes of carbon fibre composites being used their recycling has become of increasing interest. This is of importance as the off-cuts from manufacturing processes represent up to 30% of the total material used and recycling offers considerable cost savings. In addition the recycling of composite parts at end of part life has become an issue of importance. The most widely used technique to recover the carbon fibres is by pyrolysis to burn off the resin. This method requires considerable energy input and risks damaging the fibres through oxidation. A competing technique consists of removing the resin by the use of a solvent.

The recycling industry is relatively young but the numerous projects in this area show a real potential for reusing recycled carbon fibres. The carbon fibre industry is showing considerable interest in the recycling of carbon fibre and its growth can be expected to continue.

## Aramid and high-performance polymer fibres

Aramid fibres are the other type of high-performance fibre which is of major importance in advanced composites. They represent about 1% of the fibre reinforcement market in volume. As Chapter 2 reveals, they are very different from glass and carbon fibres as they can deform plastically in compression whereas they are largely elastic in tension. Although this is a limitation for their use in structures which can be subjected to compression, the high-energy absorption of composites made from these fibres finds them important markets. These fibres are used in many structures other than those which are the subject of this book, such as cables, aircraft tyres and cloths. There are two main commercial producers, DuPont™ in the USA, which also produces Kevlar®fibres in Europe and Japan, the latter as a joint venture with Toray . The other producer is Teijin Aramid based in the Netherlands. The Teijin group, which has its headquarters in Japan, produces two types of aramid fibre, the Twaron®fibre and the Technora® fibre. Toyobo , another Japanese fibre producer, must be mentioned as producing the Zylon® fibre which has the highest Young's modulus of commercially available polymer fibres. Other producers, notably in China, exist and may well become an important source of this type of reinforcement.

There are other high-performance organic fibres in production of which the most important is Dyneema® made from high-modulus polyethylene. These fibres rival aramid fibres in some markets where their lower density, toughness and high moduli are required but they are fundamentally different from aramid fibres and do not have the same high-temperature capability, as is explained in Chapter 2.

## Thermoset resin systems

To make a composite material the fibres are embedded in a matrix and in the overwhelming number of cases the matrix is a thermosetting resin or a thermoplastic. The first organic resins used were thermosets and they still represent around 60% of the overall composite market and about the same fraction of the overall market value.

Unsaturated polyester is by far the most important thermosetting resin used in composite materials but the advanced end of the market uses predominantly epoxy resins which are described in Chapter 3. Fibres can be easily impregnated with these resins and then the composite can be formed into its final shape before the thermosetting resin is cross-linked, as explained in Chapters 3 and 4. Phenol formaldehyde (phenolic) resins are used for specialist applications such as the interior of mass transport vehicles, such as trains and planes, because of its resistance in case of fire and also it is used in the facing of buildings. The finished product made of a fibre reinforced thermosetting resin cannot however be altered after manufacture and recycling off-cuts is relatively difficult as explained above for carbon fibre reinforced composites.

## Thermoplastic matrices

Recycling thermoplastic waste is much easier. The importance of thermoplastic matrices is growing and they represented, in 2020, about 40% of the market and have a higher growth rate compared to thermosetting resins. In addition, injection moulding of short fibre reinforced composites using thermoplastics is a developed industry, as is explained in Chapter 4. Long or continuous fibre reinforced thermoplastics are a developing industry and produce high-performance composite structures. Polypropylene is the preferred matrix material for many markets because of its low cost and ease of recycling however polyamide 6,6 and saturated polyester are also widely used. These and other specialist thermoplastics are described in Chapter 3.

In addition to the ability of thermoplastics to be recycled, thermoplastic matrix composites offer a number of additional advantages when compared to thermosetting resin composites. Amongst these is shorter manufacturing times and the possibility of welding parts together. The main disadvantage for thermoplastic matrices is their high viscosity when melted which requires relatively high temperatures and pressures during composite manufacture. The high viscosity makes the impregnation of the fibres difficult and several techniques exist to overcome this difficulty. Fibres can be coated with the polymer in particulate form or the polymer can be co-mingled with the fibres either during composite manufacture or as a semi-finished pre-impregnated product however this increases costs. An alternative solution is to use thermoplastics with a very low viscosity. This can be achieved by using mono- or oligomeric precursors and a polymerisation of the thermoplastic matrix in-situ; that is to say within the fibrous reinforcement. This alternative solution is called "reactive" and provides a low pressure manufacture process similar to those used with thermosets. Some composite parts, usually made with thermosetting resins are gradually being converted to take advantage of the properties of thermoplastics.

### 1.2.3   By application sectors and end-users

The markets which have developed for these organic matrix composites cover a very wide range of applications, including pleasure boats, wind turbines, storage tanks, body parts for cars, dental implants and many others. This section reveals how composite markets have developed in increasing numbers of sectors.

By adopting the JEC definition of composite industrial sectors (JEC, 2017), main industry applications are spread across nine "application sectors and/or end-users applications". Four main sectors represent around two thirds of the market in terms of share of value of total composites. As shown in Figure 1.3, these four sectors are:

- Ground transportation, i.e. automotive and road transportation, railway vehicles and infrastructure
- Building and construction
- Electrical appliances and electronics
- Aerospace and defence, i.e. aerospace and defence, security and ballistics

Key figures are provided for each of these domains from § 1.2.3 to § 1.2.3.

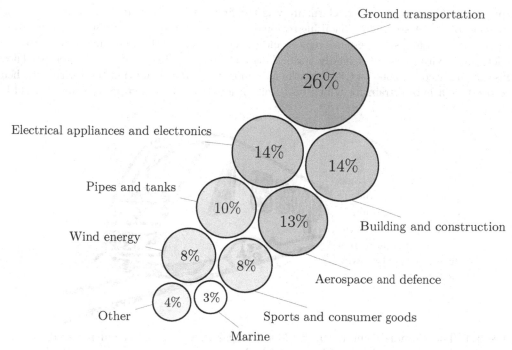

Figure 1.3: Share of value of total composites. Ground & air transportation, construction and electrical and electronic sectors represent around two thirds of the share of the value of all composites. Other sectors stand for sports, leisure and recreation, oil and gas, medical and industrial equipment.

## Ground transportation

Ground transportation covers road vehicles, monorails, tramways and trains. The automotive sector alone accounts for almost 20% of the share of value of all composites.

Trains use a lot of composite materials, nearly always reinforced with glass fibres for reasons of cost and they are found in the interior structures of the railway coaches. The need to reduce weight is important for high-speed trains and this has encouraged consideration for the use of composite materials in their primary structures. Weight saving may not at first seem a major issue with trains which spend most of the time travelling at high speed with relatively little time spent in acceleration or braking, however the very fast trains in Europe and Japan run on specially laid track. In order to make the trains more economic it is desirable to increase the numbers of passengers transported by each train. Increased numbers of passengers means an increase in weight and this poses a problem for the track. If the track is not to be replaced, at great cost, so as to take heavier trains, the weight of the trains has to be reduced. The use of aluminium for coach bodies was a first step but advanced composites are being investigated as an even better solution. Cost is the limiting factor but as the markets for advanced materials increase and material costs come down, trains will be constructed with large parts made of advanced composites. A major part which has been considered is the bogie, which is a swivelling framework with several axles connected to the wheels and springs, used to guide the powered and unpowered vehicles. Several projects to make bogies out of carbon fibre reinforced plastic have shown the feasibility of this application. An interesting development is the putting into circulation of hydrogen powered trains. Many manufacturers are lined up to develop hydrogen powered trains. Since the 2000s, many prototypes have emerged. Among the most mature projects, the first fuel cell commercial tram was deployed in China in 2015 and was built by Sifang, a subsidiary of the China South Rail

Corporation. Three years latter, Germany was the first country to benefit from a commercial circulation of the Alstom Coradia iLint regional train, see Figure 1.4. The hydrogen is stored in carbon fibre composite pressure vessels and generates energy through fuel cells to produce a quiet and environmentally friendly train. This technology is discussed in Chapter 10. These trains can be used on non-electrified tracks as a more economical solution to electrifying them or to produce a hybrid train for which both conventional and hydrogen power sources could be used.

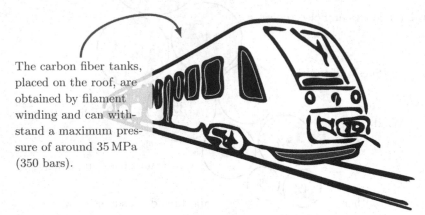

The carbon fiber tanks, placed on the roof, are obtained by filament winding and can withstand a maximum pressure of around 35 MPa (350 bars).

Figure 1.4: The Coradia iLint is the world's first passenger regional train powered by a hydrogen fuel cell. It is equipped with type-IV high-pressure cylinders, produced by Xperion, a Hexagon Composites subsidiary.

A number of countries are turning to fuel cell technology to replace traditional diesel trains in mass transit. Some manufacturers are also considering hydrogen fuelled fret trains but this poses difficulties as the weights involved are much greater than passenger trains.

Subway trains are increasingly made with low-cost composites making up an important part of the body parts of the carriages. Weight is an obvious disadvantage as the trains have to repeatedly accelerate and brake during normal use. The flexibility of manufacture that composites give is shown by the innovative designs which can be seen in many of the subway and tramway systems which have been installed in many European cities and in other parts of the world.

Cars and lorries have long attracted the attention of the composite community, above all because of the enormous volume of production which is involved. It is a well-established industry, ruled by a conservative approach to material use and above all by their cost. However, the car industry is undergoing a revolution with rejection of fossil fuels in favour of electric vehicles. This must encourage the introduction of light weight composite parts if electric cars with autonomous ranges of hundreds of kilometres are to be achieved.

Cars are made with increasing numbers of extras such as more complicated seats, air-conditioning units and safety features, all of which increase weight at a time when fuel economy is increasingly important. Another attraction of composites to the automobile industry is the possibility of making innovative body shapes as well as reducing weight to produce original vehicles with enhanced corrosion resistance. In addition, it has long been recognised that composites can absorb more energy in the case of a crash for the same weight of a metal structure. Sports cars have long since used composites whether for racing cars or sports cars for the road. The bodies of Formula 1 cars are almost all composites with an emphasis on carbon fibre composites for lightness and stiffness together with aramid fibre reinforced composites for toughness. However these are small volume productions and composites have had a harder job being accepted for the average family car. The main limitations are lack of experience for lengthy runs of manufacturing processes, recycling and cost. The costs of retooling to make composite parts are high however the tools and moulds are often cheaper than for metal parts as the forming pressures involved are much lower and they can even be made with other composite materials. Whether the composites solution is cost effective can often be a question of numbers of vehicles to be produced. Very large numbers of vehicles produced per day, say over 400, could mean that the advantages of cheaper tooling are lost as they have to be replaced frequently. Cost is certainly the biggest factor governing the use of composites in vehicles. Nevertheless, composites are making inroads in utility and family cars, particularly at the top end of the market. Low cost and long manufacturing runs for composites process technology remain a major challenge for a higher penetration rate of composite materials in the automotive market. The large numbers of cars produced by the automobile industry require innovative manufacturing processes with fast production rates which may well favour thermoplastic composites.

**Aerospace and defence**

Advanced fibre reinforced composites may not represent the greatest part of the composite market but their applications are often startling for they are used where more conventional materials cannot match their properties. This means that they find applications at the frontiers of engineering.

Space is a challenge to engineers and getting there is a competition between getting the rocket off the launch pad, the payload and gravity which means that weight is all important. Carbon fibre reinforced resin can account for 80% of the weight of the structure of a satellite because of the high specific properties of these composites. The stiffest carbon fibres, which are often pitch based, can find a role in assuring the dimensional stability of large satellite or space station constructions. In addition to their stiffness and strength, carbon fibre composites have an almost negligible coefficient of thermal expansion in the direction of the fibres. In space, the stark contrasts in temperature between the sunlit side of a structure, up to 150 °C, and the side turned away from the sun, which could be at −150 °C, means that dimensional stability is a major issue. A space telescope would lose its function if its structure varied as the temperature changed. The low atmospheric pressures encountered in near space place other constraints on composites which are made on earth. Resin matrix composites absorb water which, in low pressure conditions, will leave the composite and could condense on lenses, mirrors or electronic circuits. For these reasons, resins, such as isocyanates are used which absorb much less water than epoxies. The composites are usually in the form of panels bonded onto a honeycomb structure to form a light weight structure which is stiff in bending. This type of structure is widely used, for example, in aircraft structures but there is a difference as the honeycomb in aircraft is made of organic Nomex® paper but more often it is of aluminium in satellite structures. This again is because of concerns about water absorption which justifies the use of the heavier metal honeycomb.

The aircraft industry was the first to use fibre reinforced composites and they have become, together with aluminium, the most important class of materials used in aerospace structures. Their adoption by the industry has been spectacular and rapid and all the more impressive as the industry has some of the toughest safety regulations for the use of materials of all industrial sectors. Originally, composites were introduced into military aircraft and since the 1960s each new generation of fighter aircraft and helicopters has seen an increase in their use so that the composites share of these aircraft is approaching and surpassing 50% of the structural weight. Original applications were for secondary structures such as wing extensions to increase range, nose cones and helicopter bodies but other advanced resin matrix composites found increasing numbers of applications. Very quickly the superior fatigue resistance of composites meant that they replaced the metal blades on helicopters which needed to be changed every two thousand flying hours because of the fatigue crack growth. Composite blades last as long as the body of the helicopter and are a remarkable demonstration of composite fatigue resistance compared to that of metals. The greatest part of the composite materials used in aircraft has been carbon or aramid fibre reinforced epoxy.

Composites in military aircraft and helicopters are now used in primary structures such as wings, fuselages and tail structures and so account for such a large percentage of their structural weight. When the low density of composites is considered, compared to even light alloys, it can be seen that their percentage volume in the aircraft is even greater than 50%.

The coming of age of advanced composites in the aircraft industry must however be seen in their adoption in civil aircraft. This trend started in the 1980s, particularly with Airbus, but matured into much wider use at the beginning of the 21st century. Not only are these aircraft so much bigger than military aircraft but safety standards are even stricter. The Airbus series of aircraft have seen a gradual but a continual, step by step, increase in the use of advanced composites has been evident. An aircraft such as the largest civilian aircraft, the Airbus A380, uses composites in many structural parts such as the upper fuselage which is composed of a material called GLARE and which is a laminate of glass reinforced epoxy resin and aluminium sheets. The use of GLARE produces a 25% weight saving compared to other traditional materials. It is also used in the leading edges of the vertical and horizontal stabilisers resulting in a 20% reduction in the weight of the 14 m high tail. This hybrid material is a lower cost variation of ARALL (Aramid Reinforced ALuminium Laminate) which was developed for Fokker by the University of Delft in the Netherlands and is composed of layers of aramid reinforced epoxy resin and aluminium sheets. These materials have the deformability of aluminium but are lighter and have much superior fatigue properties. GLARE represents 3% of the weight of an Airbus A380, not counting the weight of the engines and landing gear. Most composites however are reinforced by carbon fibres and are used in the rear pressure bulkhead separating the pressurised passenger cabin from the rear, unpressurised, section of the plane, floor beams, flight control surfaces such as flaps, spoilers and ailerons, landing gear doors, most of the tail and a twelve ton centre wing box which links the two wings through the underbody of the fuselage. These advanced composite materials represent 22% of the weight of the plane without the engines and landing gear. The Airbus A380 has been an undoubted technical success but with changing market needs and volatile financial circumstances it remains to be seen if its future will not be compromised.

The other large plane maker, Boeing, is not absent in its use of composites although was slower in using them than Airbus. The large body Boeing 777 introduced in the last decade of the twentieth century has around 10% by weight of advanced composites but the middle sized Boeing 787 aircraft, the Dreamliner, capable of carrying 200 to 250 passengers is made of 50% by weight of composites. This is to be compared to 20% aluminium, 15% titanium and 10% steel. The low weight and fuel efficiency of the Dreamliner has opened up new very long distance routes so that non-stop flights right across the world are now proposed for passengers willing to sit in a plane for seventeen hours, or more, from Perth to London or Singapore to New York or fly from Toronto to New Delhi. The Airbus A350 is a direct competitor to the Dreamliner and contains marginally more carbon fibre composites in its structure (53 wt%).

This generation of aircraft has most of its fuselage and wings composed of carbon fibre laminate. This not only makes them cheaper to operate but also provides other benefits, for example windows which are 50% bigger which is possible due to the greater rigidity of carbon fibre composites compared to aluminium. It is clear that the use of advanced composites will dominate in 21st century aircraft over other types of materials.

Smaller planes, such as business jets have been able to move faster in using advanced composites. It is interesting that their introduction was not solely because of superior mechanical properties linked to lower weight but also, for some structures, because the composite solution, was cheaper than when made from aluminium. The possibility of integrating several assembly steps which are necessary with metal structures reduces overall manufacturing costs. The possibility of using composite structures for more than just mechanical purposes is attractive. This multi-functionality allows the composite material also to be used for insulation, so reducing problems of condensation, noise reduction, fire resistance, corrosion resistance and greater design flexibility. Sensors can be incorporated into the composite material turning it into what is known as an intelligent material and which can be used to monitor the state of the plane so as to reveal problems before they become acute. In addition composites offers better durability than aluminium for roughly equal cost. Nevertheless there is a continuous evaluation of the materials used in aircraft and improvements will continue to be made, not only in composites but also light alloys of aluminium and titanium. The manufacturing processes used with composites will continue to evolve with the goal of reducing costs by using automated processes and the increased use of preforming technologies drawn from the textile industry.

## Building and construction

The building sector is one of the biggest for the use of composites which, in the past, were almost invariably reinforced with glass fibres but a market for carbon fibre reinforced structures is developing. Facades in GFRP for buildings allow architects to produce novel and interesting outward appearances of new and renovated buildings. Composites offer much more however and allow imaginative architects to make entirely novel roofing shapes with large spans which could not be produced in more conventional building materials. Protection from collapse in regions prone to earthquakes can be afforded by including in the original design or by retrofitting frameworks of advanced composites which would retain their integrity in case of the building collapsing.

Much piping is also made of glass reinforced resin. For these applications, large pipes can be made by centrifugal casting as described in Chapter 4 and are used to refit large diameter drainage and sewerage. Other techniques exist which allow existing degraded pipelines to be relined in situ using preimpregnated glass felt which is cross linked once the composite cladding has been put in place.

The ease of application of composites which, as preimpregnated tapes or cloth, can be added to an existing structure and after crosslinking, act as a reinforcement, is exploited in many civil engineering structures. The renovation of subway systems, some of which have been in place for 150 years, and bridges, is facilitated by the use of high-modulus pitch based carbon fibre composite bonded onto the steel beams which need supporting. Increasingly new vehicular and pedestrian bridges are being made with composites incorporated into the structure with road decks, and struts being made from GFRP. The struts are made by pultrusion, as described in Chapter 4 whilst the decks are made in the form of lightweight sandwich or honeycomb structures.

Concrete is the main building material in civil engineering systems but, even in this type of material, fibre reinforcement is used. Historically, asbestos reinforced cement showed the way. In this type of material the role of the fibres is to hinder crack propagation in the cement. This they do by providing weak interfaces normal to the fibre crack propagation direction and also requiring pull-out for the crack to develop further. Typically fibre volume fractions are low, around 5%, so that strength or stiffness reinforcement, as seen with other composites, is not the goal. Asbestos is no longer used because of health hazards as described in Chapter 2 but other fibres, polypropylene, cellulose, steel filaments, even aramid and carbon fibres have replaced them. Such cement and concrete matrix composites show increased toughness and the result is that the thickness of say a prefabricated roof can be reduced. As mentioned above short carbon fibres used in such materials are made from spinning isotropic pitch which is then converted into a carbon fibre felt. The fibres have low moduli but resist the alkaline environment of cement which is difficult for glass fibres. Pitch based carbon fibres are discussed in Chapter 2. High-performance carbon fibres are also used to replace the steel rods in pre-stressed concrete for applications where their high resistance to corrosion or their transparency to electro-magnetic fields are important.

The growth of the global construction composites market is directly related to the growth of the underlying construction market. Composites are expected to have a higher penetration in green buildings because of the benefit they provide over traditional building materials. They are durable materials and require low maintenance. Composites are lighter than than traditional construction materials such as steel, aluminium and concrete. In addition they bring great flexibility in design to architectural structures.

### Electrical appliances and electronics

The increasing growth of digital technology is emerging as a driver of sweeping changes in the world and composites are playing a vital role in this area. These include Printed Circuit Boards, made of GFRP and advanced composites used as lightweight materials for consumer electronics such as in portable phones and computers. As an example of collaboration between different industrial sectors, it is interesting to note the long-term joint venture announced in 2013 between SGL Carbon, which is a carbon fibre producer based in Germany and Samsung the electronics giant based in Korea in the area of protective cases for laptops and portable phones.

### Pipes and tanks

The oil industry is not going to disappear in the near future despite economic and climate warming pressures and in the foreseeable future will remain a very large sector using composite materials. The Corona pandemic led to the energy market being more volatile than ever with the demand for oil and gas crashing as the world went into lock-down. The down turn in the world economics has restricted ambitions in this area so that plans for new projects in deep sea drilling may not materialise but rather there is a need for the replacement of pipes used in highly corrosive environments. Fibre reinforced plastics are obvious candidates for this type of use and they bring the added advantage of being better thermal insulators which is an advantage in retaining the flow properties of extracted oil. Nevertheless, oil and gas were amongst the industries hardest hit by the economic crisis and they face growing competition from cheaper renewable energy. The rules of the game in many ways are changing.

The use of natural gas as a fuel for buses and other mainly utility vehicles began at the beginning of the twenty-first century and foreshadowed a growing interest in using hydrogen as a fuel for a wider range of vehicles. If hydrogen is to provide acceptable ranges for these vehicles it must be highly compressed so as to provide sufficient mass of the gas to give acceptable autonomy. High-pressure gas storage vessels represent one of the biggest and fastest-growing markets for advanced composites, particularly for filament wound carbon fibre composites. This topic is discussed in Chapter 10.

### Wind energy

Concerns for the environment are encouraging the use of non-fossil fuels as techniques such as solar or wind turbine energy generation become more widely employed and the high costs of deep sea oil extraction become less attractive.

An application which is of great importance is the use of composites in wind energy generation. Wind turbines provide a valuable source of renewable energy and wind energy is thus a booming market. Wind turbines can have composite blades exceeding 100 m in length. The smaller ones are made usually from glass fibre cloth on foam or other light-weight core. Longer blades require advanced composites based on carbon fibres which allow considerable weight reduction and are being used in the hybrid structures of the bigger turbines. They provide a valuable source of renewable energy with individual wind turbines able to generate more than 5 MW of power. Plans are afoot to build wind turbines producing more than 10 MW of power.

The GE Haliade-X, one of the most powerful wind turbines at present in use, is, for example, able to power more than 16 000 European households. These very large capacity turbines are placed, often near the coast or offshore, where wind speeds are high and the wind blows almost continuously. The development of wind turbines is intimately linked to that of composite materials which offer the possibility of building ever larger structures. Figure 1.5 shows the very rapid evolution of wind turbine sizes, made possible by the increasingly large introduction of composites especially for the blades. GE Haliade-X is 260 m high with 107 m blades made from an optimised combination of carbon and glass fibre.

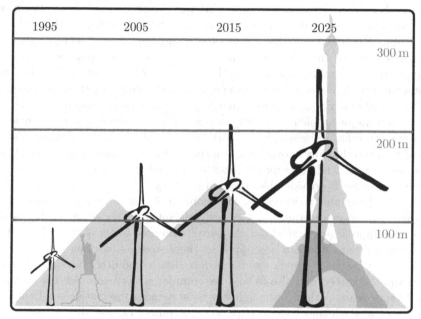

Figure 1.5: Evolution of wind turbine heights.

China has become in recent years the undisputed world leader in renewable energy, with the largest installed capacity in the world. With its long coastline and land area, China has exceptional wind energy resources. Chinese installed capacity has doubled in the last five years, while that of the EU, the second largest in the world, has doubled in the last ten years. China and the United States together account for nearly 60% of new capacity in 2019. Beyond the size and cost optimisation, future challenges are turned towards repairability and recycling of such big structures.

**Sports and consumer goods**

Lightweight and high-performance composite materials were quickly adopted in sports equipment. They are well established and have become classic materials for tennis rackets, skis, snowboards, golf clubs; pole vaulting poles; racing bicycles; fishing rods and this list could be continued. The universal adoption of composites for the sports market shows signs of stabilising because of their ubiquitous use. Carbon fibre-reinforced polymer (CFRP) composites are expected to dominate the use of composites in sports goods with glass fibre composites experiencing moderate growth as well.

An increasing trend is the customisation, particularly for well-known athletes. The development of 3D printing now enables high-end manufacturers to offer equipment specially made for the needs and desires of individuals who can afford it. Advanced composites have been used with great success for the frames of racing bikes allowing records to be broken. The first 3D-printed carbon fibre unibody production bike frame was unveiled at Euro Bike in Friedrichshafen, in 2019, by Arevo, based in the USA and the company also showed a 3D-printed thermoplastic wheel rim. In many sports applications, manufacturers are using biomaterials to enhance their environmental image by reinforcing composite structures with natural or reclaimed fibres.

## Marine

The frontier of space is being conquered by the use of composite materials and in this area they have long since been considered as standard structural materials. Another frontier however is in deep water for off-shore oil drilling. Although environmentally unfriendly, the world runs on oil and it is a diminishing resource so that the sector is obliged to look into increasingly inhospitable regions to uncover more. The off-shore oil industry involves massive resources and installations and has amassed considerable experience in drilling in depths of water using steel piping and casings. As water depths increase however the weight of the steel structures used to bring the oil to the surface becomes a serious limitation to what is feasible. At depths greater than 1500 m the weight becomes a major problem, both for transporting the parts to the off-shore site but also because the structure has to be supported by the floating platform. The density of steel is eight times that of water so that any buoyancy due to the Archimedes effect is negligible. Carbon fibre composites are obvious candidates for this frontier. Their low density means that the buoyancy effect of being in water is considerable. In addition, they can match the strength and rigidity of steel and provide greater resistance to corrosion. As if that were not enough, they also are better thermal insulators and this is not an insignificant property as the oil leaves the seabed at temperatures approaching 100 °C but the surrounding sea water is at 4 °C. As the oil cools its viscosity increases and flow can be stopped by the viscous oil blocking the riser, which is the name given to the pipe which brings it to the surface. Risers and other pipelines for transporting oil and gas from the seabed and also to land are complex structures which are required to have lifetimes of twenty years. Although the industry is by nature conservative and hesitant to adopt new materials with which they have no experience, advanced carbon fibre composites are set to make a big impact in this area. There is really no alternative for extracting oil in depths of water down to and greater than 3000 m.

Above the sea, on the platform, composites are also finding uses despite initial fears because of fire safety. The first generation of off-shore oil platforms did not use composites because of their resin matrices which were considered to be a fire hazard. Later research showed that if the composites were of a greater thickness than 8 mm they performed better than steel in a major fire. The reason is the same as why it is better to have a wooden door between you and a fire than a metal door. The latter is a good conductor of heat, and metals will soften and buckle if the temperature is great enough. Wood will first char and this will provide a protective layer to the underlying unburnt wood. Composites of sufficient thickness behave in the same way with charring and out-gassing retarding the progress of a fire. Composites are now being used in off-shore platforms and are used not only for piping but also stairways and walk ways. As with the space applications, saving weight, even on an off-shore oil platform which weighs thousands of tons, is still important. Even the mooring ropes of platforms are being made of organic fibres for the same reasons. Such ropes, made of polyester, aramid or high-modulus polyethylene fibres, are required to have breaking loads of up to 2000 t and be kilometres in length. They are replacing the much heavier chains and steel cables used in the past.

Any visitor to a marina will see hundreds of sleek and shiny pleasure boats ranging from the small to the very big, up to 40 m with their own helicopter pads. Overwhelmingly they will be made of glass reinforced polyester resin. The industry began in the late 1940s with the introduction of room temperature curing unsaturated polyester resin which could be reinforced with glass fibres. Over the years this material has become dominant in pleasure boat building.

As is described in Chapter 4, open mould manufacture of the hulls of these boats is the most common production process and has brought the cost of boats into the range of many more people than was the case before their introduction. The industry could evolve however to using more closed mould manufacturing because of the need to reduce styrene emissions.

Most high-performance yachts and catamarans, used for races, contain carbon fibre reinforced vinyl or epoxy resin in the hulls and masts. The use of composites has enabled records to be continually broken but problems, linked usually to difficulties of manufacturing limited numbers of large structures and also the reduction of safety margins, have led to some high profile and unexpected failures of masts.

Very large military boats have been produced as mine sweepers. The non-metallic nature of the hulls is the prime attraction together with their resistance to fire, but also reduced maintenance over the lifetime of the boat is an important consideration.

### 1.2.4 Conclusions

The above applications are given only as examples of the diversity of composite products. Look around and you will see many more. From fibre reinforced baths in the home to moulded bathroom units supplied as a finished product to hotels. All that is necessary is to connect it to the plumbing. Artists are using composites to make innovative sculptures and designing originally shaped furniture. Go to the local swimming pool and like as not the roof is made of laminated wooded beams. The use of hybrid beams of wood and aramid composites on the tension side allows less costly and more easily replaced wood to be used. Medical prostheses, surgical instruments and innovative filling materials for teeth all use composites. Take a plane and you can be sure to be within touching distance of advanced composites. From your mobile phone which contains composite printed circuit boards to under the bonnet of the car, composites are everywhere. If they are not there yet you can be pretty sure they are coming.

# Revision exercises

### Exercise 1.1

What makes fibres ideal for reinforcing composite materials rather than any other form of matter?

### Exercise 1.2

Explain why a fibre reinforced composite is intrinsically anisotropic.

### Exercise 1.3

To what do the specific properties of a material relate? Show that there is no advantage in the GFRP specific properties compared to conventional metals such as aluminium.

### Exercise 1.4

Despite the poor specific properties of GFRP give reasons why it is used in civil aircraft.

### Exercise 1.5

What percentage of the whole composite market do advanced composites represent? Discuss the value related to the volume of composite markets and their values.

### Exercise 1.6

How is it that natural fibres can compete as reinforcements with glass fibres?

### Exercise 1.7

What is the most widely used fibre reinforcements and matrix material?

### Exercise 1.8

Most pleasure boats are made of glass reinforced polyester made with open moulds but legislation is provoking greater use of closed moulds. Why is this?

### Exercise 1.9

Why have composites replaced metals in helicopter blades? Why is this?

### Exercise 1.10

What is approximately the percentage weight of the latest civilian airlines such as the Boeing 787 and the Airbus 350.

2

# Fibre reinforcements

## 2.1  Reinforcing fibres

Nature has developed along composite lines resulting in the plant world being made up of natural cellulose polymer fibres and the structure of animals consisting of fibres of proteins reinforcing bones and organs. The forces of nature which allowed these structures to evolve were the stresses to which they were subjected during their lifetimes and other criteria such as the need for lightweight and tough components. Man-made traditional materials have made use of natural fibres, such as natural cellulose fibres, for many textile and technical structures, including paper, which is a composite usually made from fibres taken from wood. Straw has and still is in some communities, been used to reinforce mud bricks; animal hair or silk have also been used in technical materials throughout history; some mineral fibres are often used as refractory insulation; but it is only since the end of the nineteenth century that man-made fibres have existed, despite attempts in the proceeding centuries to produce artificial silk from cellulose. The first artificial fibres exploited the long-chain molecules of the short-cellulose natural fibres by putting them into a solution and regenerating them to make continuous cellulose fibres. It was however during the second half of the twentieth century, when synthetic, purely man-made, fibres became widely available, which led to advanced fibre composites taking such a central role in advanced structures in the twenty-first century.

Fibres, whether natural, man-made by regenerating cellulose or purely synthetic are remarkable forms of matter, some of which possess properties near the limits of what physics allows. They are generally light in weight and fine, typically with diameters of the order of ten microns, sometimes a few tens of microns and sometimes just a few microns. To put that into context the micron is just about the limit of resolution of optical microscopes and the human hair is 65 μm to 80 μm in diameter. Figure 2.1 shows schematically a glass fibre and a carbon fibre, the smallest one, passing through the knot of a hair fibre. It is clear that the man-made fibres are much finer than the hair.

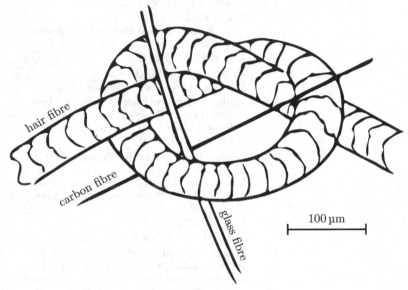

Figure 2.1: Schematic illustration of a glass fibre and a carbon fibre, the smallest one, passing through the knot of a hair. This drawing shows the very small diameters of the man-made fibres.

Fibres can be very stiff in tension but are flexible. Their flexibility in bending is a function of the reciprocal of the diameter to the fourth power $(1/\text{diameter}^4)$ and their small diameters allow them to be woven or draped over tight radii and formed into complex forms. They are usually much stronger than the same material, where it exists, in bulk form. This latter characteristic is discussed under the statistics of fibre failure later in this chapter. They are also often much stiffer in tension due to their aligned microstructures, as will also be explained. They are stiff when pulled in tension but flexible because of their small diameters. This chapter will divide fibres used for technical uses, with particular emphasis on reinforcing fibres, into Natural Fibres; Regenerated Fibres; Synthetic Polymer Fibres; High-moduli synthetic polymeric fibres; Glass Fibres; Carbon Fibres and Ceramic Fibres. For a wider treatment of the whole range of fibres and their properties the reader is referred to Bunsell (2018).

## 2.2   Natural fibres

Natural fibres can denote a filament or a filamentary cell making up certain plant or animal tissues, or even certain mineral substances. Figure 2.2 schematically illustrates these three types of fibres: Plant fibres are represented by a flax stem, animal fibres by spider silk and finally mineral fibres by asbestos.

Figure 2.2: Natural fibres cover a very wide range of material substances from plant fibres to mineral fibres via animal fibres.

## 2.2.1 Plant fibres

Plant fibres are attracting increasing interest as reinforcements for composite materials both in the industrial world and less well-developed countries (Ku et al., 2011). As the markets for composites increase and the concept of using them in an ever increasing numbers of applications is adopted in sectors which in the past were dominated by traditional materials, particularly metals, natural fibres are becoming attractive because they are renewable and sustainable, require much less energy to produce than many synthetic fibres and so are environmentally friendly, not oil based, light in weight with, in some cases, impressive properties and they can easily be recycled. An additional environmental advantage of natural fibres is that synthetic fibres, such as polyester which is the most abundant type of fibre at present, are seen increasingly as a pollutant of the natural environment. Despite these advantages there are hurdles to be overcome in using plant fibres. They have to be grown, removed, by retting or similar processes, from the parent plant and transformed into usable fibres. At each stage problems can occur and the quality of the fibres produced can be affected by the vagaries of weather, soil quality, diseases and insect attack. Variability in the radial dimensions along the fibres, due to the way they grow and also drying of the fibres which can cause the fibre to shrink transversely, is another complication. The growth of the fibres and manufacture of synthetic fibres leads naturally to directionality in their structures. This anisotropy can be used to advantage as it can enhance energy absorption but it can also be an inconvenience as it means that the strengths of many fibres are low when they are subjected to compressive forces both in the axial and radial directions. This is a trait which is shared with many fibres whether natural or man-made although an important exception is glass fibre which remains isotropic as in the bulk form. Ageing of the finished article can also be a difficulty for natural fibres as well as their propensity to take up water that can pose problems for creating effective bonds with a resin matrix. These add complications for their use. The tensile properties of natural fibres are rarely as linear, as is often the case with advanced synthetic fibres, although this can be seen, for some applications, as an advantage as natural plant fibres are generally tougher. The tensile curves of plant fibres often show marked changes in stiffness during loading with falls in stiffness as strain increases as can be seen in Figure 2.3.

Making structures with reproducible properties using natural fibres can be a challenge; however these challenges are increasingly being overcome. Usually plant fibres are seen as a competitor to glass fibres and for this reason their properties are often given with respect to their

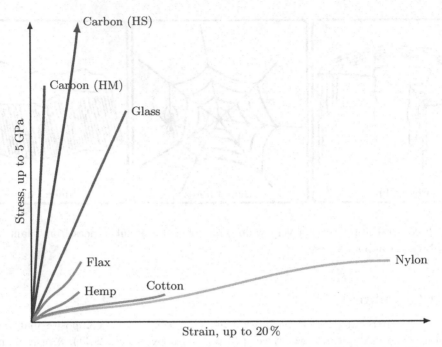

Figure 2.3: Typical stress-strain curves (schematically represented) of various fibre types.

density. These "specific properties" are obtained by dividing by the specific gravity which gives a favourable result for plant fibres as they have specific gravities around 1.2 whereas glass has a value of 2.5. The absolute values of modulus however cannot rival that of glass and many other synthetic fibres. An overall view of a number of important fibres obtained from plants is given by Ramesh (2018).

Fibres from plants are overwhelmingly made up of cellulose which is a complex natural polymer in the form of highly crystalline microfibrils, with an intrinsic elastic modulus of 140 GPa, generally embedded in a resinous matrix of lignin and non-cellulosic polysaccharides called "pectics" which are combined with cellulose in amounts which vary between different varieties of plants (Alix et al., 2009). Part of the role of pectics in the plant fibre is to provide cross-links between the cellulose microfibrils through hydrogen bonds. Cellulose and pectic compounds are the most widely found organic materials on earth.

Fibres from plants are classified by reference to where in the plant they are found.

Bast fibres such as flax, hemp and jute come from the stem of the plant. Bast fibres have higher tensile strengths than other types of plant fibres and some are of particular interest for composites (Madsen et al., 2007; Summerscales et al., 2010).

Fibres are also obtained from the hairs on seeds of plants which are from where cotton and kapok fibres are obtained although their applications are primarily for traditional textile applications. A third group, called hard fibres, come from the leaves of plants and include sisal, banana and agave as well as coir fibres obtained from the shell of coconuts (Bisanda and Ansell, 1991).

The strength and stiffness of natural plant fibres are determined by the alignment of the cellulosic micro-fibres. In fibres, such as cotton obtained from the seeds of the plants, the cellulosic micro-fibrils are arranged in a helical fashion which reverses along its length. The helical angle is important in determining the fibre strength and stiffness (Hearle and Sparrow, 1971). In bast fibres the cellulosic microfibres are generally better aligned parallel to the plant axis, so giving fibres of greater stiffness in tension. In this way natural plant fibres can be seen to be themselves

**Table 2.1**   A comparison of typical properties of plant-based natural fibres.

| Fibre | Diameter (µm) | Length (mm) | Specific Gravity | Strength (GPa) | Strain to Failure $\varepsilon$ (%) | Young's Modulus $E$ (GPa) | Specific Modulus (GPa) |
|---|---|---|---|---|---|---|---|
| Cotton | 10–35 | 10–50 | 1.5 | 0.6 | 3–10 | 8 | 5 |
| Flax | 12–24 | 10–60 | 1.5 | 0.95 | 1–3 | 60 | 40 |
| Hemp | 10–20 | 8–55 | 1.5 | 0.50 | 1–3 | 35 | 30 |
| Jute | 15–35 | 1–6 | 1.4 | 0.45 (bundle) | 1.7 | 20 | 14 |
| Bamboo | 10–20 | 0.5–5 | 1.2 | 1.5 | 4.5 | 38 | 30 |
| Sisal | 10–20 | 0.5–8 | 1.4 | 0.50 | 3.0 | 17 | 12 |
| Ramie | 10–50 | 120–150 | 1.5 | 0.75 | 2.0 | 55 | 37 |
| Kenaf | 2–20 | 2–3 | 1.4 | 0.90 | 4.5 | 19 | 13 |
| Banana | 80–200 (raw fibre) | 900 | 0.9 | 0.5 | 5 | 20 | 14 |
| Glass (E) | 5–20 | Continuous | 2.54 | 3.5 | 4.5 | 73 | 29 |

natural composite structures on all scales and their physical properties depend on the nature and alignments of the reinforcing cellulosic microfibrils.

Flax is an important textile fibre and attracts much interest as a reinforcement for composites. It belongs to the genus linum and its Latin biological name is Linum usitatissimum; the second part of the name means "most useful". It is used to produce linen and is the most often cited natural potential reinforcement for composites (Baley et al., 2018). Flax fibres are bast fibres and exhibit some of the highest values of strength and stiffness of plant fibres. They find use as reinforcements for acoustic damping panels in some high end value cars as well as being increasingly considered for other composite applications.

Many of the above mentioned natural fibres have found technical applications as twine, sailcloth, ropes and others as well as the traditional textile applications and as some forms of composites. Some of the fibres such as those from the agave Americana plant and those from banana leaves and others are used in local industries to produce some types of composite materials and can be a valuable means of generating income for these communities (Thamae and Baillie, 2007).

Cotton is the most widely used plant fibre although as a technical reinforcing fibre its use is limited. It is the second most produced fibre, after polyethylene terephthalate (PET) polyester fibre, and mainly used for textile applications. Its properties can be found in Table 2.1 together with other plant derived fibres which more easily find technical applications, as well as glass fibres which are given as a reference. With the exception of cotton and flax the literature does not give accurate values for the properties of many plant fibres which adds to the difficulty in their use. The scientific literature shows that there is a lot of scatter in the properties of plant fibres so that Table 2.1 should be taken as a guide rather than an absolute source of data. More data on the properties of plant fibres can be found in Bourmaud et al. (2018); Osorio et al. (2018); Ramesh (2018).

The scatter comes from differences in varieties of the plant, growing conditions as well as testing conditions such as relative humidity as well as difficulties in determining cross-sections of the fibres so as to convert tensile load data into stress. It also comes most probably from the definition used by different researchers as to what constitutes a fibre as plant fibres are themselves made up of microfibres and are most often obtained from some larger diameter plant stem. Table 2.1 does show however how the specific moduli of some plant fibres are impressive when compared to the most widely used reinforcing fibre which is type-E glass.

## 2.2.2   Animal fibres

Fibres from animals have been used throughout history, usually for traditional textile applications but for our purposes it is above all silk fibres which should attract our attention. Silk is different from other natural fibres in that it is very long. It is a material which has evolved in different species apparently independently and is based on fibrous proteins extruded in fibre form and having remarkable strength and strain to failure properties. It has been used for many purposes throughout its long history, from originally exclusive fine clothes for the Chinese emperor and his family; currency; stockings; sutures, ropes and in the first half of the twentieth century, parachutes. Traditional textile use of silk is increasing but also it seems that we are on the cusp of seeing the development of biomedical applications for which silk fibres are used as scaffolding for composite biomedical structures aimed at promoting tissue regeneration. This is due to the relative acceptance of silk when it is in the body.

Although many moths produce silk in the wild almost all silk, including silk used for technical purposes, is produced by the Bombyx Mori moth which has been bred over millennia, most probably from the wild Bombyx Mandarina moth which only occurs in China and like its domesticated cousin lives exclusively off mulberry leaves. The Bombyx Mori moth is blind and flightless living only to produce larvae for the next generation of silkworms. Their larvae voraciously eat the mulberry leaves and rapidly grow during their development. The silk pupae, or worms, wrap themselves in a cocoon made by extruding silk filaments, each containing two fibres, called brins, from the worm's mouth which it does so with a repeated figure of eight to-and-fro movement of its head as shown in Figure 2.4.

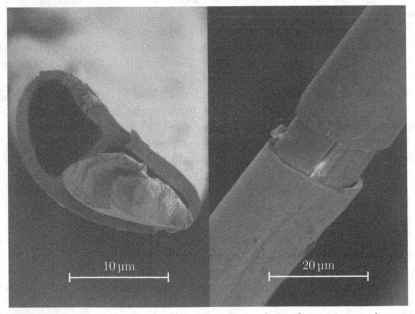

Figure 2.4: The silk worm produces a filament, called a brin, from its mouth, containing two silk fibres, called baves, by a repeated figure of eight movement of its head. SEM images taken from Colomban and Jauzein (2018).

The two fibres of insoluble fibroin encased in soluble siricin form a continuous filament called a bave with cross sectional dimensions of 15 μm to 25 μm and is made up of approximately 75 % fibron and 25 % sericin. The baves are thicker at the surface of the cocoon than inside. The silk fibres are obtained by unravelling the filaments from the cocoon after it has been dipped in boiling water to remove the sericin. The length of silk produced from each cocoon can be more

**Table 2.2** A comparison of the properties of silk fibres with synthetic organic fibres.

| Fibre | Diameter (µm) | Specific Gravity | Strength (GPa) | Strain to Failure $\varepsilon$ (%) | Young's Modulus $E$ (GPa) | Specific Modulus (GPa) | Toughness (MJ m$^{-3}$) |
|---|---|---|---|---|---|---|---|
| Spider | 3 | 1.3 | 4 | 35 | 13 | 10 | 120 |
| Bombyx mori | 10 | 1.3 | 0.6 | 18 | 7 | 8 | 70 |
| Nylon 6,6 | 25 | 1.14 | 0.95 | 18 | 5 | 4 | 80 |
| Kevlar® 49 | 14 | 1.44 | 3.6 | 2.7 | 130 | 90 | 50 |

than 1000 m. Silk fibres are not circular in cross-section, being vaguely the shape of a segment of a circle or half of an ellipse, with extreme dimensions being approximately 8.8 µm × 5 µm. Silk fibres have been shown to be composed of microfibrils embedded in a protein rich medium. Efforts to produce artificial silk similar to that produced by silk worms or other creatures have met with little success.

Another source of silk which has excited great interest is the silk produced by many spiders which are capable of producing several different types of filament depending on their function. These silk fibres, which are smooth and circular in cross section are fine, being typically around 3 µm in diameter. Spiders' long draglines have remarkable properties and have been measured as having strengths of 4 GPa, very similar to the most advanced synthetic organic fibres. However these filaments possess strains to failure which far exceed those of synthetic fibres with elongations at break of up to 35 % (Ko and Wan, 2018). This means that the energy to break of a spider's dragline is more than three times that of synthetic organic fibres such as Kevlar®. This has led to investigations into the use of such spider's silk for uses such as flak jackets or bullet proof vest but their large strains to failure would seem to work against this type of use.

Spiders produce silk from glands, called spinnerets, at the rear of the body and the chemical composition is similar to that of Nylon. Mass production of spider silk is made difficult by their small diameters and by the cannibalistic nature of spiders. There seems little willingness for spiders to collaborate amongst themselves. There have been attempts to produce man-made spider silk either by extruding silk, so emulating the natural process or by introducing some part of the spider DNA into other creatures, such as silk worms, or even mammals such as goats with the intention of spinning silk from the proteins found in the silk, called spideron, expressed in their milk. So far this has not been commercially successful even if some ersatz spider silk have been developed over the years and applied with varying degrees of success to hightech textile products.

Some averaged properties of both Bombyx mori silk and spider silk are shown in Table 2.2. The figures shown are typical values but considerable variations occur between silks produced by different types of spider (Ko and Wan, 2018; Colomban and Jauzein, 2018).

### 2.2.3 Natural mineral fibres – Asbestos

Asbestos reinforced composites predate other man-made composites by a century and became a large industry. However, as will become clear, natural mineral fibres in the form of varieties of asbestos are no longer used in making composite materials in most advanced countries. They have been used for hundreds of years in the form of heat resistant cloth as well as reinforcements for cement. Industrial exploitation of asbestos became important in the nineteenth century and remained so in many countries until the second half of the twentieth century. Many structures still exist which contain asbestos fibres and finding replacements for them has proved difficult. Asbestos is a naturally occurring mineral silicate.

There are six types of asbestos. They are all the cause of sever medical problems with chrysotile, $Mg_3(Si_2O_5)(OH)_4$, which has been widely used in the construction industry and crocidolite, known as blue asbestos, being particularly dangerous because of its friable nature. Asbestos is composed of elongated fibrous crystals which in turn break down into finer fibres of microscopic transverse dimensions. It is this friable propensity to break down into fine fibres that is the cause of serious medical problems which have led to the industry closing down. The alveolar structure of the lungs can become blocked by particles of around $1\,\mu m$ so that workers in dusty environments can suffer from pneumoconiosis and have increasing difficulties in breathing. The severity of the illness depends on the nature of the dust. Silicosis caused by inhaling particles of crystalline silica and asbestosis caused by inhaling elongated asbestos particles were recognised early in the twentieth century as being particularly dangerous having led to the deaths of possibly hundreds of thousands of people in the intervening period. The elongated form of the asbestos particle is particularly dangerous as it can become lodged in the alveoli leading to the development of macrophages as the immune system tries to deal with the foreign intrusion. This leads to difficulties in breathing and complications which includes lung cancer known as asbestosis and mesothelioma which can extend to the heart and abdomen.

A very wide use of asbestos was to reinforce cement which, for example, was used and is still being used in some countries, as corrugated roofing, typically for prefabricated building. The attraction was that the materials used, the cement and asbestos, were cheap and the asbestos reinforcement, with a volume fraction of only about $4\,\%$, hindered the growth of cracks in the cement matrix (Lenain and Bunsell, 1979). Through this mechanism the roofing could be made thinner and therefore lighter in weight. Asbestos was also used as a cheap insulation material. However, with use and abrasion fine asbestos fibres can be released from the enveloping cement or insulation and create a toxic atmosphere.

The asbestos industry has been closed down in many countries with the realisation that the material has led to many deaths. Europe had been the first region which developed the industrial use of asbestos and many European countries banned its use in the latter part of the twentieth century. The rate at which this happened, however, varied greatly from country to country despite the clear medical evidence that began to be acquired from the beginning of the twentieth century. At about the same time research and use of what are known as "whiskers", which are very fine crystalline filaments with diameters between $0.5\,\mu m$ and $1.5\,\mu m$ and which in the 1960s were of interest because of their potential very high strengths, was also abandoned because of their health implications. Today society is faced with the costs and difficulties of safely removing asbestos from buildings and these are considerable.

Asbestos is obtained by open cast mining and despite being banned in many industrialised countries it continues to be mined and used in some countries. The biggest producers are Russia, China, Kazakhstan and Brazil. Previous large producers were the Quebec Provence of Canada and South Africa but the use and mining of asbestos in these countries were both banned by 2018.

For a detailed account of the asbestos industry, its history, uses of asbestos and the medical conditions associated with its use the reader is referred to the excellent Wikipedia related article

## 2.3   Regenerated fibres

Throughout most of history there were only natural fibres available however during the eighteenth and nineteenth centuries chemists in a number of European countries examined ways of dissolving cellulose. The goal of this research was to produce artificial silk for the textile industry. The first commercialised man-made fibres were produced by the Comte H. DE CHARDONNET in France in 1892 however these cellulose nitrate fibres were highly inflammable. Two years later Charles Frederick CROSS together with Edward John BEVAN and Clayton BEADLE, in England, patented viscose silk in 1894 which dissolved the cellulose in sodium hydroxide after first reacting

**Table 2.3** Some typical properties of rayon fibres compared to other natural and synthetic fibres (Manian et al., 2018).

| Fibre | Strength (GPa) | Strain to Failure $\varepsilon$ (%) | Young's Modulus $E$ (GPa) |
|---|---|---|---|
| Viscose tyre cord | 0.7 | 11 | 15 |
| Viscose textile | 0.3 | 16 | 11 |
| Lyocell | 0.6 | 10 | 25 |
| Flax | 0.95 | 2 | 60 |

it with carbon disulphide. This gave a viscous solution from which the name "viscose cellulose" or simply "viscose" was derived. The cellulose was regenerated in dilute sulfuric acid and was to find wide use. In 1905, Courtaulds in the UK began the first commercial production of these viscose fibres. Viscose rayon found a ready market in the textile industry but also it became an important reinforcement for rubber as in tyres, pulley belting and other such applications. Development of rayon fibres was stimulated in the twentieth century by the appearance of synthetic Nylon at the end of the 1930s, then followed by polyester, and high tenacity rayon was developed in the 1940s. A technique known as the Lyocell route was developed in the 1980s by dissolving cellulose in N-methylmorpholine N-oxide (NMMO) and together with the viscose process these have become the main means of production.

The source of the cellulose for making rayon fibres is pine, spruce, or hemlock wood pulp as well as cotton residue. The fibres are dissolved in a solvent and the solution filtered to remove any particulates. The solution is then extruded through a spinneret into a bath containing a non-solvent to regenerate the cellulose. The molecular weight of the cellulose is reduced by about two-thirds during this process. The fibres are then drawn to improve molecular orientation, washed and further treated depending on their intended use. The final molecular structure of rayon fibres is semi-crystalline with crystalline fibrils forming chains oriented primarily in the fibre axis direction and linked by hydrogen bonds: the greater the degree of crystallinity the greater the fibre strength. Increases in molecular orientation parallel to the fibre axis produces improvements in tensile modulus. The fibres are usually striated and not perfectly circular in cross-section due to shrinkage induced by the loss of the solvent during manufacture. The degree of drawing imposed greatly affects the tensile behaviour of the fibres with strength and stiffness increasing with drawing at the expense of strain to failure. Strength is lowered when the fibre absorbs water. The specific gravity of viscose rayon is around 1.5 and its softening temperature is 150 °C.

Regenerated cellulosic fibres found use in the early production of carbon fibres. However this route proved not to be suitable for making high-performance carbon fibres as the low percentage of carbon available to make the fibres meant that atomic order was lost during carbonisation. This has however proved useful in slowing heat transfer through carbon-carbon heat shields so such fibres are still of some interest.

## 2.4 Synthetic polymer fibres

The 1930s saw a technical revolution which was to change the world but it happened without most people being aware of its importance. Man-made fibres were produced which did not depend on nature to create their molecular backbone. They are now ubiquitous. This was the beginning of purely synthetic polymer fibres and was to lead to ever increasing high-performance fibres of major importance for composites. It should be noted that that decade also saw the first industrial production of glass fibres. Both glass and synthetic polymer fibres would become important as reinforcements for composites.

The first truly synthetic organic fibres were developed, and commercialised in the late 1930s, independently, in the USA by E.I. DU PONT DE NEMOURS and I.G. FARBEN in Germany. These were polyamide or "Nylon" fibres and these, together with polyethylene terephthalate (PET) or polyester fibres developed in Great Britain in the 1940s, have become the most widely produced commodity synthetic fibres. These fibres, in addition to being important textile fibres find many technical applications as reinforcements for rubber, as in tyres and different types of belting. They are produced by the polycondensation of short molecules.

Polyamide 6,6 fibres, developed initially in the USA are produced by the condensation of hexamethylene diamine $H_2N-(CH_2)_6-NH_2$ and adipic acid $HOOC-(CH_2)_4-COOH$. The combination of these two products results in the elimination of water and the creation of long chain molecules with the repeat unit of $-NH-(CH_2)_6-NH-CO-(CH_2)_4-CO-$ · This gives polyamide 6,6 with the numbers referring to the two groups of six carbon atoms found in the parent molecules and which are found in the repeat unit of the molecular structure which is produced.

Polyamide 6 fibres, produced originally in Germany, are made from a single monomeric material which contains an acidic and amine function. The synthesis makes use of caprolactam which is produced from ε-aminocaproic acid with the formula $H_2N-(CH_2)_5-COOH$. The reaction of this molecule with itself and the elimination of water gives the repeat unit $-(CH_2)_5-CO-NH-$. This is the repeat unit of polycaprolactam or polyamide 6 the number indicating the number of carbon atoms in the repeat unit.

These polyamides are thermoplastics and fibres could be spun from the melt which was an advantage over the more expensive rayon fibres regenerated from solution with the necessary associated solvent recovery step. Spinning from the melt also allows faster production rates. The fibres are spun through spinnerets with holes of around 0.025 mm in diameter. The melting point of polyamide 6,6 is around 260 °C when heated in nitrogen and the molecular weight is of the order of 15 000. The melting point of polyamide 6 is around 215 °C, the molecular weight is around 20 000 and the polymer is more stable and easier to spin than polyamide 6,6. Polyamide 6,6 fibres are generally slightly stiffer and less extensible that the polyamide 6 fibres. These differences in properties open up different markets for the two types of Nylon fibre. The fibres are first spun at speeds of the order of several thousand m min$^{-1}$, cooled to around 70 °C and wound onto bobbins. The fibres can be subsequently further drawn four to five times their initial lengths so as to increase mechanical properties by a better alignment of the molecular structure, although most modern processes include this in a one-step initial operation.

Polyamide fibres are semi-crystalline, generally with a crystallinity of around 45 %. Microfibrils with diameters of around 5 nm consist of blocks of crystalline material made up of folded molecules which can pass from one block to another through a less well ordered region in the fibril. Within the fibrils and between the blocks the molecules form tie molecules which ensure the continuity of the fibrils which are themselves surrounded by other similarly oriented but not perfectly aligned molecules. The structure is anisotropic and as the molecules are not aligned continuously parallel to the fibre axis the properties of the fibres are far lower than what would be expected if the molecules were straight and aligned. The polyamide fibres contain hydrogen bonds which mean that they absorb water which changes their behaviour. This effects both their mechanical properties and their glass transition temperature which can fall dramatically from usually around 55 °C when dry to, in some cases, less than 10 °C (Humeau et al., 2018) when saturated with water. Table 2.4 shows that these fibres possess initial moduli of up to 5 GPa, It should be noted that, it is calculated, if the molecular structure were unraveled and the molecules were all aligned parallel to the fibre axis the result would be a modulus some fifty times greater. This is not possible with Nylon but it will be shown below that other, simpler, polyethylene molecules can be drawn to achieve such a morphology and the fibres do possess very high moduli.

Although work on polyesters in America was initially abandoned, because of the instability of the molecular structure which led to low melting points, it inspired work in England by J.R. WHINFIELD and J.T. DICKSON at the Calico Printers Association, later to be bought by ICI. These two chemists found that polyethylene terphthalate (PET) possessed a melting point of 260 °C. As can be seen from Table 2.4, the molecule contains an aromatic ring which confers on it thermal and chemical stability as well as interesting mechanical properties. This modification to the linear macromolecules found in polyamide defined the routes which chemists would later take to produce organic fibres with the highest tensile stiffness and which are used in high-performance composites.

The polyester fibres which were developed from PET have become, worldwide, the most widely used fibre. PET is obtained from two bifunctional reactants (C2.1);

$$HO-(CH_2)_2-OH \qquad HOOC-\langle\bigcirc\rangle-COOH \qquad (C2.1)$$

ethylene glycol    and    terephthalic acid

Poly(ethylene terephthalate) is then obtained by the reaction between the alcohol and acid groups, with elimination of water. The long chain molecules which are produced have repeat units containing ester functions giving Ch. (C2.2).

$$(C2.2)$$

Originally PET was produced as chips and polyester fibres were subsequently spun from the melt. Today it is more usual to have an integrated production process so bypassing the production of chips. Amongst other economic advantages this latter technique allows the PET to be produced more easily in a dry environment which is essential before spinning if hydrolysis is to be avoided. The water content of the polymer must be inferior by 0.005 % by weight. The PET fibres are spun in a similar manner to that of polyamide fibres at speeds of the order of several thousand m min$^{-1}$ and wound onto bobbins. The polymer is more viscous than polyamide because of the aromatic ring in the molecule and this leads to higher extrusion pressures. The fibres are subsequently further drawn, at a temperature of around 75 °C, four to five times their initial lengths so as to increase mechanical properties by a better alignment of the molecular structure. The molecular morphology of PET fibres is similar to that of polyamide fibres although crystallinity is lower. Table 2.4 shows that these fibres possess moduli of up to 17 GPa.

Polyamide (Nylon) and PET (polyester) fibres are extensively used for reinforcing rubber and these flexible composites are used in fan belts, hoses, flexible drives, moving walkways, tyres and other applications. Respectively, they account for around 25 % and 50 % of the total synthetic fibre production. Fibre reinforced rubber accounts for about the same volume of material as do all the other types of composites combined.

These fibres do not show real elastic behaviour, even at low strains, and plastic deformation as well as creep occurs when the fibre, either by itself or in an elastomeric composite, is loaded. The properties required of these fibres, for example in a tyre, are high strength, dimensional stability, fatigue resistance, thermal stability and good bonding to the matrix.

## 2.4.1   Fracture of Nylon and polyester fibres

Until the late 1960s there was no way of examining the fracture morphology of fine fibres in any detail but the development of the scanning electron microscope (SEM) changed this. Its great depth of field and ability to magnify enormously made it a most important means of examining fibres in detail. It was soon found that the fracture morphologies of Nylon and polyester fibres could be used as a diagnostic tool capable of revealing the mechanisms of failure (Hearle et al., 1998). Monotonic tensile failure in both Nylon and polyester fibres is similar, involving slow crack growth from the surface across the fibre. The crack develops and opens into a V-shape, from or near to the surface as the material ahead of the crack deforms plastically. A point is reached at which the crack becomes unstable and rapid failure occurs, now at right angles to the fibre axis. Almost identical fracture surfaces are obtained with both broken ends. It can be noted that Nylon and polyester fibres do not fibrillate, despite their anisotropic structures which often lead to splitting in other higher modulus organic fibres. However composite structures, for which organic fibres are used as reinforcements, are often subjected to dynamic loading and fatigue can be a major concern. Before the development of the SEM there were many attempts to determine whether these fibres could fail because of a fatigue process analogous to that which had been observed to occur with many metal structures. These were largely inconclusive as they sought to show statistical differences in times to failure compared to what might be expected from creep tests and the results were not clear. In addition the means of subjecting the fibres in a laboratory to cyclic loading were not ideal as their plastic behaviour led to progressive elongation of the fibres during tests which most probably failed by simple tensile processes. However in the early 1970s testing techniques were refined and with the use of the SEM revealed differences in fracture morphologies of these and other thermoplastic fibres which allowed the fatigue failure of the fibres to be identified and associated with a distinctive fracture process. Under cyclic loading, during which the maximum load each cycle is maintained constant, many thermoplastic fibres can be seen to fail by a distinct failure process leading to a characteristic fracture morphology involving longitudinal crack growth, at a small angle to the fibre axis, which gradually reduces the load bearing fibre cross section. This leads to the distinctive fatigue failure morphology which is very different from tensile or creep failure (Bunsell and Hearle, 1971). The loading conditions which lead to fatigue failure are that the cyclic load amplitude is sufficiently large but, unusually, a necessary criterion for fatigue failure is that the minimum cyclic load be nearly zero. Indeed it is possible to avoid fatigue by increasing the overall loading pattern on the fibre and if this does not raise it into a regime of creep failure, the fibre will no longer break. Work by Herrera Ramirez and Bunsell (2005) revealed that particles, used to control fibre polymerisation and spinning in the fibres, initiated fatigue failure and possibly tensile failure. At room temperature crack initiation was normally at a weak point in the fibre at the interface between the skin and core of the fibre.

The loading conditions for the fatigue failure, giving this distinctive fracture morphology, in the laboratory of individual fibres were unambiguous so for many years it remained a mystery why these breaks were difficult to find in a real structure, such as a tyre or belting which had failed during cyclic loading. Work by Le Clerc et al. (2007) showed that in many structures, such as reinforced rubber, the cyclic loading induced heating which could raise the temperature of the fibres to near or even above their glass transition temperature. Both the fibres and the surrounding rubber are good thermal insulators which means that the heat generated due to hysteresis effects cannot be dissipated and results in the rise in temperature. As single fibres were tested at increasing temperatures the initiation of fatigue failure was seen to no longer be concentrated only at particles near the surface but were associated with particles throughout the bulk of the fibre. This gave a much more complex fracture morphology of the type found in the thermoplastic fibres used to reinforce rubber composites and is possibly due to the progressive weakening of hydrogen bonds between microfibrils making up the fibres.

The development of polyester fibres showed the stabilising effect of introducing aromatic rings into the molecular structure leading to increased stiffness and higher melting points. The Nomex® fibre comes from the family of Nylon fibres but contains two aromatic rings which are joined at the base, as Table 2.4 shows. Nomex® is described as belonging to the meta-aramid group and its development foreshadowed the production of other aromatic polyamides (aramids). The meta-oriented phenylene radicals in Nomex® result in the molecule being able to rotate where

Table 2.4: Synthetic polymer fibres.

| Fibre | Repeat Unit in the Macromolecule | Specific Gravity | Maximum Young's Modulus $E$ (GPa) | Melting Point or Decomposition Temperature °C |
|---|---|---|---|---|
| Polyamide 6 (Nylon 6) IG Farben | | 1.14 | 4 | 230 |
| Polyamide 6,6 (Nylon 6,6) DuPont™ | | 1.14 | 5 | 260 |
| Polyethylene terephthalate (PET) ICI | | 1.38 | 15 | 260 |
| Poly(m-phenylenediamine-isophthalamide) (Nomex®) DuPont™ | | 1.38 | 17 | 400 |

they join resulting in a tensile stiffness which is still only about that of highly drawn PET. The fibre is however flame resistant, thermally stable to around 150 °C over the melting point of the aliphatic polyamides and PET and has good dielectric properties. As a consequence the fibre is used in apparel for which resistance to heat is required. It is also used in chopped form to make a paper material which is widely used as a honeycomb in advanced composites structures. The low specific gravity of Nomex® (1.38) compared to aluminium (2.7), which is also used for honeycomb structures, gives it an obvious advantage for applications for which weight saving is important such as in aircraft.

## 2.5   High-moduli synthetic polymeric fibres

### 2.5.1   Aramid fibres

The development of Nylon fibres first commercialised in 1938, and polyester (PET) first commercialised in 1947, was followed by ever more advanced synthetic polymeric fibres some of which are important as reinforcements for composite materials. In order to obtain high strength and above all high stiffness aramid polymers were developed and began to be commercialised in the early 1970s, containing predominantly para-oriented aromatic units. This is the case of poly(p-phenylene terephthalamide) (PPTA) which is the polymer used to make the Kevlar®fibre, made available from DuPont™ from around 1972 and shortly afterwards the Twaron®fibre produced by Akzo and since then ceded to Teijin Aramid. Since then Kevlar® and other aramid fibres have become important reinforcements for composite materials. The form of the PPTA molecule is shown in Table 2.5. It is prepared from the low temperature polycondensation of p-phenylene diamide (PPD) and terephthaloylchloride (TCl) in an appropriate solvent.

The PPTA polymer is produced at high molecular weights to give good fibre properties. The molecular weight of around 20 000 is similar to that found with polyamide 6,6. PPTA is insoluble in most solvents but a 20 % solution of the polymer in concentrated (>98 wt%) sulphuric acid is one which results in the production of a mesophase or liquid crystal solution. The rod like PPTA molecules are randomly oriented in the solution but beyond a critical concentration locally the molecules are attracted together and adopt an ordered arrangement in small domains to achieve better packing. The domains are still randomly arranged with respect to one another and the solution retains the flow properties of a liquid but in polarised light can be seen to be locally oriented. When the solution is passed through the holes of a spinneret the induced shear forces orient the molecular structure so that a highly oriented fibre can be produced without drawing. The fibres are spun through an air gap before entering a coagulating bath. The fibres are then washed and dried. The fibres can then be subsequently treated at high temperature and under tension to increase crystallinity and orientation. In this way a family of fibres can be produced.

Although the bonds are not completely straight in these fibres they are nearly so and this results in a high-tensile stiffness as well as thermal stability. The glass transition temperature of PPTA is over 375 °C and the polymer is thermally stable up to about 550 °C. The same type of fibre has been commercially produced under the name Twaron®. It was developed by Akzo in Holland and is now produced there by the Japanese company Teijin Aramid. These aramid fibres were truly remarkable as they took organic fibres from a range of stiffness of less than 20 GPa to 135 GPa.

Aramid fibres are five times as strong as steel in air for the same weight and this difference rises to thirty times when they are used in water, because of their low density and the buoyancy provided by water. It means that ropes and cables for mooring large oil tankers or platforms can be made from aramid fibres. The hydrogen transverse bonding holding the PPTA molecules together means however that they absorb water. The amount of water uptake varies with the conditions of humidity but up to 6 wt% is possible. The highly anisotropic structure of aramid fibres also results in them being weak in all other directions than simply parallel to the fibre axis. As a consequence failure of these fibres is nearly always highly fibrillar.

The structure of aramid fibres is very well ordered and aligned preferentially parallel to the fibre axis. Their highly anisotropic structure and behaviour means that the fibres are not used in primary loading structures subjected to compressive forces. It also means that the fibres split easily on failure as shown by Figure 2.5.

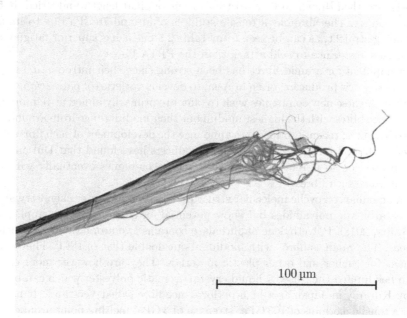

100 μm

Figure 2.5: The failure of an aramid fibre such as Kevlar® usually reveals its highly anisotropic fibrillar microstructure.

The anisotropy of aramid and similar organic fibres is difficult to quantify so that special equipment has been developed to measure their transverse properties. The transverse compression test consists of compressing diametrically a Kevlar® fibre between two parallel and rigid plates. This is known as the Single Fiber Transverse Compression Test (SFTCT) (Wollbrett-Blitz et al., 2016). The upper plate is made of sapphire, chosen for its high modulus and its transparency, so the test can be followed using a camera located on the upper side. The design of the device was conceived so as to ensure a compression cycle (loading/unloading) avoiding any drift in displacement during the experimental procedure. Modelling the results of the study revealed that the presence of a skin-core structure had to be taken into account to accurately calculate transverse properties. When this was done, the core was found to have a tensile transverse modulus of 3 GPa and the skin 0.2 GPa. This means that the transverse stiffness is about one twentieth of the longitudinal stiffness of a Kevlar® fibre and similar results could be expected from other high modulus highly oriented organic fibres.

The processes which limit the compressive properties of aramid fibres confer on them however great tenacity which means that they absorb much more energy than more brittle fibres when broken. This is because the fibre can deform plastically in compression. An elastic fibre can be bent to a minimum radius of curvature at which the tensile stresses in the convex surface attain the breaking stress of the material. When an aramid fibre is bent slippage of the molecular structure occurs on and near the concave surface leading to the development of kink bands. The result of this plastic deformation is that the neutral stress axis is displaced from the fibre axis towards the convex surface. In this way the tensile stresses on the convex surface do not attain the failure stress of the fibre, which can be bent to a zero radius of curvature. This gives the fibre great tenacity and resistance to impact loading but also means that cloth made from the fibres is difficult to cut.

The Technora® fibre is an aramid produced by Teijin, in Japan. This fibre is made from copolyamides containing PPTA and 3,4′-diphenylether. As it is produced from three basic products it is more expensive to make than the Kevlar® or Twaron® fibres but has significantly different properties. The reaction mixture is filtered and wet spun into an aqueous coagulating bath. The fibres are then drawn to 6 to 10 times their original length and dried at 500 °C to form the final product. The fibre has a tensile stiffness of around 70 GPa due to its molecular structure not being straight, as can be seen from Table 2.5 but it has superior fatigue properties and reported higher resistance to acid attack than the PPTA fibres.

Market development for aramid fibres has been strong since their introduction in the 1970s and this has tempted new producers, often in Asia, to develop varieties of para-aramid fibres. The types of fibres which these new companies wish to offer are primarily aimed at reinforcing rubber so they are not those fibres with the highest moduli but their moduli range from around 80 GPa to 115 GPa. One of the most recent producers to announce the development of such fibres is Hyosung in South Korea with a fibre called Alkex®. Other producers have found that DuPont™ has very strong patent protection so it remains to be seen which companies eventually will be able to commercially produce such fibres.

Fibres with an enhanced cyclic molecular structure based on polyester chemistry are less well known than the aromatic polyamides but show potential for some composite applications (Pegoretti and Traina, 2018). Polyethylene naphthalate contains two aromatic rings which however are not collinear. The result is fibres with moduli about double that of PET. They can be used as reinforcement for rubber and other plastic materials. They are however more expensive to produce which has limited their use. A liquid crystal aromatic polyester which can be melt spun is produced by Kuraray in Japan as a high-performance fibre called Vectran™. It has a specific gravity of 1.4; a tensile modulus of 103 GPa; strength of 3 GPa; melting point around 300 °C and high thermal stability with a degradation temperature of the polymer above 400 °C.

## 2.5.2   Heterocyclic and other high modulus polymer fibres

Other polymers with heterocyclic structures have been developed which have remarkable properties. The Zylon® fibre is made from poly-p-phenylenebenzobisoxazole, which is one of the polybenzazoles (PBO) class of polymers developed by the US Air Force as polymers to resist heat better than Kevlar®. It was first produced commercially in 1998 by Toyobo in Japan. It possesses the highest stiffness of all commercial organic fibres. This fibre is forty percent stiffer than steel for a fifth of the density. As the molecule is straight, the Zylon® fibre, as shown in Table 2.5 has a tensile stiffness twice as great as that of Kevlar®. Zylon® fibres absorb very little water (0.2 wt%) which fits into residual porosity left behind by the evaporation of the solvent. Lateral cohesion is dominated by Van der Waals bonds which explains both its hydrophobic character and also its anisotropy leading to low compressive properties and fibrillation on failure.

The scientific literature shows continuing interest in the development of even higher performance fibres but whether they will become commercially attractive is uncertain. A fibre made from polypyridobisimidazole (PIPD) has been reported. It is called M5 and was produced by the researchers from the Dutch company AkzoNobel. It was then ceded to Magellan Systems International which in turn joined the DuPont™ company. The molecule is related to PBO and the Young's modulus claimed for the fibre of 330 GPa is even higher. The producers claim that there is a highly significant difference in the behaviour of the PIPD fibre as lateral cohesion is determined by hydrogen bonds. As mentioned above, the PBO fibre relies on Van der Waals bonds for lateral cohesion which are very weak. Hydrogen bonds are ten times stronger which means that the M5 fibre has superior radial and compressive strengths. The presence of hydrogen bonds will probably mean that the fibre will absorb water however. The M5 fibre seems to be

Table 2.5: High-performance organic fibres with cyclic molecular backbones.

| Fibre | Repeat Unit in the Macromolecule | Specific Gravity | Maximum Young's Modulus $E$ (GPa) | Melting Point or Decomposition Temperature °C |
|---|---|---|---|---|
| Copoly(1,4-phenylene/3,4'-diphenylether terephthalamide (Technora®) | | 1.39 | 70 | 500 |
| Poly(p-phenylene terephthalamide) (Kevlar® 49) | | 1.45 | 135 | 550 |
| Poly(p-phenylene benzobisoxazole) (PBO, Zylon®) | | 1.56 | 280 | 650 |

very interesting from a technical view point and addresses the problem of anisotropy of aromatic fibres. It has, inevitably attracted attention for use in body armour. However, it remains unclear if this fibre will ever become commercially available.

Another avenue of research is to produce thermoplastic fibres which contain carbon fibre nano-tubes, which in the single wall form have diameters or the order of 3 nm. The extremely high Young's modulus and even more impressive theoretical strength of carbon fibre nano-tubes are seen as potential reinforcements within fine fibres. This can be seen as analogous to the structure of aramid and other similar fibres which are formed with very stiff molecules which become aligned during passage through a spinneret. So far though these fibres have not become available and scientific efforts to make them have largely been disappointing.

### 2.5.3    Ultra-high molecular weight polyethylene (UHMWPE)

The increase in mechanical properties of the fibres described above has come about by using increasingly difficult chemistry to align the molecular structure parallel to the fibre axis by making the basic molecule as stiff as possible. However it can be shown that even the simplest aliphatic polymer molecule has a very high intrinsic stiffness above 200 GPa. If this could be exploited so as to make fibres with simple polymeric molecular structures aligned along the fibre axis direction it would be expected that high-performance fibres would result. Such fibres are made from polyethylene $(C_2H_4)_n$; one of simplest polymers. This type of fibre is known as "ultra-high molecular weight polyethylene" (UHMWPE). High-modulus polyethylene fibres are produced using a dilute ($<5\%$) sol-gel, in which the polymer is dissolved in a solvent and then spun. High modulus polyethylene fibres were first produced commercially in the Netherlands in 1990 by DSM in collaboration with Toyobo in Japan under the name Dyneema® , and then in the USA by Honeywell, under the name Spectra® (Vlasblom, 2018). Since then other firms in China produce similar fibres. They have properties rivalling those of the aramid fibres, with stiffness of around 100 GPa and with a lower specific gravity of 0.97. This means that cables made from high-modulus polyethylene float which is a big advantage over other high modulus fibres when used in the off-shore industry. However, although the alignment of the molecules increases stiffness and strength, the thermal properties of the fibres are those of polyethylene. These fibres are limited in temperature to a maximum of 120 °C and suffer from creep. Nevertheless the chemistry of polyethylene and particularly high-density polyethylene makes the fibres particularly resistant to attack from chemicals such as acids and alkaline agents. Adhesion between the matrix of a composite and these fibres can be difficult however this does help in resisting ballistic impacts. The Van der Waals bonds between the long chain polyethylene molecules result in an anisotropic structure and fibrillation occurs on failure. This characteristic is found with other high-modulus polymeric fibres.

High modulus polyethylene fibres have found markets as high strength mooring ropes for large ships and oil platforms, motor cycle and military composite helmets, loud speaker cones and other composite structures, as well as in flak jackets and similar anti ballistic structures. They also find use in bio-medical applications and catheters

## 2.6    Glass, chemical vapour deposition (CVD) and carbon fibres

### 2.6.1    Glass fibres

Silica $(SiO_2)$ is the basis of glass fibres but to make them in an analogous fashion to the spinning and drawing of polymeric fibres the melting point of the glass has to be lowered. This is achieved by combining the silica with other oxides which have the effect of lowering the melting point

**Table 2.6**  Chemical composition (wt%) of glass fibres and basalt used as reinforcements

| Glass Type | E | S | Advantex® | Basalt |
|---|---|---|---|---|
| $SiO_2$ | 54 | 65 | 60 | 52 |
| $Al_2O_3$ | 15 | 25 | 13 | 18 |
| CaO | 18 | – | 22 | 5 |
| MgO | 4 | 10 | 2 | 1.3 |
| $B_2O_3$ | 8 | – | – | – |
| $F_2$ | 0.3 | – | < 0.5 | – |
| FeO | – | – | – | 2 |
| $Fe_2O_3$ | 0.3 | – | < 0.5 | 4 |
| $TiO_2$ | – | – | < 1 | 1.2 |
| $Na_2O$ | – | – | 1 | 6.4 |
| $K_2O$ | 0.4 | – | < 2 | 4.5 |
| Density | 2.54 | 2.49 | 2.62 | 2.73 |
| Strength (at 20 °C, GPa) | 3.5 | 4.65 | 3.5 | 4 |
| Elastic Modulus (at 20 °C, GPa) | 73.5 | 86.5 | 80 | 100 |
| Failure Strain (at 20 °C, %) | 4.5 | 5.3 | 4.4 | 4.5 |

from around 1750 GPa to around 1250 GPa, or less for some glasses. These additions to the silica also control its viscosity when fibres are drawn and allow speciality glasses to be produced. In this way platinum spinnerets can be used which resist the high temperatures.

Glass filaments have been formed since Roman times and before. More recently the production of fine glass filaments was demonstrated in Great Britain in the nineteenth century and used as a substitute for asbestos in Germany during the First World War. In 1931 two American firms, Owens-Illinois Glass Company and Corning Glass Works developed a method of spinning glass filaments from the melt through spinnerets. The two firms formed a separate company in 1938 called Owens-Corning Fiberglass. Since that time extensive use of glass fibres has been made. Initially the glass fibres were destined for filters and textile uses however the development of heat setting resins opened up the possibility of fibre reinforced composites and in the years following the Second World War the fibre took a dominant role in this type of material. Today, by far the greatest amount of advanced composite materials is reinforced with glass fibres.

Fibres are produced by extruding molten glass, at a temperature, usually around 1250 °C, through holes in a spinneret with diameters of one or two millimetres and then drawing the filaments to produce fibres having diameters usually between 5 μm and 20 μm. The spinnerets usually contain several hundred holes so that a strand of glass fibres is produced. The approximate compositions and properties of the most common types of glass fibres used in composites are shown in Table 2.6 (Jones and Huff, 2018). The most widely used glass for fibre reinforced composites is called E-glass; glass fibres with superior mechanical properties are known as S-glass. Advantex® is a glass fibre which has been developed by Owens-Corning Fiberglass. Other glass fibre types exist, for example, for insulation. The properties of basalt are also shown in Table 2.6. Basalt is a naturally occurring material which can be melted and spun into glass type fibres. It is discussed later in this chapter.

The strength of glass fibres depends on the size of flaws, most usually at the surface, and as the fibres would be easily damaged by abrasion, either with other fibres or by coming into contact with machinery in the manufacturing process, they are coated with a protective size. The purpose of this coating is both to protect the fibre and to hold the strand together. The size may be temporary, usually a starch-oil emulsion, to aid handling of the fibre, which is then removed and replaced with a finish to help fibre matrix adhesion in the composite. There are two main types of coupling agent used in composites. These are organometallic compounds which are typically cobalt, nickel, lead, titanium or chromium, or, more commonly, organosilanes. They are applied as an aqueous solution. The organosilanes have the general structure $X_3SiR$, in which X is either an alkoxy or a halogen group and R is an organofunctional group which reacts with the glass provoking the hydrolysis of the alkoxy groups to form silanol groups. These groups then react with the silanol groups on the glass surface creating siloxane (Si−O−Si) bonds, as is shown in Figure 2.6.

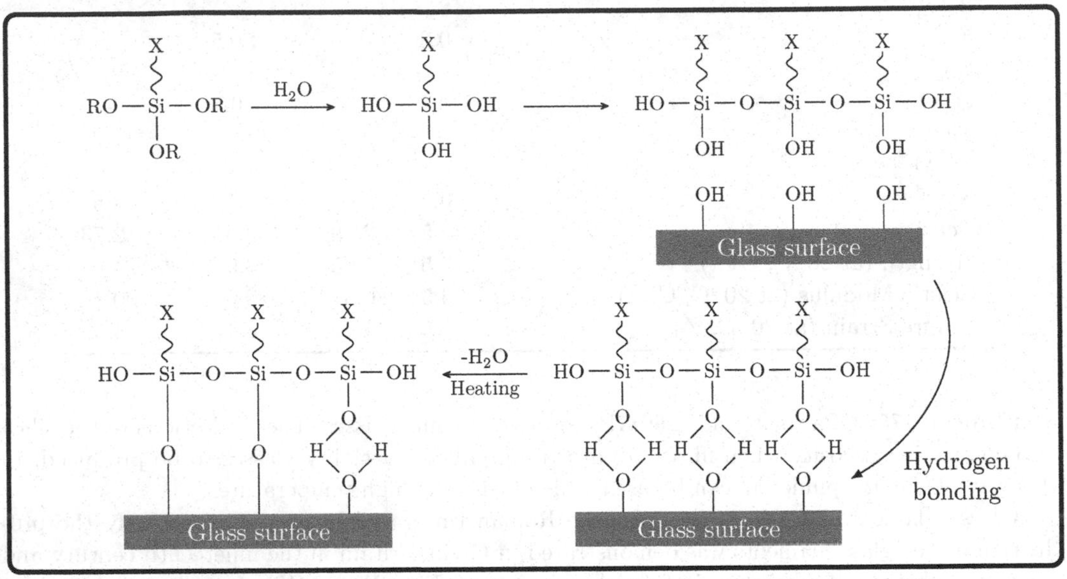

Figure 2.6: Interfacial bonding between the glass surface and an organosilane size.

The size may be of a type which has several additional functions which are to act as a coupling agent, lubricant and to eliminate electrostatic charges, as in the production of short glass fibres to be used to make sheet moulded compound and similar materials.

Continuous glass fibres may be woven, as are textile fibres; made into a non-woven mat in which the fibres are arranged in a random fashion; used in filament winding or chopped into short fibres. In this latter case the fibres are chopped into lengths of up to 5 cm and lightly bonded together to form a mat, or chopped into shorter lengths of a few millimetres for inclusion in moulding resins. Examples of the use of different forms of glass fibres are given in Chapter 4.

The structure is vitreous with no definite compounds being formed and no crystallisation taking place. The structure of glass based on silica is shown schematically in Figure 2.7.

An open network results from the rapid cooling which takes place during fibre production with the glass cooling from about 1500 °C to 200 °C in between 0.1 s to 0.3 s. Despite this rapid rate of cooling there appears to be no appreciable residual stresses within the fibre and the structure is isotropic although with a very slight reduction in density compared to the glass in bulk form. Density can evolve towards this latter state with heating. Glass is set to remain the most widely used reinforcement for general composites. They are cheaper than most other relatively high-modulus fibres and because of their flexibility do not require particularly specialised machines or techniques to handle them. The elastic modulus of glass fibres is however low when compared to many other fibres and the specific gravity of glass is relatively high. The low specific value of the mechanical properties of glass fibres means that they are not ideal for structures requiring light weight as well as high properties, although considerations such as their relatively low cost compared to other high-performance fibres and the ease of composite manufacture means that there are used in some such structures.

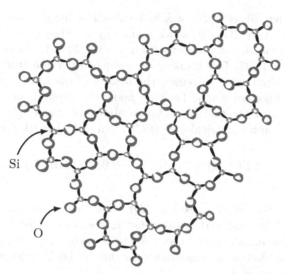

Figure 2.7: The structure of glass is based on a three-dimensional network of silica ($SiO_2$) without any long range order.

## 2.6.2 Basalt fibres

Basalt is an igneous rock which is widely found in the earth's crust. It melts in the range from 1400 °C to 1600 °C which is higher than the glasses usually used to make fibres. However its abundance makes it an attractive and cheap source of a silicate based material which, once molten, flows easily so making it attractive for fibre production. It is a naturally occurring material and Table 2.6 gives some general idea of the most frequently encountered chemical compositions to be found in basalt. Basalt is a silicate mineral and as such has much in common with glass. Its relatively high melting point has hindered its use in making fibres however continuous basalt fibres are now available made mainly by companies in the former Soviet bloc. The fibres show linear elastic behaviour in tension; they are said to be more stable, to creep at lower rates and be less likely to fail by stress corrosion than glass fibres. The fibres are circular in cross-section and usually made with diameters from 9 μm to 12 μm (Militký et al., 2018). The fibres are cheap and can be expected to compete with standard glass fibres as reinforcements in composite materials.

## 2.6.3 Boron and silicon carbide fibres made by chemical vapour deposition

Up until the early 1960s glass fibres were the stiffest filaments available but their specific moduli were only comparable to that of common metals such as steel or aluminium. The aircraft and burgeoning aerospace industries were the driving force behind the search for more advanced materials. In the early 1960s, interest in the production of fibres which combined a high modulus with low density concentrated efforts on the lightest elements in the Periodic Table. Stiffness is not related to density so some of the lightest elements were candidates. The lightest element which possesses high mechanical properties is the fourth element in the Periodic Table, beryllium, which has a high Young's modulus of over 300 GPa and a density of 1.83. However Beryllium is highly toxic, especially as an oxide and interest therefore centred on the other light elements which were the fifth, boron and the sixth, carbon. Boron fibres were produced in the USA at the beginning of the 1960s. They had remarkable properties with a Young's modulus exceeding 400 GPa as shown in Table 2.7 and quite extraordinary strength in compression. Boron is a hard material and cannot be drawn into fibres but it was found that it could be made by chemically vapour deposition (CVD) onto a core filament, which eventually was chosen to be tungsten. Fortunately tungsten filaments were made for the elements in electric light bulbs however it is very dense with a specific gravity of 19.35. This is one of the reasons why these fibres were made with large diameters, typically 140 µm so as to reduce the density of the final fibre. The fine and dense microstructure of the deposited material ensures maximum strength and Young's modulus. The CVD technique meant that the continuous boron fibres were made individually by reaction at high temperature between boron trichloride ($BCl_3$) and hydrogen ($H_2$) so that (C2.3):

$$2\,BCl_3(g) + 3\,H_2(g) \xrightarrow{t\,^{\circ}C} 2\,B(s) + 6\,HCl(g) \tag{C2.3}$$

The tungsten filaments had diameters of 12 µm and that of the final diameter of a typical boron fibre 140 µm, which produced a rather fat fibre and excluded it from being woven. At first boron fibres had no competitors and were most famously used to form the boron aluminium tubes making up the skeleton of the American space shuttle. Their high cost and lack of flexibility has limited their use however.

The primary commercial production of boron fibres is with Speciality Materials Inc. based in Lowell, Massachusetts, USA. They make a range of boron fibres with diameters from 76 µm to 279 µm and as the fibre is a composite structure the details of their properties change with the size of the diameter. Speciality Materials Inc. also uses the same CVD technique to produce silicon carbide (SiC) fibres and their composites. A company called TISICS based in Farnborough in the UK also produces large diameter SiC fibres and as the name suggests produces both the fibres and titanium matrix composite materials. Both large diameter (33 µm) carbon fibres and also tungsten wire are used in the manufacture of CVD SiC fibres (Luo and Jin, 2018). They have large diameters. In a typical process, with $CH_3SiCl_3$ as the reactant, SiC is deposited on the core as follows (C2.4):

$$CH_3SiCl_3(g) \longrightarrow SiC(s) + 3\,HCl(g) \tag{C2.4}$$

Silicon carbide fibres on cores are made with diameters from 100 µm to 140 µm and like boron fibres cannot be woven. Typical properties of these fibres are shown in Table 2.7. Although a range of applications are described for these large diameter SiC fibres the main interest is for reinforcing titanium for which they must be coated to prevent corrosion at the interface. The fibres are usually coated with SiC or $B_4C$. These titanium matrix composites are considered for gas turbine engine components. Such engines operate at very high temperatures so that heavy nickel based alloys, which withstand temperatures up to 1100 °C, are used. Even so the nickel based alloys need to be coated with a ceramic to survive in the engines. An advantageous reduction in weight, achieved by using CVD SiC fibres wound circumferentially to make up a titanium matrix composite cylinder as part of the cooler part of the body of the jet engine, would extend the use of titanium from 450 °C to 600 °C.

## 2.7  Carbon fibres

Carbon  is the sixth lightest element and the carbon-carbon covalent bond is the strongest in nature ($4000\,kJ\,mol^{-1}$), however the arrangements of the bonds and the distances between the carbon atoms can vary so giving many types of carbon, including graphite, diamond and amorphous forms. Graphite is a regular, highly anisotropic, two-dimensional array whereas a three-dimensional arrangement gives the structure of diamond. Graphite is a well defined, highly anisotropic, state of carbon in which the carbon atoms occupy planes which are separated from

Table 2.7: Typical properties of boron and silicon carbide fibres made by chemical vapour deposition

| Fibre Type | Manufacturer | Trade Mark | Composition (wt%) | Diameter (μm) | Specific Gravity | Strength (GPa) | Strain to Failure $\varepsilon$ (%) | Young's Modulus $E$ (GPa) |
|---|---|---|---|---|---|---|---|---|
| SiC | Speciality Materials Inc. | SCS-6 | SiC on carbon core | 140 | 2.7–3.3 | 3.4–4.0 | 0.8–1.0 | 427 |
| SiC | TISICS | SM3256 | SiC on W core | 140 | 3.3 | 3.8 | 0.8 | 400 |
| SiC | TISICS | SM3240 | SiC on W core | 100 | 3.4 | 3.6 | 0.95 | 380 |
| B | Speciality Materials Inc. | | B on W core | 140 | 2.48 | 4.0 | 1 | 400 |

one another by a distance of 0.34 nm. This should be borne in mind as many fibres which are described as graphite fibres contain no graphite however a close but less well ordered turbostratic carbon phase and so it is more correct to speak of carbon fibres. In 1901, T. EDISON used bamboo fibres and converted their cellulose based structures into carbon filaments for his electric light bulbs. Although T. EDISON was not the first person to attempt to produce electric light bulbs advances in vacuum technology allowed him to succeed as the use of carbon filaments required the exclusion of oxygen. He became the first person to patent carbon fibres but they proved very brittle and after three years were replaced by tungsten. Interestingly however short carbon fibres were developed in the USA in the 1950s and early 1960s using viscose fibres regenerated from cellulose (Bacon and Shalamon, 2005). This proved a slow process as the useful carbon yield from cellulose is only 24 % and mechanical properties were not high however such fibres had relatively low thermal conduction properties and are still used in carbon-carbon heat shields and brake pads. Work on converting polyacrylonitrile (PAN) had independently begun around this time in Japan and England, where as early as 1950 researchers at the Shirley Institute in Manchester, England, noted that PAN turned black when heated in air. This was the first step to converting it to carbon fibre. Akio SHINDO, working at the Government Industrial Research Institute, Osaka (GIRIO) began to work on the conversion of PAN to produce carbon fibres in the 1950s and this led to Japanese national patents showing that PAN could yield about 50 wt% carbon. Work at the Royal Aircraft Establishment by WATT and colleagues (Moreton et al., 1967) in England finally produced carbon fibres which could be commercially produced, in 1967, with properties which qualified the fibres as being truly being remarkable high-performance filaments. At this time several companies in England, of which the most significant was Courtaulds, began producing carbon fibres. A vital aspect of their work was to maintain the filaments under tension throughout carbon fibre production. This was because without being held under tension the molecular orientation induced in the PAN precursors by drawing was lost during the heat treatment to make carbon fibres during which half of the matter in the precursor was lost. This opened up the way for technical carbon fibres to be produced. Toray, in Japan, began commercial production in 1970.

The PAN fibres, held in tension, are first heated to about 250 °C in air which introduces oxygen into the polymer cross linking the structure making it infusible. Further heating under a nitrogen atmosphere leads to the loss of the oxygen and hydrogen so that at 1000 °C only carbon and about 7 wt% of nitrogen remain. Heating to above 1000 °C progressively eliminates the nitrogen and around 1500 °C to 1600 °C a fully carbon fibre is achieved. At this temperature the strength of carbon fibres reaches a maximum. The fibres are made in a continuous manner with the lengthy processes occurring with the fibres winding back and forth in the heating furnaces. Fibre production rates are of the order of $100 \, \text{m} \, \text{min}^{-1}$. The structure which results from the pyrolysis of PAN is highly anisotropic, with the basic structural units formed by the carbon atom groups aligned parallel to the fibre axis. They form imperfectly stacked layers called turbostratic layers with the interlayer spacing being greater than that of graphite. There is complete rotational disorder in the radial direction and the relatively poor stacking of the carbon atoms means that a graphite structure is never achieved. The structure contains pores which account for the density of the fibres being less than that of fully dense carbon. Table 2.8 shows some typical properties of carbon fibres; however there exists a large range of PAN based carbon fibres as their properties can be controlled by altering the pyrolysis temperature and also improvements in the precursor and spinning techniques have led to impressive increases in strength and strains to failure. Figure 2.8 shows how by varying the pyrolysis temperature several types of carbon fibres can be produced by the different producers by using the same precursor. It can be seen that there are two distinct groupings of carbon fibres. The earlier produced fibres had diameters of around 7 μm but later produced fibres had diameters of 5 μm. This change meant that the cross section of the fibres was reduced by a half which reduced the probability of finding strength reducing faults on the fibre. This is discussed at the end of this chapter. This

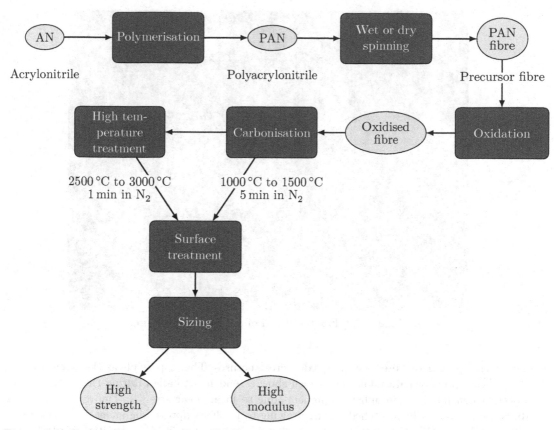

Figure 2.8: Carbon fibres made from PAN can be processed at different temperatures in order to change their properties. The filaments must be kept under tension so as to maintain atomic orientation.

resulted in a considerable increase in strength, up to 6.5 GPa, which is combined with a more efficient microstructure after pyrolysis so that the Young's modulus of the higher strength fibres also increased from 230 GPa up to 294 GPa. The fracture morphology of a T700 fibre is shown in Figure 2.9.

Carbon fibres made from pitch, which is the residue of petroleum refining or by coal tar distillation in the steel industry, were developed by Union Carbide in the USA in the early 1970s which then transferred the technology to AMOCO and in Japan in the 1980s principally by Mitsubishi Chemical Corporation. This technology is now used at Mitsubishi Rayon. The high carbon yield of pitch which approaches 90 % makes it an attractive and potentially cheap source for making carbon fibre precursors. Cost however is increased by the purification processes which are necessary for this naturally occurring material. In addition the production of high-performance carbon fibres requires that the microstructure of the precursors be highly aligned. In order to achieve this it is imperative that the polycarbon layers in the precursor fibre be aligned with the fibre axis. Some carbon fibres are made from pitch with no attempt to align their structures and the result is fibres with a low Young's modulus of around 40 GPa. These fibres, which are made from isotropic pitch are used for their chemical inertness and low cost for the reinforcement of cement and concrete. The biggest producer of this type of isotropic pitch based carbon fibre is Kureha in Japan. High-performance carbon fibres made from pitch require that the pitch be converted into a mesophase or liquid crystal solution which is then

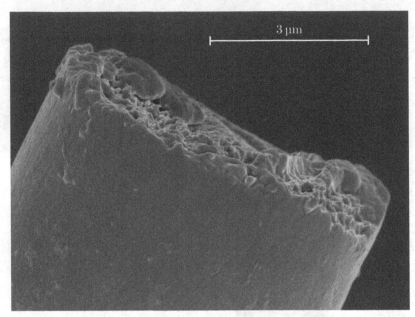

Figure 2.9: Fractured end of a T700 carbon fibre.

spun giving precursor fibres with aligned microstructures. This is possible as the pitch contains large planar polyaromatic molecules which show a tendency to align during the melt spinning process when passing through the spinneret due to local shear stresses. The mesophase pitch fibres are mechanically stretched during drawing to further improve alignment. The mesophase is formed by heating the pitch to a temperature of 400 °C to 450 °C and initially is formed of spherical liquid crystals of diameters of about 1 µm which grow until a phase inversion occurs to create a continuous nematic liquid crystalline mesophase. From this point the fabrication route is the same as that for the PAN based fibres although the final carbon fibre diameter is usually greater and usually around ten or eleven microns. Unlike PAN based fibres, there is no peak in strength at 1500 °C. Pitch based precursors heated to around 2300 °C give fibres with Young's moduli as high as those obtained with PAN based fibres at 2900 °C and heating to these higher temperatures gives even greater stiffness of up to four and a half times that of steel. Strength does not change much with temperature which means that pitch based carbon fibres become increasingly brittle as their modulus is increased. It is more economical to produce high-modulus carbon fibres from pitch than it is from PAN and high-modulus pitch based carbon fibres are made with moduli up to 935 GPa. The properties are due to a less disordered microstructure of the pitch based fibres which can become truly graphitised as the average stacking distances of the carbon atoms of such fibres made at temperatures above 2900 °C is less than 0.345 nm. The larger basic structural units in pitch based carbon fibres when compared to PAN based fibres coupled with a greater inherent anisotropy however leads to lower compressive strengths and increased difficulties and costs in producing high strength carbon fibres. The internal morphology of pitch based carbon fibres can be varied so that various producers have made fibres with a random or radial structure, which can lead to radial cracks forming during the pyrolysis process, or a circumferential arrangement of the carbon layers, which is usually preferred.

Carbon fibres made by whatever route have surfaces in which the carbon atom cycles tend to be arranged parallel to the surface which means that there are few available pendant atomic bonds with which a matrix resin can easily bond. The higher the Young's modulus of the carbon fibre the lower its tendency to interact with the matrix. For this reason carbon fibres are surface treated in order to increase interlaminar shear strength of the composites reinforced by the fibres.

**Table 2.8** Typical properties of carbon fibres made from PAN and pitch

| Fibre | Diameter (µm) | Specific Gravity | Strength (GPa) | Strain to Failure $\varepsilon$ (%) | Young's Modulus $E$ (GPa) |
|---|---|---|---|---|---|
| **Ex-PAN** | | | | | |
| High Strength (1st Generation) | 7 | 1.80 | 4.4 | 1.8 | 240 |
| High Strength (2nd Generation) | 5 | 1.82 | 7.1 | 2.4 | 294 |
| High Modulus (1st Generation) | 7 | 1.84 | 4.2 | 1.0 | 436 |
| High Modulus (2nd Generation) | 5 | 1.94 | 3.92 | 0.7 | 588 |
| **Ex-pitch** | | | | | |
| Oil derived pitch | 11 | 2.10 | 3.7 | 0.9 | 390 |
| Oil derived pitch (High Modulus) | 11 | 2.16 | 3.5 | 0.5 | 780 |
| Coal tar pitch | 10 | 2.12 | 3.6 | 0.58 | 620 |
| Coal tar pitch (High Modulus) | 10 | 2.16 | 3.9 | 0.48 | 830 |

Surface treatment results in the oxidation of the fibre surface so as to create carboxylic COOH, hydroxylic OH and other oxygen containing functional groups. There are several ways that the carbon fibre surface can be treated but the industrial process is by anodic oxidation.

In the absence of oxygen carbon fibres are the ideal reinforcement and can be used to above 3000 °C. However the fibres suffer from oxidation from around 400 °C so that they cannot be used in the presence of air for long-term applications above this temperature .

## 2.8 Ceramic fibres

### 2.8.1 Oxide fibres

Ceramic fibres and their composites seem, at first sight, an almost contradiction of terms as ceramics are hard inflexible materials and fibres are very flexible. However ceramic fibres have been developed which are used to produce ceramic matrix composites and also ceramic fibre reinforced metals (Bunsell and Berger, 1999; Bansal, 2005).

Mineral fibres produced from widely occurring rock deposits have been available and used in composite structures for many years, as the section earlier in this chapter illustrates for asbestos. The health hazards associated with the fibrillar crystalline structure of naturally occurring asbestos have led to its use being severely curtailed in many countries but other naturally occurring minerals can be melted and extruded as filaments. Amorphous aluminosilicate fibres have been produced since the late nineteenth century and used to make "mineral wool", used as a refractory insulation, including needle punched felts containing an organic binder to form furnace wall lining. The fibres contain silica ($SiO_2$) and alumina ($Al_2O_3$) together with small amounts of other oxides. Short fibres are produced by passing an electric current through the material, at around 2000 °C and blowing the molten material with compressed air. The viscous melt is formed into filaments which are broken up by turbulence in the stream of air. This is a simple and cheap technique for making these vitro-ceramic fibres. However not all of the molten material deforms in this way and this residual material is known as "shot" which is generally considered to be undesirable. This technique reveals an upper limit in the ratio of alumina to silica of 60/40. Increasing the alumina content is desirable as the Young's modulus and heat resistance increases with increasing alumina content and other techniques have been developed to produce both discontinuous and continuous fibres by sol-gel techniques which overcome this limitation.

Precursor fibres based on alumina and silica can be obtained from aluminium salts which are transformed into aluminium hydroxides. Heating the precursor fibres induces the sequential development of transition phases of alumina which if heated to a high enough temperature all convert to the most stable form which is alpha alumina. However this transformation is followed by a rapid growth of porous $\alpha$-alumina grains giving rise to weak fibres. The need therefore is to retard conversion to $\alpha$-alumina, usually by combining it with silica, or to control its phase growth to make fine grained fibres. The presence of silica in the fibres limits use to below 1000 °C due to it softening.

An extension of the earlier techniques mentioned above is the production of Saffil® fibre developed by ICI in the UK containing 4 % of silica. It is produced by the blow extrusion of partially hydrolysed solutions of some aluminium salts with a small amount of silica, in which the liquid is extruded through apertures into a high velocity gas stream. The fibre contains mainly small $\delta$-alumina grains of around 50 nm but also some $\alpha$-alumina grains of 100 nm. The widest use of the Saffil® type fibre in composites is in the form of a mat which can be shaped to the form desired and then infiltrated with molten metal, usually aluminium alloy. It is a most successful fibre reinforcement for metal matrix composites. There are now a number of producers of similar fibres.

There have been a number of continuous alumino-silicate fibres produced by different companies; the most important today is the Nextel™ series from 3M™. These fibres are not all aimed at reinforcing composites. However, there is a need for composites able to operate above 1000 °C and oxides have an obvious advantage over other systems which oxidise. Continuous fibres based on alumina have been developed by a number of companies, mainly in the USA and Japan. Although not brought to commercial production the first was Fibre FP, made by DuPont™. It was an $\alpha$-alumina fibre the microstructure of which was controlled by seeding the precursor with fine $\alpha$-alumina grains. Its diameter of 20 µm and grain size of 0.5 µm however resulted in very brittle behaviour nevertheless it showed the way to others.

A continuous $\alpha$-alumina fibre, with a diameter of 10 µm, was introduced by the 3M™ Corporation in the early 1990s with the trade-name of Nextel™ 610 fibre. The manufacturer has indicated that the fibre is composed of around 99 % alpha alumina but their more detailed chemical analysis gives 1.15 % total impurities including 0.67 % $Fe_2O_3$ used as a nucleating agent and 0.35 % $SiO_2$ as grain growth inhibitor. The fibre is polycrystalline with a grain size of 0.1 µm, five times smaller than in Fibre FP which increases strength but the greater presence of grain boundaries makes the fibre more susceptible to creep at high temperatures. The strength announced by 3M™ is 2.8 GPa, which is twice the tensile strength measured on Fibre FP. However the smaller grains and possibly the chemistry of its grain boundaries mean that when creep occurs from 900 °C the strain rates are 2 to 6 times larger than those announced for Fibre FP. A stress exponent of approximately 3 is found between 1000 °C and 1200 °C with an apparent activation energy of 660 kJ mol$^{-1}$.

There have been various attempts to improve the creep characteristics of alumina based fibres. The inclusion of zirconia has been explored but any improvement was found to be marginal. 3M™ makes a fibre called Nextel™ 720 which has a composition of around 85 wt% $Al_2O_3$ and 15 wt% $SiO_2$ giving a structure of $\alpha$-alumina and mullite. It has a specific gravity of 3.4; a modulus of 250 GPa and tensile strength of 1.9 GPa. The sol-gel route, higher processing temperatures together with the addition of seeds for $\alpha$-alumina formation have induced the growth of alumina rich mullite and alpha alumina. Unlike other alumina-silica fibres the Nextel™ 720 fibre is composed of mosaic grains of $\approx 0.5$ µm and elongated alpha alumina grains. The complex crystalline structure of the fibre provides few glide planes for creep and the fibre has excellent properties up to at least 1200 °C.

Post heat treatment leads to an enrichment of α-alumina in the fibre as mullite rejects alumina to evolve towards a 3:2 equilibrium composition. Grain growth occurs from 1300 °C. This fibre shows a dramatic reduction, of two orders of magnitude, in creep strain rates at high temperatures when compared to pure alumina fibres. The fibre exhibits the lowest creep rates of all commercial small diameter oxide fibres but it has been shown that the fibre surface is particularly reactive above 1100 °C which leads to the development of grains at the surface. Slow crack growth, initiated by these defects considerably reduces the time to failure during creep as well as the tensile strength of the fibre.

## 2.8.2  Silicon carbide fibres

Better thermomechanical stability than that obtained with the above alumina based systems can be achieved with materials having stronger bonds than the ionic/covalent bonds in alumina. For this reason SiC, which possesses covalent bonds was the first candidate for the production of reinforcements for high temperature structural materials. Small diameter ($\approx$15 µm) fibres, based on silicon carbide, have been developed through the pyrolysis of organo-silicon polycarbosilane (PCS) precursor filaments in an analogous fashion to the technique of carbon fibres produced by carbonising polyacrylonitrile (PAN) precursors. These silicon carbide based fibres allowed ceramic matrix composites to be developed and are the most widely used reinforcement for this type of composite. For these applications the fibres are often made with carbon rich surfaces which are useful in controlling the interface in composites. An alternative surface coating could be boron nitride. Complex surface coatings of SiC and pyrolytic carbon have also been prepared to control interfacial debonding.

Such fibres are produced in Japan by Nippon Carbon with the name of Nicalon™. The manufacture of Nicalon™ fibres involves the production of polycarbosilane (PCS) precursor fibres which consists of cycles of six atoms arranged in a similar manner to the diamond structure of β-SiC. The molecular weight of this polycarbosilane is low, around 1500, which makes drawing of the fibre difficult. In addition methyl groups ($-CH_3$) in the polymer are not included in the Si$-$C$-$Si chain so that during pyrolysis the hydrogen is driven off, leaving a residue of free carbon. The first series of Nicalon™ fibres involved heating the precursor fibres in air to about 300 °C to produce cross-linking of the structure. This oxidation made the fibre infusible but had the drawback of introducing oxygen into the structure which remained after pyrolysis. The ceramic fibre was obtained by a slow increase in temperature in an inert atmosphere up to 1200 °C and has a glassy appearance when observed in SEM. The fibre however contained a majority of β-SiC, of around 2 nm but also significant amounts of free carbon of less than 1 nm and excess silicon combined with oxygen and carbon as an intergranular phase. This microstructure accounts for these fibres not showing the properties of bulk silicon carbide.

The strengths and Young's moduli of Nicalon™ fibres tested in air or an inert atmosphere show little change up to 1000 °C. Above this temperature, both these properties show a slight decrease up to 1400 °C. Between 1400 °C and 1500 °C the intergranular phase begins to decompose, carbon and silicon monoxides are evacuated and rapid grain growth of the silicon carbide grains is observed. The density of the fibre decreases rapidly and the tensile properties exhibit a dramatic fall. When a load is applied to the fibres, it is found that a creep threshold stress exists above which creep occurs. The fibre is seen to creep above 1000 °C. Creep is due to the presence of the oxygen rich intergranular phase. Unloaded or at very low loads the fibres, when heated above 1000 °C, are seen to shrink. The properties and composition of these fibres are shown in Table 2.9.

A development of a second generation of Nicalon™ fibres has been produced by cross-linking the precursors by electron irradiation so avoiding the introduction of oxygen. These Hi-Nicalon™ fibres contain only 0.5 wt% oxygen. The decrease in oxygen in the Hi-Nicalon™ has resulted in an increase in the size of the SiC grains to around 10 nm and a better organisation of the free carbon.

**Table 2.9**   Properties at room temperature of the three generations of SiC fibre which are commercially available.

| SiC Fibre Generation | Trade Mark | Manufacturer | Diameter (µm) | Specific Gravity | Strength (GPa) | Young's Modulus $E$ (GPa) |
|---|---|---|---|---|---|---|
| 1st | Nicalon™ | Nippon Carbon | 14 | 2.55 | 3 | 200 |
| 2nd | Hi-Nicalon™ | Nippon Carbon | 12 | 2.74 | 2.8 | 270 |
| 3rd | Hi-Nicalon™ Type S | Nippon Carbon | 12 | 3.05 | 2.5 | 400 |
| | Tyranno SA3 | Ube Industries | 10 | 3.1 | 2.9 | 375 |
| | Sylramic™ | ATK COI Ceramics | 10 | 2.95 | 2.5 | 400 |

A significant part of the SiC is not perfectly crystallised and surrounds the ovoid β-SiC grains. Significant improvements in the creep resistance are found for the Hi-Nicalon™ fibre compared to the earlier Nicalon™. It is reported that more recent studies the National University of Defence in Changsha, China has resulted in similar fibres to Hi-Nicalon™ without the need for irradiation induced cross-linking. A third generation of the Nicalon™ series of SiC fibres is called Hi-Nicalon™ Type S and their composition is close to stoichiometry.

Another Japanese company, Ube Industries, has also marketed several types of SiC fibre. Their technology, like that of Nippon Carbon was based on the work of Yajima et al. (1976) who showed that titanium added to the PCS could be used as a precursor for the fibres. Ube Industries eventually changed the titanium for aluminium and produced several generations of SiC fibre. Their third generation fibre is called Tyranno SA3. Work in the USA at NASA in conjunction with Dow Corning® produced a fibre called Sylramic™ which is now available from ATK COI Ceramics and which extended the work by Ube Industries. The result is another source of third generation silicon carbide fibre. These fibres have the best high temperature fibres of those available both for strength and modulus retention at 1400 °C.

## 2.9   Statistical analysis of fibre strength properties

Fibres are very fine filaments of matter and in any section of a composite structure there will be found thousands and most probably millions of fibres. Such large populations lend themselves to statistical analysis and the inherent scatter in fibre properties ensures that a statistical approach to describing their properties is necessary. It is therefore not appropriate to give a single average or median value for the strength of a fibre because they show considerable scatter and the mean strength decreases with fibre length. Both the scatter and length dependence is due to the distribution of defects in the fibres and the longer the fibre the greater chance there is of it containing a sizeable defect and so being considerably weakened. For high-performance composite material applications, reliability-based designs and lifetime assessments require accurate strength data to feed into the micro-mechanical models and a more advanced statistical description of fibre strength properties is required. Failure of the fibre as a function of applied load is controlled by the random distribution of defects. When the fibres exhibit an elastic brittle behaviour, which is an acceptable assumption for the widely used carbon and glass fibres for example, fibre strengths are generally represented by using a WEIBULL distribution (Weibull, 1951).

## 2.9.1 Weibull analysis

The effect of a random distribution of a single type of defect on the strength of a solid has been described by WEIBULL who likened the failure of the solid to the breaking of a chain in which the weakest link controls failure. This reflects the behaviour of a brittle solid, under load, for which a crack, developed from the most critical defect, induces the rapid failure of the material. The assumptions of this model are: (1) the stress field in each link is considered as being uniform; (2) the failure stress of one link is independent of that of the other links; (3) the probability of a link failing is due to the random distribution of a single population of defects. The first assumption implies that local modifications of the stress field around the defect are negligible so that each link can be considered as being subjected to a uniform stress field. The second assumption means that there is no interaction between the defects. The third assumption implies that the defect distribution inside the material must be homogeneous and isotropic.

As in a chain under tensile loading, we will consider here that the stress field is uniform and uni-axial. Let's first consider one of the links in the chain subjected to a stress $\sigma$. If the failure stress of this link is denoted by $\sigma_R$, it is possible to define the failure event of the link as $\{\sigma_R < \sigma\}$ or $\{\sigma_R \leq \sigma\}$, which is the same for a continuous random variable such as $\sigma_R$. In that case, the applied stress $\sigma$ is greater or equal to the stress capacity of the link $\sigma_R$, leading to its break. It is also possible to define the survival event of the link as $\{\sigma_R > \sigma\}$, which is the complementary of the failure event. In that case, the applied stress $\sigma$ is lower than the failure stress of the link $\sigma_R$. The failure stress of a link, $\sigma_R$, is obviously a random variable that represents the possible outcomes of tensile experiments conducted on the individual chain links up to failure. The failure probability at any stress load $\sigma$, applied to a "single" link can be written $P_R(\sigma) = P\{\sigma_R \leq \sigma\}$. This probability corresponds to the cumulative distribution function of $\sigma_R$, often noted as $F(\sigma)$. In a complementary way, we can define the survival probability of a chain link as $P_S(\sigma) = P\{\sigma_R > \sigma\} = 1 - P_R(\sigma)$. This survival probability is often referred to as the reliability distribution function of $\sigma_R$, $R(\sigma)$.

Let's now consider the entire chain, consisting of $N$ links, for which we want to establish the survival probability. No need to argue that the survival of the chain for an applied stress $\sigma$ requires the survival of each of its $N$ links. It means a combination of the survival events for all the links in the chain ($1 \leq i \leq N$) and thanks to the second assumption of the model, i.e. independence of individual failure or survival events, the survival probability of the chain can be expressed as:

$$P\left(\bigcap_{i=1}^{N} \{\sigma_R - \sigma\}_i\right) = \prod_{i=1}^{N} P\{\sigma_R > \sigma\}_i = P\{\sigma_R > \sigma\}^N = P_S(\sigma)^N \tag{2.1}$$

If $P_R^{\text{Chain}}(\sigma)$ denotes the failure probability of the chain, we can write:

$$P_R^{\text{Chain}}(\sigma) = 1 - P_S(\sigma)^N \quad \Longrightarrow \quad 1 - P_R^{\text{Chain}}(\sigma) = (1 - P_R(\sigma))^N \tag{2.2}$$

This gives:

$$P_R^{\text{Chain}}(\sigma) = 1 - \exp\left[N \ln\left(1 - P_R(\sigma)\right)\right] \tag{2.3}$$

In the case of a body of volume $V$ let's consider it divided up into small volumes $V_0$ containing one defect, analogous to links in a chain. In this case $N$ is analogous to $V/V_0$:

$$P_R^{\text{Chain}}(\sigma) = 1 - \exp\left[\frac{V}{V_0} \ln\left(1 - P_R(\sigma)\right)\right] \tag{2.4}$$

In Equation (2.4), "$\ln\left[1 - P_R\left(\sigma\right)\right]$" can be replaced by any appropriate monotonic function $\sigma \mapsto \varphi\left(\sigma\right)$ decreasing from 0 when $\sigma = 0$, as $P_R\left(0\right) = 0$, to $-\infty$ when $\sigma \to \infty$ as $P_R\left(\infty\right) = 1$. Weibull took for $\varphi$ a power law equal to $-\left[\left(\sigma - \sigma_u\right)/\sigma_0\right]^m$ for $\sigma \geq \sigma_u$ and zero for $\sigma < \sigma_u$, where $\sigma$ is the applied stress, $\sigma_u$ is a threshold stress below which there is a zero probability of failure and $\sigma_0$ and $m$ are two material parameters, $\sigma_0$ is a scale parameter, and $m$, known as the Weibull shape parameter, measures the variability of the flaws in the materials. The greater the value of $m$ the less is the scatter of results.

For a defect distribution in a material following a Weibull law, the failure probability, for an applied stress $\sigma$ of a specimen, of volume $V$, is finally expressed as:

$$P_R\left(\sigma, V\right) = 1 - \exp\left[-\frac{V}{V_0}\left(\frac{\sigma - \sigma_u}{\sigma_0}\right)^m\right] \quad \text{for} \quad \sigma \geq \sigma_u \tag{2.5}$$

with $P_R\left(\sigma, V\right) = 0$ for $\sigma < \sigma_u$. The two scale factors $\sigma_0$ and $V_0$ are often found grouped into one, also called $\sigma_0$ but which, in this case, no longer has the dimensions of stress, as it replaced $\sigma_0 V_0^{1/m}$ in Equation (2.5). In this case the above expression can be rewritten more simply as:

$$P_R\left(\sigma, V\right) = 1 - \exp\left[-V\left(\frac{\sigma - \sigma_u}{\sigma_0}\right)^m\right] \quad \text{for} \quad \sigma \geq \sigma_u \tag{2.6}$$

## 2.9.2    Fibre strength distribution example for T700 carbon fibres

Fibre strengths are commonly determined using processes such as the single fibre tensile tests, fibre fragmentation tests or fibre bundle test. In the most basic form, a single fibre (or single fibre composite or fibre bundle) is subjected to a longitudinal tensile load until failure. For the following examples, the single fibre testing method, which gives the least ambiguous results, is used for determination of the fibre strength distribution.

The first practical example of the statistical analysis of fibre fracture will be given for elastic fibres with a circular cross-section, the diameter, $D$, of which can be considered not to vary. This is a reasonable assumption for glass and carbon fibres. We will assume that $\sigma_u$ can be put equal to zero and from Equation (2.6), we thus obtain Eqs. (2.7) and (2.8) with the fibre length being written $\ell$ and the constant cross section $A$:

$$P_R\left(\sigma, \ell\right) = 1 - \exp\left[-\ell A\left(\frac{\sigma}{\sigma_0}\right)^m\right] \quad \text{where} \quad A = \frac{\pi D^2}{4} \tag{2.7}$$

$$\ln\left[-\ln\left(1 - P_R\left(\sigma, \ell\right)\right)\right] = m\ln\left(\sigma\right) + \ln\left(\ell\right) + \ln\left(A\right) - m\ln\left(\sigma_0\right) \tag{2.8}$$

The statistical analysis of a series of tensile tests on single fibres can be carried out in the following way: A number, $n$, of tensile tests are performed at a given gauge length *ell* so that $P_R$ only depends on $\sigma$. Only failures can be obtained and one cannot observe an unreliability value. The results are arranged in order of increasing failure stresses, i.e. $\sigma_1 < \cdots < \sigma_i < \cdots < \sigma_n$ and an experimental cumulative failure probability $P_R\left(\sigma_i\right)$ should nevertheless be provided for each fibre of rank $i$ having failed at a stress $\sigma_i$. Unless an infinite number of fibres are tested, which is clearly impossible, the most general expression must take into account the possibility of fibres breaking at lower stresses than that of the weakest specimen as well as breaking at higher stresses than the strongest fibre tested. Rank methods determine the way an estimated unreliability is associated with each failure level. Various expressions of this probability have been proposed, depending of the shape distribution and the sample size. For sample sizes greater than 100, the problem of small sample bias becomes insignificant and the mean rank method could be used to estimate $F\left(\sigma\right)$ using $P_R\left(\sigma_i\right) = i/\left(n + 1\right)$. For smaller sample sizes, the median rank method is to be preferred to other methods since it provides positions at a specific confidence level (i.e. 50 %) and is thus best suited to some later work on confidence limits. The median rank method,

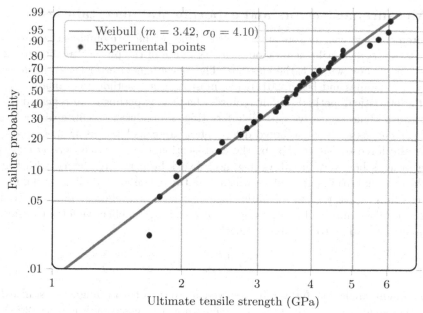

Figure 2.10: Weibull distribution obtained from T700 single fibre tests.

estimates unreliability values based on the failure order number and the cumulative binomial distribution. If the sample size is sufficient, i.e. more than 50 fibres, Benard's approximation, $P_R(\sigma_i) = (i - 0.3)/(n + 0.4)$, can be used. However, it is better to use median rank tables or directly calculate the median ranks from the cumulative binomial distribution. The median rank is, at a 50 % confidence level, the value that the true probability of failure has for the $i^{\text{th}}$ failure out of a sample of $n$ fibres. It is calculated by solving Equation (2.9) for $p$:

$$0.5 = \sum_{k=1}^{n} \binom{n}{k} p^k (1-p)^{n-k} \quad \text{with} \quad \binom{n}{k} = \frac{n!}{k!(n-k)!} \tag{2.9}$$

For the application example, let's consider 30 failure strengths obtained from single fibre tensile tests of T700 carbon fibres with a gauge length of 30 mm.

The variation of the experimental failure probability $P_R(\sigma_i)$ can be plotted as a function of the strength of the fibres as shown in Figure 2.10. A maximum likelihood estimation method can be used to obtain the Weibull distribution parameters, $m$ and $\sigma_0$. For the application example, $m$ is found equal to 3.42 and $\sigma_0$ to 4.10. In this way the distribution of fibre properties, as described by Weibull statistics, can be obtained from a series of tensile tests, usually not less than thirty depending on the scatter of results, with specimens of the same length. Nevertheless, for a given fibre type and gauge length, it could be observed from the literature that many different shape and scale values are reported. For example, the shape parameter of T700 carbon fibres given by different authors, is observed to vary between 3.5 and 12.0, which is a huge variation (Islam et al., 2019). The sources of these variations are numerous and results in the literature should therefore be used with great care, particularly in the context of lifetime assessments. Experimental procedures should always be detailed and Weibull parameters should be associated with uncertainty and a level of confidence. This is the aim of the following section.

### 2.9.3    Measurement errors and experimental uncertainties

Any measurement is wrong, sensors are not perfect and neither are humans that are using them. However measurement is still the only objective estimate of the physical reality around us. The question is always: how wrong is the measure? The measurement error has to be determined for each fibre test (Joannès and Islam, 2019). The most rigorous method to evaluate random errors is always the statistical method, but it requires that the experiment be repeated a significant number of times. This is clearly impossible for strength analysis; the fibre can only be broken once! Fibre strengths are not measured directly but are calculated using measured quantities, by dividing the force to failure, $F_R$, by the cross-sectional area of the fibre $(\pi D^2)/4$. During the testing process however, the fibres may not always be perfectly aligned with the direction in which the external load is applied, in which case the entire load applied during the test may not be successfully transferred to the fibre. The effective load applied on the fibre would depend on the angle of misalignment. For a misalignment angle, $\alpha$, the effective force transferred to the fibre is $F_R \cos(\alpha)$ leading to the fibre strength;

$$\sigma = \frac{4 F_R}{\pi D^2} \cos(\alpha) \tag{2.10}$$

The misalignment angle can further be resolved in terms of a misaligned distance $h$ and the instantaneous gauge length of the fibre specimen $L_R$ as expressed by Equation (2.11).

$$\cos(\alpha) = \frac{L_R}{\sqrt{L_R^2 + h^2}} = \left(1 + \frac{h^2}{L_R^2}\right)^{-\frac{1}{2}} \tag{2.11}$$

The instantaneous gauge length $L_R$ is further dependent on the initial gauge length $L_0$, on the elongation in fibre $L_S$ on application of tensile stress and on the compliance of the testing system $k^{-1}$ as $L_R = L_0 + L_S - k^{-1} F_R$. Substituting Equation (2.11) in Equation (2.10) leads to Equation (2.12) providing the fibre strength versus the main experimental sources of uncertainty.

$$\sigma = \frac{4 F_R}{\pi D^2} \left[1 + \left(\frac{1}{L_0 + L_S - \dfrac{F_R}{k}}\right)^2\right]^{-\frac{1}{2}} \tag{2.12}$$

All quantities necessary for calculating fibre strength, as given in Equation (2.12), have to be determined experimentally and associated with their amount of uncertainty. For example, from the multiple measurements $D_i$ of the fibre diameter (whatever the system used), the best estimate is determined by taking the average of all measurements, $\overline{D}$, and the uncertainty, $u(D)$, is provided by the corrected standard deviation divided by the square root of the number of measures as in Equation (2.13).

$$D = \overline{D} \pm u(D) \quad \text{where} \quad \begin{cases} \overline{D} = \dfrac{1}{n_D} \displaystyle\sum_{i=1}^{n_D} D_i \\[4mm] u(D) = \dfrac{1}{\sqrt{n_D}} \left[\dfrac{1}{n_D - 1} \displaystyle\sum_{i=1}^{n_D} \left(D_i - \overline{D}\right)^2\right]^{\frac{1}{2}} \end{cases} \tag{2.13}$$

An expanded uncertainty could be obtained by multiplying $u(D)$ by a coverage factor taken from the Student's T table, depending on the desired level of confidence. A 95 % confidence level is often used to report uncertainties and in that case, the coverage factor should be taken equal to 1.96 if the calculation is based on a very large number of measured diameters, i.e. $n_D \to \infty$. It should be noted that if the number of measurements is small, the coverage factor should be taken from the table for $n_D - 1$ degrees of freedom. This leads to a coverage factor of about 2.78 for $n_D = 5$ or 2.26 for $n_D = 10$.

Regarding instrument uncertainties, often the only available information from the manufacturer is that the measured value $X$ lies in a specified interval $\Delta X$. In such a case, knowledge of the quantity can be characterised by a rectangular (uniform) probability distribution leading to an uncertainty of the form $u(X) = \Delta X / \sqrt{3}$. An expanded uncertainty could be used as previously with a coverage factor equal to 1.96 for a 95 % confidence level. The combined standard uncertainty of the strength is obtained by appropriately combining the uncertainties of the input quantities using the law of propagation of uncertainty as described in the "Guide to the expression of uncertainty in measurement" (GUM, 2010). It is an estimated standard deviation of the parameter and characterises the dispersion in its values. It is obtained by taking the positive square root of the combined variance of the fibre strength. The combined standard uncertainty for fibre strength $u(\sigma)$ is obtained from the combined variance $u^2(\sigma)$ as given by Equation (2.14), where each $\partial\sigma/\partial X_i$ is a partial derivative of fibre strength for each of the quantities on which it depends.

$$u^2(D) = \sum_{i=1}^{N} \left( \frac{\partial\sigma}{\partial X_i} \right)^2 u^2(X_i) \tag{2.14}$$

Equation 2.14 assumes that variables are independent or that correlations between them can be neglected, which will be considered in the example given. Each fibre strength can now be represented as a combination of the best strength estimate and the calculated uncertainty or expanded uncertainty associated with it. Having a measurement of uncertainty means practically that if it were somehow possible to physically test a fibre again to determine its strength, the result obtained may be different. It would deviate from the best estimate due to uncertainty in accurate measurement of the quantities on which it depends. Consequently, the corresponding Weibull distribution for fibre strength would also be different, having different values of shape and scale parameters.

### 2.9.4 Estimation of uncertainty in Weibull parameters

Since it is impossible to physically test fibres that have already been tested, the estimation of uncertainty for the Weibull parameters needs to rely on simulated methods. A bootstrap method can for example be used. Measurement of uncertainty is usually assumed to follow a normal distribution and a large number of fibre strength data sets similar to the one experimentally determined could be simulated (parametric bootstrap method). Taking the experimental results as reference, fibre strength is varied around the best estimate values while choosing random values of measurement uncertainty from the uncertainty distribution. This method is applied to the T700 strength results presented previously and it is thus possible to plot the 95 % confidence bounds for the Weibull distribution estimate (Figure 2.11). With 30 tests, the wide width of the confidence interval is clearly visible and a higher number of tests is required to reduce it, combined to high resolution characterisation systems (Islam et al., 2020).

Data sets generated experimentally by conducting single fibre tensile tests are usually devoid of the strength of weak fibres and do not accurately represent the strength of all the fibres present in composite materials and structures. This is due to the effect of fibre preselection during which many fibres break before their strength can be determined. The test does not start only once the fibre is loaded, but it actually begins from the point the fibre is extracted from the bundle/roving. The prior knowledge that the fibre testing process is subjected to the preselection effect needs to be incorporated in the analysis when determining the representative fibre strength distribution. This is outside the scope of this work but is the subject of a number of publications (Berger and Jeulin, 2003; Islam, 2020).

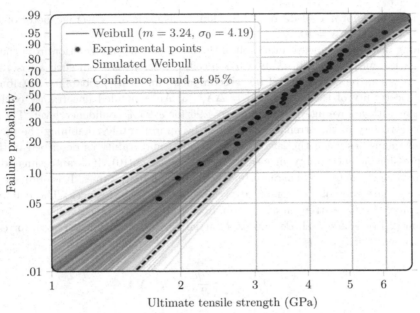

Figure 2.11: Weibull distribution obtained from T700 single fibre tests with a 95 % confidence interval on the results.

## 2.10    Conclusions

A very wide range of possible reinforcements exist which can be used to reinforce all classes of materials. Fibres which are often thought of as being for textile uses find important applications in reinforcing rubber to allow tyres and industrial belting to be made. Other organic fibres with much higher performances can be used to reinforce both thermosetting and thermoplastic matrices which can also be reinforced with carbon and glass fibres. This latter class of materials represents the most important family of composite materials.

The fineness of fibres means that a very great number are used in any application so that a statistical approach to analysing their properties is necessary. These statistical analyses require the greatest rigor in the analysis and extrapolation of the results. Only a sufficient data set allows a real decision on the distributions to be used with multiscale approaches.

# Revision exercises

## Exercise 2.1

Why can fibres be stiff in tension but easily woven?

## Exercise 2.2

What is a typical diameter of a reinforcing fibre, such as carbon and how does this compare with hair?

## Exercise 2.3

Why are the specific properties of natural fibres often given with a comparison to glass fibres?

## Exercise 2.4

Why does the molecular structure of polyethylene terephthalate give a clue to how higher performance fibres have been developed?

## Exercise 2.5

Aramid fibres are a form of polyamide so why are they so much stiffer and operate at higher temperatures than Nylon fibres?

## Exercise 2.6

What is the basic mineral to be found in glass fibres and what role do other minerals which also make up the glass play in ensuring that the fibres can be made?

## Exercise 2.7

Does a glass fibre have an anisotropic structure?

## Exercise 2.8

From what type of precursor are most carbon fibres made and how have these carbon fibres been made both stronger and stiffer since they were first developed?

## Exercise 2.9

What are the two main classes of ceramic fibre which are available as reinforcements of ceramic matrices?

## Exercise 2.10

Explain how two types of fibres could have the same mean strength but different Weibull moduli.

# 3

# Organic matrices

## 3.1  Introduction

Composite materials are most often thought of consisting of fibres embedded in a polymer matrix. This is because of the market which has developed for fibre reinforced plastics but it should not be forgotten that fibres can be used to reinforce all classes of materials. Polymers, however, have established themselves as matrix materials for composites (known as PMC) in a very broad range of applications. Market opportunities for composites with metallic (MMC) and ceramic matrices (CMC) remain much more limited

The attraction of polymers for the role of the matrix is that they are light in weight, often with a density little more than that of water and they can be used, either in solution or molten, to impregnate the fibres at pressures and temperatures which are much lower than those which would be necessary for other materials, for example metals. This results in low density of the composite material, low costs of composite manufacture and of forming tools and relative ease of fibre impregnation. The polymers are often highly resistant to corrosive environments which results in useful properties for the composites produced. The low elastic modulus of most polymers, linked to their ease of deformation, allows load transfer between fibres by shear of the matrix materials and so an effective use of the fibre properties. Certain polymers can produce composites which are particularly resistant to impact and fatigue damage.

A disadvantage of polymers as matrix materials are that they do not provide high-performance mechanical properties at right angles to the fibre directions which makes unidirectional composite materials inherently anisotropic. They also usually absorb water, which can modify the composite properties and compromise the adhesion between the fibres and the matrix, so weakening the composite. This can also lead to out-gassing of water vapour under low pressure conditions such as those found in space structures and water condensation on electronic circuitry on space mirrors could present difficulties. The absorption of water can lead to hydrolysis and matrix degradation as well as major changes in matrix properties. This is discussed in Chapter 11. Degradation of polymers by irradiation, including ultra violet light, as well as attack by atomic oxygen in low altitude space applications can be problems for some composite applications.

The polymers used as matrix materials can be divided into two primary families the natures of which depend on their molecular structures. These are thermosetting (TS) and thermoplastic (TP) resins, for example, respectively, epoxy resin and polyamide (Nylon). Thermosetting resins have been used the longest as matrices for composites and are materials which undergo a transformation of their molecular structure during the manufacture of the composite material. This is because the long macromolecules which make up thermosetting resins possess reactive bonds which can be opened by a hardener to form strong covalent lateral links with other molecules. At room temperature, the resin starts as a viscous liquid, which in the presence of the hardener, to initiate the crosslinking reaction, and usually heating, changes to a rigid solid possessing a three-dimensional molecular network. Cooling to its original temperature reveals a completely modified material which is solid. The reactions are therefore irreversible. Most people will be familiar with epoxy glues which are activated by mixing two viscous components, the resin and hardener, which then become hard. The time for such a glue to set can be modified by changing the temperature as the process is thermoactivated. Such glues are examples of thermosetting resins without reinforcements. These resins lend themselves to composite manufacture as fibre impregnation is facilitated by the viscous liquid phase of the resin which can then be made solid by the crosslinking of the molecular structure. As the crosslinking of the thermosetting resins is thermoactivated it is necessary to store the resins at low temperatures, usually $-18\,°C$, until they are required for composite manufacture. Crosslinking of many resin systems usually takes place at relatively high temperatures, for example $120\,°C$ to $180\,°C$ for epoxy resins and requires heating during the crosslinking time which can range from a few minutes to several hours. An alternative method to promote crosslinking is irradiation with ultra violet light. The time dependent heating process most often used can present difficulties for large volume production, as can the storage of resins at low temperature, the impossibility of modifying the final form after manufacture and the difficulties of recycling thermosetting materials.

Thermoplastics are polymers which undergo dramatic changes to their mechanical properties when they are heated above their glass transition temperature $(T_g)$ but these changes are reversed on lowering the temperature and no structural modification at the molecular level occurs. Unlike the thermosetting resins the macromolecules making up the thermoplastics possess no reactive lateral bonds which can link them strongly with other molecules. Lateral bounding is through secondary forces such as hydrogen and Van der Waals bonds and entanglement of the molecules. The effect of heating above the $T_g$ is to supply enough energy to the polymer to liberate the molecules from these secondary bonds and allow freer movement of the linear molecular structure. This is exploited in the forming of such thermoplastics, which can be rapid as it requires only heating the material to a high enough temperature. There is no need to store the polymers at low temperatures. Every day examples of thermoplastics can be seen in many plastic bottles, which are usually made from thermoplastic polyester, Nylon and polyester fibres, polypropylene and poly(methyl methacrylate) (PMMA), the latter which is sold under a variety of names such as Perspex and Plexiglas and which although fairly rigid at room temperature soften on heating, for example in steam; and can be deformed easily under these conditions. Cooling the thermoplastic freezes in the new structure which can be further modified at will by further heating. Thermoplastic materials such as polyamide, thermoplastic polyester and polypropylene have become widely used as matrix materials for short fibre reinforced composites made by injection moulding and more recently have been considered for long or continuous fibre reinforced materials. Thermoplastics in principle can be recycled and the final form of a structure can be modified by reheating and further forming. Processing times depend on the time to melt the thermoplastic and can be extremely rapid. Fibre impregnation is not as easy as with thermosets and is usually the responsibility of specialist material suppliers as the thermoplastic has to be molten and fibre-matrix adhesion controlled. Thermoplastic resins are gaining ground compared to thermosetting resins. These resins appeared in the 1970s and then represented only a tiny share of the composite market. Forty years later, thermoplastic resins represent almost 40 % of the composite market (in volume).

A group of resins, which can be both thermosetting or thermoplastic, are known as *thermostable* resins. Such resins retain sufficient mechanical properties to be used above 200 °C. Among highly thermostable polymers, some can be employed for limited lengths of time up to and sometimes above 400 °C. A wide variety of thermostable resins in which aromatic rings are linked to form the polymer chains has been formulated. The major high-temperature resins include bismaleimide (BMI), cyanate ester, phenolic, ... for thermosets and polyether ether ketone (PEEK), polyphenylene sulfone (PPSU), polysulfone (PSU), polyimide (PI, TPI), polyaryletherketone family (PAEK) or polyetherimide (PEI) for high-temperature thermoplastics. Such resins mainly find use in high-performances composite applications as in the aerospace industry. Many studies are nevertheless underway to improve heat-resistant properties of polymers while optimising costs and further widen the applicability of composites.

Elastomers can be seen as a lightly crosslinked thermosetting type material operating above its glass transition temperature. They form however a separate family of polymers. Elastomers, such as natural rubber, are defined by their very large deformability with essentially complete recoverability. Elastomers consist of long macromolecules which are often lightly linked together. Rubber, reinforced with a variety of fibres and used in such applications as tyres and industrial belting, is just as much a composite as other fibre reinforced materials sharing many of their characteristics and represents as big a market. The fibres are placed in specific regions of the tyre and are in the form of twisted tows, rather than aligned, or randomly arranged and represent a small overall fibre volume fraction. Rubber represents a rather special class of materials the details of which are more tied up with rubber chemistry than with composites and will not be treated in this book.

### 3.1.1 Common polymer matrix systems

Figure 3.1 shows examples of the most important families of thermosetting resin matrix systems used in composites. There can be wide variations within the thermosetting resins as they can be combined with other functional groups and crosslinking agents to give resins with different mechanical and chemical resistance properties. Thermosetting resins are primarily produced from low molecular weight monomers whereas thermoplastics have high molecular weights.

The first resin shown, in Figure 3.1 is that of an unsaturated polyester resin which has been crosslinked with styrene. Unsaturated means that there are bonds available for further chemical reactions. This type of polyester can therefore be crosslinked to give a three-dimensional molecular structure. The saturated polyester which will be mentioned later is, by contrast, a thermoplastic polymer which does not possess additional bonds for crosslinking and so remains a network of linear macromolecules. The thermosetting polyester crosslinked with styrene, shown in Figure 3.1, is the most widely used thermosetting matrix used in general composites. It is cheaper than other resins which will be mentioned but it is less tough, shrinks considerably on crosslinking and gives off dangerous smoke in case of a fire. If composite applications based on polyester resin represent the largest share in terms of volume, applications using epoxy resins represent the largest share in value (second by volume). Together, polyester and epoxy resins represent around two-thirds of the volume of thermosetting composites produced annually. Epoxy resins are used for higher performance applications as they are more expensive but tougher than polyester resins, shrink less and are generally considered to stand up to wet environments better.

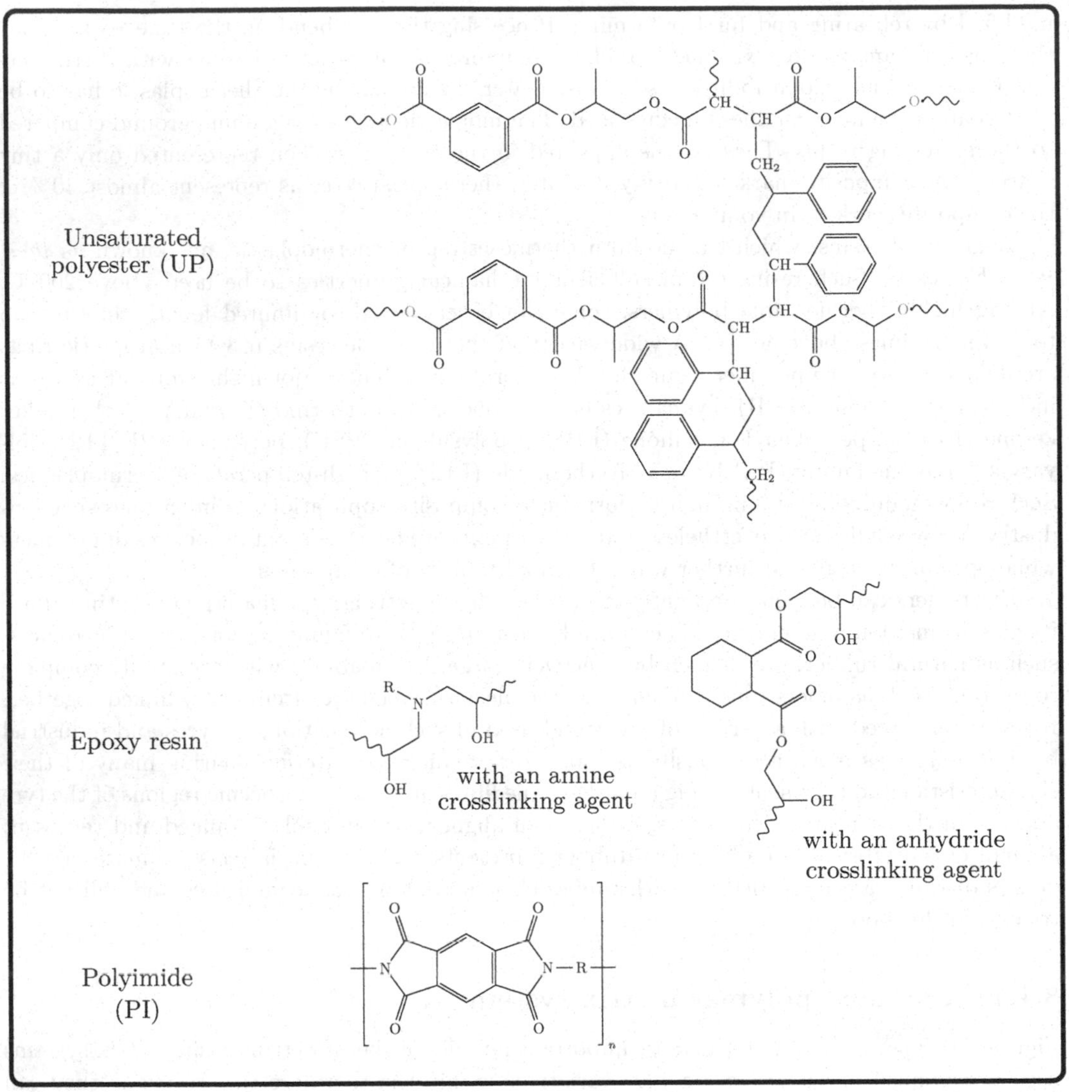

Figure 3.1: Common thermosetting resins where Rcan take different forms.

The epoxy shown in Figure 3.1 will be discussed in more detail later but it can be noted that the one crosslinked with an amine hardener is a Bisphenol A diglycidyl ether (called BADGE or DGEBA) and is the most widely employed in composite materials. A comparison of typical mechanical properties of polyester and epoxy resins is given in Table 3.1.

Polyimides can be thermoplastic but are more usually used as thermosetting resins. They are known as high-performance matrices capable of being used to higher temperatures than the two resin systems previously mentioned. They have good mechanical properties, good chemical resistance and low smoke output in case of fire.

Figure 3.2 shows the most widely used thermoplastic systems used for composite materials. These composite materials can be made by injection moulding, which is an extension of the moulding of plastic parts by the addition of short fibres. However there is an expanding market for long or continuous fibre reinforced thermoplastics as this family of matrix materials offers rapid manufacturing times and the possibility of being recycled. Polypropylene is the cheapest

**Table 3.1** A comparison of typical properties of the two most widely used families of thermosetting resins: polyester and epoxy.

| Property | Polyester | Epoxy |
|---|---|---|
| Heat deflection temperature | 85–125 °C | 155 °C |
| Tensile strength | 35–80 MPa | 90 MPa |
| Tensile modulus | 2.8–3.9 GPa | 3–4 GPa |
| Elongation | 2.8–3.9 % | 3–4 % |
| Flexural strength | 80–140 MPa | 80–140 MPa |
| Flexural modulus | 3.5 GPa | 3 GPa |

of these polymers and for that reason, coupled with its chemical inertness, is attractive as a matrix material for mass produced products. It is however difficult to produce high quality bonds between polypropylene and reinforcing fibres. The development of appropriate bonding agents either placed on the fibres' surface, in a size, or grafted onto the polymer's molecular structure overcomes many of these difficulties and polypropylene is becoming a major matrix material. Among the most recent developments is the introduction of natural fibres into the polypropylene-based composite market. Natural fibres are low-cost and eco-friendly materials and they are considered as strong competitors to replace conventional glass fibres and possibly carbon reinforcements.

Figure 3.2 shows three of the family of polyamides, which are also known as Nylon. Polyamide 6 is shown together with polyamide 6,6 and polyamide 11 so as to illustrate the significance of the numbers defining the type of polymer. The numbers refer to the number of carbon atoms which is repeated to make up the macromolecule. Polyamide 11 is interesting as it absorbs less moisture than PA66, which however is the most widely used polyamide for composite materials. Figure 3.3 shows two thermostable matrix materials which maintain their properties at temperatures above 200 °C.

## 3.2 Resin structure

If synthetic composites were not possible without the development of synthetic fibres, which were mainly developed in the second part of the twentieth century, the same is certainly true for the development of synthetic resins which are the polymers used in most composite materials. This is perhaps underlined by noting that the German chemist, H. STAUDINGER, who first described the macromolecular structure of polymers in the 1920s, received a Nobel Prize for this work in 1952.

Polymers are composed of long molecules, known as macromolecules, consisting of thousands and sometimes millions of atoms resulting in very long chain lengths. Usually the macromolecules consist of a backbone of carbon atoms as the $C-C$ bonds are particularly stable and this is the case for the resins, which are of interest for composite materials. The simplest polymer consisting of long linear macromolecules is polyethylene, which has the structure defined in Ch. (C3.1).

$$\sim\!\!\sim CH_2 - CH_2 - CH_2 - CH_2 \sim\!\!\sim \quad \text{equivalently represented as} \qquad \qquad \qquad \text{(C3.1)}$$

Polymers based on other atoms do exist such as the silicones (polysiloxanes) as represented in Ch. (C3.2).

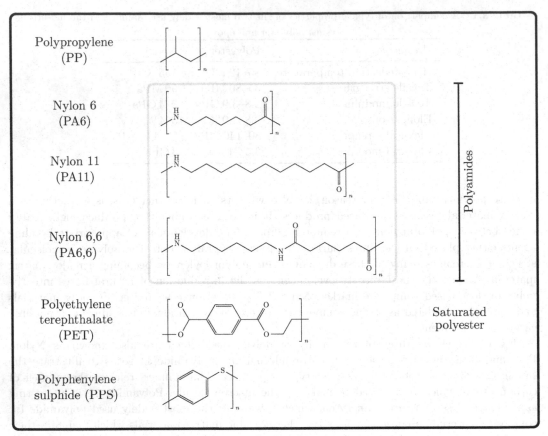

Figure 3.2: Common thermoplastic matrix systems.

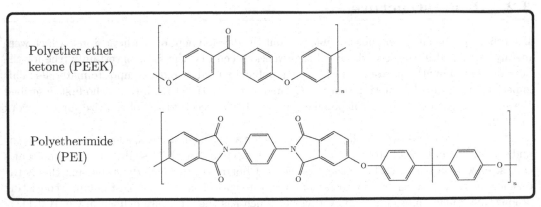

Figure 3.3: Common thermostable matrix systems.

$$\sim\!\!\text{Si}\!-\!\text{O}\!-\!\text{Si}\!-\!\text{O}\!-\!\text{Si}\!\sim \quad \text{or} \quad \left[\,\text{O}\!-\!\text{Si}\,\right]_n \qquad \qquad \text{(C3.2)}$$

The macromolecules are rarely straight or planar so that one image is that of a confused mass of intertwined long lengths of molecules, much like spaghetti on a plate. However the molecular morphology of polymers, including those used as matrix materials for composites can be much more complex than this description.

Polymers are made by the linking together of their basic units called monomers which are groups of atoms which can exist in their own right but which can be encouraged to open up so providing the opportunity to create two, three or more bonds, so permitting linkage, through covalent bonds, with other similar monomers or other chemical groups. The opening of these bonds can occur spontaneously due to molecular movement and this is further aided by raising the temperature. Opening of these bonds can also occur due to bombardment by ultraviolet light, X-rays or other forms of radiation, but the reaction is most usually encouraged by the addition of a small amount of another substance, the hardener, which usually creates free radicals which, being extremely reactive, produce a chain reaction leading to polymerisation by polyaddition. The number of bonds that a monomer can make is called its functionality. If the polymer is a combination of a number of different molecules ($n_A$, $n_B$, $n_C$, etc.) the functionality is given by the average between molecule functionalities ($F_A$, $F_B$, $F_C$, etc.) as in Equation (3.1).

$$\frac{n_A F_A + n_B F_B + n_C F_C + \ldots}{n_A + n_B + n_C + \ldots} \tag{3.1}$$

In the case of bifunctional systems, for which the functionality is 2, the resulting macromolecules are linear. Entanglements of the linear molecular chains together with secondary bonding, such as hydrogen or van der Waals bonding, give these polymers bulk cohesion which in some cases, for example cellulose, make their structure stable even to the point of resisting most solvents. Alternatively linear polymers with weak intermolecular bonding give elastomers, which in the classic case of natural rubber requires the light crosslinking of the molecules through a process known as *vulcanisation* by the addition of 6% to 8% of sulphur. Even vulcanised elastomers are characterised by great deformability, up to 1000%, due to the low density of crosslinking. Tri- or multifunctional monomers allow three-dimensional molecular networks to be created with a much greater density of crosslinking occurring between the molecular chains.

The polyaddition process occurs when the addition of another monomer leads to the repeated linking of molecular chains to create macromolecules. When only one type of monomer is involved a homopolymer is produced giving, for the monomer A, the following chain (C3.3).

$$\sim\!\!\sim A — A — A — A — A — A — A — A — A — A — A — A \sim\!\!\sim \tag{C3.3}$$

The simultaneous polymerisation of two or more different monomers, A and B for example, can lead to copolymerisation which can give several arrangements one of the simplest of which might be an alternate sequence such as in Ch. (C3.4).

$$\sim\!\!\sim A — A — B — B — A — A — B — B — A — A — B — B \sim\!\!\sim \tag{C3.4}$$

Other possible combinations are a random arrangement such as in Ch. (C3.5).

$$\sim\!\!\sim A — A — B — A — B — A — B — B — B — A — B — B \sim\!\!\sim \tag{C3.5}$$

Or arranged in sequential blocks such as in Ch. (C3.6).

$$\sim\!\!\sim AA \!\!\left[ A — BB \right]_n \!\! AA \!\!\left[ BB — A \right]_m \!\! AA \sim\!\!\sim \tag{C3.6}$$

Alternatively the copolymers can be grafted together such as in Ch. (C3.7).

$$\sim\!\!\!\sim A - A - A - A - A - A - A - A - A - A - A - A \sim\!\!\!\sim$$

$$\begin{array}{ccc} & B & B \\ & | & | \\ & B & B \\ & | & | \\ & B & B \end{array}$$

(C3.7)

Such reactions can lead to several possible molecular arrangements. If the molecules are long and linear with regular repetition of the basic units without any change in their arrangement in space they are known as isotactic. If the arrangement is regular but the groups are regularly alternatively arranged in space they are known as syndotactic. An irregular repetition of similar units is known as an atactic polymer. Polypropylene can exist in the three forms, as shown in Figure 3.4.

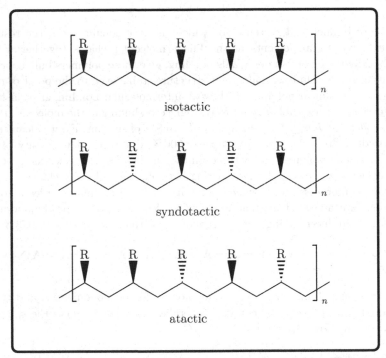

Figure 3.4: Polypropylene molecules can show three forms of tacticity.

Cyclic arrangements, in which one end of the molecule links with the other end, can also occur and are generally to be avoided. In addition three-dimensional networks and copolymerisation in which the macromolecules of the second polymer are grafted just at their ends onto the first polymer chain can also occur. If the polymer is to crystallise, the macromolecules have to be linear and regular in their arrangement, that is isotactic or syndotactic, so that they can fit alongside one another in the crystalline phase.

Polycondensation is another reaction used for producing macromolecules and differs from that described above by the loss of reaction products which are often, but not always, water molecules. To initiate such a reaction catalysers can be used but are not always necessary.

Polycondensation begins with small polyfunctional molecules which lose water either through heating or to a solvent and form, by combination, long macromolecules. If the molecules are bifunctional linear molecules are produced whereas functionality greater than two will result in three-dimensional molecular networks.

The molecular structure of three-dimensional network polymers can therefore be seen to be modelled by three main components, linear or planar, often rigid units such as benzene rings, rotational units and trifunctional bonding units whereas the structures of two-dimensional network polymers lack the trifunctional units. Three-dimensional crosslinked polymers cannot generally be sufficiently organised to crystallise however exceptions exist which do not readily spring to mind as being polymers. Examples are diamond, quartz and graphite. The two-dimensional organisation of carbon atoms gives great strength in the plane of the atoms due to the strong C−Cbond and this is exploited in polymers and for example carbon fibres. In graphite, the layers of carbon atoms are held together in a regular lattice by secondary, van der Waals bonds with a distance between the planes of 0.338 nm. This leads to the great anisotropy of graphite which in the plane of the C−Cbonds is extremely strong and rigid in tension whereas shearing of the planes is very easy, allowing graphite to be used as a lubricant.

Both thermosetting resins and thermoplastic matrices used for composite materials are made in the ways described above. Thermosetting resins can be used before three-dimensional crosslinking to impregnate the fibres which are placed in a mould or otherwise positioned in the arrangement finally desired for the composite structure and then heated, usually under pressure, to finish the curing of the resin. The result is a solid matrix with a three-dimensional molecular structure bonded together by strong covalent bonds enveloping the fibres. The process is irreversible. As the crosslinking of thermosetting resins is thermoactivated it is necessary to store the unprocessed resins at low temperature, usually around −18 °C. Other complications which can be associated with thermosetting resins are degassing during curing leading to porosity and bubbles, imperfect mixing of reacting agents leading to a heterogeneous molecular structure and excesses of uncombined hardener which can lead to degradation in the presence, for example, of water, difficulties in the control of viscosity during composite manufacture and shrinkage during crosslinking. In addition the crosslinking process is exothermic which can lead to difficulties in controlling the kinetics of the chemical reactions during composite manufacture. Thermosetting resins normally decompose rather than melt on heating to high temperatures and recycling presents major difficulties. Thermosetting resins were however the first class of polymers used as matrices for composites and represent at present approximately 60% of the organic matrix materials used (in volume).

Thermoplastic materials have found wide use for the production of fibres, such as polyamide and polyester fibres and in injection moulding processes as their ability to melt whilst retaining their macromolecular structure can be exploited in these processes. Thermoplastic materials are used as the matrix materials for both short and continuous fibre reinforced composites. The processes involved in producing short and continuous thermoplastic composites are radically different. Short fibre thermoplastic composites are usually made by injection moulding which involves a mixture of the fibres and molten polymer being injected by a system of screw threads into a mould. The injection process shears the molecules so making them more reactive to bonding with the fibres than otherwise. This shearing process is not a feature of continuous fibre reinforced composite manufacture so that the polymers must either be made more reactive by the addition of functional groups or specially adapted sizes have to be placed on the fibres to ensure good bonding.

More recently, reactive processing of fibre-reinforced thermoplastics has been developed in order to take the best amongst thermoset LCM processes and thermoplastic properties. Thermoplastic composites offer some advantages over thermosetting systems; they provide greater toughness, have faster cycle times and are intrinsically recyclable. Nevertheless, traditional melt processing used with thermoplastics is not as convenient as the liquid moulding offered by thermosets. The high-melt viscosity of thermoplastics requires high processing temperatures and pressures in order to obtain acceptable impregnation of the fibre reinforcements. This effectively limits the thickness and the size of the parts to be produced. When they are intended to be used with continuous fibres, thermoplastic composite processes often use semi-products in the form of thin pre-impregnated (also known as pre-pregs) tapes or sheets. This solution to bring the thermoplastic and the fibres in more intimate contact before the final moulding step has some drawbacks such as additional costs and poor drapability. Thermoplastic reactive processing eliminates the need of those intermediate steps by impregnating the fibres with a low viscosity mono- or oligomeric precursor. The polymerisation of the thermoplastic matrix is made *in situ*, as with the crosslinking reaction occurring with thermosets. The low pressure manufacturing processes used for thermosetting resins can in this way be used to produce thermoplastic composites.

The two-dimensional macromolecules, of which thermoplastics are composed, are made up of atoms linked together by strong covalent bonds, however interactions, other than mechanical entanglement, between macromolecules are through weak or secondary bonds. The reactions and strengths of the bonds between the macromolecules depend on the temperature so that below a critical temperature known as the glass transition temperature the bonding is sufficiently strong to result in a rigid material and the degree of bonding determines the interaction of the polymer with solvents. Above the glass transition temperature, also known as the $T_g$, the molecules have much more freedom of movement and the material becomes more rubbery in behaviour and can be deformed easily. Reducing the temperature to below the $T_g$ returns the material to its former rigid state. This behaviour is therefore reversible and is typical of thermoplastics. If the thermoplastic is atactic the polymer is amorphous and generally transparent. If the polymer is isotactic or syndotactic the macromolecules can become aligned through the influence of the secondary bonds and locally can form into a regular arrangement. These localised crystalline zones form spherulites made of folded macromolecules which can occupy a considerable percentage of the volume of the semicrystalline thermoplastic, typically 30 % or 40 %. The crystallinity of a semi-crystalline thermoplastic can be altered by melting the polymer and controlling the rate of cooling. The faster the cooling rate the lower the degree of crystallinity. When thermoplastics are drawn into fibres the crystal zones are drawn out to form microfibrils made up of both crystalline and amorphous zones. thermoplastics melt on heating to sufficiently high temperatures and this is used to impregnate the reinforcing fibres during the manufacture of thermoplastic matrix composites. Recycling is feasible with thermoplastics although practical considerations often limit its effectiveness.

The $T_g$ of amorphous thermoplastics is usually well defined whereas the mixture of phases in a semi-crystalline thermoplastic makes the identification of the $T_g$ less easy. Thermosetting resins can also be said to possess a $T_g$ above which the resin loses rigidity although this is generally less obvious than in thermoplastics.

As polymers are composed of long molecular chains made by reactions which are controlled by the local chemistry at each point in the polymer the lengths of the macromolecules are not all the same and show considerable dispersion. This dispersion can have important consequences on the behaviour of the polymer, for example, the shorter molecules can be more reactive to solvents or more easily pyrolysed than the rest of the polymer. In defining polymers it is useful therefore to define their average molecular weights and the molecular weight distributions. The "number" average molecular weight $\langle M_n \rangle$, also known as the number average molar mass, can be defined and determined by the mass of all the molecular chains divided by the number of chains such that

$$\langle M_n \rangle = \frac{\sum_i n_i M_i}{\sum_i n_i} = \frac{\sum_i w_i}{\sum_i w_i/M_i} = \frac{1}{\sum_i w_i/M_i} \tag{3.2}$$

where $n_i$ the number of molecules having a molecular weight $M_i$ and $w_i$ is the weight fraction of molecules having a molecular weight $M_i$. The molecular weight can be measured experimentally by osmometry and other techniques.

Alternatively the "mass" average molecular weight,$\langle M_w \rangle$, can be defined as the sum of all the molecular weights multiplied by their weight fractions, so that

$$\langle M_w \rangle = \sum_i w_i M_i = \frac{\sum_i w_i M_i}{\sum_i w_i} = \frac{\sum_i n_i M_i^2}{\sum_i n_i M_i} \tag{3.3}$$

The value of $M_i$ can be determined by the measurement of opacity of the resin.

It is easy to imagine the consequences of dissolving a polymer in a solvent. The movement of the macromolecules and their disentanglement in the solution will be all the easier the shorter they are so that the rheology of the solution will be affected. As a consequence a polymer can also be characterised by its viscosity in solution. If the viscosity of the solution with a certain concentration $\phi$ is $\eta$, and $\eta_0$ is the viscosity of the solvent, an expression for the intrinsic viscosity of the polymer can be written such that

$$[\eta] = \lim_{\phi \to 0} \frac{\eta - \eta_0}{\eta_0 \, \phi} \tag{3.4}$$

The intrinsic viscosity $[\eta]$ is a dimensionless number and this value is related to another expression for the average molecular weight, i.e. the "viscosity" average molecular weight $\langle M_\eta \rangle$, through

$$[\eta] = K_\eta \langle M_\eta \rangle^\alpha \quad \text{with} \quad 0.5 < \alpha < 1 \tag{3.5}$$

in which $K_\eta$ and $\alpha$ are two constants (also known as Mark-Houwink parameters (Rubinstein and Colby, 2003)) which are characteristic for each type of polymer and are found in published tables. $\langle M_\eta \rangle$ is defined by the following Equation (3.6) and is equal to $\langle M_w \rangle$ if $\alpha = 1$.

$$\langle M_\eta \rangle = \left( \frac{\sum_i n_i M_i^{\alpha+1}}{\sum_i n_i M_i} \right)^{1/\alpha} = \left( \frac{\sum_i w_i M_i^\alpha}{\sum_i w_i} \right)^{1/\alpha} = \left( \sum_i w_i M_i^\alpha \right)^{1/\alpha} \tag{3.6}$$

The values of $\langle M_n \rangle$, $\langle M_w \rangle$ and $\langle M_\eta \rangle$ are clearly related may but values may differ as a result of the methods used to evaluate them.

The molecular weight distribution can be determined by several techniques. In fractionation the solution is progressively diluted which provokes the precipitation of those parts of the polymer with progressively lower molecular weight. It is by measuring the molecular weight of each of these fractions of the polymer, $M_i$, and plotting the weight of each fraction, $w_i$, as a function of $M_i$ that the molecular weight distribution can be obtained. This technique has the disadvantage of being long to carry out and to require a considerable amount of matter to analyse.

An alternative method is by gel chromatography in which a diluted solution of the polymer passes through a chromatography column containing a porous gel consisting of a crosslinked polymer containing pores of various dimensions. The longer molecules can only traverse the gel through the largest of pores linking each side of the gel by the shortest routes so that these molecules cross the gel the fastest. The rate at which the molecules cross the gel is therefore in a decreasing order of molecular weight. The detection of the molecules can be made by the interaction of light or ultraviolet or infrared spectrophotometry and calibration of these methods is achieved by comparison with the results obtained from fractioned polymers, either the same as the one being studied or one very similar, so as to obtain molecules of a constant length.

## 3.3 Matrix thermo-mechanical behaviour

### 3.3.1 The curing of thermosetting resins

The rate of curing, or crosslinking, of a thermosetting resin system can be monitored by simply controlling the viscosity of the resin. The crosslinking process proceeds asymptotically towards the condition in which all of the resin would be combined with the hardener. This means that the process is never quite finished and optimising curing cycles can be problematic so that post curing is often performed. This involves a curing cycle which is subsequent to the initial curing process. It can also mean that the properties of the resin evolve with time and temperature. The crosslinking process is illustrated schematically in Figure 3.5.

Figure 3.5: The molecular structure is irreversibly modified by the crosslinking process as is shown from left to right. The resin is initially not crosslinked, then lightly crosslinked (at which point it gels), and finally nearly fully crosslinked.

The resin, before curing, begins as a liquid which, initially, under the effects of heat, becomes less viscous; however, as the crosslinking process progresses it becomes increasingly viscous and gels. Finally the resin becomes solid as the process approaches a fully crosslinked resin. The crosslinking process is exothermic and can generate considerable heat so that, in some manufacturing processes little or no heat is required. This can cause problems in the manufacture of thick specimens as it becomes difficult to maintain a constant temperature throughout the thickness.

The rise in temperature can however be a very relevant indicator for monitoring the crosslinking process within the composite. Several micro-sensor techniques exist and are dedicated to *in situ* temperature monitoring during curing. Other techniques based on dielectric measurements in the moulds make it possible to detect viscosity changes of the resin and thus also provide real time information about crosslinking. These manufacturing monitoring systems are also able to capture the resin arrival at strategic points of the composite part, which is useful for monitoring closed moulding processes.

The viscosity of the resin increases during crosslinking. A viscosity versus time curve can be obtained by stirring the resin with a glass rod connected to a motor and a transducer to monitor the couple necessary to keep the rod rotating during the crosslinking process. During the crosslinking, the thermoset resin changes from a low viscosity liquid, to a gel and then to a stiff solid. Such information regarding viscosity changes as a function of time of the resin-catalyst mixture helps to determine the working-life characteristics. The resin is said to have gelled when the viscosity reaches 50 Pa s (i.e. 500 poise) and this represents the beginning of the formation of a three-dimensional network through crosslinking. The "pot life" of the resin is the period between the mixing of a resin with a hardener and the gelling of the resin. This represents the useful life of the resin if it is to be used to impregnate fibres. The maximum temperature reached by reacting a thermosetting plastic composition is called the peak exothermic temperature. Gel time and peak exothermic temperature are two of the most important parameters for a thermosetting resin. These two parameters vary with the volume of material used so it is essential to specify the volume so that the results are exploitable.

The overall reaction of the resin can be monitored by a differential scanning calorimeter, or DSC, in which the temperature of a known mass of the resin positioned in a container, is compared with the temperature of an exactly similar but empty container during a continuous and slow temperature rise. As the weight and specific heat of the resin and the temperature difference are known, the heat flow, either to or from the resin, can be determined. DSC is a very widely used technique for examining polymers and determine their thermal transitions. The most important transitions are the glass transition temperature $(T_g)$, the crystallisation temperature $(T_c)$ and the melting temperature $(T_m)$.

The thermal analysis of polymers is a vast scientific field and the interested reader can find more information in Menczel and Prime (2009).

### 3.3.2 Thermoplastic and cured thermosetting resin mechanical behaviour

Thermoplastic and thermosetting resins, once cured, can be evaluated by the same procedures although their responses to mechanical loading will depend of their different molecular structures. Conceptually the simplest test is a straight tensile test often using waisted specimens so as to attempt to control the position of the failure point. Such a test can give the failure strength of matrix materials and with the use of suitable strain gauges the elastic or Young's modulus of the polymer can be obtained. However, in practice, tensile tests can present difficulties as gripping sufficiently hard to avoid slippage can damage specimens and cause failure in the grips, also many polymers are far from elastic in behaviour so that plastic deformation, creep and striction of specimens during the test complicate interpretation.

It is clear that mechanical measurements could be made at different temperatures and the variation of different characteristics determined as a function of temperature. This is effectively what happens in viscoelastic techniques although only small displacements are used so as to facilitate analysis. These techniques subject the specimens to cyclic loads which induce deformation of the material. The induced strains and applied stresses are only in phase if the material is perfectly elastic, which is rarely the case for a matrix material for a composite. The response of the material depends on the nature and time of reaction of its microstructure, which in the case of polymers varies considerably with temperature and their viscoelastic behaviour is a sensitive means of identifying the molecular mechanisms governing their characteristics. Using "Dynamic Mechanical Analysis" (Menczel and Prime, 2009) is an efficient way to evaluate the sensitivity of the material to the temperature but also to the strain rate. A measurement campaign carried out with the DMA can provide information on the temperature and speed ranges that it would be wise to characterise during tensile tests for example.

Finally, it should be noted that polymers are sensitive to hydrostatic pressure. Specific experiments are needed to characterise the dependence on hydrostatic pressure and the following papers are particularly illustrative: Boisot et al. (2011); Selles et al. (2018); Rojek et al. (2020). The dependence on hydrostatic pressure can sometimes explain the differences observed in multiscale modelling approaches. Nevertheless, in bulk form, the matrix does not necessarily behave in the same way as *in situ*, when it is confined between fibres and this constitutes another source of discrepancies. This topic is often discussed in the literature and it raises questions about the presence of interphases and their characterisation.

The mechanical characterisation of the matrix, within a composite, is a very active scientific field. Tomographic observation techniques offer today many possibilities to better understand and apprehend what is happening *in situ*.

## 3.4 Thermosetting matrix systems

### 3.4.1 Unsaturated polyester resins

This is the commercially most important resin system used with composite materials. It is not considered to be a high-performance matrix system and so is not found in advanced composites, but it is very widely used for general purpose composite applications. The basis of the unsaturated polyester resins are recurring ester groups $- CO-C -$ combined as linear molecules with

**Table 3.2**  The ingredients which can be used to make the unsaturated polyester resin backbone.

| Ingredient | | Function |
|---|---|---|
| Unsaturated acids and/or anhydrides | Maleic anhydride | Provides cure site |
| | Fumaric acid | Provides best cure site (maleic isomerises to fumaric) |
| Saturated acids and/or anhydrides | Phthalic anhydride | Low cost and hard, balance of properties |
| | Isophthalic acid | Improved strength and chemical resistance |
| | Adipic acid and homologues | Flexibility and toughness |
| | Halogenated acids/anhydrides | Flame retardance |
| Glycols | Propylene glycol | Balance of properties at lowest cost |
| | Diethylene glycol Dipropylene glycol | Flexibility and toughness |
| | Bisphenol A/PG adduct | Chemical resistance and high heat deflection temperature |
| | Neopentyl glycol | Chemical resistance and toughness |

aliphatic groups, which do not contain aromatic groups and which provide sites which can react with other unsaturated monomers such as styrene. Variation in the basic components of the polyester chain and in the ratio of the saturated and unsaturated components allows a wide range of resins to be produced to meet different performance requirements. Reactions generally occur quickly, but in a controllable manner to give a crosslinked thermoset structure.

The most widely used polyester resins, accounting for around 80 % of the market, are described as orthophthalic (ortho) resins for which orthophathalic acid is used as the saturated acid part of the backbone of the polymer. They are the cheapest form of polyester resin and are used in contact moulding, typically of large structures. A slightly more expensive resin is isophthalic (iso) polyester for which isophthalic acid is used as the saturated acid part of the molecule. Iso polyester resins are used in closed moulding processes, corrosion resistant composites and gel coats. Table 3.2 shows the ingredients which can be used to make the unsaturated polyester resin backbone.

The following chemical scheme (C3.8) shows a common polyester structure consisting of a saturated acid, an unsaturated acid and one or more glycols, this latter group being a broad range of related structures containing two hydroxyl (alcohol) groups.

(C3.8)

Saturated phthalic acid | Glycol | Unsat. maleic acid | Glycol

The molecular weight of the backbone polymer is controlled by the manufacturer but is generally in the range 1000 to 4000. The site for crosslinking is provided by the unsaturated acid or anhydride group. Commonly maleic anhydride provides cure sites for crosslinking by isomerisation to a maleate group which in turn transforms into a fumarate group, as in Ch. (C3.9).

$$(C3.9)$$

| Maleic anhydride | Maleate | Fumarate |

The isomerisation of the maleate group depends on the type of glycol used. These groups are copolymerised with other monomers which contain a reactive function. General purpose polyester resins are most often based on a saturated phthalic acid, used as phthahlic anhydride and unsaturated maleic acid, used as maleic anhydride, respectively used for the saturated and unsaturated acids with propylene glycol as the glycol. On average, one or two monomer groups link the polyester macromolecules. Polyesters with improved environmental and chemical resistance and mechanical properties are often based on maleic anhydride, isophthalic or terphthalic acids and glycols, such as neopentyl glycol. Flame resistance is achieved through the use of saturated halogenated acids or anhydrides. Such unsaturated polyester resins are liquids with low viscosity which can readily be used to impregnate fibres.

The monomers which are used to crosslink the polyester must be miscible in the polyester and capable of providing two reactive sites. The result is a relatively low viscous solution which lends itself to the easy impregnation of fibres. The general form of a crosslinking monomer is

$$(C3.10)$$

$R_1$ $R_2$

in which $R_1$ and $R_1$ are reactive sites allowing bonds to be made to the polyester. The most widely used crosslinking agent is styrene schematically represented in Ch. (C3.11), and accounts for the distinctive smell in polyester work places.

$$(C3.11)$$

Figure 3.6 shows the preparation of the polyester resin formulation and the ongoing crosslinking process is shown in Figure 3.7.

Sometimes methyl methacrylate (C3.12), mixed with styrene is used to cross link the polyester in order to provide greater transparency and external weather resistance.

$$(C3.12)$$

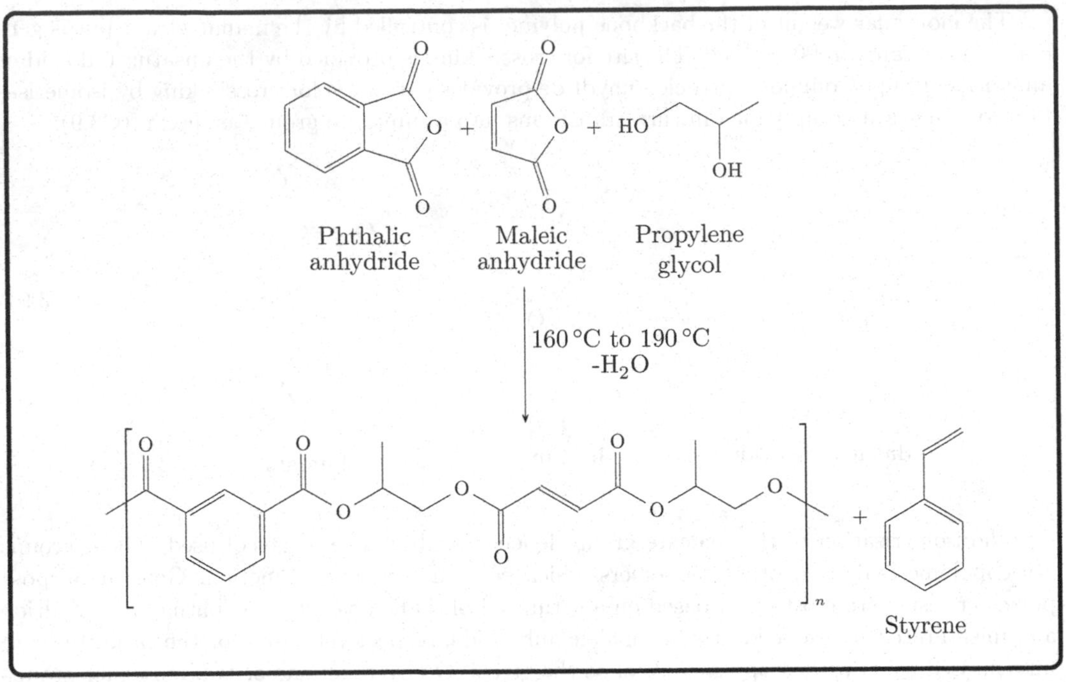

Figure 3.6: The preparation of a polyester resin.

A range of crosslinking monomers can be used to give resins with different characteristics, as shown in Table 3.3.

The crosslinking reaction has to be started and this is done by an initiator or promoter, also known, inappropriately, as a catalyst, which is often an organic peroxide which breaks down at elevated temperatures to provide free radicals. A hydroperoxide $R-O-OH$is used which decomposes when heated to provide free radicals; $R-O-OH \longrightarrow RO^{\bullet} + OH^{\bullet}$. The free radicals begin the crosslinking process.

A common peroxide which is used is methyl ethyl ketone peroxide (MEKP) which is thermally stable up to fairly high temperatures, having a half-life at $100\,°C$ of about 12 hours. Curing can occur at room temperature, however, if an accelerator is added which encourages the breakdown of the peroxide. It should be noted that the initiator and accelerator must not be mixed without the resin as an explosion will result. MEKP will readily decompose at room temperature in the presence of small amounts of transition metal ions. Cobalt naphthenate is often used as such an accelerator with a concentration of only $0.1\,\%$ of the resin weight being required.

The manufacturer of a composite structure needs to be able to control the rate of curing of the resin system to allow all the necessary manufacturing steps to take place. The rate of reaction of a polyester resin system depends on the amount of initiator, the accelerator and the temperature. Often the resins are supplied containing the accelerator so that the manufacturer can control gel time through the initiator and temperature. Inhibition of the crosslinking process through oxidation and evaporation of the styrene can cause difficulties during manufacture. This is sometimes minimised by the addition of small amounts of wax dissolved in the resin and which, during curing, exudes to the surface to form a film which keeps oxygen out and the styrene in.

Typically the failure stresses of the polyester resins are in the range $40\,MPa$ to $100\,MPa$, their elastic moduli between $2\,GPa$ to $4\,GPa$ and the $T_g$ of most polyester falls in the range $80\,°C$ to $120\,°C$. The heat deflection temperature is often a property given for polyester resins and is a short-term indication of the ability of the resin structure to resist the thermally induced molecular

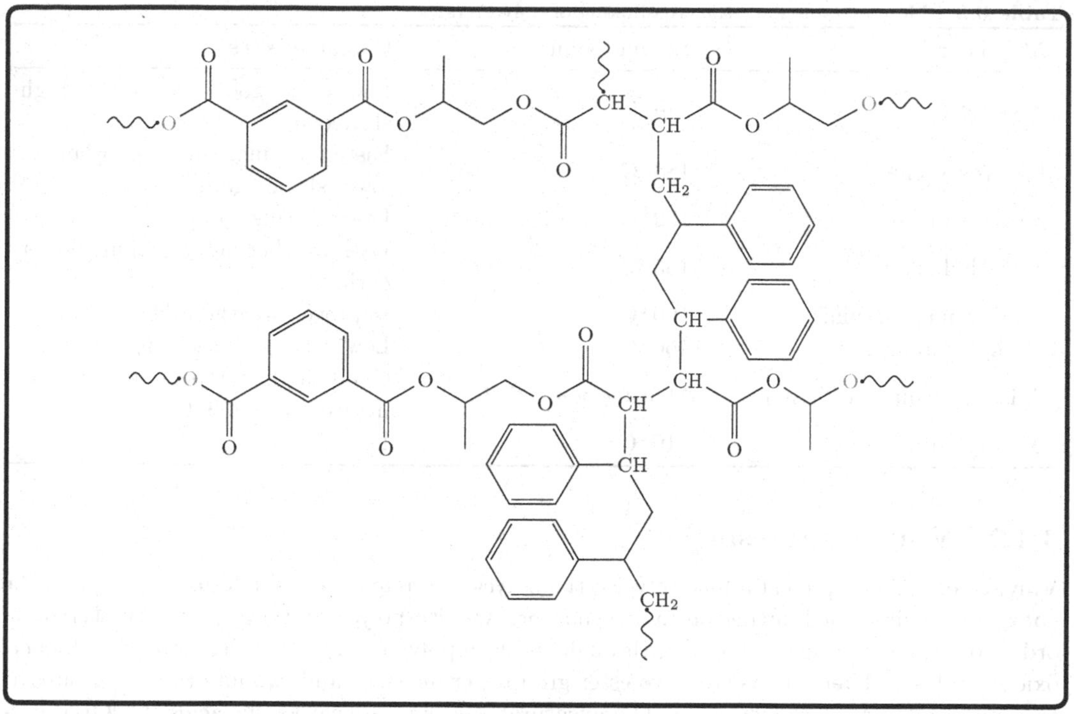

Figure 3.7: The crosslinking (or curing) of a polyester resin. The process is shown as being ongoing as the bottom carbon atom in the diagram, has an unused bond which is available for further crosslinking.

movements. The properties of some typical polyester systems in the form of cast unreinforced specimens are shown in Table 3.4.

Flammability and smoke production are factors which can determine the choice of a resin system for a particular application. Fire redundancy is achieved using halogenated resin systems or with the use of an additives which is often alumina trihydrate. Smoke density depends in part on the crosslinking monomer. Styrene, the most widely used product, gives the highest smoke output and methyl methacrylate the lowest. The glycol which gives the highest smoke output is the widely used propylene whereas increasing the unsaturated acid content for isophthalic-maleic resins decreases smoke production. Certain fire retardant additives such as antimony trioxide increase smoke production however the presence of alumina trihydrate has the opposite effect.

Polyester resins are generally resistant to short-term exposure to common acids, bases solvents and water however long-term exposure especially at elevated temperatures can lead to degradation. Hydrolysis can occur with long-term exposure to water and this is accelerated in the presence of $H^+$ or $OH^-$ ions.

The unsaturated polyesters represent a large family of resins and the environmental resistance of each system clearly depends on the exact chemistry used and the conditions in which the resin is to be used. Orthophthalic polyester resin does not resist well to boiling water however it is widely and satisfactorily used in composite boat hulls. The quality of the fibre-matrix interface is often an important factor in determining resin and composite degradation as a poor interface allows rapid ingress of the environment into the composite.

**Table 3.3** Monomers used for the crosslinking of polyesters.

| Monomer | Boiling point | Characteristics |
|---|---|---|
| Styrene | 145 °C | Low cost, good reactivity, high shrinkage |
| Chlorostyrene | 188 °C | Faster curing, high exotherms, lower shrinkage |
| Vinyl toluene | 172 °C | Lower shrinkage |
| α-Methyl styrene | 165 °C | Less exothermic reaction slower curing |
| Methyl methacrylate | 100 °C | Improved weatherability |
| Diallyl phtalate | 160 °C | Low volatility, useful in prepregs |
| Triallyl cyanurate (TAC) | Melting point 27 °C | Good high temperature performance up to 200 °C |
| T-butyl styrene | 219 °C | Low volatility |

### 3.4.2   Vinyl ester resins

Vinyl ester resins represent a link between the polyester group of resins described above and the epoxy resins described further on in the chapter. As with polyester, they react with styrene in order to form a crosslinked structure through addition polymerisation which is promoted by peroxide catalysts. There are usually two ester groups per molecule and two unsaturated groups at the end of an epoxy polymer chain. The unsaturated groups are carboxylic acids (COOH), usually methacrylic acid but also acrylic acid. The resins can be cured at temperatures above 10 °C or at elevated temperatures using peroxide initiators and the results are resins with properties which are intermediate between those of polyester and epoxy resins.

The chemical scheme (C3.13) shows a typical chemical structure of a general purpose vinyl ester based on a bisphenol A epoxy backbone which results in a combination of corrosion resistance, high modulus and strength, toughness, uniformity of the cured structure and reduced internal stresses.

Typical properties of such a resin are shown in Table 3.5. Increasing the molecular weight of the epoxy backbone increases toughness and chemical resistance although this is accompanied with some loss in heat deflection temperature.

Most vinyl ester resins used in composite materials employ methacrylates as the unsaturated end groups as they are more resistant to base hydrolysis. As the reactive sites occur only at the ends of the molecules, the total number of ester linkages is lower than in a typical unsaturated polyester resin, which results in comparatively improved resistance to caustic degradation. In addition the reaction of the acid with epoxy is complete in vinyl esters so that no excess glycol or similar reactive groups remain available to promote processes such as osmosis which causes blistering of gel coats, used as protective coatings of composite structures and of unsaturated polyester composite laminates.

The hydroxyl groups in the vinyl ester chain promote good wetting and adhesion to polar surfaces such as those of glass fibres.

Specialised vinyl ester resins are produced for particular purposes. Acrylic vinyl ester resins can be cured by exposure to ultraviolet light. Vinyl esters based on multifunctional epoxy back-

Table 3.4: Typical characteristics of common unreinforced thermosetting resins used in composite materials.

| Resin | Physical properties | | | | | | Chemical resistance properties | | | |
|---|---|---|---|---|---|---|---|---|---|---|
| | Flexural strength (MPa) | Tensile strength (MPa) | Tensile modulus (GPa) | Failure strain (%) | Heat deflection temperature (°C) | Normal maximum temperature limit (°C) | Water | Solvent | Acid | Alkali |
| **Unsaturated polyester** | | | | | | | | | | |
| Ortho-phthalate | 100–135 | 50–75 | 3.2–4.0 | 1.2–4.0 | 55–100 | 80–100 | ∨ | ≫ | ∨ | ≫ |
| Iso-phthalate | 100–140 | 55–90 | 3.0–4.0 | 0.8–2.8 | 100–125 | 100–130 | ◯ | ∨ | ◯ | ∨ / ≫ |
| Modified bisphenol type | 125–135 | 65–75 | 3.2–3.8 | 0.9–2.6 | 130–180 | 130–180 | ∧ | ∨ | ◯ | ◯ / ∨ |
| **Epoxy (bisphenol)** | | | | | | | | | | |
| Aliphatic polyamide cure | 85–125 | 50–70 | 3.5 | 1.0–3.5 | 60–90 | 100 | ◯ | ∨ / ∧ | ◯ / ∧ | ◯ / ∨ |
| Boron trifluoride complex | 110 | 85 | 3.0–4.0 | 1.0–2.5 | 120–190 | 90–150 | ◯ | ∨ / ∧ | ◯ | ◯ / ∧ |
| Aromatic amine cure | 80–130 | 60–75 | 3.0–3.5 | 1.5–3.5 | 85–170 | 120–180 | ≪ / ∨ | ◯ / ∧ | ◯ | ∧ / ≫ |
| Aromatic anhydride cure | 90–130 | 80–105 | 2.65–3.5 | 2.0–2.5 | 130–200 | 150–220 | ∨ / ≫ | ∨ / ≫ | ◯ | |
| **Vinyl ester** | | | | | | | | | | |
| Polyimide | 110–130 | 70–85 | 3.3 | 1.0–4.0 | 90–125 | 90–125 | ◯ | ◯ / ∨ | — | ◯ |
| Friedel–Crafts | 75–130 | 50–120 | 3.1–4.7 | 2.0–3.5 | 250–360 | 250–360 | ∨ | — | — | ∨ |
| Phenolic furane | 110–120 | 95–110 | 4.1 | 1.5–3.0 | 160–240 | 150–300 | ≪ | ◯ | ◯ | ◯ / ∨ |
| Silicone | 100–120 | 60–75 | 2.5–3.5 | 0.5–1.0 | 180–220 | 250–300 | ◯ | ≪ | ◯ | ≫ |

≫ poor, ∨ fair, ◯ good, ∧ very good, ≪ excellent.

**Table 3.5**   Typical properties of a vinyl ester resin based on bisphenol A epoxy.

| Liquid properties | | Cast properties | |
|---|---|---|---|
| Specific gravity | 1.4 | Tensile strength | 80 MPa |
| Gel time at 25 °C | 28 min | Failure strain | 6 % |
| Gel time at 80 °C | 12 min | Flexural modulus | 3.1 GPa |
| | | Heat deflection temperature | 100 °C |

bones which results in a high density of crosslinking can be made which enhances resistance to solvents and increases $T_g$. Fire retardant properties can be achieved by the use of a backbone structure of tetrabromobisphenol A which also improves fatigue resistance. Impact resistance is improved through rubber toughening by incorporating carboxy terminated acrylonitrile butadiene.

The uses of vinyl esters are many and include storage tanks and piping because of their resistance to corrosive environments and parts for reasonably high temperature applications. The resin system is suitable for most manufacturing processes including hand lay-up, projection techniques, filament winding, resin transfer moulding and pultrusion.

## 3.4.3   Epoxy resins

Epoxy resins are generally considered to possess more attractive properties than polyester resins for high-performance composite materials as they provide a resin which has low shrinkage, high adhesive strength, excellent mechanical strength and provided that the correct system is used, chemical resistance.

The molecular structure of epoxy resins is based on the epoxy (or oxirane) group, consisting of two carbon atoms and an oxygen atom within a molecular chain so that $R_1$ and $R_1$ below, (C3.14), represent the continuations of the chain on each side of the group.

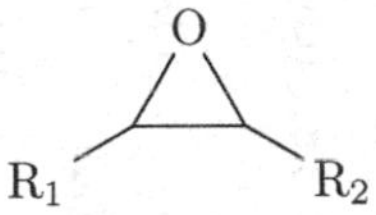

$$(C3.14)$$

Epoxide group

The epoxy cycle is strained which accounts for its high reactivity. A type of epoxy used in composites is known as cycloaliphatic epoxy resins in which the oxygen atom is attached to a six member carbon ring, as shown in Ch. (C3.15).

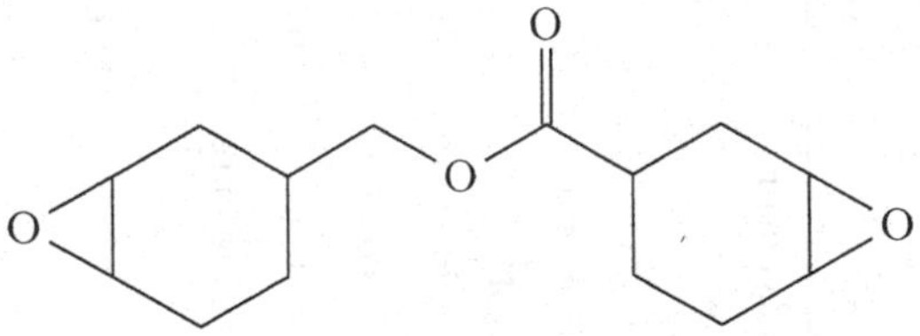

$$(C3.15)$$

Cycloaliphatic resin

This type of resin is recommended for use in combination with bisphenol A epoxy resin for filament winding to increase their heat distortion temperature and to reduce viscosity of the resin. However, the epoxide group which is most often used is glycidyl ether as

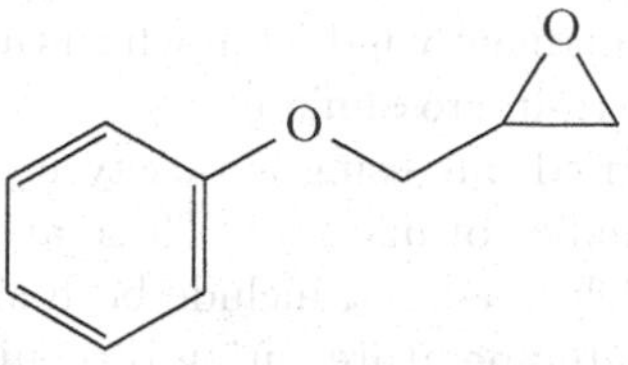

(C3.16)

Glycidyl ether

The commercially most important epoxy resin known as bisphenol A diglycidyl ether (DGEBA) in which the epoxy group is attached to a hydrogen atom ( $R_1$ ) and to glycidated polyhydroxyphenols ( $R_1$ ), accounts for 90 % of epoxy resins produced. They are synthesised from epichlorohydrin and bisphenol A in the presence of sodium hydroxide. The synthesis of DGEBA epoxy resin is shown in Figure 3.8.

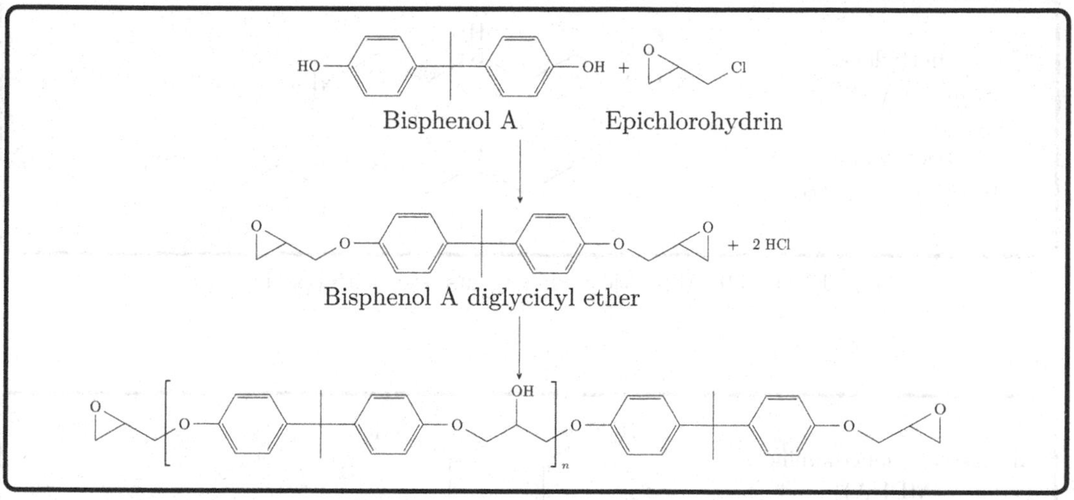

Figure 3.8: Synthesis of bisphenol A diglycidyl ether (DGEBA) epoxy resin.

The value of $n$, in Figure 3.8 indicates the number of times the structure within the brackets is repeated and this can be varied so that the density of crosslinking can be altered. In commercially available resins $n$ falls in the range 0 to 25. As $n$ increases so does the number of hydroxyl groups. The epoxy resins with low values of $n$ usually require curing agents that react with the epoxy group, whereas resins with higher $n$ values are cured through the hydroxyl groups. In any given resin there will be a range of chemical structures so that the value of $n$ is an average value and so can be less than 1. Liquid epoxy resins, such as those used in composite materials, have values of 0 to 1 with values greater than 2 being solid.

Many applications including encapsulation, casting and filament winding which require fluid resins epoxies with $n$ values less than 1 are used as they offer the best balance of handling, reactivity and performance properties. Epoxy resins with $n$ values higher than 1 can be used for composite manufacture if they are heated so as to bring down their viscosity. This is the case for some resins cured with aromatic amines; however, increasing the temperature also increases the rate of reaction so that the pot life or useable lifetime for the resin is reduced. Solvents which do take part in the crosslinking process can also be used to reduce the viscosity of some resins, and this is applicable in the manufacture of some laminates in which the solvent can be removed by evaporation before curing. If the solvent is not completely removed during composite manufacture its presence can seriously degrade the composite properties so that such systems are not used in processes such as filament winding in which thick layers of fibre may be produced inhibiting solvent removal before resin crosslinking.

Curing of epoxies can be carried out using a variety of agents. However, the most widely used systems consist of three families of material. These are amines, anhydrides and catalytic agents. Amines are the most widely used and include both aliphatic and aromatic amines. The aliphatic amines are used if room temperature curing is required as they are more reactive than anhydride amines. For this reason they are used for contact moulding processes for which their short curing time is useful, but they are not used for prepegging for which a slower reaction is required. The high volatility of the aliphatic amines can lead to handling problems including skin irritation and for this reason they can be combined with other agents to reduce their reactivity. Aromatic amines require raised temperatures to produce crosslinking of epoxides and so can be used for processes such as filament winding and prepegging. Figure 3.9 and Figure 3.10 show some aliphatic and aromatic curing agents.

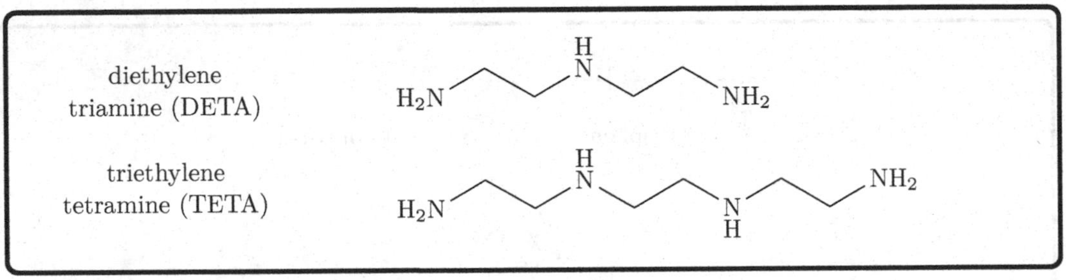

Figure 3.9: Aliphatic curing agents used with epoxides.

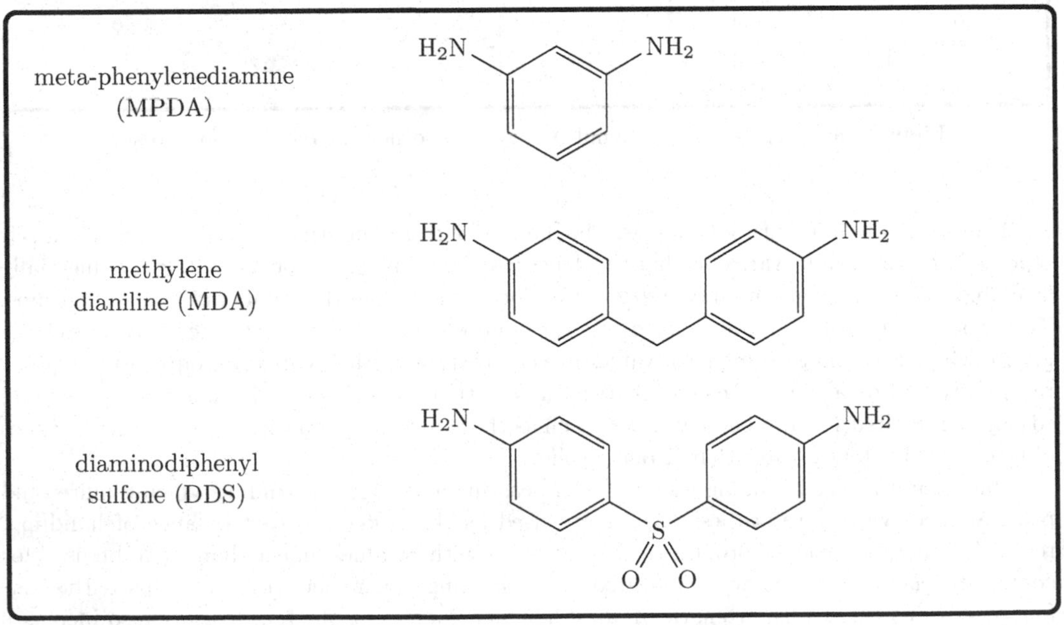

Figure 3.10: Aromatic curing agents used with epoxides.

Anhydrides are widely used to cure epoxies as they give long pot lives and low exotherms. They require high curing temperatures, typically 180 °C for periods of several hours to achieve crosslinking and the cured resin have good high temperature and chemical stability, except to caustic soda, NaOH. They are used both for prepreg and filament winding processes. It should be noted, however, that anhydride cured epoxies are susceptible to degradation by water uptake, particularly if the water is warmer than around 40 °C. Figure 3.11 shows some anhydride systems used to cure epoxy resins.

nadic methy anhydride (NMA)

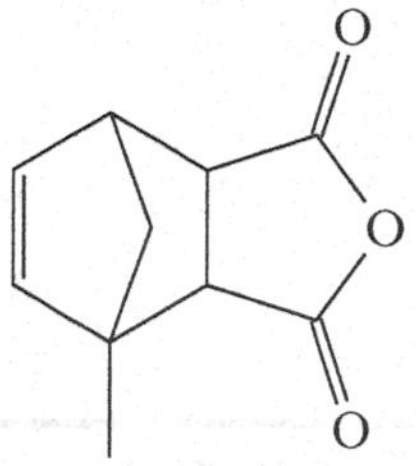

hexahydrophthalic anhydride (HHPA)

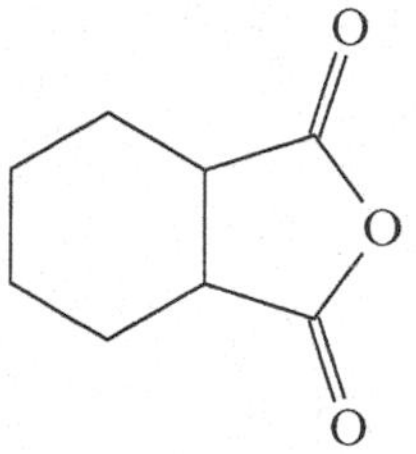

phthalic anhydride (PA)

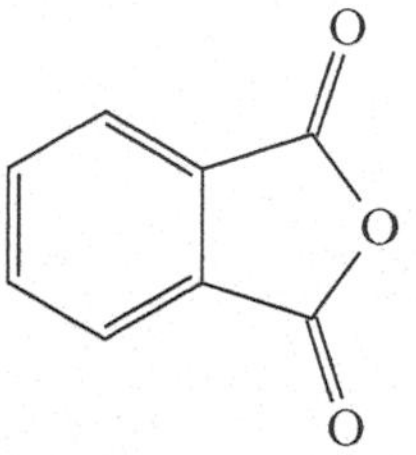

Figure 3.11: Anhydride curing agents for epoxides.

Catalytic hardeners promote homopolymerisation of the epoxy to achieve cure. They belong to a group of agents known as Lewis acids or bases and require curing temperatures of around 180 °C. Such agents are shown in Figure 3.12.

The process of crosslinking is very dependent on the type of curing agent used. One of the most used systems is shown in Ch. (C3.17) in which an anhydride hardener is used.

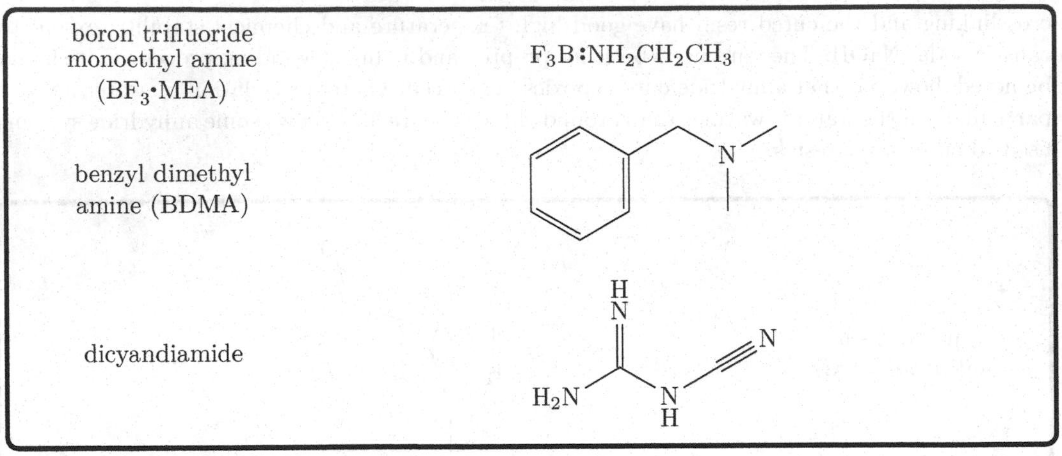

Figure 3.12: Catalytic curing agents for epoxides.

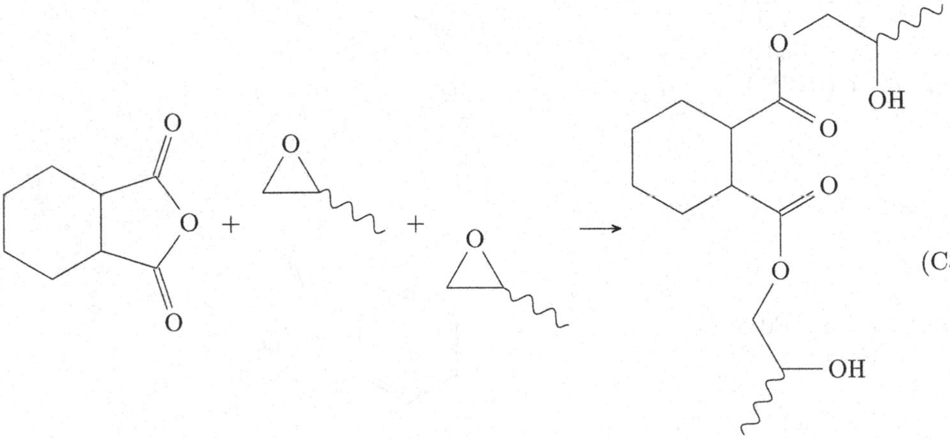

(C3.17)

An amine cured epoxy system is illustrated by the chemical scheme (C3.18). The reaction involves the epoxide group opening to combine with the hydrogen atom in the amine. In the case of aliphatic amines curing can occur at room temperature but higher temperatures around 150 °C to 180 °C are necessary in the case of aromatic amines. The latter system is known as a B-staged resin.

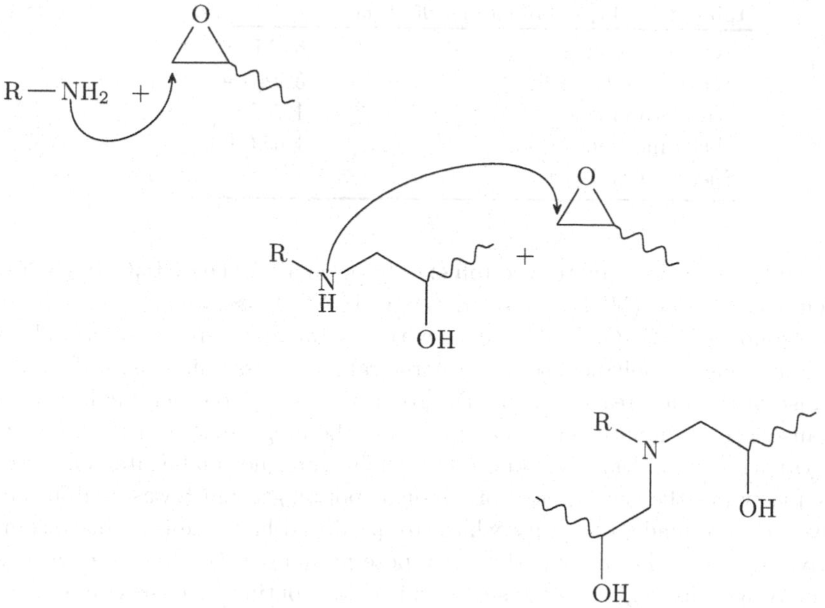

$$R-NH_2 +$$

(C3.18)

The above discussion is primarily concerned with the most widely available and used epoxy system based on bisphenol A, which is restricted in use at temperatures below 150 °C. Higher temperature stability can be obtained using epoxy systems which have functionalities greater than 2. A commonly used system of this type is the tetrafunctional resin tetetraglycidylamine (TGMDA), shown in Ch. (C3.19). This type of resin is gradually being overtaken by more complex systems with higher $T_g$ and greater resistance to water. Such epoxy systems extend the long term-temperature range of epoxies to 175 °C.

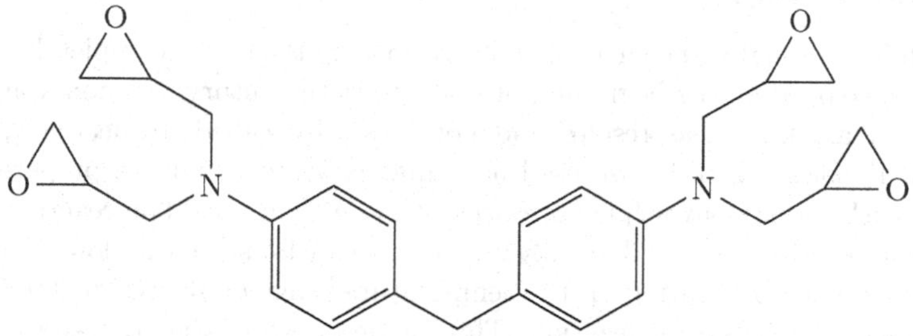

(C3.19)

tetraglycidyl-4,4'-diaminodiphenylmethane
(TGMDA)

### 3.4.4 Polyurethane resins

These resins represent a large family of resin systems most of which are made into foams, elastomers and surface coatings. They are not usually associated with composite materials; however, there are some exceptions which are finding use in reinforced reaction injection moulding (RRIM) and continuous manufacturing processes such as pultrusion. These manufacturing techniques are described in Chapter 4.

Polyurethane are polymers which are made by a reaction of an organic isocyanate with a polyol, such as an alcohol, which is any compound containing multiple hydroxyl groups. The general formula for an isocyanate is $R-N=C=O$ in which $R$ is a reactive, usually large aromatic

**Table 3.6**   Typical properties of phenolic resole resins.

| | |
|---|---|
| Tensile strength | 61 MPa |
| Tensile modulus | 3.9 GPa |
| Strain to failure | 1.7 % |
| Shear modulus | 1.65 GPa |
| Specific gravity | 1.26 |

group. The most widely used isocyanates are toluene diisocyanate (TDI) $CH_3C_6H_3(NCO)_2$ and methylene diphenyl isocyanate (MDI) $OCNC_6H_4CH_2C_6H_4NCO$. Isocyanates contain the extremely reactive group $-N=C=O$. It will react with any hydroxy group $-OH$ and can be combined with many different polymers so that a large range of possibilities exist. It should be noted that, because of the high reactivity of this group, it poses a considerable health risk as inhalation will cause a reaction with the $-OH$ groups in the lung tissue, causing a respiratory illness called isocyanate asthma. Damage can be very debilitating and can be fatal. One variation of isocyanate chemistry was the development of phosgene poison gas which was used in the First World War. However, once made, the resins which are produced have the isocyanate combined with other reactive agents so that after curing they pose no danger. For example, they can be combined with epoxy systems to give resin systems which have continuous in-service temperature capabilities of 200 °C.

These resins are of interest because they allow rapid crosslinking, usually with times of a few minutes and at temperatures which are low, from 25 °C to 60 °C. The resins are generally tough, due to their high impact strength and strain to failure which is greater than most other systems. They adhere well to fibres. The composites which are produced with these resins absorb much less water than say epoxy resins and therefore are of interest for applications for which this could present a problem.

## 3.4.5   Phenolic resins

Phenol formaldehyde, usually known as phenolic resins, are the oldest completely synthetic polymers and were developed at the beginning of the twentieth century. Phenolics are divided into two classes which are known as resoles and novalacs, also called, respectively, one-step and two-step resins. Resoles are the most used as a matrix system and are prepared under alkaline conditions with formaldehyde/phenol ratios of more than one. The reaction is stopped by cooling to give a reactive and soluble polymer which can be stored at low temperatures and subsequently crosslinked by heating, in the temperature range of 130 °C to 200 °C, as a continuation of the initial condensation reaction. This is the origin of the "one-step" nomenclature. Resoles are used to impregnate papers or fabrics for both electrical and structural applications. Novolacs are not often used as a matrix for composite structures. They are produced under acid conditions with formaldehyde/phenol ratios of less than one. The reaction is carried to completion to give an unreactive thermoplastic which in a second step can be cured by the addition of hexamethylenetetramine. See Table 3.6 for properties of phenolic resins.

## 3.4.6   Polyimides

High temperature stability is achieved with the use of polyimide resins, which can be used as matrix materials up to temperatures around 300 °C. For this reason polyimides are attractive not only for structural composite applications but also for electrical insulation. Polyimides can be made both in thermosetting and thermoplastic forms. These resins are generally made in two stages, consisting of the polycondensation of a dianhydride and an aromatic diamine which gives an acid which is soluble in polar solvents. Heating, usually above 250 °C to remove water results in the formation of the polyimide cycle as shown in Figure 3.13. Typical properties of polyimide resins are shown in Table 3.7.

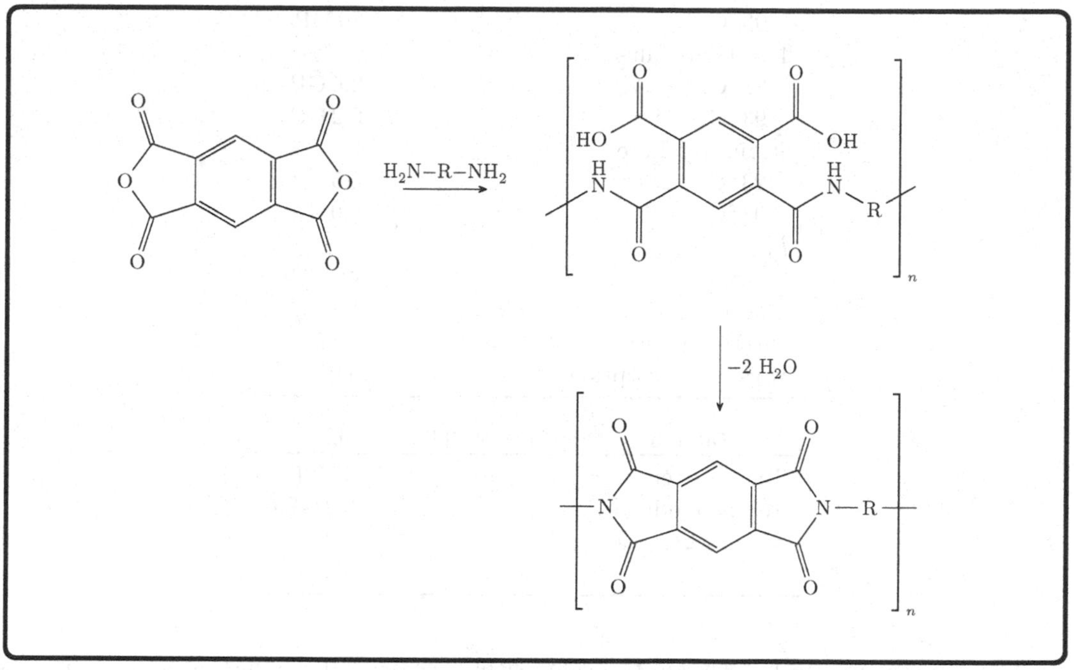

Figure 3.13: The formation of polyimide where Rcan take different forms.

A variety of routes exist for making polyimides so that polymerisation by addition is also possible. Generally it is difficult to control the chemistry of the reactions taking place to create polyimides, so problems associated with the stability of the reactants, their volatility, the molecular weight distribution and porosity can occur. The latter can be a particular problem in thick structures and can lead to bubbles and delamination of the structure. The manufacture of composites using polyimide resin systems usually requires long curing cycles.

Bismaleimides are thermosetting resins which are based on the polyimide chemistry but made by addition curing and can be more easily processed than most resins in the family. The chemistry of their manufacture is shown in Figure 3.14, resulting from the reaction of a maleic anhydride with methylene dianiline (MDA). However, the MDA is increasingly replaced by other aromatic diamines as it presents a health risk. The reaction on heating is with a linear function with a carbon backbone. The resulting resin is very brittle, but this can be countered by the addition of an aromatic diamine which is copolymerised with the bismaleimide. Processing occurs at high temperatures, above 175 °C, and leads to crosslinking of the resin.

Bismaleimides have an upper temperature use of around 225 °C and can be processed almost as easily as epoxies. In addition they are inexpensive and require only low pressures during composite manufacture. However, in general, bismaleimides are more brittle than epoxies, showing poor fracture toughness and as a consequence suffer from microcracking in laminates.

In situ polymerisation of monomeric reactants (PMR) gives the most widely used resins with the highest in-service temperature capabilities. PMR-15 is a thermosetting polyimide with temperature capabilities up to 316 °C for at least 1500 hours in air. It is made by impregnating the fibres with a low molecular weight polymer consisting of nadic ester (NE), 4,4'-diaminodiphenyl methane (MDA) and benzophenone tetracarboxylic acid diethyl ester (BTDE). The MDA component is carcinogenic which has led to much interest in producing resins with similar high

**Table 3.7**   Typical properties of a polyimide resins.

| | |
|---|---|
| Tensile strength | |
| 23 °C | 100 MPa |
| 93 °C | 80 MPa |
| Tensile modulus | |
| 23 °C | 2.5 GPa |
| 93 °C | 2.2 GPa |
| Strain to failure | |
| 23 °C | 6.5 % |
| 93 °C | 5.0 % |
| $T_g$ | |
| Dry | 210 °C |
| Wet | 190 °C |
| Specific gravity | 1.37 |
| Moisture absorption | 0.9 % |

**Table 3.8**   Properties of PMR-15 resin.

| | |
|---|---|
| Tensile strength | 55 MPa |
| Tensile modulus | 3.2 GPa |
| Specific gravity | 1.3 |
| $T_g$ | 350 °C |

temperature capabilities but without this component. The thermal crosslinking addition process for producing PMR-15 is shown in Figure 3.15.

PMR-15 has a molecular weight of 1500 which gives it its name. It is brittle and is of primary interest to the aerospace industry for high temperature uses. It is destined most probably to be displaced with modified polyimides which should be tougher and less toxic during its processing. Properties of PMR-15 are shown in Table 3.8.

Most resins which are said to retain their properties to above 300 °C suffer, however, from a fall in $T_g$ to around 225 °C when subjected to temperature cycles in a humid atmosphere.

## 3.5   Thermoplastic matrix materials

The plastic industry produces far more thermoplastics than it does thermosetting plastics, approximately in the ratio of 4 : 1; however, this ratio is not maintained in the area of composite materials which represent about 3 % of the total plastic industry. Approximately 60 % of the production of composites comes from thermosets, so about 40 % comes from thermoplastics. The rate of growth of thermoplastic matrix composites is however considerably higher than that of the thermosetting composites. This, in part, is because of the widening use of composite parts and the need for faster production rates with thermoplastics than are possible with most thermosetting resins. Thermoplastics can simply be melted and formed with no delay, as occurs with thermosets while they crosslink. In addition thermoplastics do not need to be stored at low temperatures, which reduces costs and handling problems, and they are inherently tougher than thermosetting resin. Costs are further reduced as there are fewer scrapped parts as thermoplastics can be reprocessed and recycled. The natures of the two types of composites are, in the main, different, thermoplastics originally being primarily used for short fibre injection moulded composites but are now finding their way into long or continuous fibre composites. This difference is accounted for partly by historical reasons, with thermosets being employed for impregnating fibre structures well before the use of thermoplastics, but above all for reasons of processing.

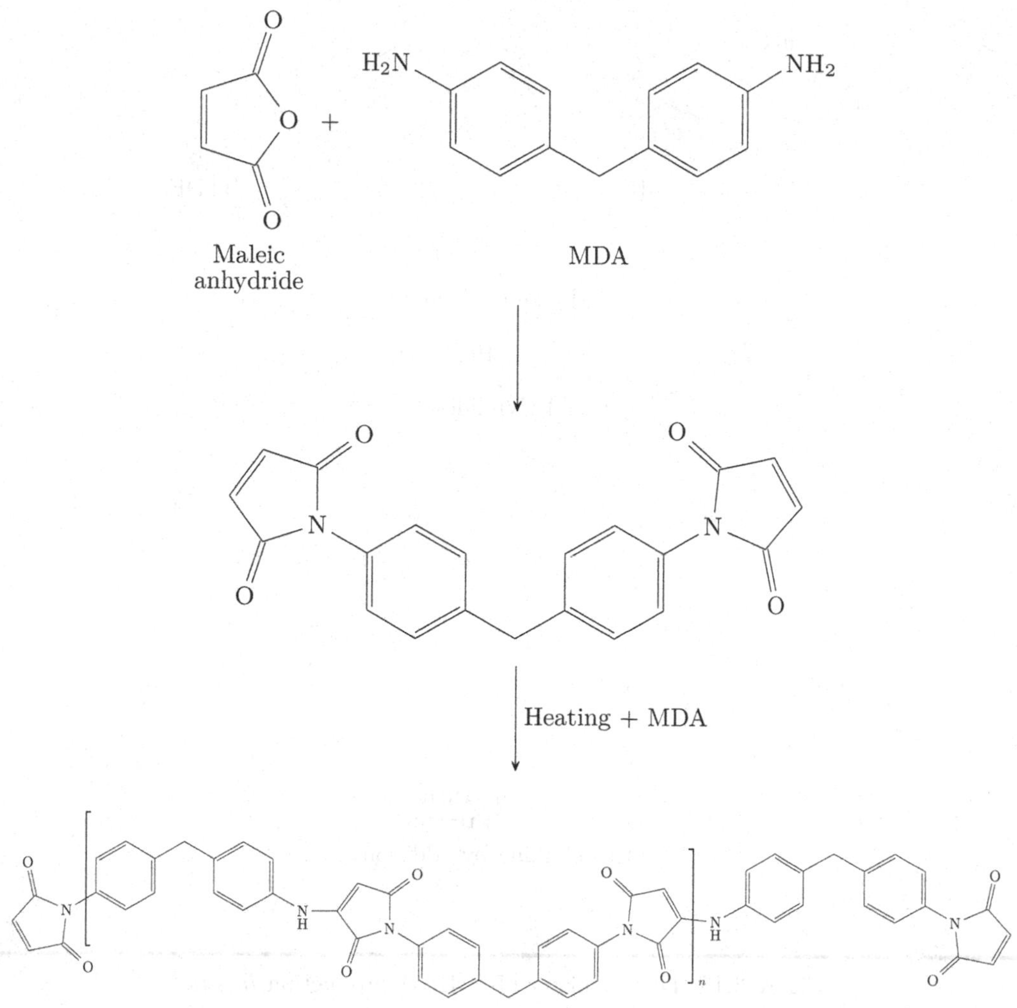

Figure 3.14: The chemistry of bismaleimide.

The thermosetting resins can be easily used to impregnate fibre tows as they are either liquid or easily put into solution when in their non crosslinked state. The thermoplastics are solids so that they have to be put into intimate contact with the reinforcements when in a molten state. This clearly is the case for injection moulded composites and in the case of continuous fibres, analogous to most thermosetting systems, the thermoplastics are placed in contact with the fibres either in the form a powder or in the form of filaments. This intimate combination is then heated so as to melt the thermoplastic matrix material. Adhesion with the fibres can be a difficulty with some thermoplastic matrices, such as polypropylene, which have to be modified for good fibre-matrix bonding to occur. The high viscosity of molten thermoplastics and their

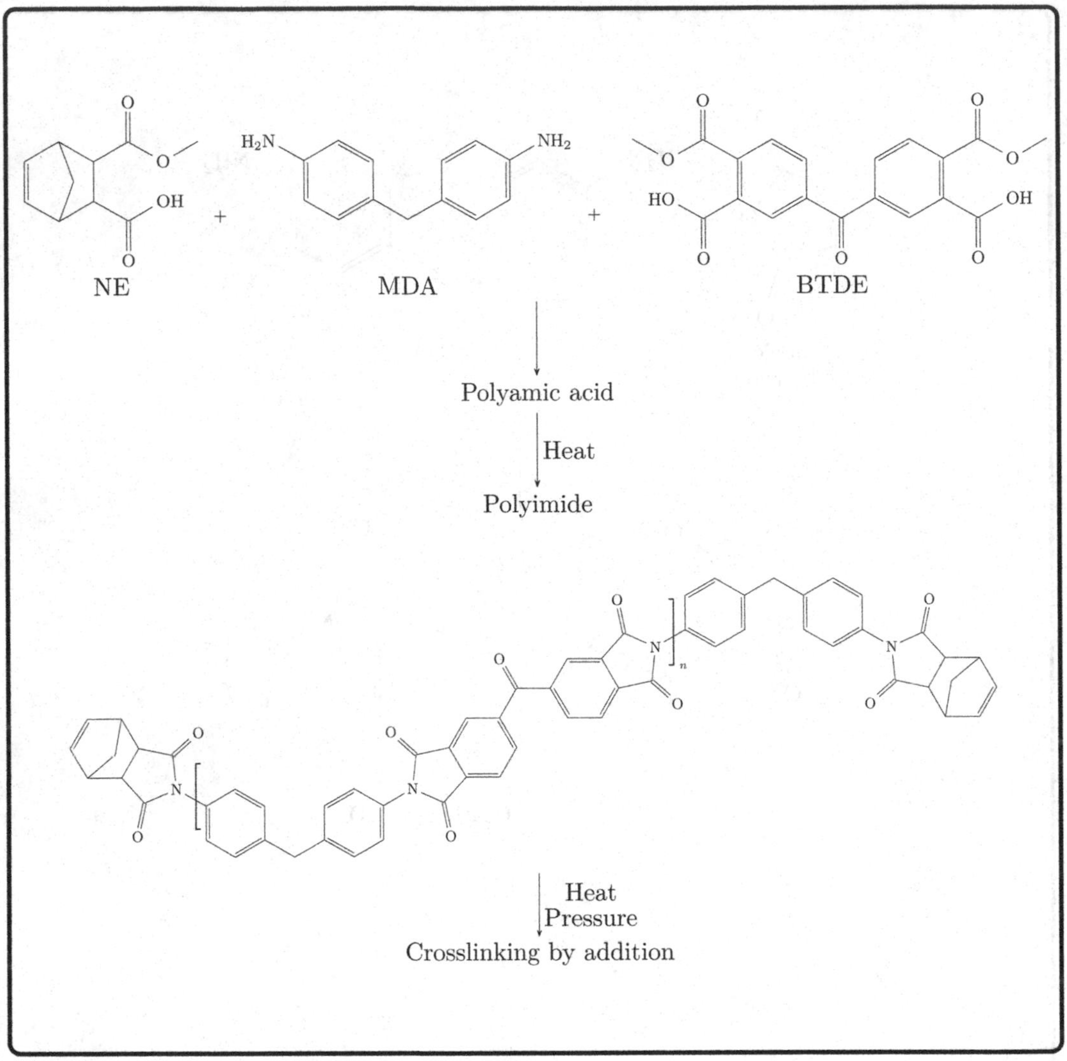

Figure 3.15: The chemistry of PMR-15 production *in situ.*

lack of available free bonds are the causes for processing difficulties and the difficulty of creating good adhesion with fibres. This is more of a difficulty with continuous fibre composites than with injection moulded parts as the shearing which occurs in the injection process can create bonding sites in the matrix material.

These differences between thermosets and thermoplastics tend to disappear with the emergence of new industrial solutions. On the one hand, it is about exploiting thermoplastic reactive processes greatly facilitating impregnation (van Rijswijk and Bersee, 2007). On the other hand, thermosets benefit from increasingly shorter crosslinking times. Some processes, historically dedicated to thermosets are now accesible to thermoplastics. This is particularly the case with RTM and pultrusion.

Thermoplastic composites can be based with readily available materials, such as polypropylene, polyamide (Nylon) and saturated polyethylene terephthatlate (polyester) but also other more exotic and high-performance thermoplastics are used such as polyether ether ketone (PEEK). Reactive solutions are not yet possible for all thermoplastics but very low viscosity

**Table 3.9**  Typical properties of undrawn bulk polypropylene.

| | |
|---|---|
| Density | $905\,\mathrm{kg\,m^{-3}}$ |
| Melting point | 165–170 °C |
| Elastic modulus | 1.0–1.4 GPa |
| Yield stress | 25–38 MPa |
| Strain to failure | 300 % |
| Coefficient of linear expansion | $175 \times 10^{-6}\,°\mathrm{C}^{-1}$ |
| Processing temperature | 190–285 °C |

grades are also being developed.

## 3.5.1  Polypropylene

Polypropylene, together with polyethylene and polybutadiene, make up the important polyolefine group of polymers. Polypropylene is a commodity plastic and together with polyvinylchloride (PVC) and polysulfone, which are also commodity plastics, they represent 70 % of the unreinforced thermoplastic market. These three plastics share the attribute of being of relatively low cost but polypropylene is interesting as a composite matrix material as it is more rigid and stronger than PVC, more easily worked than polysulfone and has the lower density ($900\,\mathrm{kg\,m^{-3}}$). Polypropylene is semi crystalline when in the isotactic, most usual form and has a crystallinity of about 68 %. The polymer can therefore be seen as consisting of an amorphous phase, which is in the rubbery phase at room temperature, in which are embedded the spherulitic crystals. Polypropylene is a linear polymer based on a backbone of C−Cbonds with lateral methyl groups, as shown in Ch. (C3.20).

(C3.20)

Polypropylene

The physical and mechanical properties of unreinforced polypropylene are shown in Table 3.9. The low melting point should be noted. The polymer shows several transition temperatures at which changes to the structure occur. The most important of these transitions is the glass transition temperature ($T_g$). As polypropylene is semicrystalline there is a $T_g$ for the amorphous regions which is around 5 °C and one for the crystalline regions, around 90 °C. Above these temperatures the molecular structure acquires sufficient energy to allow much freer movement. There is another, secondary transition temperature around −80 °C corresponding to localised movements in the amorphous phase. Such transitions can be revealed by viscoelastic tests, as are described above, in which the energy absorbed during cyclic loading leads is revealed by the phase difference between the applied strain and the induced load in the material. At such transitions the energy absorbed is a maximum.

Polypropylene is not very reactive because of the non-polar nature of the molecule and this leads to problems in forming bonds between the matrix and the fibres in a composite. Two approaches are used to improve bonding. The more usual technique is to incorporate into the size on the fibres a coupling agent which can bond both to the glass and to the polymer. Such coupling agents are often based on silanes which contain the $-\mathrm{Si(OH)_3}$ group and which can bond to the silanol $-\mathrm{Si-O}-$ groups on the glass fibre surface, as described in Chapter 2. The silane can then bond to the polymer matrix so achieving adhesion. The lack of functionality of pure polypropylene means that to achieve bonding the polymer is often modified by the addition

**Table 3.10**   Typical properties of Polyamide 6,6 .

| | |
|---|---|
| Density | $1140 \, \mathrm{kg \, m^{-3}}$ |
| Melting point | $264 \, °\mathrm{C}$ |
| Elastic modulus | $1.5–2.5 \, \mathrm{GPa}$ |
| Yield stress | $60–75 \, \mathrm{MPa}$ |
| Strain to failure | $40–80 \, \%$ |
| Coefficient of linear expansion | $90 \times 10^{-6} \, °\mathrm{C}^{-1}$ |
| Processing temperature | $260–325 \, °\mathrm{C}$ |

of reactive groups to the polymer chain. Bonding of the polypropylene matrix to the glass fibres takes place in two steps, during which the macroalkyl radicals are formed by shearing of the polymer chains making up the matrix in the injection moulding process.

An adduct is then produced by grafting a modifier or functionalised group to the activated chain. The modifier must be bifunctional with one end group reacting with the size on the fibres and the other with the polymer. The addition of such modifying agents as acrylic acid or maleic anhydride, as shown in Ch. (C3.9), leads to improved bonding as the matrix is able to form strong bonds with the size or directly with the glass.

### 3.5.2   Polyamide 6,6

The polyamides are a group of polymers often referred to as Nylon which is a commercial name. They are engineering plastics which mean that they possess some useful properties, such as strength, stiffness, toughness and resistance to chemical attack when in the unreinforced state. One of the most widely used polyamide, both in fibre and bulk form, is polyamide 6,6 which, as its name indicates, consists of linear molecules based on repeated units of two groups of six carbon atoms, as shown in Ch. (C3.21), extract from Figure 3.2.

Polyamide 6,6                                                                                                    (C3.21)

Polyamide 6,6 is made by the copolymerisation of hexamethylene diamine and adipic acid, both of which contain six atoms of carbon. It is semi-crystalline, ranging in crystallinity from 25 % to 50 %. It has one of the highest melting points of the polyamide family, 264 °C, and is typically processed around this temperature, but in common with other polyamides has relatively low viscosity in the molten form which aids processing. It adheres well to reinforcing fibres and shows good resistance to most chemical environments, although it is sensitive to strong acids. In its unreinforced form it is used in many light engineering structures from fans and door handles in cars and curtain fittings which make use of the low coefficient of friction of polyamides. As a short glass fibre reinforced composite it finds uses around car engines due to its excellent resistance to oil. Some typical properties of bulk polyamide 6,6are shown in Table 3.10.

Polyamide 6,6 absorbs significant amounts of water which causes large falls in glass transition temperature. For example a completely dry specimen will show a $T_g$ of around 70 °C but as it absorbs water the $T_g$ will fall as the water acts as a lubricant to the molecular structure. This can increase toughness as it reduces stress concentrations, however, water absorption, at saturation, can reach 6 % of the dry weight of an immersed specimen at which point the $T_g$ is around 0 °C.

**Table 3.11**    Typical properties of PET.

| | |
|---|---|
| Density | $1350\,\mathrm{kg\,m^{-3}}$ |
| Melting point | $260\,°C$ |
| Elastic modulus | $2.8\,\mathrm{GPa}$ |
| Yield stress | $80\,\mathrm{MPa}$ |
| Strain to failure | $70\,\%$ |
| Coefficient of linear expansion | $70 \times 10^{-6}\,°C^{-1}$ |
| Processing temperature | $225–350\,°C$ |
| Water uptake at $100\,\%$ RH ($20\,°C$) | $0.5\,\%$ |

This effect is largely reversible if the polymer is dried. Polyamide 11 and polyamide 12 have extra methylene groups between the amide groups which make these forms of Nylon more flexible, tougher and also reduces the amount of water they absorb so that for some marine applications these polymers could be considered as matrix materials.

Polyamides lend themselves well to the reactive processes, i.e. *in situ* polymerisation, as described in van Rijswijk and Bersee (2007).

### 3.5.3    Polyester thermoplastic matrix

All polyesters contain the ester group $-\,CO-C\,-$. This group can be hydrolysed at temperatures around $150\,°C$. For this reason it is most important to ensure that the polymer is dry when processing it and making composites. The most widely used polyester is polyethylene terephthalate (PET), which is shown in Figure 3.2 and also reproduced in Ch. (C3.22).

Polyethylene terephthalate (PET)

(C3.22)

It is a semicrystalline homopolymer made by the polycondensation of terephthalic acid ethylene glycol and with the crystalline phase typically making up $35\,\%$ of the material. Its wide use is due to it being the lowest cost and highest volume polyester produced. It is melt blown for such products as drink bottles and spun into fibres. It has a low viscosity which helps the processing of composites. The presence of the aromatic ring means that the PET matrix is stiffer, stronger and can operate at higher temperatures than say polypropylene. It is used in many of the same types of applications as polyamide 6,6. Compared to this material it absorbs less water and its $T_g$, which is around $75\,°C$, does not decrease with water uptake. The effect of water at temperatures above the $T_g$ is to broaden the peak revealed by DSC measurements as hydrolysis sets in and cuts the long PET molecules. This enables the molecules to reassemble and increase crystallinity up to as much as $55\,\%$ and broaden the range of crystallite sizes in the matrix. The effect of increased crystallinity is to reduce toughness which if taken to extreme values can induce cracking in the matrix. Typical properties of unreinforced PET are shown in Table 3.11.

As described in van Rijswijk and Bersee (2007), "macrocyclic oligomers can be obtained through cyclodepolymerisation of linear PET and subsequently repolymerised". PET can thus be used in a reactive way leading to liquid route processes, usually intended for thermosets.

**Table 3.12**    Typical properties of PEI.

| | |
|---|---|
| Density | $1260\,\mathrm{kg\,m^{-3}}$ |
| Melting point | $349\,^{\circ}\mathrm{C}$ |
| Elastic modulus | $2.7\text{--}4.0\,\mathrm{GPa}$ |
| Yield stress | $105\,\mathrm{MPa}$ |
| Strain to failure | $5\text{--}90\,\%$ |
| Coefficient of linear expansion | $50 \times 10^{-6}\,^{\circ}\mathrm{C}^{-1}$ |
| Processing temperature | $335\text{--}370\,^{\circ}\mathrm{C}$ |
| Water uptake at $100\,\%$ RH ($20\,^{\circ}\mathrm{C}$) | $1.25\,\%$ |

### 3.5.4    Polyimides

This class of matrix material is interesting because of its high transition temperatures how-ever, partly as a result of this, processing is difficult. Linear polyimides are thermoplastics but they degrade before reaching their melting point so that they are made from solution or con-verted into a polymer *in situ*. Most polyimides are viscous and this complicates processing. Ch. (C3.23) (from Figure 3.3) shows polyetherimide (PEI) which is marketed by GE Plastics under the name of ULTEM.

Polyetherimide (PEI)                                                                 (C3.23)

This matrix material is derived from phthalic anhydride, bisphenol A and Meta-phenyllenediamine (MPD). The inclusion of an ether bond in the molecular backbone allows it to be processed from the melt. The molecular is irregular in shape, as can be seen from Ch. (C3.23) and for this reason it is amorphous.

Polyetherimide is a general purpose moulding resin with excellent processing characteristics, excellent resistance to chemical attack and hydrolysis. It has high mechanical strength, high heat deflection temperature and excellent flame resistance. PEI is used in composite applications be-cause of its high in-service temperature, toughness, good interfacial adhesion and nonflamability. Typical properties of polyimide resins are shown in Table 3.12.

### 3.5.5    Sulphur-containing polymer matrix

Sulphur (S) and the sulfone group of compounds containing the radical $SO_2^-$ united with two hydrocarbon radicals can be placed in the backbone of aromatic polymers to give useful high temperature matrix systems. The chemical formulae of several such polymers are given in Fig-ure 3.16.

The presence of sulphur in the backbone increases flexibility whereas sulfone in more rigid. Polyphenylene sulfide (PPS) is a homopolymer and is crystalline. It is used as a composite matrix because of its chemical resistance, good mechanical properties and low moisture uptake. It has a high melting point and is non-flammable. Typical properties are shown in Table 3.13.

Both polysulfone and polyethersulfone are amorphous. PS has a $T_g$ in the range of $180\,^{\circ}\mathrm{C}$ to $265\,^{\circ}\mathrm{C}$ and resists thermal oxidation well. They are therefore used in applications such as food trays which require repeated sterilisation. These amorphous polymers are very resistant to

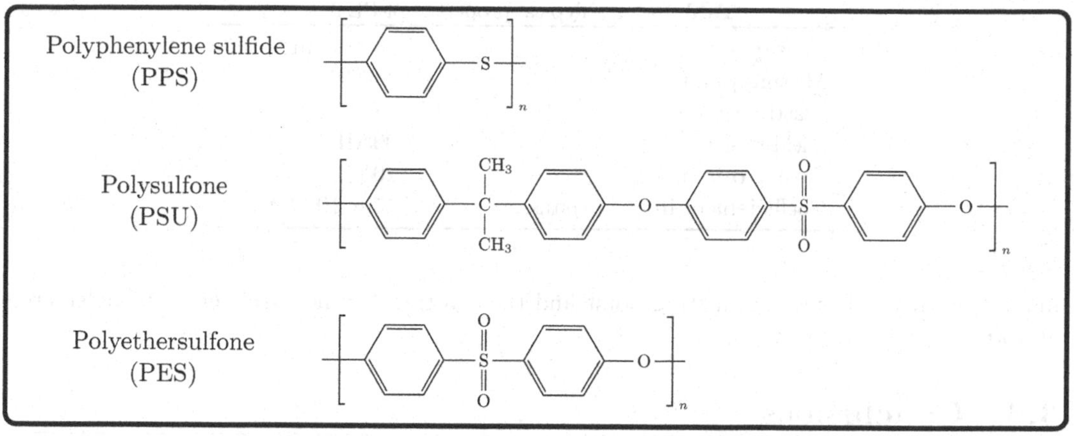

Figure 3.16: Sulphur-containing matrix systems.

**Table 3.13**    Typical properties of sulphur-containing matrix systems.

| Polymer | Density $(kg\,m^{-3})$ | Elastic modulus (GPa) | Failure stress (MPa) | Failure strain (%) | Coefficient of expansion $(°C^{-1})$ | Processing temperature (°C) |
|---|---|---|---|---|---|---|
| PPS | 1350 | 3.35 | 65 | 2.5 | $35 \times 10^{-6}$ | 330 |
| PSU | 1240 | 2.60 | 67 | 75 | $55 \times 10^{-6}$ | 360 |
| PES | 1370 | 2.60 | 85 | 60 | $50 \times 10^{-6}$ | 350 |

hydrolysis as well as to aqueous acids and bases however it is susceptible to chlorinated solvents and most highly polar organic solvents. They can be used continuously from 150 °C to 220 °C. Reactive processing of such high-performance plastics is detailed in van Rijswijk and Bersee (2007).

## 3.5.6   Polyethers

Polyethers contain oxygen bonded to carbon in their backbone. The ether group introduces flexibility into the molecule which makes melt processing easier. There are many polyether thermoplastic resins available and some are used for composite materials. However the best known is polyether ether ketone (PEEK) which is semi-crystalline, possesses a melting point around 380 °C and can be used for high temperature applications. Its molecular formula can be seen in Ch. (C3.24).

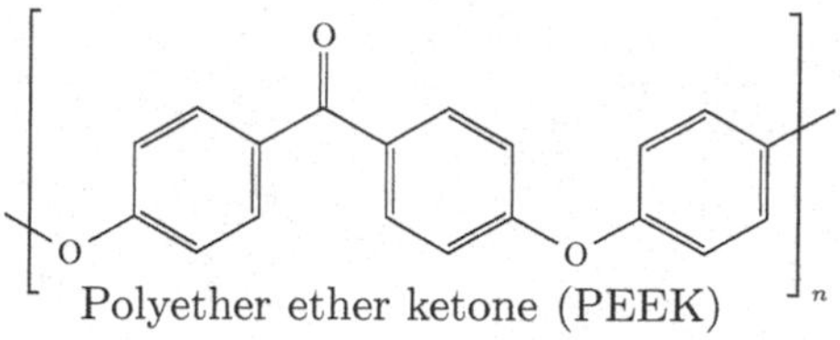

(C3.24)

Polyether ether ketone (PEEK)

It possesses high mechanical properties, as can be seen in Table 3.14, as well as being highly resistant to chemical attack and being nonflammable. It shows excellent adhesion to carbon fibres and gives very good interlaminar shear strengths for the composites for which it is used as a matrix. It shows transcrystalline growth from the irregularities on the carbon fibres surfaces. In order to control the microstructure of the PEEK it is recommended to consolidate structures

**Table 3.14**    Typical properties of PET.

| | |
|---|---|
| Density | $1300 \, \text{kg m}^{-3}$ |
| Melting point | $340 \, °\text{C}$ |
| Elastic modulus | $3.6 \, \text{GPa}$ |
| Yield stress | $90 \, \text{MPa}$ |
| Strain to failure | $20 \, \%$ |
| Coefficient of linear expansion | $45 \times 10^{-6} \, °\text{C}^{-1}$ |

made from it at around the melting point and then to transfer the specimen to a cooler press at 150 °C.

## 3.6    Conclusions

Organic resins find wide uses as matrices for composite materials. They can be either thermosetting resins or thermoplastic polymers. Their properties depend intimately on their chemistry as does their cost and ease of incorporation as a matrix material. Polyester resins are most widely used in general purpose composites but epoxy resins represent the most important group of resin systems used in high-performance composites. Other resin systems exist which have better high temperature properties than these common systems but their chemistry is generally more complex. Thermoplastic resins offer an adaptability after manufacture which is lacking in thermosetting systems and so are attractive both for the speed of composite manufacture that they allow but also as there should be less problems associated with recycling. Whilst thermosetting resin systems represent about 60 % of the matrix systems used in composites thermoplastics show two or three times their growth rate as composites find increasing uses in an ever widening field of industrial sectors.

# Revision exercises

### Exercise 3.1

Explain the difference between a thermosetting resin and a thermoplastic. What happens if they are heated? Give examples of each type of material.

### Exercise 3.2

The macromolecules on which resins are built are derived from monomers. Discuss how these monomers can be made into long chain macromolecules . Explain how a bifunctional system results in a thermoplastic and how cross-linking can occur with other systems.

### Exercise 3.3

Polyesters are the most commonly used thermosetting resin system but polyesters are also used in the production of fibres. Discuss and explain the signification of this observation.

### Exercise 3.4

Explain the significance of glass transition temperature.

### Exercise 3.5

Discuss the phenomena involved in the cross-linking of a thermosetting resin, explaining what is meant by gel time and how shrinkage can be a problem with these resins.

### Exercise 3.6

What is the most common cross-linking agent used with thermosetting polyesters and what are its disadvantages?

### Exercise 3.7

There are several families of curing agents which are used with epoxy resins but one is particularly sensitive to degradation by water. Which one is it and why?

### Exercise 3.8

Give examples of resin systems which can be used at higher temperatures than those which are possible with epoxy systems.

### Exercise 3.9

What is the chemical name of Nylon 6,6 and what do the numbers signify?

### Exercise 3.10

What is the melting point of saturated polyester (PET) and why does this mean that it is important to prepare the polymer under dry conditions.

# 4

# Composite manufacturing processes

## 4.1 Introduction

Manufacturing processes used for composite materials are often very different from those used with conventional structural materials such as metals, although in some cases similarities can be found with the processing of polymers. The materials we are considering are based on fine filaments, which are very long with respect to their diameters, so that there are many aspects of composite manufacture which are common with textile processing. Metal structures are most often made by first producing bulk metal and then machining it into the final form, however advanced composite materials are most often two-dimensional materials in which the fibres lie parallel to the plane of the material. A carbon fibre pressure vessel can be made by winding the filaments around a preform which can become an impermeable liner. An aircraft wing may be a three-dimensional structure but it can be made by stretching a two-dimensional composite skin over a honeycomb form. The fibres in the skin may be laid up in discrete layers of parallel fibres with each layer having fibres arranged in different directions or the fibres can be woven. The directions in which the fibres are placed control the behaviour of the final product. In some cases techniques borrowed from the textile industry can place fibres at right angles to the plane of the fabric by such means as needle punching, sewing or knitting so as to improve interlaminar shear behaviour. The flexibility of fibres allows them to be woven or wound onto mandrels so that complex forms can be produced, which when impregnated with resins, which are then cured, can lead directly to the final product which requires little or no extra finishing. In this way the total cost of manufacturing a component with a composite material can be reduced, compared to traditional materials. Another technique borrowed by the composite industry from traditional fibre processing is braiding. Braiding is rather like plaiting hair so that, by using a number of spools of fibres, virtually continuous tubes or profiles can be made at high speed. Pultrusion also produces virtually continuous tubes or profiles by combining the simultaneous pulling of the fibres through a die, which determines the cross- sectional shape, and the extrusion of the resin through the same die. The final product is generally unidirectional but subsequent turning the profile after the die or over-winding fibres around it can overcome this limitation. Lower cost products may be made by injection or closed moulding techniques which resemble those use in the plastics industry. In these processes the fibres are of a short length ranging from a few millimetres to one or two centimetres and are arranged randomly either in the plane of a sheet of the composite or in three dimensions so as to give a material which can be formed in a closed mould. Alternatively fibres can be chopped and sprayed together with the resin onto an open mould. This technique is useful for large structures including boat hulls but is also used for other smaller low-cost structures. By far the greatest number of composite materials are made with polymeric matrices which can be either thermosetting or thermoplastics. To this statement can be added elastomers in the form of natural and synthetic rubber used by the tyre industry and which represent a very large volume of materials.

All manufacturing processes require the material to be formed into its final shape. An advantage with composites is that the final shape can often be achieved in one operation with little need for further expensive finishing processes whereas, apart from the use of additive processes, metals often have to be machined into the final form required. However shrinkage of the composite due to the crosslinking process can be considerable, if not countered. A polyester resin will shrink by around 8 % in volume which will lead to a linear shrinkage of around 2 %. Nevertheless the presence of reinforcements as well as appropriate fillers can reduce this to a fraction of a percent in volume. Often the composite is made in a mould but as the pressures and temperatures which are necessary for forming composites are low compared to say metal casting the moulds can be made of cheaper materials. For many fabrication methods the moulds which are used are themselves made from composite materials. The cost of such moulds made in metal would typically be ten times higher. This explains why composites can be competitive with metal structures for modest fabrication runs but if the rate of production is increased they may become less attractive as moulds have to be replaced more frequently or more expensive, longer lasting moulds have to be used. For example a production run, of, say, one hundred thousand composite structures per year may be competitive but doubling this rate may mean that the composite has to be replaced by metal.

### 4.1.1  Composite manufacturing processes and their classification

As mentioned previously, there exist numerous technologies for manufacturing composite structures. The choice in using one technique or another is conditioned by cost constraints related to the number of parts to be made and technical considerations linked to the materials used including surface finish required, dimensions, sustainability etc.

As a consequence, many possibilities exist to classify the composites manufacturing processes. The most common is based on the nature of the constituents. It is then possible to distinguish thermoplastic processes from thermosetting ones (or even those reserved for elastomers). If this distinction was obvious twenty years ago, the emergence of reactive thermoplastic process solutions is somewhat confusing the cards and the processes historically dedicated to thermosets are now possible with thermoplastics. While remaining in the field of constituents, it is also possible to distinguish continuous fibre processes from those with staple fibres, short or long. Regarding raw materials, several preparation steps are sometimes necessary before being able to combine the fibre reinforcement with the matrix. This topic is described in more detail in the following § 4.1.2. Wet processing route or a two-stage dry route: this is another way to class composite manufacturing processes. The wet route allows both the manufacture and impregnation of the structure; the dry route consists of the use of a semi-finished product with the final manufacturing step occurring afterwards as a function of requirements. Another possibility of classification is based on the moulding techniques, i.e. closed moulding versus open moulding techniques. For health and safety reasons, the trend is clearly to favour so-called closed-mould techniques, but many process ways still exist with open moulds. It is also possible to classify the processes in relation to the quantity of parts that it is possible to produce, this comes down to considering the cycle time but above all the tools to be used to move towards more automation. The use of robotics in the manufacture of composites is growing very rapidly and is probably the most promising technology for the future.

In this book, the choice has turned to this last mentioned classification, i.e. from manual processes to full automated ones or from small series to large series.

## 4.1.2  Preparation of raw materials

### Matrices

As many produced polymers do not possess intrinsically advanced properties such as high strength, oxidation resistance and other properties which need to be enhanced for their final use they need to be further processed through a formulation stage. In this stage, chemical additives are added to the pure polymer in a bid to improve its characteristics. Additives are added to improve their characteristics and to allow the polymer to be plasticized and processed without oxidation or degradation. During this stage other special features can sometimes be added to the polymer such as colour; UV resistance; fire and smoke retardancy; lubricants and others.

As intimated above, plastics in their raw material stage are available in several forms and some of these are granules and plastic pellets, which are large-sized particles (small granules, 1 mm to 5 mm in diameter) that can be separated by hand.

Depending on the viscosity of the matrix, other semi-finished products can be useful to facilitate impregnation: powders, fibres or sheets. If the use of semi-finished products is technically attractive, it is generally accompanied by a significant additional cost which can sometimes be redundant. Large series production is generally associated with fully integrated solutions.

### Reinforcements

Fibre reinforcements come in different forms and glass fibres present the widest range of these forms. Glass fibres are typically 14 µm in diameter, which is a fifth of the diameter of a human hair, so although glass is a brittle elastic material the fineness of the fibres makes them very flexible and they can be treated in much the same way as continuous textile fibres. They can be woven or filament wound around a mandrel. The fibres are however sensitive to abrasion so that they are coated with a resin or size which serves to protect them, to facilitate their conversion into finished products by acting as a lubricant, to hold them together as a tow to aid handling and also to bond them to the composite matrix and in some cases as an antistatic agent. Woven cloth of glass fibres can be used to form composite parts with complex shapes as the weave holds the fibres in position. The use of a woven cloth means also that the position of the fibres is controlled so that they can be placed so as to make optimum use of their properties. Fabrics can have different weaves, from canvas to twill to satin. It is also possible to use NCF, i.e. non-crimp fabrics, which corresponds to multiple layers of unidirectional fibres, with each ply placed in a different orientation; the whole being held by a sewing thread. NCF are mainly used in high-performance composite materials.

It is sometimes possible to combine these textile structures to achieve a dry – semi-finished – product which is called a preform. The preform corresponds to the fibrous skeleton of the part which will then be impregnated with resin in the mould. The preforming techniques, which are not detailed in this book, are numerous and make it possible to produce in larger series while optimizing the use of production tools. Many books dedicated to technical textiles detail the manufacture of all these fabrics.

Some fibres, usually carbon but also glass and aramid fibres, are available preimpregnated in the form of a sheet and this material is known as prepreg. The continuous unidirectional fibres are coated with the uncured thermosetting resin or the thermoplastic polymer. The thickness of the prepreg sheets is around 0.3 mm and they can easily be cut to shape with a cutter and then stacked in a mould for processing. The layers of prepreg can be arranged in different directions so as place fibres in the numbers and directions to give optimum properties to the finished composite.

Continuous or long fibres can be used to form non-woven mats of randomly arranged fibres. The size is used to hold the fibres together before the mat is cut to shape and impregnated with the uncured thermosetting resin before composite manufacture. Cutting out the weaving process reduces cost but means that the fibres cannot be aligned preferentially so as to make optimum use of their properties.

Glass and sometimes other fibres, such as carbon fibres, are often chopped into short lengths varying from 5 mm to 75 mm for use in a variety of manufacturing processes. The widest use is in the form of sheet moulded compound (SMC) in which the fibres are cut to short lengths and allowed to fall on a moving sheet of uncured thermosetting resin. The resulting mixture of short fibres in a viscous resin can then be cut to shape, placed in a mould for a cheap method of making a composite part. Short fibres must also be used in making composite parts by injection moulding into a closed mould. The fibres can be added separately from the thermoplastic resin in the screw extruder or they can be first prepared in the form of granules. In this latter case, the continuous fibres are first extruded with the molten polymer to form a continuous ribbon of unidirectional material which is then cut into short length of around 20 mm. These granules can then be used in the extrusion process with better control over final length.

## 4.2   Small series – Manual processes

Manual production processes are the simplest for producing composite structures but they are also the slowest. They can be adapted for the production of large structures but are only of interest for short production runs. Most commonly the reinforcements which are used are glass fibres and the resin is unsaturated polyester although higher performance fibres and epoxy resin are both sometimes employed but, as these are more costly, other techniques are usually preferred for these latter systems. The techniques often involve the use of open moulds which means that the quality of the finish of only one surface can usually be controlled as it is formed against the mould. Also the emission of styrene, which is used in the crosslinking of polyester resins, can be high. Although no direct link between styrene and health problems has been conclusively demonstrated the distinctive smell associated with its use leads to much speculation as to its effects. Legislation varies enormously with factors of five existing between different industrialised countries. A tightening of international regulations on styrene emission can be expected.

At the laboratory scale, the simplest method of producing a high-performance composite flat plate is to use a process called hot pressing and makes use of a female mould of the dimensions of the desired plate. The clean mould is coated with a release agent, such as PTFE or Teflon spray so as to allow the finished product to be removed. The fibres are taken from a bobbin, cut into lengths equal to the length of the mould which has the dimensions of the desired plate, placed in the mould and then impregnated with uncured resin, which has been premixed with the hardener, by stippling the resin into the fibres with a brush and closing the mould. The mould is designed so as to have escape holes around the edge through which excess resin can escape. The shape of the mould controls the final dimensions of the composite plate and the quantity of fibres placed in the mould determines the fibre volume fraction. In such a process the mould is placed between the heated platens of a press, closed and taken through the heat cycle for curing the resin. If a solvent is used to control the viscosity of the resin a dwell time to allow the solvent to evaporate may be necessary. On heating, the resin first becomes very liquid and the pressure applied forces it around the fibres. Excess resin is forced out of the mould. As the resin begins to crosslink its viscosity increases dramatically and eventually the resin gels. Crosslinking proceeds asymptotically towards complete crosslinking, as discussed in Chapter 3.

As described, the resin flows towards the edge of the mould, taking with it any entrapped bubbles of air or solvent but also dragging fibres so that the finished composite plate often shows fibre misalignment, particularly in the centre. The use of a bleeder cloth placed over the composite, within the mould, allows the excess resin to be absorbed and provides a much shorter channel across the thickness of the specimen for the resin to be removed. This largely eliminates the displacement of the fibres. At the end of the curing process the mould is opened and the composite plate removed.

The process described above allows us to examine several important factors which must be taken into account when making a composite. Firstly a release agent must be used so as to be able to remove the finished product. At the laboratory level a Teflon spray can be used however this is an expensive product and more often a wax, such as bees' wax or a synthetic wax is used, often diluted with a solvent. Silicone sprays are usually avoided so as not to pollute the composite and because the finish obtained is not always smooth. Several layers of wax may be necessary but need not be replaced after each fabrication cycle. An alternative method is to use a thermoplastic film such as thermoformable polyester, PVC (polyvinyl chloride) or cellulosic films. Not all thermoplastic films are suitable as they are attacked by the styrene monomer used in the crosslinking process of polyester resins. Other films can be deposited onto the mould. These are PVA, polyvinyl alcohol dissolved in water which has the advantage of allowing the mould to be easily cleaned.

The surface finish of the composite which is made depends on the quality of the surfaces of the mould. Many structures have a gel-coat which is a surface layer of resin, usually containing other products which act both as fillers and modify the properties of the resin. Such a gel-coat takes on the finish of the mould and, if this is very smooth, can provide an excellent surface finish. Most pleasure boats and some bigger boats such as mine sweepers, are made by applying the composite to an open mould which has the shape of the hull. The attractive appearance of many of these crafts is due to the quality of the moulds and the gel-coat which ensures a smooth surface finish. The gel-coat can be coloured to give the composite material a particularly desired appearance and is also used as a protective layer against abrasion and water ingress, although this latter characteristic is not as obvious as it is often claimed as water can diffuse through such a resin layer. Environmental ageing is discussed in Chapter 11 which discusses long-term behaviour. The gel-coats vary as a function of the composite system used but are often of the same type of resin as used in the matrix. They should not be diluted with acetone as the presence of this solvent can cause problems due to outgassing, forming bubbles and porosities, so producing an unacceptable finish. In the case of polyester resins, the gel-coat can be diluted with styrene monomer. They can be applied to the inside of the mould with a brush or projected onto it. The gel-coat has to be left to gel before the composite can be added to mould and the mould closed.

After the manufacture of the composite it may be desirable to post cure it which may enhance the crosslinking process and in the case of polyester resins, for example, remove excess styrene left over from the manufacturing process.

### 4.2.1   Contact moulding

This technique is simple, suitable for large as well as small structures but it is slow and only suitable for small series at a slow rate of production of up to 500 units per year. The most widely used composite system which is used with this method is glass fibre reinforced polyester.

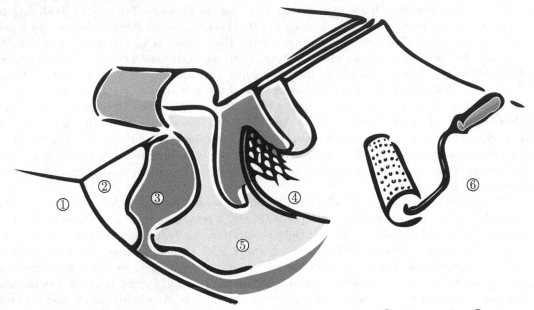

Figure 4.1: Schematic illustration of the contact moulding process. ① rigid mould, ② release agent, ③ gel-coat, ④ reinforcement, ⑤ manual impregnation using a brush or a roller ⑥.

As shown in Figure 4.1, a rigid mould with a hard surface ① is used and is most often made with glass fibre reinforced polyester or epoxy resin, concrete, wood or plywood, plaster, aluminium or steel sheet. The fibres, in the form of mat or woven roving, are placed inside a female mould or on a male mould, depending on the desired finished article, which has previously been coated with a release agent ② and if required, a gel-coat ③ with a thickness of up to 1 mm, which, if required, can be coloured. The reinforcement ④ is then manually ⑤ impregnated with the resin to which an appropriate catalysing agent has been added to promote crosslinking. The resins used for this process are most commonly unsaturated polyesters, vinyl esters and epoxies, although the latter are used less frequently than the other two. The laminate is then consolidated with a brush or roller ⑥ to eliminate air bubbles, as shown in Figure 4.1. To produce a thick composite the process is repeated the number of times which is necessary with a dwell time between each layer as the layers should begin to gel but not to cure completely between each application. Finally the composite is allowed to fully cure, usually at room temperature. The finished product has a smooth appearance on the surface which is in contact with the mould however the other surface will usually be rough and reveal the fibre reinforcements. This process can be used to make very large structures such as large boats.

### 4.2.2   Spray-up moulding

This technique has much in common with contact moulding and the two are often combined in alternating sequences. Figure 4.2 presents a schematic illustration of this process.

The principle of the technique is to use compressed air to simultaneously project the fibres and the resin onto the surface of the mould. The continuous fibres are brought to the spray

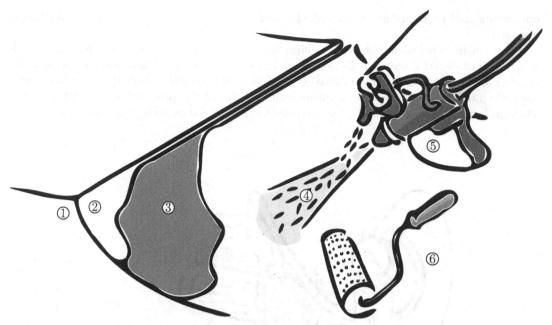

Figure 4.2: Schematic illustration of the spray-up moulding process. ① rigid mould, ② release agent, ③ gel-coat, ④ chopped fibres projected with the resin by a spray-gun ⑤, ⑥ roller.

gun ⑤, which is often hand held, from bobbins and are then chopped, to lengths between 25 mm and 75 mm, by a rotary chopping head before being projected together with the resin ④, as shown in Figure 4.2. The fibres are randomly distributed on the surface of the mould. After projection the laminate has then to be rolled ⑥ to eliminate entrapped air and ensure good consolidation. Successive layers can be sprayed to obtain the desired thickness and if required, mat or woven layers can be included in the thickness.

Both contact moulding and spray moulding may be combined in the production of a composite structure, for example a boat hull, as the latter technique is unable to provide the fibre content of the former and this determines mechanical properties. Although spray lay-up can be automated, if the structure is simple, such as cylindrical tubes and pipes, and is the cheaper and more versatile process, it is restricted to a glass fibre weight content of around 35 % whereas around double is possible with hand lay-up. In addition hand lay-up allows greater control over fibre alignment which again is important for mechanical properties.

## 4.2.3 Vacuum bag and infusion moulding

The previous techniques do not use any applied pressure during crosslinking of the composite. Bag moulding allows pressure to be applied evenly over the composite structure. A flexible sheet, which can be rubber, is laid over the composite lay-up, sealed at the edges and the air sucked from beneath the sheet. Vacuum bag moulding can be used with wet-layered laminates (stacked and impregnted by hand lay-up) or prepreg composites. The depressurisation needs only to be relatively slight as too great a suction will extract styrene from polyester systems which is needed for the crosslinking process. Alternatively, pressure in the range of 1 to 3 atmospheres is applied over the whole area of the sheet. Often both air removal and pressure is applied. The appearance of the two surfaces of the composite is better than in the two previously described techniques. Further curing can take place after this initial process by heating in an oven. This is usually only considered necessary in the case of advanced composite systems for which prepreg sheets are used.

Vacuum infusion moulding mainly differs from the vacuum bag moulding in terms of impregnation of the reinforcement. Unlike the previous process for which the atmospheric pressure only serves to compress the impregnated stack, the present technique uses the vacuum to drive the resin into a dry fibre preform. A schematic illustration of this process is shown in Figure 4.3. Vacuum infusion moulding can be used to manufacture very large structures like boat hulls.

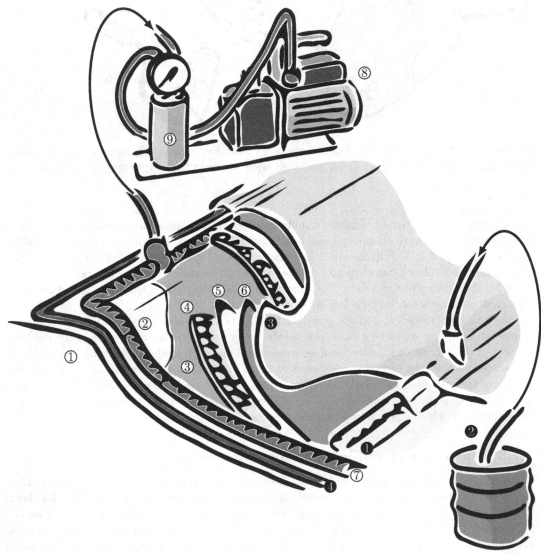

Figure 4.3: Schematic illustration of the vacuum infusion moulding process. ① rigid mould, ② release agent, ③ gel-coat, ④ dry preform, ⑤ peel ply treated with release agent, ⑥ breather/bleeder fabric, ⑦ vacuum line, ⑧ vacuum pomp and its resin trap ⑨, ❶ resin flow channel, ❷ feedline and resin, ❸ bag film and ❹ sealant tape.

As for the previous small series manufacturing processes, a mould tool ① is covered with a release agent ② and optionally with a gel-coat ③ (Figure 4.3). The dry preform ④ is placed inside the mould and is covered with a peel-ply ⑤ which is intended to be peeled off after the process is finished. This layer, treated with release agent, is a woven fabric of Nylon, glass or other synthetic materials and it absorbs some of the excess resin. When removed, the peel-ply fabric also provides a textured surface particularly suitable for joining operations such as bonding. A bleeder fabric ⑥ is then stacked over the peel-ply. Its role is to improved control over the resin distribution. A vacuum line ⑦ is placed around the perimeter of the part and is connected to the vacuum pump ⑧ passing through a resin trap ⑨. A resin flow channel ❶ is connected to the feedline bringing the resin ❷. A bag film ❸ is placed above the assembly and is sealed to the outer perimeter by using a sealant tape ❹.

Many variations of these processes exist and the jungle of acronyms sometimes makes it difficult to understand the very essence of the manufacturing methods that we are trying to highlight in this book (van Oosterom et al., 2019; Hindersmann, 2019).

### 4.2.4 Autoclaving

Moulding in an autoclave is used primarily for the production of small numbers, of up to 100, high-performance composite structures, as higher pressures can be obtained for consolidation but, in principle, has much in common with the bag moulding techniques. The process is time consuming with manufacturing cycles lasting one to three hours. It is therefore expensive. The composite to be made is laid up as previously described for bag moulding and placed in the autoclave. Often a peel ply, treated with release agent, is placed on top or under the composite and is removed after fabrication for final surface finishing such as bonding or painting. A bleeder material, usually of glass fibre, is also used to absorb excess resin during the manufacturing cycle. There is then a barrier film of material such as PTFE. Then comes a breather material which allows a uniform application of the pressure by allowing entrapped air or volatiles to be removed during cure. It may be a loosely woven fabric or felt which can be easily draped over the composite to take its form. Then comes the vacuum bag. The air in the bag is removed so that it collapses onto the composite to be consolidated and allows the solvents to be removed. The curing cycle and applied pressure can then be automatically controlled. The autoclave is usually pressurised with nitrogen or a mixture of nitrogen and air in order to avoid a fire risk. The autoclave is designed so as to allow gas circulation to achieve a uniform temperature. This manufacturing process is of little interest for large volume production but is widely used in the aerospace industry.

## 4.3 Medium series moulding

Manual techniques are slow and are rarely used in industrial processing except for very large structures such as boat hulls. Medium series processes uses techniques for which the impregnation of the fibres is achieved automatically and the finish of the structure is more reproducible than in the above techniques.

### 4.3.1 Vacuum moulding

This technique requires little investment, particularly as the mould costs are lower than in the straight RTM method as they do not have to resist the higher pressures involved in that method.

A closed mould is used so that the final product has well-finished surfaces but, this time both the fibres and the uncured resin are placed in the female part of the mould which is then closed and the air extracted through the male part. The air is extracted from points at the extremity of the composite item being made and a reservoir of resin is connected to the mould at its lowest point to ensure that enough resin is provided to completely impregnate the reinforcements. The two parts of the mould are forced together by atmospheric pressure and impregnation of the fibres is achieved. The points at which air is extracted are closed when resin begins to be seen. Styrene emission can be high. The moulding cycles are slow with this technique and there is a danger that the reinforcements will be displaced by the movement of the viscous resin. Fibre volume fractions achieved are typically in the range 0.25 to 0.40 and the composite can contain some porosities due to trapped air or solvent.

### 4.3.2 RTM light – light resin transfer moulding

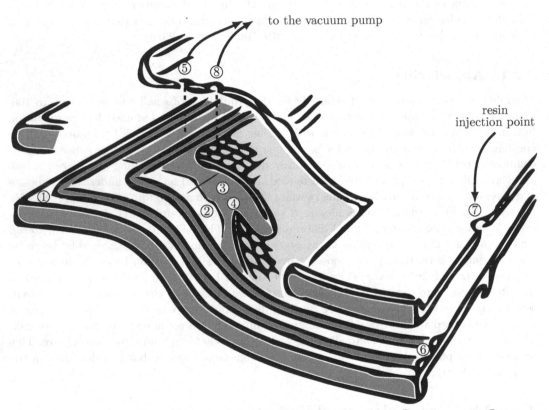

Figure 4.4: Schematic illustration of the light RTM moulding process. ① rigid mould, ② release agent, ③ gel-coat, ④ dry preform, ⑤ to the vacuum pump in order to close the mould, ⑥ silicone seals, ⑦ injection point, ⑧ vacuum injection assistance.

This technique is shown in Figure 4.4. A closed mould ① is covered with a release agent ② and, optionally, a gel-coat ③. A dry preform ④ is placed inside the mould which is hermetically sealed thanks to a vacuum pomp ⑤ and silicone seals at the periphery ⑥. The resin is injected ⑦ at low pressure into the mould thanks to an injection machine. The air is pumped out ⑧ of the mould at locations removed from the injection points. The design of the mould is more complex than in the two previous techniques and this increases cost. However the manufactured products have very low porosity and surface finish is good. The uniform quality and lack of porosity means that the technique is used to produce composite parts for applications for which their good dielectric qualities are used. The emission of styrene is low. The fibre volume fraction which can be achieved is in the range of 0.25 to 0.55.

Light RTM has the advantage of requiring less rigid moulds than the RTM process and, as a consequence, can be both cheaper and lighter.

## 4.4 Large series methods

Many composite manufacturers are concerned with small series production, for which manual or other slow processing techniques may be appropriate however as soon as large numbers of similar items, with reproducible characteristics, need to be made, some of the processes so far described are not ideal. The differences may not be of quality as some industrial sectors require much lower production rates than others as a rapid consideration of the aerospace and automobile industries illustrates. The manufacturing processes which are described below are those which are more suited for high production numbers, in particular because of the investments to be made.

### 4.4.1 Hydraulic press moulding

One of the first investments that is necessary to increase production rates is the vertical press. It is easy to understand that if we want to increase the pressure, for example in the case of the RTM light process, we will also have to increase the closing force on the mould. Press moulding, which is widely used by the composites industry, can be carried out "cold" or "hot".

This industrial process employs male and female moulds fitted to a simple hydraulic press. Large structures such as lorry cabs, roofs for caravans and mobile homes as well as large panels, baths, shower units and containers of various types can be produced by this method. As the moulds have to withstand greater pressures, they are usually made of metallic materials.

Regarding cold press moulding, the cost of the moulds is not high as they are not heated and only low pressures in the range of 0.354 atmospheres (35 kPa to 400 kPa) are used. The dry reinforcement, such as continuous glass mat, is placed in the mould and the catalysed resin is poured onto it. Polyester resins of higher reactivity are used than in contact moulding. The faster rate of crosslinking allows a relatively higher production rate. The manufacture of items at rates of five to ten per hour is possible. The only heating which occurs is due to the exothermic reaction of the resin curing process but temperatures of up to 50 °C can be developed.

A variation on this technique is thermoforming in which the composite, generally of glass reinforced polyester is sandwiched between two thermoplastic acrylic sheets of around 0.6 mm thickness. The acrylic sheets are first shaped, either by at room temperature by vacuum forming or, for higher production rates, at temperature and then the reinforcements and resin, combined with a room temperature curing catalyst, are sandwiched between the acrylic sheets and the press closed during curing. The finished product has the high quality surface appearance of acrylic sheets and can be produced with a range of colours at a lower cost than if a paint cycle were necessary. The fibre volume fraction is around 0.25.

It is also possible to use semi-finished products based on thermoplastics. Fibre reinforced stampable thermoplastic sheet, incorporating the reinforcement and the matrix, should be heated before being placed in the mould which is quickly closed. This technique also makes it possible to produce sandwich parts.

Hot compression moulding of composites is often used in preference to the cold techniques described above for the manufacture of large series of an article, typically up to and above 20 000 units per year. The moulds which are heated in this process are made of steel and the pressures are of the order of twenty to two hundred atmospheres (2 MPa to 20 MPa). The starting materials can be in the form of chopped or continuous fibre mat, the latter being more common, woven cloth or fibre preforms, often made by spraying the fibres onto a pre-shaped metal mesh. The fibres are held onto the preform by suction and bonded lightly together. Alternatively short chopped fibres, most often glass, can be mixed with the resin to form a dough or sheet which can then be moulded into a finished article.

Several processes are based on the hot compression moulding process, starting with RTM which is described below and is one of the flagship processes for the transformation of composites. Wet or dry compression moulding, the latter corresponding for example to the widely used SMC, are also available.

## 4.4.2   Resin transfer moulding (RTM)

This is one of the most widely used composite manufacturing techniques used in industry. It can be used to produce quite large series of complex composite structures. The principle of the technique is to inject resin into a closed mould which contains the reinforcements which have previously been placed there. With high pressure RTM, the pressures to force the resin into the mould could be as high as 150 bar (15 MPa). The number of pieces to be made will determine the type of material used for the mould. Composite moulds with a polyester resin can be used to produce up to 2000 items but better finishes and larger production runs of up to 4000 items are obtained with epoxy composite moulds.

Stamped steel moulds will allow several tens of thousands of composite structures to be made although the forms which can be made can only be simple. Composite moulds with a projected surface finish of zinc bismuth or zinc aluminium can give composite surfaces which can be either mat or shiny and allow production runs of up to 10 000 units. For runs of up to 50 000 pieces, nickel copper electro-plated onto epoxy composite moulds can be used and give excellent surface finishes.

As shown in Figure 4.5, the reinforcement ④ is placed inside the mould ① which is then closed and the resin subsequently injected ⑦ into it under low to relatively high pressure which must nevertheless avoid disturbing the placement of the fibres. The resins are of low viscosity, typically in the range of 0.1 Pa s to 1 Pa s, fast curing resins such as polyester and epoxy are used, although bismaleimide and phenolic based resins also find uses for high temperature applications; and the reinforcement is chopped strand or continuous glass fibre mat. If mould closing pressures used are low, in the range of one to six atmospheres, 100 kPa to 600 kPa, it means that the tonnage of the presses need not be high ($20 \, \text{t m}^{-2}$) which keep the cost of equipment low. Presses of up to 3000 t are nevertheless used if the pressures or the sizes increase. Permeability of the mat is therefore important. Air is driven out as the resin impregnates the fibres so that escape holes for the air have to be provided in the mould. The resin spreads through the fibre reinforcements. If the permeability of the mat is low a bleeder material can be used to increase the speed of resin transfer. Crosslinking occurs without further heating of the composite. Styrene emissions are low. The surfaces of the mould must be non-deformable and if this is the case an excellent finish for both surfaces is achieved. Fibre volume fractions can be in the range of 0.20 to 0.55.

RTM is now possible with ultra-fluid thermoplastic matrices or by a reactive route. This process then bears the name T-RTM (Ageyeva et al., 2019).

## 4.4.3   Wet compression moulding

Wet compression molding's main advantages over RTM is the ability to apply resin outside of the mold, which speeds resin impregnation for reduced cycle time.

In this process the fibres are positioned onto a heated mould, which has been preheated to the temperature at which the crosslinking of the resin will take place, the catalysed resin poured onto the fibres and the mould closed. The predetermined manufacturing cycle depends on the reactivity of the resin and the thickness of the laminate. The resins are usually filled with additives which counter shrinkage of the composite article due to the crosslinking process and higher temperatures used in manufacture. The moulds, which are simple in form, are usually

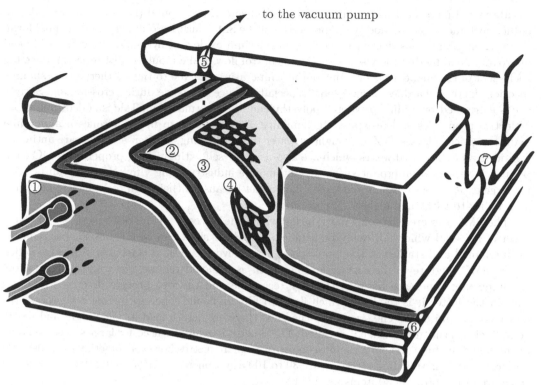

to the vacuum pump

Figure 4.5: Schematic illustration of the RTM moulding process. ① steel mould, ② release agent, ③ the gel-coat, not shown here, can be replaced by a post-curing painting operation, ④ dry preform, ⑤ to the vacuum pump, ⑥ silicone seals, ⑦ injection point.

heated to 100 °C to 140 °C and the pressures applied are most often between 20 to 50 atmospheres (2 MPa to 5 MPa). A production rate of up to thirty articles per hour can be achieved with this technique. Both surfaces of the finished product have excellent appearances, porosity is generally very low, although flash has to be removed and the complexity of the finished shapes which it is possible to produce is limited.

## 4.4.4   Dry compression moulding

This technique uses one of the most common forms of pre-impregnated composite material called SMC which is an abbreviation of "sheet moulded compound". It is made by chopping continuous glass fibre tow into lengths, typically from 25 mm to 50 mm. The chopped fibres fall continuously onto a moving belt onto which a polyethylene film has been laid and onto which the catalysed but uncured resin has been poured.

The result is an almost completely random arrangement of the glass fibres, in the plane of the belt, which become impregnated with the resin. The fibre content is usually in the range of 25 % to 40 %, by weight. The resin may be coloured if required. In this continuous process the fibre resin mixture is then covered with another polyethylene film to produce a sheet of pre-impregnated fibre resin, cut to length and usually around 6 mm thick. The resins used are most commonly polyesters with styrene or acrylic used as crosslinking agents. The acrylic monomers are used to obtain low shrinkage and vinyl can be used for higher temperature strength. Peroxides such as benzoyl peroxide (BPO) and t-butyl perbenzoate for higher temperatures are used as catalysts. Fillers, such as calcium carbonate and sometimes ground (kaolin) clay are added to reduce cost and others to modify the properties of the SMC during moulding and in its final form. Filler agents, known as thickeners, such as magnesium oxide and hydroxide, calcium oxide and hydroxides, can modify the viscosity so as to control flow in the mould whilst others, reduce the coefficient of thermal expansion, control the temperature rise due to the exothermic crosslinking reaction, increase hardness, rigidity and especially reduce shrinkage during crosslinking. Shrinkage is controlled by adding powdered polyethylene to the formulation. Talc is added to improve dielectric properties and resistance to humidity. As resistance to fire is a concern aluminium hydrate, which releases 35 % by weight of water during heating, is often added as are antimony trioxide and chlorinated waxes which are also used for self-extinguishing properties. SMC compounds have also been produced for the automobile industry using vinylester and can be used for large exterior body panels and under the bonnet because of their high resistance to corrosion and ability to withstand higher temperatures.

The SMC is stored for some days during which the viscosity of the resin increases to give a sheet of material which can easily be handled. The SMC sheet can be cut into shapes required by the mould and stacked in the mould so as to achieve the desired thickness. Stearates added to the SMC migrate to the surface and act both as a mould release agent and pigmentation. An important difference between this dry technique and the wet process described above is that in the latter it is the resin which flows to fill the mould whereas in this dry process the entire compound flows. The size of the charge is therefore determined by the weight and has to completely fill the mould when it is closed. The process is generally used for large series moulding of small and medium size parts of up to one or two square metres however sometimes it is used for larger mouldings. Moulding pressures are 30 to 100 atmospheres (3 MPa to 10 MPa). Production rates are of the order of 30 items per hour.

As the fibre volume fraction is not very high and the short fibres are randomly arranged in the plane of the composite the mechanical properties are far less than in some other types of composite material but the ease of manufacture and low cost make the process attractive.

Other forms of this type of compound exist such as HMC, which is a compound with a high glass fibre content of 50 % to 60 % by weight. To achieve this volume fraction the fibres are usually chopped to lengths of 25 mm. TMC describes a form of SMC, which however is made in thicker sheets, hence the T in the initials and XMC which is a unidirectional reinforced SMC produced by a winding technique with a high glass content of 60 % to 70 % by weight.

A three-dimensional distribution of glass fibres in the matrix to give a sort of dough which can be compression moulded is known as DMC. Not surprisingly this stands for "dough moulded compound". This compound is made by the fibres, which are chopped to lengths of 6 mm to 10 mm, the resin, catalysts, pigments and fillers being mixed by a twin bladed mixer. The three-dimensional arrangement of the fibres naturally leads to fewer fibres being arranged in any given direction, when compared to SMC, so that mechanical properties of DMC are lower than those of SMC. A form of DMC with higher fibre volume fraction is BMC which is short for "bulk moulded compound". The fibres in BMC are chopped to shorter lengths than in DMC and are between 4 mm and 10 mm. Such compounds can be used in the production of small size pieces even if of thick and complicated shapes.

The moulds are similar to those used in the wet process and the description of the manufacturing process is reminiscent of cold press moulding but for one important difference. In the latter process the movement of fibres is unwanted whereas it is a requisite of the hot press technique. The SMC, or DMC, flows under the effect of the applied pressure to fill the mould and as it does so the fibres also move. If the mould is of a complex shape additional pre-cut SMC may be positioned in particular parts of the mould where filling by the flow of the compound might be difficult. Moulding pressures are in the range of 30 to 100 atmospheres (3 MPa to 10 MPa) and mould temperatures are between 140 °C and 160 °C. Manufacturing rates are in the range of 20 to 40 items per hour, perhaps less for DMC pieces. The investment cost for this process are higher than in many other techniques due to the higher pressures used but the higher production rates makes it attractive.

## 4.4.5 Injection moulding

It is possible to inject short fibres mixed in an uncured resin into moulds in much the same way as many unreinforced plastics are moulded. Compounds, with fibre lengths up to 30 mm are used and are fed manually or automatically, from a roll or bulk form of the material. The steel moulds are heated in a temperature range of 140 °C to 160 °C. The technique is suitable for making small thick specimens and it is possible to produce up to 60 per hour. Variations on the technique exist which differ essentially in the manner by which the compound is injected into the mould. The compound can be transferred to the principal injection chamber by a piston and then injected into the mould by another piston. In this way the fibres are not broken down to shorter lengths which is of great importance in maintaining optimum mechanical properties in the final composite product. Alternatively the injection can be achieved, as with the injection moulding of unreinforced plastics using a screw thread injection system, as shown in Figure 4.6.

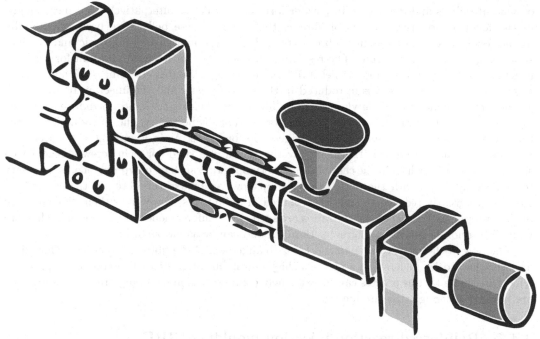

Figure 4.6: Schematic illustration of the injection moulding process using a screw thread injection system. The material is pushed by the screw into the mould which is shown on the left of the illustration.

This is known as ZMC which have much longer fibres than the conventional BMC. Injection pressures are high, 50 MPa to 100 MPa and the parts which can be made are limited by the dimensions of the mould and the force necessary to hold it closed. The compound is introduced into the injection chamber by a piston and then a screw thread operating on the Archimedes screw principle injects it into the mould. No pressure is applied during this transportation stage and as the feed rate is kept constant, with a low rotation rate, generally between 50 rpm and 90 rpm, a minimum of damage to the fibres occurs. The injection channel is heated with the temperature increasing from 30 °C at the arrival point for the compound to 80 °C to 90 °C at the entrance to the mould, which is typically at a temperature of 150 °C to 200 °C. One difficulty with this technique is that the injection of the compound has to be stopped once the mould is filled. This can be by simply stopping the movement of the screw thread or by the use of a valve which closes off the entrance to the mould. Both these actions can introduce some variability into the quantity of compound placed in the mould and also can cause the breakage of the fibres. Overall however the screw thread offers advantages over a piston as production cycle times can be reduced and a greater control over product quality can be achieved.

### 4.4.6   Injection moulding of fibre reinforced thermoplastics

The production of parts made by the injection of a molten thermoplastic into a, usually, cold mould, which after cooling of the part can be opened and the part ejected, is a rapid and well established production route suitable for high-production rates. When fibres are added to the plastic the extruder has to be made so as to resist the added wear due to abrasion which occurs. Injection moulding has been adapted for fibre reinforced thermosetting resins as described above although in this case the mould has to be heated to induce crosslinking. In the case of injection moulding of fibre reinforced thermoplastics the starting materials are solid and can be in the form of separate fibres and the matrix in powder form or granules or alternatively, and increasingly, in the form of preformed pellets of fibre in the matrix. These pellets are produced initially in the form of continuous aligned fibres extruded with the molten thermoplastic and cut into short lengths of around 25 mm. The resulting fibre lengths can be short (0.1 mm for separate mixing and 4 mm to 6 mm for the pellets). The moulding pressures are some 30 % higher than when no fibre reinforcement is introduced in the thermoplastic and the injection temperatures are generally higher to reduce viscosity. The fibres and thermoplastic are introduced into the injection moulding machine, either together or, if thought necessary, the introduction of the fibres can be nearer the mould so as to reduce fibre breakage. As in Figure 4.6 the material to be injected is moved towards the mould by the action of a single or twin screw thread which shear the material, which in the case of some thermoplastics, such as polypropylene, is necessary to achieve adhesion with the fibres. This action however breaks the fibres so that their final lengths are greatly reduced to lengths measured in millimetres, however as their diameters are of the order of 0.01 mm their aspect ratios are such that reinforcement is still achieved. The fibre volume fraction is generally no greater than 0.4 and can be as low as 0.25.

The flow of the matrix in the mould leads to alignment of the fibres parallel to the flow lines and this has to be understood so as to achieve optimum properties. The composite products which are made by this process can be extremely complex in shape and high rates of production are possible although sizes are limited.

### 4.4.7   Reinforced reaction injection moulding (RRIM)

Many composite manufacturing processes are too slow for really large volume fabrication such as might be needed in the automotive industry and so reinforced reaction injection moulding is of particular interest as it allows large parts to be made quickly. The composite is made in a closed heated mould into which are injected two highly reactive monomers, one of which is also used to carry the short milled glass fibres into the mould. Milled or broken fibres are generally preferred to cut fibres because of lower cost. The monomers are carefully metered and brought together in a mixing head and injected under low pressure. The interaction between the two monomers produces a cross-linked polymer which fills the mould and which is reinforced with between 8 % and 25 % by weight of the fibres. The reaction is generally between a liquid polyol and an isocyanate, such as diphenylmethane 4,4′-diisocyanate (MDI), to produce a polyurethane, as described in Chapter 3. It is important to avoid exposure of the monomers to water as this can produce gas during manufacture of the composite and poor results. Isocyanates are particularly sensitive to water which will produce a hard precipitate so hindering processing and reduce the characteristics of the composite. For this reason the isocyanates are generally stored under nitrogen. Storage of the monomers is usually controlled to maintain a temperature between 24 °C and 30 °C in order to keep them liquid. Flourocarbon blowing agents can be used in small quantities to produce a microcellular structure at the centre of the composite. This means that the composite is less dense and allows for better packing of the mould. Filling the mould occurs in seconds and the entire process, including curing and removal from the mould, usually takes only two minutes. The process is cheap as the low pressures involved allow less expensive moulds to be used and little heat input is needed as the reaction is exothermic. The process can be automated and fast moulding times are possible. However polyurethanes have high coefficients of thermal expansion and poor dimensional stability at raised temperatures. The presence of the fibres however considerably reduces these problems. Tensile strengths in the range of 30 MPa and above are attainable, with flexural moduli being around 1.4 GPa. Strains to failure are high and in the region of 25 %.

A variation of RRIM combined with the RTM process involves injecting the polyols and isocyanate into a mould prefilled with glass fibre mat. This allows the fibres to be placed in particular areas of the mould rather than relying on the polyol to transport them there.

## 4.5  Rotational moulding

There are several processes which involve placing the fibre reinforcements, together with the matrix material, onto a rotating mandrel in order to fabricate the composite part. Both short fibre reinforcement and continuous reinforcements can be used but they produce very different end products.

### 4.5.1  Centrifugal casting

A low-cost process for producing cylinders and pipes involves placing both resin and fibres into the interior of a rotating circular mould. The fibres can be in the form of mats or woven rovings or chopped fibres can be projected into the rotating mould. The mould rotates at a sufficiently high speed for the centrifugal force to cause the glass fibres to be well impregnated with the resin and induce good compaction. This technique is the only process which gives a good finish on both sides of the composite product without the use of matching moulds and gives a very smooth finish to both surfaces. The dimensions of the products which can be made do not have an upper limit but sprayed chopped fibres cannot be used for cylinders of small diameter as a gap of 250 mm to 300 mm has to be left between the spray head and the mould so as to ensure even fibre distribution. The process can be used to make cylindrical or slightly tapered pipes.

### 4.5.2  Filament winding

Filament winding of continuous filaments onto a rotating mandrel after first being impregnated with uncured thermosetting resin or coated with thermoplastic is a means of making high-performance, highly reproducible, composite structures. The simplest structures which can be made by this technique are tubes, using a cylindrical rotating mandrel ① onto which the continuous fibres ② & ⑥ are wound from a head ③ which repeatedly sweeps from one end of the mandrel to the other (Figure 4.7). Such cylinders can be very large.

Wet winding in which the fibres pass first through a bath ④ of uncured thermosetting resin is the most common. The fibres, impregnated with the resin, are then wound onto the mandrel and subsequently cured either at room temperature or in an oven. There is generally no external pressure applied during manufacture, except that provided by the tension in the fibres, although,

Figure 4.7: Schematic illustration of the wet filament winding process. ① cylindrical rotating mandrel, ② continuous fibres, ③ sweeping head, ④ bath of uncured resin, ⑤ separator combs, ⑥ continuous strand roving.

in some cases, curing can take place under pressure in an autoclave. It is also possible to use prepreg rovings which can then be wound onto the mandrel without the need of a resin bath. The tackiness of the material is ensured by control of the temperature at which the winding takes place. Occasionally the fibres can be wound onto the mandrel and the resin applied afterwards.

If a thermoplastic matrix is required the fibres must first be processed to produce a ribbon impregnated with the thermoplastic. This can be achieved by passing the tow of fibres through the thermoplastic in the form of a fine powder so that it adheres to the fibres by electrostatic forces or the fibres can be co-extruded with the molten polymer. In both cases the ribbons are wound onto the mandrel as in the wet process but at the point of contact the ribbon is both heated to remelt the thermoplastic and a pressure is applied locally to achieve bonding. In situ polymerisation is an increasingly interesting technique which allows the monomers making up a thermoplastic to be directly applied to the fibres, emulating the use of thermosetting resins, so making manufacture much easier.

The mandrels can either be dismantled or dissolved away after composite manufacture or most often remain and serve a useful role as a liner providing an impervious barrier for whatever fluid, liquid or gas, is to be held in the structure. Typically such liners are in aluminium or polyethylene.

Filament winding machines can be extremely complex and possess up to seven degrees of movement so as to enable complex shapes to be wound. Polar winding, which allows spheres to be produced, involves the mandrel remaining stationary whilst the fibre feed arm rotates about the longitudinal axis which is inclined at the necessary winding angle. In both helical and polar windings the fibre tows can be arranged so as to sit perfectly next to one another so as to give a complete covering in one ply or alternatively several plies can be used to cover the mandrel entirely.

The fibres are generally wound on the mandrel following geodesic paths, which are the shortest distances between two points on a three-dimensional surface and this ensures that no slippage occurs and no friction is required to hold the fibres in place. A geodesic path can be described by the relation $r \sin \alpha = $ constant, in which $r$ is the radius at a given point on the surface and $\alpha$ is the winding angle. It follows that the fibres must experience only tensile forces so that the following expressions can be used to calculate the strength of a filament wound tube.

Where $S$ is the unit strength of the wrapped tow, $S_H$ is the strength of the cylinder in the hoop direction, $S_A$ is the strength of the cylinder in the axial direction, $t$ is the wall thickness, $\alpha$ is the winding angle, $W$ is the width of the tow and $L$ is defined as the ratio $W/\sin \alpha$, $F_1$ is the component of the strength of the tow in the $\alpha$ direction, and $F_2$ is the component of the strength of the tow in the hoop direction, we can write:

$$F_1 = \frac{S\,W\,t}{2} \tag{4.1}$$

$$F_2 = \frac{S\,W\,t\,\sin \alpha}{2} \tag{4.2}$$

$$S_H = \frac{2\,F_2}{L\,t} = \frac{2\,S\,W\,t\,\sin \alpha}{2\,W\,t\,(\sin \alpha)^{-1}} = S \sin^2 \alpha \tag{4.3}$$

Similarly $S_A = S \cos^2 \alpha$. This means that for a cylinder with a 2 : 1 ratio of hoop strength to axial strength:

$$\frac{S_H}{S_A} = \frac{2}{1} = \frac{S \sin^2 \alpha}{S \cos^2 \alpha} = \tan^2 \alpha \tag{4.4}$$

This gives a value for the winding angle $\alpha$ of 54.75° and coincides with a minimum value, given by a more detailed analysis, of the shear forces in the composite and of the stresses tending to separate the fibres.

Filament wound composite pressure vessels require at least one end fitting so that they can be filled. These fittings are incorporated into the structure during winding. The fibres which wrap around the ends provide the necessary strength to support the end forces due to the internal pressure however they are inevitably at an angle with respect to the hoop axis of the vessel over the central cylindrical part of the structure. This means that the winding angles of a pressure vessel are not those of a simple filament wound tube so that typically the longitudinal fibres could be placed at ±20° whilst the circumferential windings are at approximately 90° to the main axis. Passing around the ends of the vessel can produce a too thick layer of fibre as several windings are made. This is avoided by slightly changing the winding angle for subsequent windings. The fretting circumferential windings are wound around the central cylindrical section to provide the required hoop strength.

## 4.6   Automated composites manufacturing

Automated composite processes date back to the 1970s, when the first developments took place. Since the 1980s, commercial pultrusion and fibre placement machines have existed. 3D printing is the most recent example of automated composite manufacturing that is clearly bringing composites into the Fourth Industrial Revolution.

### 4.6.1   Pultrusion

This technique integrates fibre and resin mixing, composite forming and curing in one continuous process. The fibres, which are generally glass but can be of any continuous type, are brought together from a number of spools or bobbins ⑥ and impregnated with a, usually, thermosetting resin ④ such as polyester. The impregnated fibres then pass through a die, heated between 90 °C and 150 °C, which crosslinks the resin sufficiently for a sufficiently strong composite profile to be pulled from the die. The process allows profiles of extremely complex cross section to be made at speeds in the range of $0.6 \, \text{m} \, \text{min}^{-1}$ to $1.2 \, \text{m} \, \text{min}^{-1}$.

Figure 4.8 illustrates the pultrusion process with the profile being pulled through the die ③ and heating oven by two constantly rotating, cleated belts ② between which the cured profile is pulled and then cut ① to length. The cleated caterpillar belts pulling system has the drawback of requiring the belts to be shaped to maintain contact with the profile being drawn. Single clamps can be used which pull the required length of profile through the die. Improved production rates can be achieved by using a radio frequency (RF), typically at 70 MHz, as a means of dielectric heating the resin. This accelerates curing and allows the profile to be pulled through the die quicker.

Profiles can be either solid or hollow and can be made of all one type of reinforcement or a combination of continuous fibres and fibreglass mat. In this latter case the continuous fibre rovings are sandwiched between two layers of mat as the mat would not be strong enough to pass through the resin bath without the roving. This type of material is used to make flat sections or thin-walled shapes.

Thermoplastics can be used as the matrix material however the higher viscosity of the polymers is an added complication compared to thermosetting resins which have viscosities two or three orders of magnitude lower, around $1 \, \text{Pa s}$ to $10 \, \text{Pa s}$. Despite this difficulty the thermoplastics have one big advantage in that they can be remelted or shaped after fabrication of the profile. Initially the process was developed for high-performance polymers such as PEEK, PEI and PPS, however the process has evolved to less costly polymers such as saturated polyester.

The profiles produced by pultrusion are intrinsically unidirectional even though a variation of the technique can induce some twist into the profile by turning the pulling mechanism. The technique however can be developed into a pullwinding process in which the pultruded unidirectional profile is wrapped after emerging from the pultruder by filaments which overcomes the anisotropy of the unidirectional structure.

### 4.6.2   Automated fibre placement (AFP)

Fibre placement is an automated composite manufacturing process that, as the name suggests, allows the reinforcement to be placed in the most suitable directions and proportions for the part. In this sense, the filament winding presented previously also falls into this category, but the automatic fibre placement (AFP) mainly designates a set of processes where the quality, stability and efficiency of the placement make it possible to work on complex geometries.

Fibre placement, initially reserved for aerospace applications, now finds applications in many other fields. Fibre usually comes in the form of prepreg tows, whether based on thermosets or thermoplastics. The placement of fibres can be used to improve the stiffness and/or the

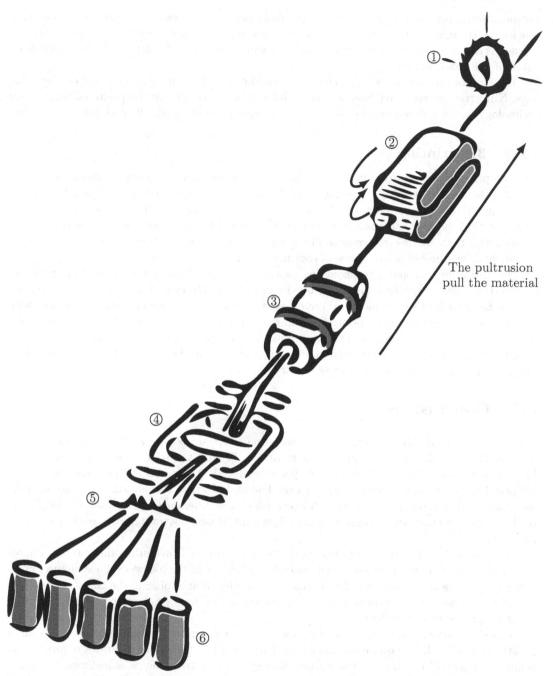

The pultrusion
pull the material

Figure 4.8: Schematic illustration of the pultrusion process. ① cut-off saw, ② pull mechanism, ③ forming and curing die, ④ bath of uncured resin, ⑤ separator combs, ⑥ continuous strand roving.

resistance of structures, or to locally reinforce composite parts. This process can also be used to obtain a complex load-strain response that precisely meets a specification. The fibre steering process is designed for this purpose in order to drastically improve conventional AFP processes by allowing curved placement of fibres. In all cases, AFP technologies make extensive use of optimization algorithms to determine the optimal orientation of reinforcements. Soundboards of musical instruments have been produced in this way as well as satellite structures; for the latter, it was necessary to ensure that various modes of vibration of the structure were decoupled from those of the space launcher.

The core of the method is essentially based on the placing of the head mounted on the robot axes. Numerous patents have been and are still being proposed for this composite manufacturing technology which still has great development potential (Crosky et al., 2015; Zhang et al., 2020).

### 4.6.3    3D printing

The 3D-printing sector is booming, manufacturers are increasingly using composite materials because they allow very high mechanical performance to be achieved. Most Fused Deposition Modeling (FDM) type printers on the market can print using short fibre reinforced filaments provided they have a hardened steel nozzle. However, there are 3D printers based on other technologies that are able to combine fibres and matrix during printing; this makes it possible to benefit from the full performance of continuous fibres.

One of the most popular technologies, uses a fast-curing thermosetting resin to impregnate a strand of continuous fibres deposited by the print head. Other systems allow the continuous fibre to be combined with a molten thermoplastic. Many suppliers are entering the market with equipment ranging from office printers to industrial printing benches capable of manufacturing pieces measuring several meters. Thermwood has for example introduced a Large Scale Additive Manufacturing (LSAM) machine able to print 4 m parts for the aircraft industry. With composite 3D printing, we are only at the start of a new era.

## 4.7    Conclusions

As we have seen in this chapter, the manufacturing processes of composites are numerous and are constantly evolving. Developments in materials, IT and transducers are all elements that help improve the quality and quantity of parts produced. Among all the processes presented, 3D printing can be seen as a move towards the Fourth Industrial Revolution. Although initially seen as something from science fiction this technology has developed rapidly from the beginning of the present century and represents a paradigm shift in manufacturing processes of composite materials.

Stimulated by the Internet of Objects this fourth industrial revolution aims at a metamorphosis of production techniques and will extend to all stages of the life cycle of a structure. From there it is only a small step to arrive at the ultimate objective of placing the client at the heart of the technico-industrial economy via a personalisation of mass production. 3D printing pulls up other processes in this direction.

The development of more intelligent and more rapid transformation processes will allow greater reliability, better quality of the finished article whilst respecting the rates of production required. This will become possible by the increasing use of transducers, sometimes integrated into intelligent structures, which will allow continuous monitoring of their physical state and allowing continuous tracing from manufacture to end of life by being connected, often remotely, to monitoring systems. During manufacture this will allow decisions to be taken in real time during the forming process such as the laying down of fibre reinforcements, their spacial arrangements and the advancement of the curing cycle of the resin. This will optimise the manufacturing process and ensure that any variation from the optimum is corrected.

Finally, this does not mean that more manual processes will disappear because the repair sector is also booming. The variety of repairs to be undertaken requires and will require manual intervention for many years to come. Good practices still need to be "disseminated" to the smaller organizations entering the sector for environmental, health and safety reasons.

# Revision exercises

### Exercise 4.1

Which well-established industry is often a source of manufacturing processes for composite materials?

### Exercise 4.2

Glass fibres are used in many forms and are therefore adaptable to many manufacturing processes. Give examples of the forms that glass fibres can obtained and briefly indicate some corresponding processes in which they are used.

### Exercise 4.3

Composite structures made from glass mat do not give as high properties in the final structure as woven glass fibre cloth. Why is this?

### Exercise 4.4

Describe manufacturing with an open mould. Why is a bleeder cloth desirable in this type of process? Approximately what rate of manufacture is possible with such a technique? Give an example of a product made in this way.

### Exercise 4.5

What is an autoclave? What pressures are generally involved with this process? What is the approximate rate of production with this technique?

### Exercise 4.6

Describe what the RTM process is. Discuss the choice of moulds which can be used and explain why different materials are used. What sort of pressures are used?

### Exercise 4.7

What advantages are there in vacuum assisted RTM? What would be a typical fibre volume fraction obtained with this technique?

### Exercise 4.8

Cold press moulding is used in industry to produce reasonably large pieces. How does this technique differ from RTM and what sort of production rate is possible? What advantage does hot press moulding have over cold press moulding?

### Exercise 4.9

Thermoplastic matrix composites are often made by injection moulding. Explain this technique. How does it compare to the injection of non-reinforced thermoplastics? What does the Fourth Industrial Revolution mean and how will it transform manufacturing of composites?

**Exercise 4.10**

Tubes and other structures can be made by the filament winding technique. Describe this technique for manufacturing with both thermosetting and thermoplastic matrix materials. Why are the fibres automatically laid down following geodesic paths? Derive the equation for the hoop strength of a filament wound cylinder.

# 5

# Constitutive relations

## 5.1  Introduction

A mathematical treatment of the behaviour of materials, including composite materials, is necessary in order to design even the simplest structures. The models have to take into account all the possible variables which determine the behaviour of the material and to be applied require that the variables are first determined experimentally as accurately as possible. Writing a constitutive law is needed to solve the equations of a continuum mechanics problem. The local equations introduce six unknown stresses $\underset{\approx}{\sigma}$ and three unknown displacements $u_1$, $u_2$ and $u_3$. That means there are nine unknown variables related by only three available equilibrium equations. Six other equations are therefore needed to determine all the nine variables. These equations which relate the behaviour between stresses and strains are called constitutive equations. As we have six stresses and strains, we get six new equations which allow the studied medium, the material, to be characterised.

The material is considered to be made up of regions which possess its characteristics, the sum of which makes up the whole material. In fact, we have to describe the response of the *Representative Elementary Volume* (REV) subjected to a given loading. We shall take loading to mean all events which can lead to changes in the material response. Of course this means the effects of applying mechanical forces but also other factors such as environmental conditions of humidity and temperature. The choice of the REV, which will be further discussed in Chapter 6, is very important because it will determine the nature of the constitutive law which will be applied. This means that the REV has to be statistically representative of the medium we wish to characterize. In order to make this choice to establish a constitutive law the following steps are carried out.

- The first step is an experimental analysis of samples which are sufficiently representative of the REV. As the observable physical phenomena are well known, the aim of this step is to identify non-observable phenomena which are internal and dissipative in the material. These observations need both destructive and non-destructive tools to evaluate all physical events, appreciate their relative importance and so decide which are the most important for modelling.

- The second step is a theoretical one. As the physical events to be considered have been selected, we have to formulate the most appropriate mathematical model to describe each variable state and its evolution. This formulation has to respect some theoretical frameworks which we are going to develop.

- Finally the third step is an identification procedure. Like the first one, it is an experimental phase carried out on representative samples of the material. As a mathematical law describes different physical phenomena, we have to identify the different coefficients which characterize this law and which are dependent of the material itself. In order to do this, we have to select appropriate experiments to determine specific coefficients. It is most important that the results used to measure these coefficients be as accurate and reproducible as possible.

This chapter will pay particular attention to the second step described above. The general principles which have to be considered when formulating constitutive relations have to be coherent with the physics of all bodies in motion and be able to describe the effects of motion which differ according to the nature of the material which makes up the body. A general theory of constitutive relations has to be able to describe an infinite number of classes of materials, distinguished by properties of symmetry and invariance. We can classify these general principles as follows.

- *Principle of Material Frame-Indifference.* This means that we regard materials properties as being indifferent to the choice of the coordinate system. Since constitutive equations are designed to express idealized material properties, we require that they shall be frame-indifferent. In popular terms this means that the responses of a material and its microstructure depend neither on their location in the world nor on the position of different observers.

- *Respect of the geometrical microstructure.* This means that the constitutive relations have to satisfy material symmetry during any mapping or change of coordinate system.

- *To be thermodynamically admissible during any motion combined with acting forces.* Continuum mechanics provide a framework which will be considered in order to describe dynamic process as a function of time $t$.

Considering these three last requirements, it is first necessary to answer the following question: to state the constitutive law of a REV for a given point at a time $t$, is it sufficient to consider only its own history or is it necessary to take into account the history of all those neighbouring with their own histories too? This answer will be given by the *Principle of Determinism* and the *Principle of Local Action* which we shall now study.

## 5.2   Principle of determinism

Modelling the response at a material point $M$ at time $t$ when the material is submitted to a stress field, requires answers to the following questions :

- How can the local stress field be taken into account?
- What is the influence of the surroundings on the behaviour of the studied particle.

The *Principle of Determinism* answers these two questions by assuming that the stress at the place occupied by a material point $M$ at time $t$ is determined by the history of the motion of the point $M$ itself, and also of all other points surrounding $M$ up to time $t$. That means the history of all points making up the body must be taken into account. The concept of material defined here reflects the common observation that many natural bodies exhibit a memory of their past experiences, sometimes continuing to respond to the effect by a change of form long after the stress field has been applied. In this case the constitutive relationships describe the past and present positions of points in the material as a function of forces acting on the studied material. These mathematical expressions have to include the variables which have been used to model observable phenomena (generally, stress, $\sigma$, strain, $\varepsilon$ or temperature, $T$) and internal variables, $V_i$ (plasticity, damage, ageing, rupture, ... ). Changes in these mechanical or physical variables are consequences of the applied loading. In this way the constitutive expressions are able to describe the state of all the points in the material at any time up to time $t$.

## 5.3    Principle of Local Action

In the principle of determinism the motions of body-points that lie far away from $M$ are allowed to affect the stress at $M$. This principle is too general and also too difficult to be applied. Accordingly, we assume a second constitutive axiom: only the motion of body-points at a finite distance from $M$ and in the neighbourhood of $M$ are allowed to affect the stress state at $M$. Sometimes we limit this principle to the point $M$ itself.

## 5.4    Principle of Material Frame-Indifference

The question is now to know if the constitutive laws and the associated variables are indifferent to the choice of the coordinate system during motion of the body. Considering that the material behaviour has to be objective, constitutive laws have to be universal and then indifferent to the choice of the coordinate system. That leads to the use of invariants of the stress and strain tensors.

## 5.5    Thermodynamically admissible processes

All the above mentioned conditions can be verified in the frame of continuum mechanics and thermodynamics. Once the variables describing the state considered are defined, we can postulate the existence of a thermodynamic potential from which the laws of state can be derived. This potential is both a function of observable variables and internal variables, such as

$$\psi = \psi\big(\varepsilon,\, T,\, V_1,\, \ldots,\, V_I,\, \ldots,\, V_N\big) \tag{5.1}$$

where $\varepsilon$ is the strain tensor and $T$ is the temperature. The refinement of the description of physical phenomena controlling the material characteristics will determine the number of variables $V_I$ and their significance. According to the Principle of Local Action, these variables, at a point $M$ and time $t$, will only depend on the point considered and its particular history up to time $t$. The evolution of the material is thermodynamically admissible if obeying the inequality which characterized the first and second law of thermodynamics. In the case where there is no variation of the temperature $T$ during the transformation, i.e. $\dot{T} = 0$, and considering a uniform temperature $\underline{\mathrm{grad}}\,(T) = 0$, we obtain

$$\left(\sigma - \rho\frac{\partial\psi}{\partial\varepsilon}\right) : \dot{\varepsilon} - \rho\sum_{I=1}^{N}\frac{\partial\psi}{\partial V_I}\dot{V_I} \geq 0 \tag{5.2}$$

where $\rho$ is the density of the material.

If we now imagine particular transformations in which all variables are zero except one, we obtain the law of state associated with this variable. If $A_I$ is defined as the thermodynamic force associated with the internal variable $V_I$, we have

$$A_I = \rho \frac{\partial \psi}{\partial V_I} \tag{5.3}$$

These relations which constitute the laws of state give very interesting information about the behaviour of the material, but do not describe the dissipation processes. To describe the evolution of the internal variables, a complementary formalism is needed. These laws which are able to describe these dissipation processes are called "complementary laws".

We are not going to discuss in detail the construction of these complementary laws as this chapter is only concerned with linear elastic media, but we can say that dissipation can be described by the introduction of a dissipation potential (or pseudo-potential) expressed within the same constraints as the thermodynamic potential: the principle of material frame-indifference and the material symmetries of the media. The other condition is that the dissipation process of a body in motion given by Equation (5.2) is never negative. With this condition that the thermodynamic processes are admissible, the complementary laws describe dissipative phenomena and the variables introduced in these dissipation potentials depend on the position of point $M$ and its own history up to time $t$.

In this way the thermo-mechanical behaviour at a point $M$ is determined with a potential and associated dissipation processes. We shall mention some of these processes when discussing damage in composite materials.

## 5.6    Anisotropic linear elastic materials

For reversible (or elastic) phenomena, at each instant $t$, the state only depends on observable variables. If we only consider isothermal transformations, the unique observable variable is the elastic strain equal, in that case, to the total strain $\underset{\sim}{\varepsilon}$. Then the thermodynamic potential can be written as

$$\psi = \psi\left(\underset{\sim}{\varepsilon}\right) \tag{5.4}$$

and the unique law of state is given by

$$\underset{\sim}{\sigma} = \rho \frac{\partial \psi}{\partial \underset{\sim}{\varepsilon}} \tag{5.5}$$

We set the following equation to define the thermodynamic potential

$$\rho\psi\left(\underset{\sim}{\varepsilon}\right) = \frac{1}{2}\underset{\sim}{\varepsilon} : \underset{\approx}{C} : \underset{\sim}{\varepsilon} \tag{5.6}$$

where $\underset{\approx}{C}$ corresponds to the elastic stiffness tensor, leading to Hooke's law

$$\underset{\sim}{\sigma} = \underset{\approx}{C} : \underset{\sim}{\varepsilon} \tag{5.7}$$

By inverting the above relation, the linear strain-stress relation is written

$$\underset{\sim}{\varepsilon} = \underset{\approx}{S} : \underset{\sim}{\sigma} \tag{5.8}$$

where $\underset{\approx}{S}$ is the elastic compliance tensor. $\underset{\approx}{C}$ and $\underset{\approx}{S}$ are fourth-order tensors, the components of which are respectively $C_{ijkl}$ and $S_{ijkl}$. These tensors depend on 81 ($= 3^4$) coefficients since each indices varies from 1 to 3. Further, as the stress and the strain tensors are symmetrical, their number reduces to 6 independent components instead of 9 and the linear relationship between stress and strain can be written in a general tensorial way, $\sigma_{ij} = C_{ijkl}\,\varepsilon_{kl}$, with 36 material coefficients all referred to an orthogonal Cartesian coordinate system fixed in the body and defined in the basis $\left(\underline{e}_1,\ \underline{e}_2,\ \underline{e}_3\right)$.

At this point, the introduction of *Voigt notation* is particularly useful. This notation is a powerful way to represent a symmetrical tensor by reducing its order; it enables, for example, a fourth-order tensor to be represented by a $6 \times 6$ matrix. Equation (5.7) can be rewritten as

$$\{\sigma\} = [C]\{\varepsilon\} \tag{5.9}$$

which corresponds to the condensed form of the generalized Hooke's law for a linear, anisotropic, elastic continuum material. $[C]$ is the stiffness matrix the components of which are the elastic moduli which may depend on the temperature. It must be clear that the linear strain-stress relation (5.8) becomes

$$\{\varepsilon\} = [S]\{\sigma\} \tag{5.10}$$

where $[S]$ is the elastic compliance matrix, the condensed form of $\underset{\approx}{S}$.

Nomenclature may vary according to what is traditional in the field of application. In this book, the most common rule is applied in writing the stress $\{\sigma\}$ and strain $\{\varepsilon\}$ in condensed forms, provided as follows

$$\{\sigma\} = {}^t\{\sigma_{11},\ \sigma_{22},\ \sigma_{33},\ \sigma_{23},\ \sigma_{13},\ \sigma_{12}\}$$
$$\{\varepsilon\} = {}^t\{\varepsilon_{11},\ \varepsilon_{22},\ \varepsilon_{33},\ 2\,\varepsilon_{23},\ 2\,\varepsilon_{13},\ 2\,\varepsilon_{12}\}$$

Variants exist as to the introduction of the coefficients for the strain part and the order of the indices. For memorizing Voigt notation, write down the stress (or strain) tensor in matrix form and draw the arrow from $\sigma_{11}$ to $\sigma_{12}$ as follows

$$\begin{bmatrix} \sigma_{11} & \sigma_{12} & \sigma_{13} \\ & \sigma_{22} & \sigma_{23} \\ \text{sym.} & & \sigma_{33} \end{bmatrix} \qquad \begin{bmatrix} \sigma_{11} & \sigma_{12} & \sigma_{13} \\ & \sigma_{22} & \sigma_{23} \\ \text{sym.} & & \sigma_{33} \end{bmatrix} \tag{5.11}$$

The generalized Hooke's law is then written in the following form

$$\begin{Bmatrix} \sigma_{11} \\ \sigma_{22} \\ \sigma_{33} \\ \sigma_{23} \\ \sigma_{13} \\ \sigma_{12} \end{Bmatrix} = \begin{bmatrix} C_{1111} & C_{1122} & C_{1133} & C_{1123} & C_{1113} & C_{1112} \\ C_{2211} & C_{2222} & C_{2233} & C_{2223} & C_{2213} & C_{2212} \\ C_{3311} & C_{3322} & C_{3333} & C_{3323} & C_{3313} & C_{3312} \\ C_{2311} & C_{2322} & C_{2333} & C_{2323} & C_{2313} & C_{2312} \\ C_{1311} & C_{1322} & C_{1333} & C_{1323} & C_{1313} & C_{1312} \\ C_{1211} & C_{1222} & C_{1233} & C_{1223} & C_{1213} & C_{1212} \end{bmatrix} \begin{Bmatrix} \varepsilon_{11} \\ \varepsilon_{22} \\ \varepsilon_{33} \\ 2\,\varepsilon_{23} \\ 2\,\varepsilon_{13} \\ 2\,\varepsilon_{12} \end{Bmatrix} \tag{5.12}$$

The tensors $\underset{\approx}{C} \rightarrow [C_{ijkl}]_{i,j,k,l\,\in\{1,2,3\}}$ and $\underset{\approx}{S} \rightarrow [S_{ijkl}]_{i,j,k,l\,\in\{1,2,3\}}$ are symmetrical tensors so that $C_{ijkl} = C_{ijlk} = C_{jikl} = C_{jilk}$ and $S_{ijkl} = S_{ijlk} = S_{jikl} = S_{jilk}$. In this way the number of independent coefficients is reduced to 21, which means that the most general anisotropic elastic continuum medium can be characterized with 21 coefficients.

A single subscript notation is often used for stresses and strains which is called the contracted notation with the following correspondences:

$$\underline{\sigma} \rightarrow \{\sigma\} \equiv \begin{Bmatrix} \sigma_1 \\ \sigma_2 \\ \sigma_3 \\ \sigma_4 \\ \sigma_5 \\ \sigma_6 \end{Bmatrix} = \begin{Bmatrix} \sigma_{11} \\ \sigma_{22} \\ \sigma_{33} \\ \sigma_{23} \\ \sigma_{31} \\ \sigma_{12} \end{Bmatrix} \quad \text{and} \quad \underline{\varepsilon} \rightarrow \{\varepsilon\} \equiv \begin{Bmatrix} \varepsilon_1 \\ \varepsilon_2 \\ \varepsilon_3 \\ \varepsilon_4 \\ \varepsilon_5 \\ \varepsilon_6 \end{Bmatrix} = \begin{Bmatrix} \varepsilon_{11} \\ \varepsilon_{22} \\ \varepsilon_{33} \\ 2\,\varepsilon_{23} \\ 2\,\varepsilon_{31} \\ 2\,\varepsilon_{12} \end{Bmatrix} = \begin{Bmatrix} \varepsilon_{11} \\ \varepsilon_{22} \\ \varepsilon_{33} \\ \gamma_{23} \\ \gamma_{31} \\ \gamma_{12} \end{Bmatrix} \tag{5.13}$$

In this case, Equation (5.12) can be written

$$\{\sigma\} = [C]\{\varepsilon\} \quad \Leftrightarrow \quad \begin{Bmatrix} \sigma_1 \\ \sigma_2 \\ \sigma_3 \\ \sigma_4 \\ \sigma_5 \\ \sigma_6 \end{Bmatrix} = \begin{bmatrix} C_{11} & C_{12} & C_{13} & C_{14} & C_{15} & C_{16} \\ & C_{22} & C_{23} & C_{24} & C_{25} & C_{26} \\ & & C_{33} & C_{34} & C_{35} & C_{36} \\ & & & C_{44} & C_{45} & C_{46} \\ & \text{sym.} & & & C_{55} & C_{56} \\ & & & & & C_{66} \end{bmatrix} \begin{Bmatrix} \varepsilon_1 \\ \varepsilon_2 \\ \varepsilon_3 \\ \varepsilon_4 \\ \varepsilon_5 \\ \varepsilon_6 \end{Bmatrix} \quad (5.14)$$

In the same way, the inverse strain-stress relations are given by (with $[S] = [C]^{-1}$)

$$\{\varepsilon\} = [S]\{\sigma\} \quad \Leftrightarrow \quad \begin{Bmatrix} \varepsilon_1 \\ \varepsilon_2 \\ \varepsilon_3 \\ \varepsilon_4 \\ \varepsilon_5 \\ \varepsilon_6 \end{Bmatrix} = \begin{bmatrix} S_{11} & S_{12} & S_{13} & S_{14} & S_{15} & S_{16} \\ & S_{22} & S_{23} & S_{24} & S_{25} & S_{26} \\ & & S_{33} & S_{34} & S_{35} & S_{36} \\ & & & S_{44} & S_{45} & S_{46} \\ & \text{sym.} & & & S_{55} & S_{56} \\ & & & & & S_{66} \end{bmatrix} \begin{Bmatrix} \sigma_1 \\ \sigma_2 \\ \sigma_3 \\ \sigma_4 \\ \sigma_5 \\ \sigma_6 \end{Bmatrix} \quad (5.15)$$

It should be noted that, due to the Voigt notation adopted, the coefficients of the above mentioned matrix $[S]$ are related to $S_{ijkl}$ with a multiplier 1, 2 or 4. Indeed, from the strain-stress tensorial relation $\varepsilon_{ij} = S_{ijkl}\,\sigma_{kl}$, we get, for example

$$\begin{aligned} \varepsilon_{12} = &S_{1211}\,\sigma_{11} + S_{1212}\,\sigma_{12} + S_{1213}\,\sigma_{13} \\ &S_{1221}\,\sigma_{21} + S_{1222}\,\sigma_{22} + S_{1223}\,\sigma_{23} \\ &S_{1231}\,\sigma_{31} + S_{1232}\,\sigma_{32} + S_{1233}\,\sigma_{33} \end{aligned} \quad (5.16)$$

which is equivalent to the following equation due to the symmetries of $\underset{\approx}{S}$ and $\underset{\sim}{\sigma}$

$$\begin{aligned} \varepsilon_{12} = &S_{1211}\,\sigma_{11} + S_{1212}\,\sigma_{12} + S_{1213}\,\sigma_{13} \\ &S_{1212}\,\sigma_{12} + S_{1222}\,\sigma_{22} + S_{1223}\,\sigma_{23} \\ &S_{1213}\,\sigma_{13} + S_{1223}\,\sigma_{23} + S_{1233}\,\sigma_{33} \end{aligned} \quad (5.17)$$

thus

$$\varepsilon_{12} = S_{1211}\,\sigma_{11} + S_{1222}\,\sigma_{22} + S_{1233}\,\sigma_{33} + 2\,S_{1223}\,\sigma_{23} + 2\,S_{1213}\,\sigma_{13} + 2\,S_{1212}\,\sigma_{12} \quad (5.18)$$

leading to

$$\underbrace{2\,\varepsilon_{12}}_{\varepsilon_6} = \underbrace{2\,S_{1211}\,\sigma_{11}}_{S_{61}\,\sigma_1} + \underbrace{2\,S_{1222}\,\sigma_{22}}_{S_{62}\,\sigma_2} + \underbrace{2\,S_{1233}\,\sigma_{33}}_{S_{63}\,\sigma_3} + \underbrace{4\,S_{1223}\,\sigma_{23}}_{S_{64}\,\sigma_4} + \underbrace{4\,S_{1213}\,\sigma_{13}}_{S_{65}\,\sigma_5} + \underbrace{4\,S_{1212}\,\sigma_{12}}_{S_{66}\,\sigma_6} \quad (5.19)$$

The converse to Equation (5.12) should be written

$$\begin{Bmatrix} \varepsilon_{11} \\ \varepsilon_{22} \\ \varepsilon_{33} \\ 2\,\varepsilon_{23} \\ 2\,\varepsilon_{13} \\ 2\,\varepsilon_{12} \end{Bmatrix} = \begin{bmatrix} S_{1111} & S_{1122} & S_{1133} & 2\,S_{1123} & 2\,S_{1113} & 2\,S_{1112} \\ S_{2211} & S_{2222} & S_{2233} & 2\,S_{2223} & 2\,S_{2213} & 2\,S_{2212} \\ S_{3311} & S_{3322} & S_{3333} & 2\,S_{3323} & 2\,S_{3313} & 2\,S_{3312} \\ 2\,S_{2311} & 2\,S_{2322} & 2\,S_{2333} & 4\,S_{2323} & 4\,S_{2313} & 4\,S_{2312} \\ 2\,S_{1311} & 2\,S_{1322} & 2\,S_{1333} & 4\,S_{1323} & 4\,S_{1313} & 4\,S_{1312} \\ 2\,S_{1211} & 2\,S_{1222} & 2\,S_{1233} & 4\,S_{1223} & 4\,S_{1213} & 4\,S_{1212} \end{bmatrix} \begin{Bmatrix} \sigma_{11} \\ \sigma_{22} \\ \sigma_{33} \\ \sigma_{23} \\ \sigma_{13} \\ \sigma_{12} \end{Bmatrix} \quad (5.20)$$

This remark does not apply to $[C]$, indeed, from $\sigma_{ij} = C_{ijkl}\,\varepsilon_{kl}$, we get, for example

$$\begin{aligned} \sigma_{12} = &C_{1211}\,\varepsilon_{11} + C_{1212}\,\varepsilon_{12} + C_{1213}\,\varepsilon_{13} \\ &C_{1221}\,\varepsilon_{21} + C_{1222}\,\varepsilon_{22} + C_{1223}\,\varepsilon_{23} \\ &C_{1231}\,\varepsilon_{31} + C_{1232}\,\varepsilon_{32} + C_{1233}\,\varepsilon_{33} \end{aligned} \quad (5.21)$$

which condenses to

$$\underbrace{\sigma_{12}}_{\sigma 6} = \underbrace{C_{1211}\,\varepsilon_{11}}_{C_{61}\,\varepsilon_1} + \underbrace{C_{1222}\,\varepsilon_{22}}_{C_{62}\,\varepsilon_2} + \underbrace{C_{1233}\,\varepsilon_{33}}_{C_{63}\,\varepsilon_3} + \underbrace{2\,C_{1223}\,\varepsilon_{23}}_{\substack{C_{64}\,\varepsilon_4 \\ C_{1223}=C_{64}}} + \underbrace{2\,C_{1213}\,\varepsilon_{13}}_{\substack{C_{65}\,\varepsilon_5 \\ C_{1213}=C_{65}}} + \underbrace{2\,C_{1212}\,\varepsilon_{12}}_{\substack{C_{66}\,\varepsilon_6 \\ C_{1212}=C_{66}}} \qquad (5.22)$$

## 5.7 Material symmetry

As we have seen previously the stiffness matrix and the compliance matrix are determined using 21 material coefficients. This is the general case for an anisotropic material, described as being "triclinic" and displaying no symmetry. The response of such a characteristic REV submitted to a given loading depends on the loading direction.

The question which is now raised is whether, when the REV is loaded in different directions, some of them give the same response. If it is the case, the identical responses characterize the material symmetry. According to the set of material symmetry, which characterises the material, its behaviour is said to be monoclinic, orthotropic, transversally isotropic or isotropic. It is mainly the geometry of the microstructure which induces material symmetry (single crystal, fibre reinforced composite, textile, ...). In addition other causes can modify the microstructure or change material symmetry (pre-stresses, damage,...) and modify the life-time of the material.

The constitutive law has to be written to respect the material symmetry and then the number of independent elastic coefficients can be reduced. To respect these physical symmetries, we choose a thermodynamic potential defined with polynomial quantities of state variables which remain invariable with the material plane symmetry. We can show that the number of these invariants is finite and describe the "integrity basis" (Dieudonné and Carrel, 1971; Derksen and Kemper, 2002). Further, the definition of a thermodynamic potential written with variables according to material symmetries allows the material behaviour to be correctly described. A summary of the form taken by $[S]$ or $[C]$ depending on the material symmetry is provided in Table 5.2.

## 5.8 Monoclinic material

A monoclinic material is a material with only one plane of symmetry. The stiffness matrix or compliance matrix do not have to be modified by transformation with respect to a symmetry of this plane. In the case where the plane of symmetry is $(\underline{e}_1, \underline{e}_2)$, the stiffness matrix is given by

$$\begin{bmatrix} C_{11} & C_{12} & C_{13} & 0 & 0 & C_{16} \\ & C_{22} & C_{23} & 0 & 0 & C_{26} \\ & & C_{33} & 0 & 0 & C_{36} \\ & & & C_{44} & C_{45} & 0 \\ & \text{sym.} & & & C_{55} & 0 \\ & & & & & C_{66} \end{bmatrix} \qquad (5.23)$$

The compliance matrix is similar. The number of independent coefficients reduces to 13.

## 5.9 Orthotropic material

An orthotropic material has three orthogonal planes of material symmetry; each plane is normal to the others. The existence of two of them implies the existence of the third plane. The stiffness matrix is then obtained by adding another plane of symmetry to the monoclinic material. Respecting invariance towards these three planes of material symmetry, requires a thermodynamic potential to be written as a function of seven invariant polynoms of strains in the "integrity basis":

- Degree 1 with strains: $I_1 = \varepsilon_1$, $I_2 = \varepsilon_2$, $I_3 = \varepsilon_3$
- Degree 2 with strains: $I_4 = \varepsilon_4$, $I_5 = \varepsilon_5$, $I_6 = \varepsilon_6$
- Degree 3 with strains: $I_7 = \varepsilon_4\,\varepsilon_5\,\varepsilon_6$

A strain second-order thermodynamic potential can then be written as

$$\psi = \psi(I_1, I_2, I_3, I_4, I_5, I_6) \tag{5.24}$$

The law of state gives the stiffness matrix as

$$\begin{bmatrix} C_{11} & C_{12} & C_{13} & 0 & 0 & 0 \\ & C_{22} & C_{23} & 0 & 0 & 0 \\ & & C_{33} & 0 & 0 & 0 \\ & & & C_{44} & 0 & 0 \\ & \text{sym.} & & & C_{55} & 0 \\ & & & & & C_{66} \end{bmatrix} \tag{5.25}$$

The number of independent coefficients is reduced to 9. If the material coordinate system is taken to be a Cartesian coordinate system, defined in the basis $\left(\underline{e}_1,\ \underline{e}_2,\ \underline{e}_3\right)$ in which the direction given by $\underline{e}_1$ is parallel to the fibre longitudinal axis, the compliance matrix can be written as

$$\begin{Bmatrix} \varepsilon_1 \\ \varepsilon_2 \\ \varepsilon_3 \\ \varepsilon_4 \\ \varepsilon_5 \\ \varepsilon_6 \end{Bmatrix} = \begin{bmatrix} 1/E_1 & -\nu_{21}/E_2 & -\nu_{31}/E_3 & 0 & 0 & 0 \\ -\nu_{12}/E_1 & 1/E_2 & -\nu_{32}/E_3 & 0 & 0 & 0 \\ -\nu_{13}/E_1 & -\nu_{23}/E_2 & 1/E_3 & 0 & 0 & 0 \\ 0 & 0 & 0 & 1/\mu_{23} & 0 & 0 \\ 0 & 0 & 0 & 0 & 1/\mu_{13} & 0 \\ 0 & 0 & 0 & 0 & 0 & 1/\mu_{12} \end{bmatrix} \begin{Bmatrix} \sigma_1 \\ \sigma_2 \\ \sigma_3 \\ \sigma_4 \\ \sigma_5 \\ \sigma_6 \end{Bmatrix} \tag{5.26}$$

Where $E_1$, $E_2$, $E_3$ are Young's moduli in $\underline{e}_1$, $\underline{e}_2$ and $\underline{e}_3$ material directions, respectively, $\nu_{ij}$ is Poisson's ratio, defined as the ratio of transverse strain in the $j$th direction to the axial strain in the $i$th direction when stressed in the $i$ direction, and $\mu_{23}$, $\mu_{13}$, $\mu_{12}$ are the shear moduli in the $\left(\underline{e}_2,\ \underline{e}_3\right)$, $\left(\underline{e}_1,\ \underline{e}_3\right)$ and $\left(\underline{e}_1,\ \underline{e}_2\right)$ planes, respectively. The symmetry of the behaviour implies that the following reciprocal relations hold

$$\frac{\nu_{ij}}{E_i} = \frac{\nu_{ji}}{E_j} \quad \text{(no sum on } i,j) \text{ i.e.} \quad \begin{cases} \dfrac{\nu_{12}}{E_1} = \dfrac{\nu_{21}}{E_2} \\[2mm] \dfrac{\nu_{13}}{E_1} = \dfrac{\nu_{31}}{E_3} \\[2mm] \dfrac{\nu_{23}}{E_2} = \dfrac{\nu_{32}}{E_3} \end{cases} \tag{5.27}$$

## 5.10    Transversally isotropic material

A continuum material is said to be transversally isotropic for a given axis, if its properties remain invariant or isotropic with respect to a change in direction orthogonal to this axis. Often fibre-reinforced lamina are characterized as being transversely isotropic. If the fibre direction is parallel to the direction given by $\underline{e}_1$, we experimentally observe that perpendicular to this axis all the properties are isotropic. We can demonstrate that the "integrity basis" of such a material is characterized by four independent polynoms of strains:

- $I_1 = \varepsilon_1$
- $I_2 = \varepsilon_2 + \varepsilon_3$
- $I_3 = \varepsilon_4^2 - \varepsilon_2 \varepsilon_3$
- $I_4 = \varepsilon_5^2 + \varepsilon_6^2 - \varepsilon_1 \varepsilon_2 - \varepsilon_1 \varepsilon_3$

Thus the constitutive law can be written with five independent material coefficients

$$
\begin{Bmatrix} \sigma_1 \\ \sigma_2 \\ \sigma_3 \\ \sigma_4 \\ \sigma_5 \\ \sigma_6 \end{Bmatrix} = \begin{bmatrix} C_{11} & C_{12} & C_{12} & 0 & 0 & 0 \\ & C_{22} & C_{12} & 0 & 0 & 0 \\ & & C_{22} & 0 & 0 & 0 \\ & & & \dfrac{C_{22} - C_{23}}{2} & 0 & 0 \\ & \text{sym.} & & & C_{66} & 0 \\ & & & & & C_{66} \end{bmatrix} \begin{Bmatrix} \varepsilon_1 \\ \varepsilon_2 \\ \varepsilon_3 \\ \varepsilon_4 \\ \varepsilon_5 \\ \varepsilon_6 \end{Bmatrix} \tag{5.28}
$$

$$
\begin{Bmatrix} \varepsilon_1 \\ \varepsilon_2 \\ \varepsilon_3 \\ \varepsilon_4 \\ \varepsilon_5 \\ \varepsilon_6 \end{Bmatrix} = \begin{bmatrix} S_{11} & S_{12} & S_{12} & 0 & 0 & 0 \\ & S_{22} & S_{12} & 0 & 0 & 0 \\ & & S_{22} & 0 & 0 & 0 \\ & & & \dfrac{S_{22} - S_{23}}{2} & 0 & 0 \\ & \text{sym.} & & & S_{66} & 0 \\ & & & & & S_{66} \end{bmatrix} \begin{Bmatrix} \sigma_1 \\ \sigma_2 \\ \sigma_3 \\ \sigma_4 \\ \sigma_5 \\ \sigma_6 \end{Bmatrix} \tag{5.29}
$$

Young's moduli in direction $\underline{e_2}$ and $\underline{e_3}$ are the same, and so are the shear moduli in planes $\left(\underline{e_1}, \underline{e_2}\right)$ and $\left(\underline{e_1}, \underline{e_3}\right)$. Only 5 independent coefficients are required and we obtain

$$
\begin{cases} S_{11} = \dfrac{1}{E_1} \\[2mm] S_{22} = S_{33} = \dfrac{1}{E_2} = \dfrac{1}{E_3} \\[2mm] S_{44} = \dfrac{2\left(1 + \nu_{23}\right)}{E_2} \\[2mm] S_{55} = S_{66} = \dfrac{1}{\mu_{12}} = \dfrac{1}{\mu_{13}} \end{cases} \quad \text{and} \quad \begin{cases} S_{12} = S_{13} = \dfrac{-\nu_{12}}{E_1} = \dfrac{-\nu_{21}}{E_2} \\[2mm] S_{23} = \dfrac{-\nu_{23}}{E_2} \end{cases} \tag{5.30}
$$

The coefficients of $[C]$, although analytically definable, are more easily obtained by the inversion of $[S]$. Of course, depending on the Poisson's ratio, $C_{11}$ can be close to $E_1$ but it is not equal to $E_1$, a mistake sometimes made...

$$
C_{11} = \frac{E_1^2 \left(1 - 2\nu_{23}\right)}{-E_1 \left(1 - 2\nu_{23}\right) + 2\nu_{12}E_2} = \frac{E_1 \left(1 - 2\nu_{23}\right)}{1 - \nu_{23} - 2\nu_{12}\nu_{21}} \tag{5.31}
$$

Instead of using indices 1, 2 and 3, another – more explicit – notation, is often used in the case of transverse isotropy. $L$ denotes in that case the longitudinal direction and $T$, $T'$ the transverse ones (defining the isotropy plane). The five independent coefficients can be

$$
E_L, \quad E_T = E_{T'}, \quad \nu_{TT'}, \quad \nu_{LT} = \nu_{LT'}, \quad \mu_{LT} = \mu_{LT'} \tag{5.32}
$$

The transverse shear modulus, $\mu_{TT'}$, is equal to

$$
\frac{E_T}{2\left(1 + \nu_{TT'}\right)} \tag{5.33}
$$

## 5.11    Isotropic material

A material is said to be isotropic if, when stressed in all the directions, the responses remain identical. The mechanical properties are the same whatever the considered direction and so the stiffness and compliance matrices have to be invariant for any orthogonal frame transformation. In that case, we have two basic strain invariants:

- $I_1 = \text{Tr}\left(\underline{\varepsilon}\right) = \varepsilon_1 + \varepsilon_2 + \varepsilon_3$
- $I_2 = \dfrac{1}{2}\,\text{Tr}\left(\underline{\varepsilon}^2\right)$

This requires that the thermodynamic potential $\psi$ be a quadratic invariant of the strain tensor as a linear combination of the square of $I_1$ and $I_2$

$$\psi = \psi\left(I_1{}^2,\, I_2\right) \tag{5.34}$$

The number of independent coefficients reduces to 2 leading to the stiffness matrix

$$
\begin{bmatrix}
C_{11} & C_{12} & C_{12} & 0 & 0 & 0 \\
 & C_{11} & C_{12} & 0 & 0 & 0 \\
 & & C_{11} & 0 & 0 & 0 \\
 & & & \dfrac{C_{11}-C_{12}}{2} & 0 & 0 \\
 & \text{sym.} & & & \dfrac{C_{11}-C_{12}}{2} & 0 \\
 & & & & & \dfrac{C_{11}-C_{12}}{2}
\end{bmatrix}
\tag{5.35}
$$

Generally, the index stiffness coefficients are also replaced by Lamé's two coefficients, $\lambda$ and $\mu$

$$
\begin{cases}
\lambda = C_{12} \\[2mm]
\mu = \dfrac{C_{11}-C_{12}}{2}
\end{cases}
\tag{5.36}
$$

and so $C_{11} = \lambda + 2\,\mu$. The stress-strain relation becomes

$$
\begin{Bmatrix}
\sigma_1 \\ \sigma_2 \\ \sigma_3 \\ \sigma_4 \\ \sigma_5 \\ \sigma_6
\end{Bmatrix}
=
\begin{bmatrix}
\lambda+2\mu & \lambda & \lambda & 0 & 0 & 0 \\
 & \lambda+2\mu & \lambda & 0 & 0 & 0 \\
 & & \lambda+2\mu & 0 & 0 & 0 \\
 & & & \mu & 0 & 0 \\
 & \text{sym.} & & & \mu & 0 \\
 & & & & & \mu
\end{bmatrix}
\begin{Bmatrix}
\varepsilon_1 \\ \varepsilon_2 \\ \varepsilon_3 \\ \varepsilon_4 \\ \varepsilon_5 \\ \varepsilon_6
\end{Bmatrix}
\tag{5.37}
$$

By inverting the above relation, or by differentiating a dual potential, we find the compliance matrix given as

$$
\begin{Bmatrix}
\varepsilon_1 \\ \varepsilon_2 \\ \varepsilon_3 \\ \varepsilon_4 \\ \varepsilon_5 \\ \varepsilon_6
\end{Bmatrix}
=
\begin{bmatrix}
1/E & -\nu/E & -\nu/E & 0 & 0 & 0 \\
 & 1/E & -\nu/E & 0 & 0 & 0 \\
 & & 1/E & 0 & 0 & 0 \\
 & & & 1/\mu & 0 & 0 \\
 & \text{sym.} & & & 1/\mu & 0 \\
 & & & & & 1/\mu
\end{bmatrix}
\begin{Bmatrix}
\sigma_1 \\ \sigma_2 \\ \sigma_3 \\ \sigma_4 \\ \sigma_5 \\ \sigma_6
\end{Bmatrix}
\tag{5.38}
$$

The Young's modulus $E$ and the poisson's ratio $\nu$ are related to Lamé's coefficients according to Table 5.1. Coefficients $E$ and $\nu$ are generally mostly used because they can directly be obtained from experiments (tensile tests, for example).

Table 5.1: Relationships between isotropic material coefficients

|  | $\lambda,\mu$ | $\lambda,E$ | $\lambda,\nu$ | $\lambda,K$ | $\mu,E$ | $\mu,\nu$ | $\mu,K$ | $E,\nu$ | $E,K$ | $\nu,K$ |
|---|---|---|---|---|---|---|---|---|---|---|
| $\lambda=$ | $\lambda$ | $\lambda$ | $\lambda$ | $\lambda$ | $\dfrac{\mu(2\mu-E)}{E-3\mu}$ | $\dfrac{2\mu\nu}{1-2\nu}$ | $\dfrac{3K-2\mu}{3}$ | $\dfrac{E\nu}{(1+\nu)(1-2\nu)}$ | $\dfrac{3K(3K-E)}{9K-E}$ | $\dfrac{3K\nu}{1+\nu}$ |
| $\mu=$ | $\mu$ | $\dfrac{\mathcal{A}-3\lambda+E}{4}$ | $\dfrac{\lambda(1-2\nu)}{2\nu}$ | $\dfrac{3(K-\lambda)}{2}$ | $\mu$ | $\mu$ | $\mu$ | $\dfrac{E}{2(1+\nu)}$ | $\dfrac{3KE}{9K-E}$ | $\dfrac{3K(1-2\nu)}{2(1+\nu)}$ |
| $E=$ | $\dfrac{\mu(3\lambda+2\mu)}{\lambda+\mu}$ | $E$ | $\dfrac{\lambda(1+\nu)(1-2\nu)}{\nu}$ | $\dfrac{9K(K-\lambda)}{3K-\lambda}$ | $E$ | $2\mu(1+\nu)$ | $\dfrac{9K\mu}{3K+\mu}$ | $E$ | $E$ | $3K(1-2\nu)$ |
| $\nu=$ | $\dfrac{\lambda}{2(\lambda+\mu)}$ | $\dfrac{\mathcal{A}-E-\lambda}{4\lambda}$ | $\nu$ | $\dfrac{\lambda}{3K-\lambda}$ | $\dfrac{E-2\mu}{2\mu}$ | $\nu$ | $\dfrac{3K-2\mu}{2(3K+\mu)}$ | $\nu$ | $\dfrac{3K-E}{6K}$ | $\nu$ |
| $K=$ | $\dfrac{3\lambda+2\mu}{3}$ | $\dfrac{\mathcal{A}+3\lambda+E}{6}$ | $\dfrac{\lambda(1+\nu)}{3\nu}$ | $K$ | $\dfrac{\mu E}{3(3\mu-E)}$ | $\dfrac{2\mu(1+\nu)}{3(1-2\nu)}$ | $K$ | $\dfrac{E}{3(1-2\nu)}$ | $K$ | $K$ |

$\mathcal{A}=\sqrt{E^2+2\lambda E+9\lambda^2}$

The transverse bulk modulus, $k$, not mentioned in this table, should not be confused with the bulk modulus $K$; i.e. $k = K + \dfrac{\mu}{3} = \dfrac{\mu}{1-2\nu}$

**Table 5.2**    Form of the $[S]$ and $[C]$ matrices

| | |
|---|---|
| • | Zero component |
| ○ | Non-zero and independent components |
| ● | Dependent components (standard ones) |
| ○—○ ●—● | Equal components |

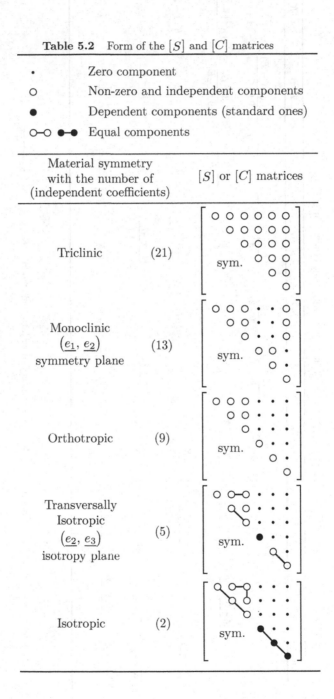

| Material symmetry with the number of (independent coefficients) | $[S]$ or $[C]$ matrices |
|---|---|
| Triclinic            (21) | |
| Monoclinic $(\underline{e_1}, \underline{e_2})$ symmetry plane            (13) | |
| Orthotropic            (9) | |
| Transversally Isotropic $(\underline{e_2}, \underline{e_3})$ isotropy plane            (5) | |
| Isotropic            (2) | |

# Revision exercises

### Exercise 5.1

Give the generalized Hooke's law using three different notations.

### Exercise 5.2

Give the matrix relationship between $\{\varepsilon\}$ and $\{\sigma\}$ without using the condensed Voigt notation for $\{\varepsilon\}$.

### Exercise 5.3

Give the matrix relationship between $\{\sigma\}$ and $\{\varepsilon\}$ without using the condensed Voigt notation for $\{\varepsilon\}$.

### Exercise 5.4

Can you explain the differences between the two previous relationships?

### Exercise 5.5

Under the isotropic assumption, are the variables $E$, $\mu$ and $\nu$ independent?

### Exercise 5.6

The ply behaviour is generally said to be transversally isotropic for the fibre axis. In this case, what is the number of independent constants which are necessary to describe this behaviour.

### Exercise 5.7

Explain what orthotropy is.

### Exercise 5.8

What relationship can be written between these four variables: $E_1$, $E_2$, $\nu_{12}$ and $\nu_{21}$?

### Exercise 5.9

Rewrite relations Eqs. (5.28) and (5.29) in the case where the fibre direction is parallel to the direction given by $\underline{e_3}$.

### Exercise 5.10

Are are the terms "bulk modulus" and "transverse bulk modulus" synonymous?

# Micromechanical models for composite materials

## 6.1   Introduction

The appearance of composite materials of the type reinforcement/matrix gives rise to problems, in terms of Mechanics, which were almost unknown until the mid-1970s. The small sizes of the constitutions making up the microstructure compared to the dimensions of the structures in which they were used, and the need to access damage phenomena which appeared, directly causing the ruin of the structure, required a detailed knowledge of the description of the stresses and strains close to the reinforcements. But, the computers were far from being sufficiently powerful to solve such a problem. Then, multi-scale approaches were developed. Now, these techniques are widely used, also for metallic materials, because of the considerable advances they have made in terms of structural calculations.

The objective of this chapter is first to show that the inner constitution of the composite materials induces difficulties for solving apparently simple mechanical problems. After that, the multi-scale process involving the homogenisation and localisation steps are described, in a general way, leading to the Effective Moduli Method and the Periodic Method, using the averaging technique. Finally, all of these concepts are applied in the case of unidirectional fibre / resin composites. The conclusion is that we must be very careful about the use of these methods which can give, depending on the case, excellent results but also significantly misleading results.

## 6.2   Dumb script convention (DSC) or Einstein convention

### 6.2.1   Definition

We consider an algebraic expression, i.e. a symbolic mathematical expression formed by monomials (themselves made up of numbers, letters, symbols) separated by an addition or subtraction operation (it is thus a polynomial expression). It is assumed that the letters and symbols present in the monomials can carry numeric and/or literal underscripts/upperscripts. In order to simplify the notations here, the word script means either an underscript or upperscript.

Definitions – Literal scripts appearing twice in a monomial form a dumb script, literal scripts appearing once in a monomial form, each, a frank script.

Fundamental rules – A literal script should never be repeated more than twice in a monomial, you can debaptize a dumb script by replacing it with any other letter (respecting the first rule), you cannot debaptize a frank script.

Dumb script convention – By omitting to write the summation sign (usually denoted $\Sigma$) and provided that the scripts always vary in the same way, if a script is dumb in a monomial it means that we have a summation of this script.

Notes – This explains why it is possible to change the letter of a dumb script. If a frank script is present on both sides of an equality, it must be designated by the same name.

Example – If an algebraic expression is:

$$E_k = \sum_{i=1}^{3} u_i v_i w_k$$

$k$ is a straightforward script, he appears with the same name in both sides of the equality, $i$ is a dumb script. In the framework of the DSC, this expression takes the following form (admitting that the script evolves from 1 to 3):

$$E_k = u_i v_i w_k$$

We can also write: $E_k = u_i v_i w_k = u_j v_j w_k$.

## 6.2.2 Example with the ellipticity condition on the fourth-order tensor of rigidity. Flexibility tensor and identity tensor

We note $a = (a_{ijkh})_{i,j,k,h=1,2,3}$ the fourth-order tensor of rigidity of a material which possesses the usual properties of symmetry and ellipticity. The symmetry conditions are:

$$
\begin{cases}
a_{ijkh} = a_{jikh} \\
a_{ijkh} = a_{ijhk} \\
a_{ijkh} = a_{khij}
\end{cases}
$$

In each of the previous equalities, $i$, $j$, $k$ and $h$ are straightforward scripts. The ellipticity condition is:

$$
\exists \alpha \in \mathbb{R}_+^* \ / \ \forall \varepsilon = (\varepsilon_{ij} = \varepsilon_{ij})_{i,j=1,2,3} \in \mathbb{R}^6,
$$

$$
\sum_{i=1}^3 \sum_{j=1}^3 \sum_{k=1}^3 \sum_{h=1}^3 a_{ijkh}\varepsilon_{ij}\varepsilon_{kh} \geq \alpha \sum_{i=1}^3 \sum_{j=1}^3 \varepsilon_{ij}\varepsilon_{ij}
$$

In the first part of the definition, $i$ and $j$ are straightforward scripts. In the second part of the definition, $i$, $j$, $k$ and $h$ are dumb scripts. Then, in the framework of the DSC, the definition can be written:

$$
\exists \alpha \in \mathbb{R}_+^* \ / \ \forall \varepsilon = (\varepsilon_{ij} = \varepsilon_{ij})_{i,j=1,2,3} \in \mathbb{R}^6,
$$

$$
a_{ijkh}\varepsilon_{ij}\varepsilon_{kh} \geq \alpha\varepsilon_{ij}\varepsilon_{ij}
$$

The fourth-order tensor of flexibility is denoted as $A = (A_{ijkh})_{i,j,k,h=1,2,3}$. It is obtained, starting from $a$, by writing:

$$
A_{ijlm}a_{lmkh} = T_{ijkh} = \frac{1}{2}\left(\delta_{ik}\delta_{jh} + \delta_{jk}\delta_{ih}\right)
$$

and then, solving a linear system. It can be shown that $A$ possesses the same proprerties as $a$. Using the description of second-order tensors, we can say that $a$ and $A$ are inverses with respect to each other.

# 6.3 Justification of the necessity of micromechanical approaches for composite materials

## 6.3.1 The problem to solve. Example with a pressure vessel

Let's consider a structure $S$ assumed to be made of an inhomogeneous medium consisting of several $(n)$ constituents $C^{(I)}$ $(I = 1, \ldots, n)$ (phases and inclusions) that are themselves homogeneous. This implies that the characteristic sizes of the inclusions are small compared to the characteristic size of the structure and we assume that they are numerous and approximately regularly distributed all over $S$. Moreover, we assume that the adhesion between phases and inclusions is perfect, which is to say that there is perfect bonding with no possibility of dissipative phenomena at the interfaces of the phases/inclusions.

We wish to calculate the stress and strain fields in each point of $S$ in a way so as to predict its failure. It is well known that, in general, the origin of the failure of the structure takes place around inclusions. In order to obtain a good prediction of the failure of the structure, it is therefore necessary to know the stress and strain fields around the inclusions.

As an example, we shall consider a pressure vessel made with long fibres embedded in a resin. The constitutive medium of the vessel consists of one phase (the resin) and it can be considered that the fibres are inclusions in the resin, even if they are numerous. The characteristic size of the vessel (axial length) is $\approx 2 \times 10^1$ m and the characteristic size of the fibres (the diameter) is $\approx 7 \times 10^{-6}$ m. The ratio of the orders of magnitude is very large. For this type of structure, it is well known that the failure is due to the fibre breakage. It is therefore necessary to know the stress and strain fields in the fibres to arrive at an accurate prediction of the burst pressure of the vessel.

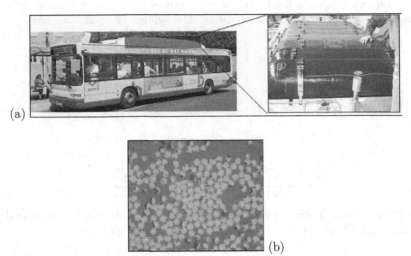

(a)

(b)

Figure 6.1: Pressure vessels typical of those used on buses (a) made by fibres embedded in resin (b). The ratio between the orders of magnitude of the characteristic size of the vessel and the characteristic size of the fibre is very large (Blassiau, 2005). The failure of the vessel is due to fibre breakage.

## 6.3.2   The most general framework and data of the problem to be solved

Let's consider the Euclidian affine space $\varepsilon^3$ modelling the physical space, for which the associated vector space is $\mathbb{R}^3$. This affine space is referred to the framework $R = (O, b)$ for which the orthonormal basis is $b = (\vec{x}_1, \vec{x}_2, \vec{x}_3)$ and $O$ the origin. The vector of the coordinates of a point $M$ relative to this framework is given by $\vec{x} = (x_i)_{i=1,2,3}$. The relationship of a function $q$ with respect to the variables $(x_i)_{i=1,2,3}$, that is to say with respect to the point $M$, is noted by $q(x)$ where $x = (x_1, x_2, x_3)$ or $q(M)$. The dependence of $q$ with respect to $x$ or $M$ can be omitted as all the values implicitly considered are related to it, except if this information brings necessary information and clarity to the problem. The fields of displacement, stress (supposed symmetrical) and strain (linearised) are respectively noted as $\vec{u} = (u_i)_{i=1,2,3}$, $\sigma = (\sigma_{ij})_{i,j=1,2,3}$ and $e = (e_{ij})_{i,j=1,2,3}$ where $e_{ij} = \frac{1}{2}(\frac{\partial u_i}{\partial x_j} + \frac{\partial u_j}{\partial x_i})$.

The material system $S$ studied, assuming the Hypothesis of Small Perturbations and quasi-static conditions, is being displaced and coincides in time with the domain $D = \Omega \cup \partial\Omega$. $D$ is a continuous path-connected bounded and closed set in $\varepsilon^3$, $\Omega$ designating its interior and $\partial\Omega$ its boundary which is considered to be regular. The outer unit vector at a point $M$ of $\partial\Omega$ is noted as $\vec{n}(M)$.

We assume that the behaviour of each constituent $\mathcal{C}^{(I)}$ making up the constitutive media of the structure is linearly elastic and possibly anisotropic. We note $a^{(I)} = (a_{ijkh}^{(I)})_{i,j,k,h=1,2,3}$ the fourth-order tensor of rigidity of the constituent $\mathcal{C}^{(I)}$ which possesses the usual properties of symmetry and ellipticity:

$$
\begin{cases}
a_{ijkh}^{(I)} = a_{jikh}^{(I)} \\
a_{ijkh}^{(I)} = a_{ijhk}^{(I)} \\
a_{ijkh}^{(I)} = a_{khij}^{(I)} \\
\exists \alpha \in \mathbb{R}_+^* \ / \ \forall \varepsilon = (\varepsilon_{ij} = \varepsilon_{ij})_{i,j=1,2,3} \in \mathbb{R}^6, a_{ijkh}^{(I)} \varepsilon_{ij} \varepsilon_{kh} \geq \alpha \varepsilon_{ij} \varepsilon_{ij}
\end{cases}
$$

The fourth-order tensor of flexibility is denoted as $A^{(I)} = (A_{ijkh}^{(I)})_{i,j,k,h=1,2,3}$ and possesses the same properties as $a^{(I)}$.

The loadings applied to $S$ can be remote actions acting on $\Omega(t)$ induced by a volume density of force denoted as $\rho(M)\vec{f}(M,t)$. Here we assume that this volume density of force is induced by gravity $(\vec{g})$, assumed to be constant in the domain of the space where the structure is situated: $\rho(M)\vec{f}(M,t) = \rho(M)\vec{g}$. The loadings applied to the domain can be also contact actions on $\partial\Omega(t)$. To construct a general framework for the contact actions it is necessary to divide the frontier of the domain into three distinct parts $\partial\Omega_U$, $\partial\Omega_F$ and $\partial\Omega_{FU}$. It is supposed that this division is independent of time and that it forms a partition of $\partial\Omega(t)$, which is to say that summing the different parts reforms the totality of $\partial\Omega(t)$ and that their intersection two by two is equal to the empty set:

$$
\partial\Omega(t) = \partial\Omega_U \cup \partial\Omega_F \cup \partial\Omega_{FU}
\qquad
\begin{cases}
\partial\Omega_U \cap \partial\Omega_F = \emptyset \\
\partial\Omega_F \cap \partial\Omega_{FU} = \emptyset \\
\partial\Omega_U \cap \partial\Omega_{FU} = \emptyset
\end{cases}
$$

One or two of these parts can be equal to the empty set. But, when a part is not equal to the empty set, it is a requirement that its measure (its area) is strictly positive. Starting from this division, we define the following general boundary conditions:

- on $\partial\Omega_F$, the surface density of force $\vec{F}(M,t)$ is given;
- on $\partial\Omega_U$, the displacement field $\vec{U}_d(M,t)$ is given;
- on $\partial\Omega_{FU}$, a local vector of information, for which one component can be a component of a surface density of force or a component of a displacement field, is given. For instance:

$$
\begin{pmatrix}
F_1'(M,t) \\
U_{d2}'(M,t) \\
F_3'(M,t)
\end{pmatrix}
$$

where $F_1'(M,t)$ and $F_3'(M,t)$ are the first and third components of a surface density of force $\vec{F}'(M,t)$ and $U_{d2}'(M,t)$ the second component of a displacement field $\vec{U}_d{}'(M,t)$. These boundary conditions are those of a standard problem. Associated with linear elastic behaviour (as considered here) they are those of a standard problem of linear elasticity. In this case, theorems exist that give the existence and uniqueness of the solutions for this problem, in terms of stress and strain fields. Also in this case, theorems exist that give the existence and uniqueness or otherwise of the displacement field. The uniqueness or not depends on the type of the boundary conditions. Generally the displacement field solution of the problem has the form of a particular solution in which a movement of a rigid body is added for which all, several or no components are equal to zero, depending on the boundary conditions. In the first case, the displacement field solution of the problem is unique. For other cases, the displacement field solution of the problem is not unique.

In the following, we assume that any boundary condition depends on time. This does not affect the generality of what is explained here.

### 6.3.3　Solving the problem to find the stress and strain fields around the inclusions: theoretically possible but technically impossible

**The equations**

The problem, denoted as $(\mathcal{P}_1)$, which has to be solved consists of finding, at each $M$ point of the domain, the following unknowns:

- the displacement vector $\vec{u}(M)$;
- the stress tensor $\sigma(M)$,

verifying:

- the local equation of equilibrium (including the hypotheses given):

$$\overrightarrow{div}\,\sigma(M) + \rho(M)\vec{g} = \vec{0}\ \forall M \in \Omega$$

- the behaviour law (the constituents are homogeneous and linearly elastic)

  · state law:

$$\left\{ \begin{array}{l} \sigma(M) = a(M)\varepsilon(\vec{u}(M)) \Longleftrightarrow \sigma_{ij}(M) = a_{ijkh}(M)\varepsilon_{kh}(\vec{u}(M)) \\ \Longleftrightarrow \\ \varepsilon(\vec{u}(M)) = A(M)\sigma(M) \Longleftrightarrow \varepsilon_{ij}(\vec{u}(M)) = A_{ijkh}(M)\sigma_{kh}(M) \end{array} \right\}$$

  or, equivalently:

$$\left\{ \begin{array}{l} \sigma(M \in \mathcal{C}^{(I)}) = a^{(I)}\varepsilon(\vec{u}(M \in \mathcal{C}^{(I)})) \Longleftrightarrow \sigma_{ij}(M \in \mathcal{C}^{(I)}) = a_{ijkh}^{(I)}\varepsilon_{kh}(\vec{u}(M \in \mathcal{C}^{(I)})) \\ \Longleftrightarrow \\ \varepsilon(\vec{u}(M \in \mathcal{C}^{(I)})) = A^{(I)}\sigma(M \in \mathcal{C}^{(I)}) \Longleftrightarrow \varepsilon_{ij}(\vec{u}(M \in \mathcal{C}^{(I)})) = A_{ijkh}^{(I)}\sigma_{kh}(M \in \mathcal{C}^{(I)}) \end{array} \right\}$$

  · no evolution law (no dissipative phenomenon)

- boundary conditions:

$$\left\{ \begin{array}{l} \text{on } \partial\Omega_F \ : \ \sigma(M \in \partial\Omega_F) \times \vec{n}(M \in \partial\Omega_F) = \vec{F}(M) \\[2ex] \text{on } \partial\Omega_U \ : \ \vec{u}(M \in \partial\Omega_U F) = \vec{U}_d(M) \\[2ex] \text{on } \partial\Omega_{FU} \ : \ \left\{ \begin{array}{l} \{\sigma(M \in \partial\Omega_{FU}) \times \vec{n}(M \in \partial\Omega_{FU})\}_1 = F_1'(M) \\ \{\vec{u}(M \in \partial\Omega_{FU})\}_2 = U_{d2}'(M) \\ \{\sigma(M \in \partial\Omega_{FU}) \times \vec{n}(M \in \partial\Omega_{FU})\}_3 = F_3'(M) \end{array} \right. \end{array} \right.$$

**Theorems of existence and uniqueness of the solutions exist. However, technically impossible to solve**

Solving this problem is theoretically possible. Its writing as a standard problem of linear elasticity guaranties that the solution in terms of stress exists and is unique. But, solving this problem cannot be made analytically. So solving this problem requires a numerical solution by, for example, the Finite Element Method. However, it is easy to understand that the discretisation of this problem is so big (it should be made with elements of sizes equal to or less than the characteristic length of the inclusion) that no computer could solve this problem. Another way of tackling the problem has to be found.

　　To do that, it must be pointed out that, in the present problem considered no RVE has been defined for the medium making up the structure. However, it is possible, by using the given hypotheses and by accepting that the inclusions are numerous, to assume that an RVE can be defined. Consequently, the notion of scale, that has been, until now, not really necessary, has to be defined.

**Reminder of energy theorems**

The set of the kinematic admissible fields is:

$$U_{ad} = \left\{ \begin{array}{l} \vec{v}(M) = (v_i(M))_{i=1,2,3} \,/ \\ * \; v_i(M) \text{ is a regular function} \\ * \text{ On } \partial\Omega_U, \; \vec{v}(M) = \vec{U}(M) \\ * \text{ On } \partial\Omega_{FU}, \; v_2(M) = U_2'(M) \end{array} \right\}$$

The set of the statically admissible fields is:

$$\Sigma_{ad} = \left\{ \begin{array}{l} \tau(M) = (\tau_{ij}(M))_{i,j=1,2,3} \,/ \\ * \; \tau_{ij}(M) \text{ is a regular function} \\ * \; \tau_{ij}(M) = \tau_{ji}(M) \\ * \; \vec{div}\tau(M) + \rho(M)\vec{f}(M) = \vec{0} \\ * \text{ On } \partial\Omega_F, \; \tau(M)\vec{n}(M) = \vec{F}(M) \\ * \text{ On } \partial\Omega_{FU}, \; \tau_{1j}(M)n_j(M) = F_1'(M), \tau_{3j}(M)n_j(M) = F_3'(M) \end{array} \right\}$$

The variational formulation in term of displacement gives:

$$(\mathcal{P}_1) \Longleftrightarrow (\mathcal{P}_2) \left\{ \begin{array}{l} \vec{u}(M) \in U_{ad}, \; b(\vec{u}, \vec{v} - \vec{u}) = l(\vec{v} - \vec{u}) \; \forall \vec{v} \in U_{ad} \\ \text{with} \\ * \; b(\vec{u}, \vec{v}) = \int_\Omega a_{ijkh}(M)\varepsilon_{ij}(\vec{u}(M))\varepsilon_{kh}(\vec{v}(M)) \\ * \; l(\vec{v}) = \int_\Omega \rho(M)f_i(M)v_i(M) + \int_{\partial\Omega_F} F_i(M)v_i(M) + \ldots \\ \ldots \int_{\partial\Omega_{FU}} (F_1'(M)v_1(M) + F_3'(M)v_3(M)) \end{array} \right\}$$

$$(\mathcal{P}_2) \Longleftrightarrow (\mathcal{P}_3) \left\{ \begin{array}{l} \vec{u}(M) \in U_{ad}, \; I(\vec{v}) \geq I(\vec{u}) \; \forall \vec{v} \in U_{ad} \\ \text{with } I(\vec{v}) = \frac{1}{2}b(\vec{v}, \vec{v}) - l(\vec{v}) \end{array} \right\}$$

$I(\vec{v})$ is called the potential energy of the kinematic admissible field $\vec{v}$.

The variational formulation in term of stress gives:

$$(\mathcal{P}_1) \Longleftrightarrow (\mathcal{P}_4) \left\{ \begin{array}{l} \sigma(M) \in \Sigma_{ad}, \; B(\sigma, \tau - \sigma) = L(\tau - \sigma) \; \forall \tau \in \Sigma_{ad} \\ \text{with} \\ * \; B(\sigma, \tau) = \int_\Omega A_{ijkh}(M)\sigma_{ij}(M)\tau_{kh}(M) \\ * \; L(\tau) = \int_{\partial\Omega_U} \tau_{ij}(M)n_j(M)U_i(M) + \int_{\partial\Omega_{FU}} \tau_{2j}(M)n_j(M)U_2'(M) \end{array} \right\}$$

$$(\mathcal{P}_4) \Longleftrightarrow (\mathcal{P}_5) \left\{ \begin{array}{l} \sigma(M) \in \Sigma_{ad}, \; J(\tau) \geq J(\sigma) \; \forall \tau \in \Sigma_{ad} \\ \text{with } J(\tau) = \frac{1}{2}B(\tau, \tau) - L(\tau) \end{array} \right\}$$

$-J(\vec{v})$ is called the complementary energy of the statically admissible field $\tau$.

Additionally, it can be demonstrated that if $(\sigma(M), \vec{u}(M))$ are solutions of $(\mathcal{P}_1)$;

$$\forall \vec{v} \in U_{ad}, \forall \tau \in \Sigma_{ad}$$

$$I(\vec{v}) \geq I(\vec{u}) = -J(\sigma) \geq -J(\tau)$$

## 6.3.4 Solving the problem to find the stress and strain fields around the inclusions using two scales: highlighting the concept of homogenisation and localisation

**Definitions of two scales: macroscopic and microscopic scales**

We assume now that a RVE has been found for the medium making up the considered structure $S$. As previously considered, the medium is made of several $(n)$ constituents $\mathcal{C}^{(I)}$ $(I = 1, \ldots, n)$

(phases and inclusions) that are themselves homogeneous. Moreover, it is assumed that the adhesion between phases and inclusions is perfect, which is to say that there is perfect bonding with no possibility of dissipative phenomena at the interfaces of the phases/inclusions. We assume also that the behaviour of each constituent $\mathcal{C}^{(I)}$ is linearly elastic and possibly anisotropic. We note $a^{(I)} = (a^{(I)}_{ijkh})_{i,j,k,h=1,2,3}$ the fourth-order tensor of rigidity of the constituent $\mathcal{C}^{(I)}$ which possesses the usual properties of symmetry and ellipticity. This RVE allows the definition of a mechanical behaviour law for the medium for which it has been defined. Without any other consideration except invoking that each constituent of the medium making up $S$ shows linear elastic behaviour and that the link between phases and inclusions is perfect (so, no dissipative phenomena can appear), we state that the behaviour of the RVE is linearly elastic (possibly anisotropic). The rigidity tensor is denoted as $a^H$ and the flexibility tensor is denoted as $A^H$.

It is therefore understood that now, the medium making up the considered structure $S$ is taken to be homogeneous and that its behaviour is characterised by $a^H$ and $A^H$. It is also understood that the phases and inclusions have been merged into a continuum. Obviously, that does not suit us because we should recall that we wish to evaluate the stress and strain fields around the inclusions, so as to predict accurately the failure of the structure $S$. To avoid this difficulty, we are going to solve two problems:

- one at a scale where we do not see phases and inclusions. This scale is denoted as the macroscopic scale;
- and afterwards, one at a the scale where we can see phases and inclusions in the RVE, defined as the microscopic scale.

The framework for the macroscopic scale which is $R = (O, b)$ for which the orthonormal basis is $b = (\vec{x}_1, \vec{x}_2, \vec{x}_3)$ and $O$ the origin. The vector of the coordinates of a point $M$ relative to this framework is given by $\vec{x} = (x_i)_{i=1,2,3}$. The relationship of a function $q$ with respect to the variables $(x_i)_{i=1,2,3}$, that means, with respect to the point $M$, is noted by $q(x)$ where $x = (x_1, x_2, x_3)$ or $q(M)$. The dependence of $q$ with respect to $x$ or $M$ can be omitted as all the values implicitly considered are related to it, except if this information brings necessary information and clarity to the problem. The fields of displacement, stress (supposed symmetrical) and strain (linearised) are respectively noted as $\vec{U} = (U_i)_{i=1,2,3}$, $\Sigma = (\Sigma_{ij})_{i,j=1,2,3}$ and $E = (E_{ij})_{i,j=1,2,3}$ where $E_{ij} = \frac{1}{2}(\frac{\partial U_i}{\partial x_j} + \frac{\partial U_j}{\partial x_i})$.

The microscopic scale is associated with a direct framework $R_y = (O_y, b_y)$, $b_y = (\vec{y}_1, \vec{y}_2, \vec{y}_3)$ designating its direct orthonormal basis and $O_y$ as its origin. The vector of the coordinates of a point $P$ relative to this framework is given by $\vec{y} = (y_i)_{i=1,2,3}$. The relationship of a function $q$ with respect to the variables $(y_i)_{i=1,2,3}$, which means, with respect to the point $P$, is noted by $q(y)$ where $y = (y_1, y_2, y_3)$ or $q(P)$. The dependence of $q$ with respect to $y$ or $P$ can be omitted as all the values implicitly considered are related to it, except if this information brings necessary information and clarity to the problem. The fields of displacement, stress (supposed symmetrical) and strain (linearised) are respectively noted as $\vec{u}^m = (u^m_i)_{i=1,2,3}$, $\sigma^m = (\sigma^m_{ij})_{i,j=1,2,3}$ and $\varepsilon^m = (\varepsilon^m_{ij})_{i,j=1,2,3}$ where $\varepsilon^m_{ij} = \frac{1}{2}(\frac{\partial u^m_i}{\partial y_j} + \frac{\partial u^m_j}{\partial y_i})$.

**The problem at the macroscopic scale. Access to the stress and strain fields acting on the RVE. Macroscopic stress and strain fields**

The problem $(\mathcal{P}^M)$ which has to be solved at the macroscopic scale (defined as the macroscopic problem), consists of finding, at each $M$ point of the domain the following unknowns:

- the displacement vector $\vec{U}(M)$;
- the stress tensor $\Sigma(M)$,

verifying:

- the local equation of equilibrium (including the hypotheses given):

$$\overrightarrow{div}\,\Sigma(M) + \rho(M)\vec{g} = \vec{0}\;\forall M \in \Omega$$

- the behaviour law (the medium is homogeneous and linearly elastic)
  - state law:

$$\left\{\begin{array}{l} \Sigma(M) = a^H E(\vec{U}(M)) \Longleftrightarrow \Sigma_{ij}(M) = a^H_{ijkh}E_{kh}(\vec{U}(M)) \\ \Longleftrightarrow \\ E(\vec{U}(M)) = A^H\Sigma(M) \Longleftrightarrow E_{ij}(\vec{u}(M)) = A^H_{ijkh}\Sigma_{kh}(M) \end{array}\right\}$$

  - no evolution law (no dissipative phenomenon)
- boundary conditions:

$$\left\{\begin{array}{l} \text{on } \partial\Omega_F \;:\; \Sigma(M \in \partial\Omega_F) \times \vec{n}(M \in \partial\Omega_F) = \vec{F}(M) \\[2mm] \text{on } \partial\Omega_U \;:\; \vec{U}(M \in \partial\Omega_U) = \vec{U}_d(M) \\[2mm] \text{on } \partial\Omega_{FU} \;:\; \left\{\begin{array}{l} \{\Sigma(M \in \partial\Omega_{FU}) \times \vec{n}(M \in \partial\Omega_{FU})\}_1 = F'_1(M) \\ \left\{\vec{U}(M \in \partial\Omega_{FU})\right\}_2 = U'_{d2}(M) \\ \{\Sigma(M \in \partial\Omega_{FU}) \times \vec{n}(M \in \partial\Omega_{FU})\}_3 = F'_3(M) \end{array}\right. \end{array}\right.$$

A way to solve this problem $a^H$ or $A^H$ has to be identified. How can we gain access to $a^H$ or $A^H$? One possibility is experimental. But, a non experimental possibility is available: the homogenisation process. This process is linked to the following defined in the next section.

**The problem at the microscopic scale. Access to the stress and strain fields around the inclusions: theoretically possible however with no unique solution available. Microscopic stress and strain fields**

The material system considered here is the RVE situated at point $M$, where it exists at the macroscopic scale $\vec{U}(M)$, $\Sigma(M)$ and $E(M) = E(\vec{U}(M))$ coming from $(\mathcal{P}^M)$. The domain defined by the RVE is the domain $D^m = \Omega^m \cup \partial\Omega^m$ where $\Omega^m$ designates its interior and $\partial\Omega^m$ its boundary. The outer unit vector in a point $P$ of $\partial\Omega^m$ is noted as $\vec{n}^m(P)$.

The problem $(\mathcal{P}^m)$ which has to be solved at the microscopic scale (defined as the microscopic problem), to access to the stress and strain tensor around the inclusion existing in the RVE submitted to the macroscopic fields at point $M$, consists of finding the following unknowns:

- the displacement vector $\vec{u}^m(P)$;
- the stress tensor $\sigma^m(P)$,

verifying:

- the local equation of equilibrium:

$$\overrightarrow{div}\,\sigma^m(P) = \vec{0}\;\forall P \in \Omega^m$$

- the behaviour law (the constituents are homogeneous and linearly elastic)
  - state law:

$$\left\{\begin{array}{l} \sigma^m(P) = a(P)\varepsilon^m(\vec{u}^m(P)) \Longleftrightarrow \sigma^m_{ij}(P) = a_{ijkh}(P)\varepsilon^m_{kh}(\vec{u}^m(P)) \\ \Longleftrightarrow \\ \varepsilon^m_{kh}(\vec{u}^m(P)) = A(P)\sigma^m(P) \Longleftrightarrow \varepsilon_{ij}(\vec{u}^m(P)) = A_{ijkh}(P)\sigma^m_{kh}(P) \end{array}\right\}$$

or, equivalently:

$$\left\{\begin{array}{l} \sigma^m(P \in \mathcal{C}^{(I)}) = a^{(I)}\varepsilon(\vec{u}^m(P \in \mathcal{C}^{(I)})) \Longleftrightarrow \sigma_{ij}^m(P \in \mathcal{C}^{(I)}) = a_{ijkh}^{(I)}\varepsilon_{kh}(\vec{u}^m(P \in \mathcal{C}^{(I)})) \\ \Longleftrightarrow \\ \varepsilon(\vec{u}^m(P \in \mathcal{C}^{(I)})) = A^{(I)}\sigma^m(P \in \mathcal{C}^{(I)}) \Longleftrightarrow \varepsilon_{ij}(\vec{u}^m(P \in \mathcal{C}^{(I)})) = A_{ijkh}^{(I)}\sigma_{kh}^m(P \in \mathcal{C}^{(I)}) \end{array}\right\}$$

      · no evolution law (no dissipative phenomenon)

- boundary conditions on $\partial\Omega^m$: giving boundary conditions so that $E(M) =< \varepsilon^m(\vec{u}^m) >$ or $\Sigma(M) =< \sigma^m >$,

with $< q >= \frac{1}{|D|} \int_D q(y)dy$ is the average of $q$ over $D$ with a size $|D|$.

Accepting for now that this problem can be solved, even if the boundary conditions are not clearly defined, it allows us to obtain, at the microscopic scale, the stress ($\sigma^m$) and strain ($\varepsilon^m$) tensors in each point $P$ of the RVE situated at the point $M$ of the macroscopic scale subjected to the macroscopic fields $E(M)$ or $\Sigma(M)$. In particular, we access the microscopic stress and strain field around the inclusions: this is the goal we had to predict to calculate the failure of $S$ with good precision. This process which allows the microscopic fields to be obtained, starting from the macroscopic level, is called the localisation step.

### 6.3.5    Definition of a multi-scale process

Both the homogenisation process and localisation process define the two steps of a multi-scale process. It will be shown that the homogenisation and localisation processes are closely linked. More precisely, we will see that the homogenisation process which needs the clear definition of the boundary conditions of ($\mathcal{P}^m$) induces the localisation process. This is the reason why, in the following of this chapter we mainly discuss the homogenisation process and therefore we present the associated localisation process.

The homogenisation process is not unique and can be classified into different families. In fact, from a mathematical point of view, ($\mathcal{P}^m$) is a problem badly posed in the sense that the existence and the uniqueness of its solutions cannot be demonstrated as for ($\mathcal{P}^M$).

Posing correctly the definition of particular boundary conditions gives answers to the queries of how to find $a^H$ or $A^H$ (the homogenisation step) and clearly defines the boundary conditions (the localisation step) of the microscopic problem ($\mathcal{P}^m$). This allows also families of homogenisation methods to be defined.

### 6.3.6    Objective of the chapter. Average technique homogenisation method: effective moduli type and periodic type

Our objective here is not to present all the existing homogenisation methods (and their associated localisation methods) but mainly to analyse some of them and to point out their particularities and their differences. In particular, we shall expose only the homogenisation methods for the following cases:

- boundary conditions of ($\mathcal{P}^m$) imply $E(M) =< \varepsilon^m(\vec{u}^m) >$ or $\Sigma(M) =< \sigma^m >$. This gives a family of homogenization methods called an average technique homogenization method. If the boundary conditions imply $E(M) =< \varepsilon^m(\vec{u}^m) >$, the approach is the so-called strain approach. If the boundary conditions imply $\Sigma(M) =< \sigma^m >$, the approach is the so-called stress approach;
- the behaviour of the constituents of the RVE is considered to be linearly elastic;
- the arrangement of constituents inside the RVE is or is not periodic.

The Effective Moduli Method is defined by the average technique homogenisation method without a periodicity condition. The Periodic Average Technique Method (also called here, Periodic Method) is defined by the average technique homogenisation method with periodicity conditions.

Strictly speaking it is clear that the periodic case is very rare because of the irregular distribution of the components in the composite due to the manufacturing process, or to local fluctuations which lead to a scatter in properties throughout the material (Figure 6.2). However, the periodic arrangement can be see as the configuration which is, on average, representative of all others. Moreover, this assumption allows theorems concerning the existence and the uniqueness of the solutions of $(\mathcal{P}^m)$ and strict equivalence between stress and strain approach to be obtained. This is not the case for the Effective Moduli Method. This is one of the important disadvantages of this method rather than that of Periodic Method.

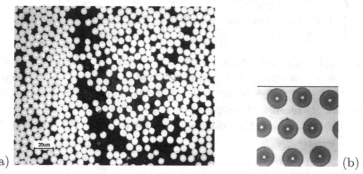

(a)                (b)

Figure 6.2: Unidirectional long fibres / matrix material and fibre periodicity hypothesis. The periodicity of the arrangement of fibres is often assumed. The reality can be far from this ideal configuration (a, unidirectional carbon fibres / epoxy resin material (Chou et al., 2015)) or very close to a periodic pattern (b, unidirectional silicon carbide fibres / titanium resin material (Thionnet and Renard, 1998)).

## 6.4   Effective moduli method

### 6.4.1   The homogenised behaviour defined as a relationship between values averaged on the RVE

In the framework of homogenisation using the averaging technique, the so-called homogenised behaviour law is the relation which exists between the average on the RVE of the microscopic fields of strain and stress solutions of $(\mathcal{P}^m)$. We recall that in this framework no geometric periodicity exists in the RVE of the considered media we want to homogenise.

### 6.4.2   Homogenisation step

**Strain approach: trying to find $a^H$**

Consider that the boundary conditions are $(\mathcal{P}^m)$ so that $E = < \varepsilon^m(\vec{u}^m) >$ is given. As the $(\mathcal{P}^m)$ problem is linear, its solutions are then linearly dependent of its data. We can then write:

- $\sigma_{ij}^m = s_{ij}^{kh-\varepsilon} E_{kh}$ ;
- $u_i^m = v_i^{kh-\varepsilon} E_{kh}$.

The behaviour law and second relation give:

$$\sigma_{ij}^m = a_{ijpq}\varepsilon_{pq}^m(\vec{u}^m) = a_{ijpq}\varepsilon_{pq}^m(\vec{v}^{kh-\varepsilon} E_{kh}) = a_{ijpq}\varepsilon_{pq}^m(\vec{v}^{kh-\varepsilon})E_{kh}$$

Consequently, with the first relation:

$$\sigma_{ij}^m = a_{ijpq}\varepsilon_{pq}^m(\vec{v}^{kh-\varepsilon})E_{kh} = s_{ij}^{kh-\varepsilon}E_{kh}$$

and then:

$$s_{ij}^{kh-\varepsilon} = a_{ijpq}\varepsilon_{pq}^m(\vec{v}^{kh-\varepsilon})$$

Finally:

$$< \sigma_{ij}^m > = < s_{ij}^{kh-\varepsilon}E_{kh} > = < s_{ij}^{kh-\varepsilon} > E_{kh} \Longrightarrow a_{ijkh}^{H-\varepsilon} = < s_{ij}^{kh-\varepsilon} >$$

We can then find the tensor $a^{H-\varepsilon}$ by solving the six following $(\mathcal{P}^{kh-\varepsilon})$ problems:

$$(\mathcal{P}^{kh-\varepsilon}) \begin{cases} \text{Find } (s^{kh-\varepsilon}, \vec{v}^{kh-\varepsilon})/ \\ \vec{div}\, s^{kh-\varepsilon} = \vec{0}, \forall P \in \Omega^m \\ s^{kh-\varepsilon} = a(y)\varepsilon^m(\vec{v}^{kh-\varepsilon}) \Longleftrightarrow \varepsilon^m(\vec{v}^{kh-\varepsilon}) = A(y)s^{kh-\varepsilon} \\ \text{Boundary conditions on } \partial\Omega^m\text{- Giving boundary conditions so that:} \\ < \varepsilon_{ij}^m(\vec{v}^{kh-\varepsilon}) > = E_{ij}^{kh} = \frac{1}{2}(\delta_{ik}\delta_{jh} + \delta_{ih}\delta_{jk}) \end{cases}$$

where the terms $(\delta_{ij})_{i,j=1,2,3}$ define the Kronecker's numbers and the six tensors $(E^{kh})_{i,j=1,2,3}$ define a basis of the vector space of symetrical second-order tensors. The second-order tensors $s^{kh-\varepsilon}$ and $\varepsilon^m(\vec{v}^{kh-\varepsilon})$ $(k, h = 1, 2, 3)$ are called the stress and strain localisation tensors or the stress and strain density tensors. We can also demonstrate that $a^{H-\varepsilon}$ obeys the usual symmetry and ellipticity properties. We can then obtain a flexibility tensor $A^{H-\varepsilon}$ starting from the rigidity tensor $a^{H-\varepsilon}$.

## Stress approach: trying to find $A^H$

Consider that the boundary conditions are $(\mathcal{P}^m)$ so that $\Sigma = < \sigma^m) >$ is obtained. As the $(\mathcal{P}^m)$ problem is linear, its solutions are then linearly dependent of its data. We can then write:

- $\sigma_{ij}^m = s_{ij}^{kh-\sigma}\Sigma_{kh}$ ;
- $u_i^m = v_i^{kh-\sigma}\Sigma_{kh}$.

The behaviour law, the first and second relations give:

$$\varepsilon_{ij}^m(\vec{u}^m) = \varepsilon_{ij}^m(\vec{v}^{kh-\sigma}\Sigma_{kh}) = \varepsilon_{ij}^m(\vec{v}^{kh-\sigma})\Sigma_{kh} = A_{ijpq}\sigma_{pq}^m = A_{ijpq}s_{pq}^{kh-\sigma}\Sigma_{kh}$$

and then:

$$\varepsilon_{ij}^m(\vec{v}^{kh-\sigma}) = A_{ijpq}s_{pq}^{kh-\sigma}$$

Finally:

$$< \varepsilon_{ij}^m(\vec{u}^m) > = < \varepsilon_{ij}^m(\vec{v}^{kh-\sigma}\Sigma_{kh}) > = < \varepsilon_{ij}^m(\vec{v}^{kh-\sigma})\Sigma_{kh} > = < \varepsilon_{ij}^m(\vec{v}^{kh-\sigma}) > \Sigma_{kh}$$

$$\Longrightarrow A_{ijkh}^{H-\sigma} = < \varepsilon_{ij}^m(\vec{v}^{kh-\sigma}) >$$

We can then find the tensor $A^{H-\sigma}$ by solving the six following $(\mathcal{P}^{kh-\sigma})$ problems:

$$(\mathcal{P}^{kh-\sigma}) \begin{cases} \text{Find } (s^{kh-\sigma}, \vec{v}^{kh-\sigma})/ \\ \vec{div}\, s^{kh-\sigma} = \vec{0}, \forall P \in \Omega^m \\ s^{kh-\sigma} = a(y)\varepsilon^m(\vec{v}^{kh-\sigma}) \Longleftrightarrow \varepsilon^m(\vec{v}^{kh-\sigma}) = A(y)s^{kh-\sigma} \\ \text{Boundary conditions on } \partial\Omega^m \text{ - Giving boundary conditions so that:} \\ < s^{kh-\sigma} > = \Sigma_{ij}^{kh} = \frac{1}{2}(\delta_{ik}\delta_{jh} + \delta_{ih}\delta_{jk}) \end{cases}$$

where the terms $(\delta_{ij})_{i,j=1,2,3}$ define the Kronecker's numbers and the six tensors $(\Sigma^{kh})_{i,j=1,2,3}$ define a basis of the vector space of symetrical second-order tensors. The second-order tensors $s^{kh-\sigma}$ and $\varepsilon^m(\vec{v}^{kh-\sigma})$ $(k, h = 1, 2, 3)$ are called the stress and strain localisation tensors or the stress and strain density tensors. We can also demonstrate that $A^{H-\sigma}$ obeys to the usual symmetry and ellipticity properties. We can then obtain a rigidity tensor $a^{H-\sigma}$ starting from the flexibility tensor $A^{H-\sigma}$.

**Did we get $a^H$ and $A^H$? Equivalence between strain and stress approach?**

Recall that the fourth-order tensors $a^H$ and $A^H$ define the macroscopic behaviour law, which demonstrates the relationship between the macroscopic stress tensor $\Sigma$ and the macroscopic strain tensor $E$ which is present in the $(\mathcal{P}^M)$ problem:

$$\Sigma = a^H E \Longleftrightarrow E = A^H \Sigma$$

with the usual symmetry and ellipticity properties. The strain approach tries to calculate $a^H$ and we get $a^{H-\varepsilon}$. The stress approach tries to calculate $A^H$ and we obtain $A^{H-\sigma}$. Then important questions have to be resolved.

The first ones concern $a^H = a^{H-\varepsilon}$? $A^H = A^{H-\varepsilon}$? $A^H = A^{H-\sigma}$? $a^H = a^{H-\sigma}$? The answers are: this is not correct and only approximations of $a^H$ and $A^H$ can be obtained. It should be understood that the difference between these approximations and $a^H$ and $A^H$ can be large, small or equal to zero depending on the boundary conditions of $(\mathcal{P}^{kh-\sigma})$ and $(\mathcal{P}^{kh-\varepsilon})$. However for particular ways of applying the boundary conditions, results can be positive.

The second ones are do $a^{H-\varepsilon} = a^{H-\sigma}$? $A^{H-\varepsilon} = A^{H-\sigma}$? The answers are: no. This means that the strain approach and the stress approach are not equivalent. However, as previously shown, it should be understood that the difference between these tensors can be large, small or equal to zero depending on the boundary conditions of $(\mathcal{P}^{kh-\sigma})$ and $(\mathcal{P}^{kh-\varepsilon})$. In this way it can be seen that for particular ways of applying the boundary conditions, the relations can be validated.

It will be shown later that in case where there exists a geometrical periodicity in the RVE, this implies that, in the framework of the Periodic Method, all the previous questions have a positive answer.

## 6.4.3 Localisation step

Coming back to the multi-scale process, the problem at the macroscopic scale $(\mathcal{P}^M)$ is assumed to be solved. That means that at each point $M$ of the macroscopic scale, the macroscopic stress and strain tensors $\Sigma(M)$ and $E(M)$ are known. Each of them can be decomposed according to the basis of the symmetrical tensors as the following way:

$$\Sigma(M) = (\Sigma_{kh}(M))_{k,h=1,2,3} = \Sigma_{kh}(M)\Sigma^{kh}$$

and

$$E(M) = (E_{kh}(M))_{k,h=1,2,3} = E_{kh}(M)E^{kh}$$

Assuming, the problems $(\mathcal{P}^{kh-\varepsilon})$ and/or $(\mathcal{P}^{kh-\sigma})$ have been solved, the density tensors $s^{kh-\varepsilon}$, $\varepsilon^m(\vec{v}^{kh-\varepsilon})$ and/or $s^{kh-\sigma}$, $\varepsilon^m(\vec{v}^{kh-\sigma})$ are known. Then, by linearity, we obtain the microscopic stress and strain tensors $\sigma^m(P)$, $\varepsilon^m(\vec{v}^m(P))$ at each point $P$ of the RVE situated at the point $M$ of the macroscopic scale:

$$\sigma^m(P) = \Sigma_{kh}(M)s^{kh-\varepsilon}(P) \text{ or } \Sigma_{kh}(M)s^{kh-\sigma}(P)$$

$$\varepsilon^m(\vec{v}^m(P)) = E_{kh}(M)\varepsilon^m(\vec{v}^{kh-\varepsilon}(P)) \text{ or } E_{kh}(M)\varepsilon^m(\vec{v}^{kh-\sigma}(P))$$

Then, it appears that, once the homogenisation step has been solved (by the stress or strain approach), it allows easy access, without any new calculations (except the sum for the linearity principle) at the localisation step using the localisation tensors.

### 6.4.4   Explicit boundary conditions for the $(\mathcal{P}^m)$ problem

Until now, the detailed and precise boundary conditions for the $(\mathcal{P}^m)$ problem (and also $(\mathcal{P}^{kh-\varepsilon})$ and $(\mathcal{P}^{kh-\sigma})$) have not been explicitly written. The reason is that the main results can be obtained without these conditions. However, when the homogenisation step has to be solved, these condition are necessary. Pointing out that there is an infinite number of ways to verify the average conditions (there is an infinite number of regular real functions that have the same average value in a given interval), it is evident that we cannot have an exhaustive presentation of them all. It should also be pointed out that as previously mentioned, the result of the homogenisation step can be very far from or very close to the real value of the homogenised behaviour, depending on the way these average values are obtained with the boundary conditions imposed. Then, the most natural ones are presented here.

First, assume we impose:

$$\vec{u}^m(P \in \partial\Omega^m) = E(M)\overrightarrow{O_y P} = E(M)\vec{y} \Longrightarrow u_i^m(P \in \partial\Omega^m) = E_{ik}(M)y_k$$

Let's verify that it allows to verify: $< \varepsilon_{ij}^m(\vec{u}^m(y)) >= E_{ij}(M)$. Then:

$$< \varepsilon_{ij}^m(\vec{u}^m(y)) > = \frac{1}{|\Omega^m|} \int_{\Omega^m} \varepsilon_{ij}^m(\vec{u}^m(y))dy = \frac{1}{|\Omega^m|} \int_{\Omega^m} \frac{1}{2}\left( \frac{\partial u_i^m(y)}{\partial y_j} + \frac{\partial u_j^m(y)}{\partial y_i} \right) dy$$

$$= \frac{1}{|\Omega^m|} \int_{\partial\Omega^m} \frac{1}{2}\left( u_i^m(y)n_j^m + u_j^m(y)n_i^m \right) dy$$

$$= \frac{1}{|\Omega^m|} \int_{\partial\Omega^m} \frac{1}{2}\left( E_{ik}(M)y_k n_j^m + E_{jk}(M)y_k n_i^m \right) dy$$

$$= \frac{E_{ik}(M)}{2} \frac{1}{|\Omega^m|} \int_{\partial\Omega^m} y_k n_j^m dy + \frac{E_{jk}(M)}{2} \frac{1}{|\Omega^m|} \int_{\partial\Omega^m} y_k n_i^m dy$$

$$= \frac{E_{ik}(M)}{2} \frac{1}{|\Omega^m|} \int_{\Omega^m} \frac{\partial y_k}{\partial y_j} dy + \frac{E_{jk}(M)}{2} \frac{1}{|\Omega^m|} \int_{\Omega^m} \frac{\partial y_k}{\partial y_i} dy$$

$$= \frac{E_{ik}(M)}{2} \frac{1}{|\Omega^m|} \int_{\Omega^m} \delta_{kj} dy + \frac{E_{jk}(M)}{2} \frac{1}{|\Omega^m|} \int_{\Omega^m} \delta_{ki} dy$$

$$= \frac{E_{ik}(M)}{2}\delta_{kj} + \frac{E_{jk}(M)}{2}\delta_{ki} = \frac{1}{2}\left( E_{ij}(M) + E_{ji}(M) \right)$$

$$= E_{ij}(M)$$

Secondly, assume that we impose:

$$\sigma^m(P \in \partial\Omega^m)\vec{n}^m(P \in \partial\Omega^m) = \Sigma(M)\vec{n}^m(P \in \partial\Omega^m)$$

$$\Longrightarrow \sigma_{ik}^m(P \in \partial\Omega^m)n_k^m(P \in \partial\Omega^m) = \Sigma_{ik}(M)n_k^m(P \in \partial\Omega^m)$$

Let's verify that it allows to verify: $< \sigma_{ij}^m(y)) >= \Sigma_{ij}(M)$. We have:

$$\frac{\partial \sigma_{ik}^m y_j}{\partial y_k} = \frac{\partial \sigma_{ik}^m}{\partial y_k} y_j + \sigma_{ik}^m \frac{\partial y_j}{\partial y_k} = \sigma_{ik}^m \frac{\partial y_j}{\partial y_k} = \sigma_{ik}^m \delta_{jk} = \sigma_{ij}^m$$

because the local equilibrium gives: $\frac{\partial \sigma^m_{ik}}{\partial y_k} y_j = 0$. Then:

$$< \sigma^m_{ij}(y) > = \frac{1}{|\Omega^m|} \int_{\Omega^m} \sigma^m_{ij}(y) dy = \frac{1}{|\Omega^m|} \int_{\Omega^m} \frac{\partial \sigma^m_{ik}(y) y_j}{\partial y_k} dy = \frac{1}{|\Omega^m|} \int_{\partial\Omega^m} \sigma^m_{ik}(y) y_j n^m_k dy$$

$$= \frac{1}{|\Omega^m|} \int_{\partial\Omega^m} \sigma^m_{ik}(y) n^m_k y_j dy = \frac{1}{|\Omega^m|} \int_{\partial\Omega^m} \Sigma_{ik}(M) n^m_k y_j dy$$

$$= \Sigma_{ik}(M) \frac{1}{|\Omega^m|} \int_{\partial\Omega^m} n^m_k y_j dy = \Sigma_{ik}(M) \frac{1}{|\Omega^m|} \int_{\Omega^m} \frac{\partial y_j}{\partial y_k} dy$$

$$= \Sigma_{ik}(M) \frac{1}{|\Omega^m|} \int_{\Omega^m} \delta_{jk} dy = \Sigma_{ik}(M) \delta_{jk}$$

$$= \Sigma_{ij}(M)$$

The $(\mathcal{P}^m)$ problem for which the boundary conditions are those giving the macroscopic strain is denoted as $(\mathcal{P}^{m-\varepsilon})$. The $(\mathcal{P}^m)$ problem for which the boundary conditions are those giving the macroscopic stress is denoted as $(\mathcal{P}^{m-\sigma})$. At the following section, we try to solve both problems using the usual theorems of energy.

## Solving $(\mathcal{P}^{m-\varepsilon})$. Voigt upper bounds of the rigidity tensor

The $(\mathcal{P}^m)$ problem for which the boundary conditions are those giving the macroscopic strain is denoted as $(\mathcal{P}^{m-\varepsilon})$:

$$(\mathcal{P}^{m-\varepsilon}) \begin{cases} \text{Find } (\sigma^{m-\varepsilon}, \vec{u}^{m-\varepsilon})/ \\ \vec{div}\, \sigma^{m-\varepsilon} = \vec{0} \text{ in } \Omega^m \\ \sigma^{m-\varepsilon} = a(y)\varepsilon^m(\vec{u}^{m-\varepsilon}) \Longleftrightarrow \varepsilon^m(\vec{u}^{m-\varepsilon}) = A(y)\sigma^{m-\varepsilon} \\ \text{Boundary conditions on } \partial\Omega^m \text{ - } \vec{u}^{m-\varepsilon} = E\vec{y} \Longrightarrow u^{m-\varepsilon}_i = E_{ik} y_k \end{cases}$$

It should be pointed out that:

- $< \varepsilon^m(\vec{u}^{m-\varepsilon}) >= E$;
- $< \sigma^{m-\varepsilon} >= \Sigma^\varepsilon \neq \Sigma$ in general;
- $\Sigma^\varepsilon = a^{H-\varepsilon} E$, that says, $a^{H-\varepsilon} \neq a^H$ in general;
- with this approach, $E$ is given, $\Sigma^\varepsilon$ is calculated and $a^{H-\varepsilon}$ is obtained;
- we note $E = A^{H-\varepsilon} \Sigma^\varepsilon$, that says, $A^{H-\varepsilon} \neq A^H$ in general.

The set of the kinematically admissible fields of $(\mathcal{P}^{m-\varepsilon})$ is:

$$U^{m-\varepsilon}_{ad} = \begin{Bmatrix} \vec{v} = (v_i)_{i=1,2,3} / \\ * \ v_i \text{ is a regular function} \\ * \text{ on } \partial\Omega^m, \ v_i = E_{ik} y_k \end{Bmatrix}$$

The set of the statically admissible fields of $(\mathcal{P}^{m-\varepsilon})$ is:

$$\Sigma^{m-\varepsilon}_{ad} = \begin{Bmatrix} \tau = (\tau_{ij})_{i,j=1,2,3} / \\ * \ \tau_{ij} \text{ is a regular function} \\ * \ \tau_{ij} = \tau_{ji} \\ * \ \vec{div}\tau = \vec{0} \text{ in } \Omega^m \end{Bmatrix}$$

It can be show that:

$$(\mathcal{P}^{m-\varepsilon}) \Longleftrightarrow \begin{Bmatrix} \vec{u}^{m-\varepsilon} \in U^{m-\varepsilon}_{ad}, \ I(\vec{v}) \geq I(\vec{u}^{m-\varepsilon}) \ \forall \vec{v} \in U^{m-\varepsilon}_{ad} \\ \text{with } I(\vec{v}) = \frac{1}{2} \int_\Omega a_{ijkh}(y)\varepsilon^m_{ij}(\vec{v})\varepsilon^m_{kh}(\vec{v}) \\ = \frac{1}{2}|\Omega^m| < \varepsilon^m(\vec{v})a(y)\varepsilon^m(\vec{v}) > \end{Bmatrix}$$

where $I(\vec{v})$ is the potential energy of the kinematically admissible field $\vec{v}$. $I(\vec{v})$ can be calculated for $\vec{u}^{m-\varepsilon}$ (using the properties that has been demonstrated in the local appendix section):

$$I(\vec{u}^{m-\varepsilon}) = \frac{1}{2}|\Omega^m| < \varepsilon^m(\vec{u}^{m-\varepsilon})a(y)\varepsilon^m(\vec{u}^{m-\varepsilon}) >$$

$$= \frac{1}{2}|\Omega^m| < \varepsilon^m(\vec{u}^{m-\varepsilon})\sigma^{m-\varepsilon} >= \frac{1}{2}|\Omega^m|E < \sigma^{m-\varepsilon} >= \frac{1}{2}|\Omega^m|Ea^{H-\varepsilon}E$$

It can also be shown that:

$$(\mathcal{P}^{m-\varepsilon}) \Longleftrightarrow \left\{ \begin{array}{l} \sigma^{m-\varepsilon} \in \Sigma_{ad}^{m-\varepsilon}, \; J(\tau) \geq J(\sigma^{m-\varepsilon}) \; \forall \tau \in \Sigma_{ad}^{m-\varepsilon} \\ \text{with } J(\tau) = \frac{1}{2}\int_{\Omega^m} A_{ijkh}(y)\tau_{ij}\tau_{kh} - \int_{\partial\Omega^m} \tau_{ij}n_j^m E_{ik}y_k \\ = \frac{1}{2}|\Omega^m| < \tau A(y)\tau > -|\Omega^m|E < \tau > \end{array} \right\}$$

because:

$$\int_{\partial\Omega^m} \tau_{ij}n_j^m E_{ik}y_k = E_{ik}\int_{\partial\Omega^m} \tau_{ij}n_j^m y_k = E_{ik}\int_{\Omega^m} \frac{\partial\tau_{ij}y_k}{\partial y_j}dy$$

$$= E_{ik}\int_{\Omega^m} \left( \frac{\partial\tau_{ij}}{\partial y_j}y_k + \tau_{ij}\frac{\partial y_k}{\partial y_j} \right) dy$$

$$= E_{ik}\int_{\Omega^m} \tau_{ij}\frac{\partial y_k}{\partial y_j}dy \; (\frac{\partial\tau_{ij}}{\partial y_j}y_k = 0)$$

$$= E_{ik}\int_{\Omega^m} \tau_{ij}\delta_{jk}dy = E_{ik}\int_{\Omega^m} \tau_{ik}dy = E_{ik}|\Omega^m| < \tau_{ik} >$$

$$= |\Omega^m|E < \tau >$$

where $-J(\tau)$ is the complementary energy of the statically admissible field $\tau$. $J(\tau)$ can be calculated for $\sigma^{m-\varepsilon}$ (using the properties that has been demonstrated in the local appendix section):

$$J(\sigma^{m-\varepsilon}) = \frac{1}{2}|\Omega^m| < \sigma^{m-\varepsilon}A(y)\sigma^{m-\varepsilon} > -|\Omega^m|E < \sigma^{m-\varepsilon} >$$

$$= \frac{1}{2}|\Omega^m| < \sigma^{m-\varepsilon}\varepsilon^m(\vec{u}^{m-\varepsilon}) > -|\Omega^m|E < \sigma^{m-\varepsilon} >$$

$$= \frac{1}{2}|\Omega^m| < \sigma^{m-\varepsilon} > E - |\Omega^m|E < \sigma^{m-\varepsilon} >$$

$$= -\frac{1}{2}|\Omega^m| < \sigma^{m-\varepsilon} > E = -\frac{1}{2}|\Omega^m|Ea^{H-\varepsilon}E$$

Recall that:

$$\forall \vec{v} \in U_{ad}^{m-\varepsilon}, \forall \tau \in \Sigma_{ad}^{m-\varepsilon}$$

$$I(\vec{v}) \geq I(\vec{u}^{m-\varepsilon}) = -J(\sigma^{m-\varepsilon}) \geq -J(\tau)$$

We verify, with the previous results, that the equality is trivial and verified. Now, taking $\vec{v} \in U_{ad}^{m-\varepsilon}$ equals to $v_i = E_{ij}y_j$:

$$I(\vec{v}) = \frac{1}{2}|\Omega^m| < \varepsilon^m(\vec{v})a(y)\varepsilon^m(\vec{v}) >= \frac{1}{2}|\Omega^m|E < a(y) > E$$

And then:

$$I(\vec{v}) \geq I(\vec{u}^{m-\varepsilon}) \Longrightarrow E \left( < a(y) > -a^{H-\varepsilon} \right) E \geq 0$$

This gives the Voigt upper bounds of the rigidity tensor.

## Solving $(\mathcal{P}^{m-\sigma})$. Reuss lower bounds of the flexibility tensor

The $(\mathcal{P}^m)$ problem for which the boundary conditions are those giving the macroscopic stress is denoted as $(\mathcal{P}^{m-\sigma})$:

$$(\mathcal{P}^{m-\sigma}) \left\{ \begin{array}{l} \text{Find } (\sigma^{m-\sigma}, \vec{u}^{m-\sigma})/ \\ \vec{div}\,\sigma^{m-\sigma} = \vec{0} \text{ in } \Omega^m \\ \sigma^{m-\sigma} = a(y)\varepsilon^m(\vec{u}^{m-\sigma}) \Longleftrightarrow \varepsilon^m(\vec{u}^{m-\sigma}) = A(y)\sigma^{m-\sigma} \\ \text{Boundary conditions on } \partial\Omega^m \text{ - } \sigma^{m-\sigma}\vec{n}^m = \Sigma\vec{n}^m \Longrightarrow \sigma_{ik}^{m-\sigma} n_k^m = \Sigma_{ik} n_k^m \end{array} \right.$$

It should be pointed out that:

- $<\sigma^{m-\sigma}> = \Sigma$;
- $<\varepsilon^m(\vec{u}^{m-\sigma})> = E^{m-\sigma} \neq E$ in general;
- $E^{m-\sigma} = A^{H-\varepsilon}\Sigma$ that says, $A^{H-\sigma} \neq A^H$ in general;
- with this approach, $\Sigma$ is given, $E^{m-\sigma}$ is calculated and $A^{H-\sigma}$ is obtained;
- we note $\Sigma = a^{H-\sigma}E^\varepsilon$, that says, $a^{H-\sigma} \neq A^H$ in general.

The set of the kinematically admissible fields of $(\mathcal{P}^{m-\sigma})$ is:

$$U_{ad}^{m-\sigma} = \left\{ \begin{array}{l} \vec{v} = (v_i)_{i=1,2,3} / \\ * \; v_i \text{ is a regular function} \end{array} \right\}$$

The set of the statically admissible fields of $(\mathcal{P}^{m-\sigma})$ is:

$$\Sigma_{ad}^{m-\sigma} = \left\{ \begin{array}{l} \tau = (\tau_{ij})_{i,j=1,2,3} / \\ * \; \tau_{ij} \text{ is a regular function} \\ * \; \tau_{ij} = \tau_{ji} \\ * \; \vec{div}\tau = \vec{0} \text{ in } \Omega^m \\ * \text{ on } \partial\Omega^m, \tau_{ik}^m n_k^m = \Sigma_{ik} n_k^m \end{array} \right\}$$

It can be show that:

$$(\mathcal{P}^{m-\sigma}) \Longleftrightarrow \left\{ \begin{array}{l} \vec{u}^{m-\sigma} \in U_{ad}^{m-\sigma}, I(\vec{v}) \geq I(\vec{u}^{m-\sigma}) \; \forall \vec{v} \in U_{ad}^{m-\sigma} \\ \text{with } I(\vec{v}) = \frac{1}{2} \int_\Omega a_{ijkh}(y)\varepsilon_{ij}^m(\vec{v})\varepsilon_{kh}^m(\vec{v}) - \int_{\partial\Omega^m} \Sigma_{ik} n_k^m v_i \\ = \frac{1}{2}|\Omega^m| < \varepsilon^m(\vec{v})a(y)\varepsilon^m(\vec{v}) > -|\Omega^m|\Sigma < \varepsilon^m(\vec{v}) > \end{array} \right\}$$

because:

$$\int_{\partial\Omega^m} \Sigma_{ik} n_k^m v_i = \Sigma_{ik} \int_{\partial\Omega^m} n_k^m v_i$$

$$= \Sigma_{ik} \int_{\Omega^m} \frac{\partial v_i}{\partial y_k}$$

$$= \frac{\Sigma_{ik}}{2} \int_{\Omega^m} \frac{\partial v_i}{\partial y_k} + \frac{\Sigma_{ki}}{2} \int_{\Omega^m} \frac{\partial v_k}{\partial y_i}$$

$$= \frac{\Sigma_{ik}}{2} \int_{\Omega^m} \frac{\partial v_i}{\partial y_k} + \frac{\Sigma_{ik}}{2} \int_{\Omega^m} \frac{\partial v_k}{\partial y_i}$$

$$= \Sigma_{ik} \int_{\Omega^m} \frac{1}{2}\left( \frac{\partial v_i}{\partial y_k} + \frac{\partial v_k}{\partial y_i} \right)$$

$$= \Sigma_{ik}|\Omega^m| < \varepsilon_{ik}^m(\vec{v}) > = |\Omega^m|\Sigma < \varepsilon^m(\vec{v}) >$$

where $I(\vec{v})$ is called the potential energy of the kinematically admissible field $\vec{v}$. $I(\vec{v})$ can be calculated for $\vec{u}^{m-\sigma}$ (using the properties that has been demonstrated in the local appendix):

$$
\begin{aligned}
I(\vec{u}^{m-\sigma}) &= \frac{1}{2}|\Omega^m| < \varepsilon^m(\vec{u}^{m-\sigma})a(y)\varepsilon^m(\vec{u}^{m-\sigma}) > -|\Omega^m|\Sigma < \varepsilon^m(\vec{u}^{m-\sigma}) > \\
&= \frac{1}{2}|\Omega^m| < \varepsilon^m(\vec{u}^{m-\sigma})\sigma^{m-\sigma} > -|\Omega^m|\Sigma < \varepsilon^m(\vec{u}^{m-\sigma}) > \\
&= \frac{1}{2}|\Omega^m| < \varepsilon^m(\vec{u}^{m-\sigma}) > \Sigma - |\Omega^m|\Sigma < \varepsilon^m(\vec{u}^{m-\sigma}) > \\
&= -\frac{1}{2}|\Omega^m| < \varepsilon^m(\vec{u}^{m-\sigma}) > \Sigma = -\frac{1}{2}|\Omega^m|\Sigma A^{H-\sigma}\Sigma
\end{aligned}
$$

It can be shown that:

$$
(\mathcal{P}^{m-\sigma}) \Longleftrightarrow \left\{
\begin{array}{l}
\sigma^{m-\sigma} \in \Sigma_{ad}^{m-\sigma}, \ J(\tau) \geq J(\sigma^{m-\sigma}) \ \forall \tau \in \Sigma_{ad}^{m-\sigma} \\
\text{with } J(\tau) = \frac{1}{2}\int_{\Omega^m} A_{ijkh}(y)\tau_{ij}\tau_{kh} \\
\qquad\quad = \frac{1}{2}|\Omega^m| < \tau A(y)\tau >
\end{array}
\right\}
$$

where $-J(\tau)$ is called the complementary energy of the statically admissible field $\tau$. $J(\tau)$ can be calculated for $\sigma^{m-\sigma}$ (using the properties that have been demonstrated in the local appendix):

$$
\begin{aligned}
J(\sigma^{m-\sigma}) &= \frac{1}{2}|\Omega^m| < \sigma^{m-\sigma} A(y)\sigma^{m-\sigma} > \\
&= \frac{1}{2}|\Omega^m| < \sigma^{m-\sigma}\varepsilon^m(\vec{u}^{m-\sigma}) > \\
&= \frac{1}{2}|\Omega^m|\Sigma < \varepsilon^m(\vec{u}^{m-\sigma}) >= \frac{1}{2}|\Omega^m|\Sigma A^{H-\sigma}\Sigma
\end{aligned}
$$

Recall that:

$$
\forall \vec{v} \in U_{ad}^{m-\sigma}, \forall \tau \in \Sigma_{ad}^{m-\sigma}
$$
$$
I(\vec{v}) \geq I(\vec{u}^{m-\sigma}) = -J(\sigma^{m-\sigma}) \geq -J(\tau)
$$

We verify, with the previous results, that the equality is trivially verified. Now, taking $\tau \in \Sigma_{ad}^{m-\sigma}$ equals to $\tau_{ij} = \Sigma_{ij}$:

$$
J(\tau) = \frac{1}{2}|\Omega^m| < \tau A(y)\tau >= \frac{1}{2}|\Omega^m|\Sigma < A(y) > \Sigma
$$

And then:

$$
-J(\sigma^{m-\sigma}) \geq -J(\tau) \Longrightarrow \Sigma \left( < A(y) > - A^{H-\sigma} \right) \Sigma \geq 0
$$

This gives the Reuss lower bounds of the flexibility tensor.

### 6.4.5   Local appendix: Hill-Mandel Lemma and other considerations

Recall the different admissible sets defined for the $(\mathcal{P}^{m-\varepsilon})$ and $(\mathcal{P}^{m-\sigma})$ problems:

$$
U_{ad}^{m-\varepsilon} = \left\{
\begin{array}{l}
\vec{v} = (v_i)_{i=1,2,3} \ / \\
* \ v_i \text{ is a regular function} \\
* \text{ on } \partial\Omega^m, \ v_i = E_{ik}y_k
\end{array}
\right\}
\quad
\Sigma_{ad}^{m-\varepsilon} = \left\{
\begin{array}{l}
\tau = (\tau_{ij})_{i,j=1,2,3} \ / \\
* \ \tau_{ij} \text{ is a regular function} \\
* \ \tau_{ij} = \tau_{ji} \\
* \ \vec{div}\tau = \vec{0} \text{ in } \Omega^m
\end{array}
\right\}
$$

$$
U_{ad}^{m-\sigma} = \left\{
\begin{array}{l}
\vec{v} = (v_i)_{i=1,2,3} \ / \\
* \ v_i \text{ is a regular function}
\end{array}
\right\}
\quad
\Sigma_{ad}^{m-\sigma} = \left\{
\begin{array}{l}
\tau = (\tau_{ij})_{i,j=1,2,3} \ / \\
* \ \tau_{ij} \text{ is a regular function} \\
* \ \tau_{ij} = \tau_{ji} \\
* \ \vec{div}\tau = \vec{0} \text{ in } \Omega^m \\
* \text{ on } \partial\Omega^m, \ \tau_{ik}^m n_k^m = \Sigma_{ik}n_k^m
\end{array}
\right\}
$$

- Property - $\forall \tau \in \Sigma_{ad}^{m-\varepsilon}$ or $\forall \tau \in \Sigma_{ad}^{m-\sigma}$, $\frac{\partial \tau_{ik} y_j}{\partial y_k} = \tau_{ij}$.

Proof - Reminding that $\frac{\partial \tau_{ik}^m}{\partial y_k} = 0$:

$$\frac{\partial \tau_{ik} y_j}{\partial y_k} = \frac{\partial \tau_{ik}}{\partial y_k} y_j + \tau_{ik} \frac{\partial y_j}{\partial y_k} = \tau_{ik} \frac{\partial y_j}{\partial y_k} = \tau_{ik} \delta_{jk} = \tau_{ij} \ \square$$

- Property - $\forall \vec{v} \in U_{ad}^{m-\varepsilon}$ and $\forall \tau \in \Sigma_{ad}^{m-\varepsilon}$ or $\forall \vec{v} \in U_{ad}^{m-\sigma}$ and $\forall \tau \in \Sigma_{ad}^{m-\sigma}$

$$\tau_{ij} \varepsilon_{ij}^m(\vec{v}) = \tau_{ij} \frac{\partial v_i}{\partial y_j}$$

Proof - $\tau_{ij} \frac{\partial v_i}{\partial y_j} = \tau_{ji} \frac{\partial v_j}{\partial y_i} = \tau_{ij} \frac{\partial v_j}{\partial y_i} = \tau_{ij} \frac{1}{2} \left( \frac{\partial v_i}{\partial y_j} + \frac{\partial v_j}{\partial y_i} \right) = \tau_{ij} \varepsilon_{ij}^m(\vec{v}) \ \square$

- Property - $\forall \vec{v} \in U_{ad}^{m-\varepsilon}$ and $\forall \tau \in \Sigma_{ad}^{m-\varepsilon}$ or $\forall \vec{v} \in U_{ad}^{m-\sigma}$ and $\forall \tau \in \Sigma_{ad}^{m-\sigma}$

$$\tau_{ik} \frac{\partial v_j}{\partial y_k} = \frac{\partial \tau_{ik} v_j}{\partial y_k}$$

Proof - Reminding that $\frac{\partial \tau_{ik}}{\partial y_k} = 0$:

$$\frac{\partial \tau_{ij} v_i}{\partial y_j} = \frac{\partial \tau_{ij}}{\partial y_j} v_i + \tau_{ij} \frac{\partial v_i}{\partial y_j} = \tau_{ij} \frac{\partial v_i}{\partial y_j} \ \square$$

- Property - $\forall \vec{v} \in U_{ad}^{m-\varepsilon}$ and $\forall \tau \in \Sigma_{ad}^{m-\varepsilon}$ or $\forall \vec{v} \in U_{ad}^{m-\sigma}$ and $\forall \tau \in \Sigma_{ad}^{m-\sigma}$

$$< \tau_{ij} \varepsilon_{ij}^m(\vec{v}) > = \frac{1}{|\Omega^m|} \int_{\partial \Omega^m} \tau_{ij} n_j^m v_i dS$$

Proof -

$$< \tau_{ij} \varepsilon_{ij}^m(\vec{v}) > = \frac{1}{|\Omega^m|} \int_{\Omega^m} \tau_{ij} \varepsilon_{ij}^m(\vec{v}) dy = \frac{1}{|\Omega^m|} \int_{\Omega^m} \tau_{ij} \frac{\partial v_i}{\partial y_j} dy = \frac{1}{|\Omega^m|} \int_{\Omega^m} \frac{\partial \tau_{ij} v_i}{\partial y_j} dy$$

$$= \frac{1}{|\Omega^m|} \int_{\partial \Omega^m} \tau_{ij} v_i n_j^m dS = \frac{1}{|\Omega^m|} \int_{\partial \Omega^m} \tau_{ij} n_j^m v_i dS \ \square$$

- Property - $\forall \vec{v} \in U_{ad}^{m-\varepsilon}$, $< \varepsilon_{ij}^m(\vec{v}) > = E_{ij}$

Proof -

$$< \varepsilon_{ij}^m(\vec{v}) > = \frac{1}{|\Omega^m|} \int_{\Omega^m} \varepsilon_{ij}^m(\vec{v}) dy = \frac{1}{|\Omega^m|} \int_{\Omega^m} \frac{1}{2} \left( \frac{\partial v_i}{\partial y_j} + \frac{\partial v_j}{\partial y_i} \right) dy$$

$$= \frac{1}{|\Omega^m|} \int_{\partial \Omega^m} \frac{1}{2} \left( v_i n_j^m + v_j^m n_i^m \right) dS$$

$$= \frac{1}{|\Omega^m|} \int_{\partial \Omega^m} \frac{1}{2} \left( E_{ik} y_k n_j^m + E_{jk} y_k n_i^m \right) dS$$

$$= \frac{E_{ik}}{2} \frac{1}{|\Omega^m|} \int_{\partial \Omega^m} y_k n_j^m dS + \frac{E_{jk}}{2} \frac{1}{|\Omega^m|} \int_{\partial \Omega^m} y_k n_i^m dS$$

$$= \frac{E_{ik}}{2} \frac{1}{|\Omega^m|} \int_{\Omega^m} \frac{\partial y_k}{\partial y_j} dy + \frac{E_{jk}}{2} \frac{1}{|\Omega^m|} \int_{\Omega^m} \frac{\partial y_k}{\partial y_i} dy$$

$$= \frac{E_{ik}}{2} \frac{1}{|\Omega^m|} \int_{\Omega^m} \delta_{kj} dy + \frac{E_{jk}}{2} \frac{1}{|\Omega^m|} \int_{\Omega^m} \delta_{ki} dy$$

$$= \frac{E_{ik}}{2} \delta_{kj} + \frac{E_{jk}}{2} \delta_{ki} = \frac{1}{2} \left( E_{ij} + E_{ji} \right)$$

$$= E_{ij} \ \square$$

- Property - $\forall \tau \in \Sigma_{ad}^{m-\sigma}, \; < \tau_{ij} >= \Sigma_{ij}$

Proof -

$$
\begin{aligned}
< \tau_{ij}) > &= \frac{1}{|\Omega^m|} \int_{\Omega^m} \tau_{ij} dy = \frac{1}{|\Omega^m|} \int_{\Omega^m} \frac{\partial \tau_{ik} y_j}{\partial y_k} dy = \frac{1}{|\Omega^m|} \int_{\partial \Omega^m} \tau_{ik} y_j n_k^m dS \\
&= \frac{1}{|\Omega^m|} \int_{\partial \Omega^m} \sigma_{ik} n_k^m y_j dS = \frac{1}{|\Omega^m|} \int_{\partial \Omega^m} \Sigma_{ik} n_k^m y_j dS \\
&= \Sigma_{ik} \frac{1}{|\Omega^m|} \int_{\partial \Omega^m} n_k^m y_j dS = \Sigma_{ik} \frac{1}{|\Omega^m|} \int_{\Omega^m} \frac{\partial y_j}{\partial y_k} dy \\
&= \Sigma_{ik} \frac{1}{|\Omega^m|} \int_{\Omega^m} \delta_{jk} dy = \Sigma_{ik} \delta_{jk} \\
&= \Sigma_{ij} \;\square
\end{aligned}
$$

- Strain-Hill-Mandel Lemma –

$$
\forall \vec{v} \in U_{ad}^{m-\varepsilon} \; \forall \tau \in \Sigma_{ad}^{m-\varepsilon} \quad < \tau_{ij} \varepsilon_{ij}^m(\vec{v}) >= E_{ij} < \tau_{ij} >
$$

Proof –

$$
\begin{aligned}
< \tau_{ij} \varepsilon_{ij}^m(\vec{v}) > &= \frac{1}{|\Omega^m|} \int_{\partial \Omega^m} \tau_{ij} n_j^m v_i dS = \frac{1}{|\Omega^m|} \int_{\partial \Omega^m} \tau_{ij} n_j^m E_{ik} y_k dS \\
&= E_{ik} \frac{1}{|\Omega^m|} \int_{\partial \Omega^m} \tau_{ij} n_j^m y_k dS = E_{ik} \frac{1}{|\Omega^m|} \int_{\Omega^m} \frac{\partial \tau_{ij} y_k}{\partial y_j} dy = E_{ik} \frac{1}{|\Omega^m|} \int_{\Omega^m} \tau_{ik} dy \\
&= E_{ik} < \tau_{ik} > \;\square
\end{aligned}
$$

- Stress-Hill-Mandel Lemma –

$$
\forall \vec{v} \in U_{ad}^{m-\sigma} \; \forall \tau \in \Sigma_{ad}^{m-\sigma} \quad < \tau_{ij} \varepsilon_{ij}^m(\vec{v}) >= \Sigma_{ij} < \varepsilon_{ij}^m(\vec{v}) >
$$

Proof –

$$
\begin{aligned}
< \tau_{ij} \varepsilon_{ij}^m(\vec{v}) > &= \frac{1}{|\Omega^m|} \int_{\partial \Omega^m} \tau_{ij} n_j^m v_i dS = \frac{1}{|\Omega^m|} \int_{\partial \Omega^m} \Sigma_{ij} n_j^m v_i dS \\
&= \Sigma_{ij} \frac{1}{|\Omega^m|} \int_{\partial \Omega^m} n_j^m v_i dS = \Sigma_{ij} \frac{1}{|\Omega^m|} \int_{\Omega^m} \frac{\partial v_i}{\partial y_j} dy = \Sigma_{ij} \frac{1}{|\Omega^m|} \int_{\Omega^m} \varepsilon_{ij}^m(\vec{v}) dy \\
&= \Sigma_{ij} < \varepsilon_{ij}^m(\vec{v}) > \;\square
\end{aligned}
$$

## 6.5 Periodic method

Concerning the concepts of periodic homogenisation we implicitly refer to the works of Suquet (1982), Léné (1984), Dumontet (1990) and Sanchez-Hubert and Sanchez-Palencia (1992). It is possible to find in these references all the demonstrations and results discussed here.

### 6.5.1 Reminder of the general hypotheses and framework. Additional periodicity hypothesis

We assume that a RVE has been found for the medium making up the considered structure $S$. But now, moreover, we assume that this RVE has a periodic geometry which is small compared to the overall dimensions of $S$. What does that mean and why are these hypotheses necessary? The periodicity indicates that the RVE of the considered material is made from a repeated pattern called the (periodic elementary) cell $Y$. It should be noted that the geometry of $Y$ is not unique in that it offers several choices for the pattern used to reconstruct the RVE by periodicity. Nevertheless the results obtained by the Periodic Method are independent of the cell chosen. Even if we do not explicitly need it we shall give to the $Y$ cell the geometry of a parallelepiped volume. We can then ask ourselves if this particular choice is restrictive for a general solution. The response is clearly that it is not: it is only necessary so as to evoke the fact that all elementary cells are the transformations of parallelepipeds through a regular transformation. Concerning the hypothesis of the small size of the cell, it is necessary to justify the calculation of the exact effective characteristics of the equivalent material which is achieved by extrapolating the size of the cell to zero.

We assume that the cell is made with several $(n)$ constituents $\mathcal{C}^{(I)}$ $(I = 1, \ldots, n)$ (phases and inclusions, but no voids) that are themselves homogeneous. Moreover, we assume that the adhesion between phases and inclusions is perfect, which is to say that there is perfect bonding with no possibility of dissipative phenomena at the interfaces of the phases/inclusions. At least, we assume also that the behaviour of each constituent $\mathcal{C}^{(I)}$ is linearly elastic and possibly anisotropic. We note $a^{(I)} = (a_{ijkh}^{(I)})_{i,j,k,h=1,2,3}$ the fourth-order tensor of rigidity of the constituent $\mathcal{C}^{(I)}$ which possesses the usual properties of symmetry and ellipticity.

We define a scale of work associated with the $Y$ cell which represents the microstructure of the material. This scale is the microscopic scale. The $Y$ cell is constituted from its interior noted $\Omega^Y$ and from its boundary $\partial \Omega^Y$.

The framework for the macroscopic scale which is $R = (O, b)$ for which the orthonormal basis is $b = (\vec{x}_1, \vec{x}_2, \vec{x}_3)$ and $O$ the origin. The vector of the coordinates of a point $M$ relative to this framework is given by $\vec{x} = (x_i)_{i=1,2,3}$. The relationship of a function $q$ with respect to the variables $(x_i)_{i=1,2,3}$, that means, with respect to the point $M$, is noted by $q(x)$ where $x = (x_1, x_2, x_3)$ or $q(M)$. The dependence of $q$ with respect to $x$ or $M$ can be omitted as all the values implicitly considered are related to it, except if this information brings necessary information and clarity to the problem. The fields of displacement, stress (supposed symmetrical) and strain (linearised) are respectively noted as $\vec{U} = (U_i)_{i=1,2,3}$, $\Sigma = (\Sigma_{ij})_{i,j=1,2,3}$ and $E = (E_{ij})_{i,j=1,2,3}$ where $E_{ij} = \frac{1}{2}(\frac{\partial U_i}{\partial x_j} + \frac{\partial U_j}{\partial x_i})$.

The microscopic scale is associated with a direct framework $R^{(Y)} = (O^{(Y)}, b^{(Y)})$, $b^{(Y)} = (\vec{y}_1, \vec{y}_2, \vec{y}_3)$ designating its direct orthonormal basis and $O^{(Y)}$ as its origin. The vector of the coordinates of a point $P$ relative to this framework is given by $\vec{y} = (y_i)_{i=1,2,3}$. The relationship of a function $q$ with respect to the variables $(y_i)_{i=1,2,3}$, which means, with respect to the point $P$, is noted by $q(y)$ where $y = (y_1, y_2, y_3)$ or $q(P)$. The dependence of $q$ with respect to $y$ or $P$ can be omitted as all the values implicitly considered are related to it, except if this information brings necessary information and clarity to the problem. The fields of displacement, stress (supposed symmetrical) and strain (linearised) are respectively noted as $\vec{u}^m = (u_i^m)_{i=1,2,3}$, $\sigma^m = (\sigma_{ij}^m)_{i,j=1,2,3}$ and $\varepsilon^m = (\varepsilon_{ij}^m)_{i,j=1,2,3}$ where $\varepsilon_{ij}^m = \frac{1}{2}(\frac{\partial u_i^m}{\partial y_j} + \frac{\partial u_j^m}{\partial y_i})$.

## 6.5.2 The homogenised behaviour defined as a relationship between values averaged on the cell $Y$

In the framework of the periodic homogenisation using the averaging technique, the so-called homogenised behaviour law is the relationship which exists between the average on the cell $Y$ of the microscopic fields of strain and stress solutions of $(\mathcal{P}^Y)$ problem written in the following. This problem have a form close to the $(\mathcal{P}^m)$ problem, including the periodicity conditions.

### 6.5.3    Homogenisation step

**Strain approach finding $a^H$ equivalent to the stress approach finding $A^H$ and independence regarding the geometry of the cell $Y$ used**

In the case of the present periodicity condition, the previous $(\mathcal{P}^m)$ problem is adapted in a way to take into account this condition.

Recall that the fourth-order tensors $a^H$ and $A^H$ define the macroscopic behaviour law, that gives, the relationship between the macroscopic stress tensor $\Sigma$ and the macroscopic strain tensor $E$ which is present in the $(\mathcal{P}^M)$ problem:

$$\Sigma = a^H E \Longleftrightarrow E = A^H \Sigma$$

with the usual symmetry and ellipticity properties.

Compared to the case without the periodicity condition, we have here a very important new result: contrary to the case without the periodicity condition, here the strain approach exactly gives $a^H$ and not an approximation, and the stress approach exactly gives $A^H$ and not an approximation. It is then equivalent to solving the strain approach which could be formulated in a $(\mathcal{P}^{Y-\varepsilon})$ problem as well as the stress approach which could be formulated in a $(\mathcal{P}^{Y-\sigma})$ problem.

This is the reason why here we only consider the strain approach coming from the $(\mathcal{P}^m)$ problem adapted with the periodicity condition, called the $(\mathcal{P}^Y)$ problem (not indicating the type of the approach), and consisting of finding:

- the displacement vector $\vec{u}^m(P)$;
- the stress tensor $\sigma^m(P)$,

verifying:

- the local equation of equilibrium:

$$\overrightarrow{div}\,\sigma^m(P) = \vec{0}\ \forall P \in \Omega^Y$$

- the behaviour law (the constituents are homogeneous and linearly elastic)

  . state law:

$$\left\{ \begin{array}{l} \sigma^m(P) = a(P)\varepsilon^m(\vec{u}^m(P)) \Longleftrightarrow \sigma_{ij}^m(P) = a_{ijkh}(P)\varepsilon_{kh}^m(\vec{u}^m(P)) \\ \Longleftrightarrow \\ \varepsilon_{kh}^m(\vec{u}^m(P)) = A(P)\sigma^m(P) \Longleftrightarrow \varepsilon_{ij}(\vec{u}^m(P)) = A_{ijkh}(P)\sigma_{kh}^m(P) \end{array} \right\}$$

  or, equivalently:

$$\left\{ \begin{array}{l} \sigma^m(P \in \mathcal{C}^{(I)}) = a^{(I)}\varepsilon(\vec{u}^m(P \in \mathcal{C}^{(I)})) \Longleftrightarrow \sigma_{ij}^m(P \in \mathcal{C}^{(I)}) = a_{ijkh}^{(I)}\varepsilon_{kh}(\vec{u}^m(P \in \mathcal{C}^{(I)})) \\ \Longleftrightarrow \\ \varepsilon(\vec{u}^m(P \in \mathcal{C}^{(I)})) = A^{(I)}\sigma^m(P \in \mathcal{C}^{(I)}) \Longleftrightarrow \varepsilon_{ij}(\vec{u}^m(P \in \mathcal{C}^{(I)})) = A_{ijkh}^{(I)}\sigma_{kh}^m(P \in \mathcal{C}^{(I)}) \end{array} \right\}$$

  . no evolution law (no dissipative phenomenon)

- boundary conditions on $\partial\Omega^Y$:

  . $\sigma^m$ and $\varepsilon^m(\vec{u}^m)$ are Y-periodic;

  . $E(M) = <\varepsilon^m(\vec{u}^m)>$.

It can be shown that the solution of $(\mathcal{P}^Y)$ in term of stress is unique. It can also be demonstrated that there is an infinite number of displacement fields that are solutions of this problem. The difference between two solutions is a translation movement of rigid body.

Moreover, it can be shown that these results do not depend on the cell $Y$ used: every cell $Y$ gives exactly the same results.

As previously shown, the $(\mathcal{P}^Y)$ problem is split into six elementary problems because of its linearity, giving that its solutions are linearly dependent of its data. We can then write:

- $\sigma_{ij}^m = s_{ij}^{kh} E_{kh}$ ;
- $u_i^m = v_i^{kh} E_{kh}$.

The behaviour law and the second relation give:

$$\sigma_{ij}^m = a_{ijpq}\varepsilon_{pq}^m(\vec{u}^m) = a_{ijpq}\varepsilon_{pq}^m(\vec{v}^{kh} E_{kh}) = a_{ijpq}\varepsilon_{pq}^m(\vec{v}^{kh}) E_{kh}$$

Consequently, with the first relation:

$$\sigma_{ij}^m = a_{ijpq}\varepsilon_{pq}^m(\vec{v}^{kh}) E_{kh} = s_{ij}^{kh} E_{kh}$$

and then:

$$s_{ij}^{kh} = a_{ijpq}\varepsilon_{pq}^m(\vec{v}^{kh})$$

Finally:

$$< \sigma_{ij}^m >=< s_{ij}^{kh} E_{kh} >=< s_{ij}^{kh} > E_{kh} \Longrightarrow a_{ijkh}^H =< s_{ij}^{kh} >$$

We can then found the tensor $a^H$ (and not an approximation) by solving the six following cellular problems $(\mathcal{P}^{kh-Y})$:

$$(\mathcal{P}^{kh-Y}) \begin{cases} \text{Find } (s^{kh}, \vec{v}^{kh}) / \\ \vec{div}\, s^{kh} = \vec{0}, \forall P \in \Omega^Y \\ s^{kh} = a(y)\varepsilon^m(\vec{v}^{kh}) \Longleftrightarrow \varepsilon^m(\vec{v}^{kh}) = A(y)s^{kh} \\ \text{Boundary conditions on } \partial\Omega^Y - \\ s^{kh} \text{ and } \varepsilon^m(\vec{v}^{kh}) \text{ are } Y\text{-periodic} \\ < \varepsilon_{ij}^m(\vec{v}^{kh}) >= E_{ij}^{kh} = \frac{1}{2}(\delta_{ik}\delta_{jh} + \delta_{ih}\delta_{jk}) \end{cases}$$

where the terms $(\delta_{ij})_{i,j=1,2,3}$ define the Kronecker's numbers and the six tensors $(E^{kh})_{i,j=1,2,3}$ define a basis of the vector space of symetrical second-order tensors. The second-order tensors $s^{kh}$ and $\varepsilon^m(\vec{v}^{kh})$ $(k, h = 1, 2, 3)$ are called the stress and strain localisation tensors or the stress and strain density tensors. We can also demonstrate that $a^H$ obeys to the usual symmetry and ellipticity properties. We can then obtain the flexibility tensor $A^H$ (and not an approximation) starting from the rigidity tensor $a^H$.

### 6.5.4 New formulations of the elementary cellular problems and a new expression of the homogenised coefficients

The difficulty in the process of periodic homogenisation resides in solving the six cellular problems and notably in the application of the periodicity conditions. The calculation codes are generally written in terms of displacement and then, to impose as boundary condition that the displacement field solution of the problem must be with a periodic deformation, that is to say to impose conditions on the spatial derivatives of the field, is not easy. Depending on the method employed for imposing this periodicity condition the resolution of the problems $(\mathcal{P}^{kh-Y})$ is either exact or approximate. We therefore obtain either an exact value or an approximate value of the homogenised behaviour tensor. In the case of the approximate resolution the method of homogenisation is no longer of a periodic type but qualified as Effective Moduli Method. These methods offer the advantage of being quick to resolve. They do have, however, the disadvantage of being strongly dependent on the choice of geometry of the $Y$ cell (for which it should be recalled does not have a unique form) and also of the way used to impose the periodicity of the deformation (which is also not unique). In other terms the results obtained for $a^H$ depend strongly on the geometry and the boundary conditions imposed. All these disadvantages disappear when the $(\mathcal{P}^{kh-Y})$ problems are formulated in an exact manner. However, as the imposition of the periodicity condition for the deformation is difficult to carry out the cellular problems $(\mathcal{P}^{kh-Y})$ are reformulated by a change of variable on which the periodicity condition is directly written (rather than its derivatives). Thus we write:

$$\vec{\chi}^{kh} = -\vec{v}^{kh}+ < grad(\vec{v}^{kh}) > \vec{y} \Longleftrightarrow \chi_i^{kh} = -v_i^{kh}+ < \frac{\partial v_i^{kh}}{\partial y_j} > y_j$$

We then deduce:

$$\varepsilon_{ij}^m(\vec{\chi}^{kh}) = \frac{1}{2}\left(\frac{\partial \chi_i^{kh}}{\partial y_j} + \frac{\partial \chi_j^{kh}}{\partial y_i}\right) = -\frac{1}{2}\left(\frac{\partial v_i^{kh}}{\partial y_j} + \frac{\partial v_j^{kh}}{\partial y_i}\right) + \frac{1}{2}\left(< \frac{\partial v_i^{kh}}{\partial y_j} > + < \frac{\partial v_j^{kh}}{\partial y_i} >\right)$$

$$= -\frac{1}{2}\left(\frac{\partial v_i^{kh}}{\partial y_j} + \frac{\partial v_j^{kh}}{\partial y_i}\right) + < \frac{1}{2}\left(\frac{\partial v_i^{kh}}{\partial y_j} + \frac{\partial v_j^{kh}}{\partial y_i}\right) >= -\varepsilon_{ij}^m(\vec{v}^{kh})+ < \varepsilon_{ij}^m(\vec{v}^{kh}) >$$

$$\Longleftrightarrow \varepsilon^m(\vec{v}^{kh}) = -\varepsilon^m(\vec{\chi}^{kh})+ < \varepsilon^m(\vec{v}^{kh}) >= -\varepsilon^m(\vec{\chi}^{kh}) + E^{kh}$$

Therefore:

$$s^{kh} = a(y)\varepsilon^m(\vec{v}^{kh}) = a(y)\left(-\varepsilon^m(\vec{\chi}^{kh}) + E^{kh}\right) = -a(y)\varepsilon^m(\vec{\chi}^{kh}) + a(y)E^{kh}$$

We note:

$$\sigma^{kh} = a(y)\varepsilon^m(\vec{\chi}^{kh})$$

Furthermore:

$$s^{kh} = -\sigma^{kh} + a(y)E^{kh} \Longleftrightarrow s_{ij}^{kh} = -\sigma_{ij}^{kh} + a_{ijpq}(y)E_{pq}^{kh} = -\sigma_{ij}^{kh} + a_{ijkh}(y)$$

With the new variables $(s^{kh}, \vec{\chi}^{kh})$, the cellular problems $(\mathcal{P}^{kh-Y})$ change their form and become the following $(\overline{\mathcal{P}}^{kh-Y})$ cellular problems (in the sens of the distributions theory):

$$(\overline{\mathcal{P}}^{kh-Y}) \begin{cases} \text{Find } (\sigma^{kh}, \vec{\chi}^{kh})/ \\ \frac{\partial \sigma_{ij}^{kh}}{\partial y_j} = \frac{\partial a_{ijkh}(y)}{\partial y_j}, \forall M \in \Omega^Y \\ \sigma^{kh} = a(y)\varepsilon^m(\vec{\chi}^{kh}) \\ \vec{\chi}^{kh} \text{ is Y-periodic} \end{cases}$$

The homogeneised coefficients take therefore the following form:

$$a_{ijkh}^H =< a_{ijkh}(y) > - < a_{ijpq}(y)\varepsilon_{pq}^m(\vec{\chi}^{kh}) >$$

Finally, we can demonstrate that the homogeneised tensor $a^H$ possesses the usual properties of symmetries and ellipticity:

$$\begin{cases} a_{ijkh}^H = a_{jikh}^H \\ a_{ijkh}^H = a_{ijhk}^H \\ a_{ijkh}^H = a_{khij}^H \\ \exists \alpha \in \mathbb{R}_+^* \ / \ \forall \varepsilon = (\varepsilon_{ij} = \varepsilon_{ij})_{i,j=1,2,3} \in \mathbb{R}^6, a_{ijkh}^H \varepsilon_{ij}\varepsilon_{kh} \geq \alpha \varepsilon_{ij}\varepsilon_{ij} \end{cases}$$

The tensor $a^H$ can therefore be qualified as the tensor of rigidity of the equivalent homogeneous material and can be associated with the tensor of flexibility $A^H$ so as to finally write the homogenised behaviour as is done for the usual linear elastic medium:

$$\left\{\Sigma = a^H E \Longleftrightarrow \Sigma_{ij} = a_{ijkh}^H E_{kh}\right\} \Longleftrightarrow \left\{E = A^H \Sigma \Longleftrightarrow E_{ij} = A_{ijkh}^H \Sigma_{kh}\right\}$$

## 6.5.5 Localisation step

Coming back to the multi-scale process, the problem at the macroscopic scale $(\mathcal{P}^M)$ is assumed to be solved. That is says at each point $M$ of the macroscopic scale, the strain tensor $E(M)$ is known. It can be decomposed according to the basis of the symmetrical tensors in the following manner:

$$E(M) = (E_{kh}(M))_{k,h=1,2,3} = E_{kh}(M)E^{kh}$$

Assuming, the problems $(\mathcal{P}^{kh-Y})$ have been solved, the density tensors $s^{kh}$, $\varepsilon^m(\vec{v}^{kh})$ are known. Then, by linearity, we obtain the microscopic stress and strain tensors $\sigma^m(P)$, $\varepsilon^m(\vec{v}^m(P))$ at each point $P$ of the RVE situated at the point $M$ of the macroscopic scale:

$$\sigma^m(P) = E_{kh}(M)_{kh}(M)s^{kh}(P)$$

$$\varepsilon^m(\vec{v}^m(P)) = E_{kh}(M)\varepsilon^m(\vec{v}^{kh}(P))$$

## 6.5.6 Proof of the equivalence between the strain approach finding $a^H$ and the stress approach finding $A^H$

The $(\mathcal{P}^m)$ problem adapted with the periodicity condition and the stress approach, always called the $(\mathcal{P}^Y)$ problem (not indicating the type of the approach), consists of finding:

- the displacement vector $\vec{u}^m(P)$;
- the stress tensor $\sigma^m(P)$,

verifying:

- the local equation of equilibrium:

$$\overrightarrow{div}\,\sigma^m(P) = \vec{0}\;\forall P \in \Omega^Y$$

- the behaviour law (the constituents are homogeneous and linearly elastic)

  . state law:

$$\left\{ \begin{array}{l} \sigma^m(P) = a(P)\varepsilon^m(\vec{u}^m(P)) \Longleftrightarrow \sigma^m_{ij}(P) = a_{ijkh}(P)\varepsilon^m_{kh}(\vec{u}^m(P)) \\ \Longleftrightarrow \\ \varepsilon^m_{kh}(\vec{u}^m(P)) = A(P)\sigma^m(P) \Longleftrightarrow \varepsilon_{ij}(\vec{u}^m(P)) = A_{ijkh}(P)\sigma^m_{kh}(P) \end{array} \right\}$$

  or, equivalently:

$$\left\{ \begin{array}{l} \sigma^m(P \in \mathcal{C}^{(I)}) = a^{(I)}\varepsilon(\vec{u}^m(P \in \mathcal{C}^{(I)})) \Longleftrightarrow \sigma^m_{ij}(P \in \mathcal{C}^{(I)}) = a^{(I)}_{ijkh}\varepsilon_{kh}(\vec{u}^m(P \in \mathcal{C}^{(I)})) \\ \Longleftrightarrow \\ \varepsilon(\vec{u}^m(P \in \mathcal{C}^{(I)})) = A^{(I)}\sigma^m(P \in \mathcal{C}^{(I)}) \Longleftrightarrow \varepsilon_{ij}(\vec{u}^m(P \in \mathcal{C}^{(I)})) = A^{(I)}_{ijkh}\sigma^m_{kh}(P \in \mathcal{C}^{(I)}) \end{array} \right\}$$

  . no evolution law (no dissipative phenomenon)

- boundary conditions on $\partial\Omega^Y$:

  . $\sigma^m$ and $\varepsilon^m(\vec{u}^m)$ are Y-periodic;

  . $\Sigma(M) = <\sigma^m_{ij}>$.

As previously indicated, the $(\mathcal{P}^Y)$ problem is split into six elementary problems because of its linearity, giving that its solutions are linearly dependent of its data. We can then write:

- $\sigma^m_{ij} = c^{kh}_{ij}\Sigma_{kh}$ ;
- $u^m_i = w^{kh}_i\Sigma_{kh}$.

The behaviour law with the first relation give:

$$\varepsilon_{ij}^m(\vec{u}^m) = A_{ijpq}\sigma_{pq}^m = A_{ijpq}c_{pq}^{kh}\Sigma_{kh}$$

Consequently, with the second relation:

$$\varepsilon_{ij}^m(\vec{u}^m) = \varepsilon_{ij}^m(\vec{w}^{kh}\Sigma_{kh}) = \varepsilon_{ij}^m(\vec{w}^{kh})\Sigma_{kh}$$

and then:

$$\varepsilon_{ij}^m(\vec{w}^{kh}) = A_{ijpq}c_{pq}^{kh}$$

Finally:

$$< \varepsilon_{ij}^m(\vec{u}^m) >=< \varepsilon_{ij}^m(\vec{w}^{kh})\Sigma_{kh} >=< \varepsilon_{ij}^m(\vec{w}^{kh}) > \Sigma_{kh} \implies A_{ijkh}^{H-\sigma} =< \varepsilon_{ij}^m(\vec{w}^{kh}) >$$

We can then find the tensor $A^H$ (and not an approximation.) by solving the six following cellular problems $(\mathcal{P}^{kh-Y})$:

$$(\mathcal{P}^{kh-Y}) \begin{cases} \text{Find } (c^{kh}, \vec{w}^{kh})/ \\ c^{kh} = a(y)\varepsilon^m(\vec{w}^{kh}) \iff \varepsilon^m(\vec{w}^{kh}) = A(y)c^{kh} \\ \text{Boundary conditions on } \partial\Omega^Y - \\ c^{kh} \text{ and } \varepsilon^m(\vec{w}^{kh}) \text{ are Y-periodic} \\ < c_{ij}^{kh} >= \Sigma_{ij}^{kh} = \frac{1}{2}(\delta_{ik}\delta_{jh} + \delta_{ih}\delta_{jk}) \end{cases}$$

where the terms $(\delta_{ij})_{i,j=1,2,3}$ define the Kronecker's numbers and the six tensors $(\Sigma^{kh})_{i,j=1,2,3}$ define a basis of the vector space of symetrical second-order tensors. The second-order tensors $c^{kh}$ and $\varepsilon^m(\vec{w}^{kh})$ ($k, h = 1, 2, 3$) are called the stress and strain localisation tensors or the stress and strain density tensors.

Now, to demonstrate that both strain and stress approaches are strictly equivalent (recall that we do not mark the difference in the notation of the microscopic problems denoted independently of the considered approach as $(\mathcal{P}^{kh-Y})$), it should be demonstrated that $a^H$ found with the strain approach and $A^H$ found with the stress approach are so that:

$$A_{ijkh}^H a_{khlm}^H = I_{ijlm} = \frac{1}{2}(\delta_{il}\delta_{jm} + \delta_{im}\delta_{jl})$$

Proof -

$$\begin{aligned} A_{ijkh}^H a_{khlm}^H &= A_{khij}^H a_{khlm}^H =< \varepsilon_{kh}^m(\vec{w}^{ij}) >< s_{kh}^{lm} > \\ &=< A_{khpq}c_{pq}^{ij} >< s_{kh}^{lm} >=< A_{khpq}c_{pq}^{ij}s_{kh}^{lm} >=< A_{pqkh}s_{kh}^{lm}c_{pq}^{ij} > \\ &=< A_{pqkh}s_{kh}^{lm} >< c_{pq}^{ij} >=< \varepsilon_{pq}^m(\vec{v}^{lm}) >< c_{pq}^{ij} > \\ &= \frac{1}{2}(\delta_{pl}\delta_{qm} + \delta_{pm}\delta_{ql})\frac{1}{2}(\delta_{pi}\delta_{qj} + \delta_{pj}\delta_{qi}) \\ &= \frac{1}{2}(\delta_{li}\delta_{mj} + \delta_{lj}\delta_{mi}) \end{aligned}$$

as we recall that:

- $< \varepsilon_{ij}^m(\vec{v}^{kh}) >= \frac{1}{2}(\delta_{ik}\delta_{jh} + \delta_{ih}\delta_{jk})$;
- $< c_{ij}^{kh} >= \frac{1}{2}(\delta_{ik}\delta_{jh} + \delta_{ih}\delta_{jk})$. $\square$

## 6.6 Application. Homogenisation of a unidirectional fibre/matrix composite in case of periodic assumption using Periodic Method and Effective Moduli Method

We consider the case of a unidirectional fibre/matrix composite in the case of the periodic assumption in which the centre of the fibres is regularly placed according to a square network.

Two cells can be considered (Figure 6.3). The first, denoted as $Y_{1F}$, considers that its cross section is a square in which one fibre appears. The second, denoted as $Y_{4QF}$, considers that its cross section is a square in which four quarters of four fibres appear. The lengths of the cells are not important.

Periodic Method, which does consider the periodicity of the RVE has been used, that says: the problems $(\mathcal{P}^{kh-Y})$ / $(\overline{\mathcal{P}}^{kh-Y})$ which give $a^H$ and $A^H$ have been solved, for both $Y_{1F}$ and $Y_{4QF}$ cells (Figure 6.4, Figure 6.5, Figure 6.6, Figure 6.7, Table 6.1). It should be pointed out that these problems have been solved using a strain approach. But, they can be also solved by a dual approach by imposing the periodicity conditions on the stress tensor and a value for the macroscopic stress tensor. The two approaches can be demonstrated to be equivalent. This has been done here, and the results obtained confirm this property.

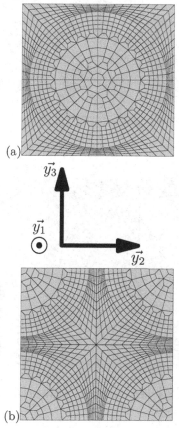

Figure 6.3: Two possible periodic $Y$ cells. (a) $Y_{1F}$ cell. (b) $Y_{4QF}$ cell.

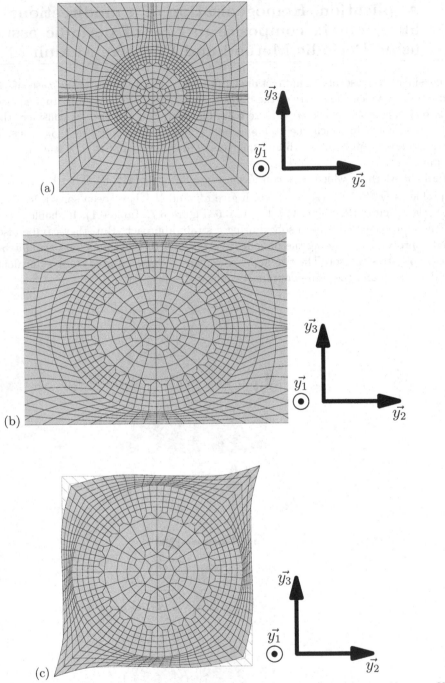

Figure 6.4:  Periodic Method / $Y_{1F}$ cell. Solutions of the cellular problem. (a) $\vec{\chi}^{11}$. (b) $\vec{\chi}^{22}$. (c) $\vec{\chi}^{23}$.

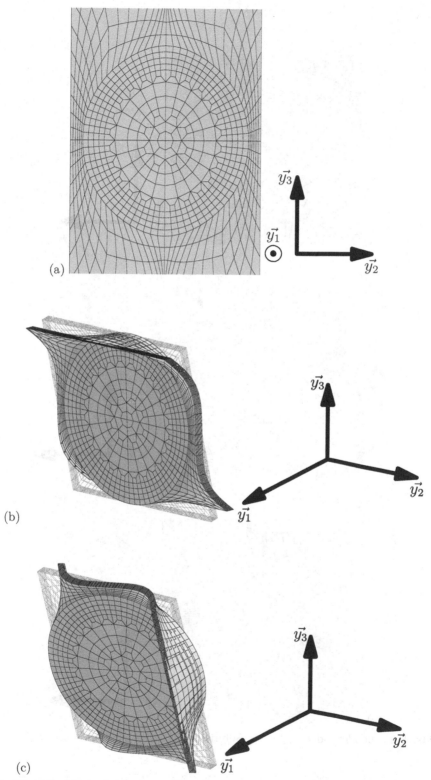

Figure 6.5:  Periodic Method / $Y_{1F}$ cell. Solutions of the cellular problem. (a) $\vec{\chi}^{33}$. (b) $\vec{\chi}^{13}$. (c) $\vec{\chi}^{12}$.

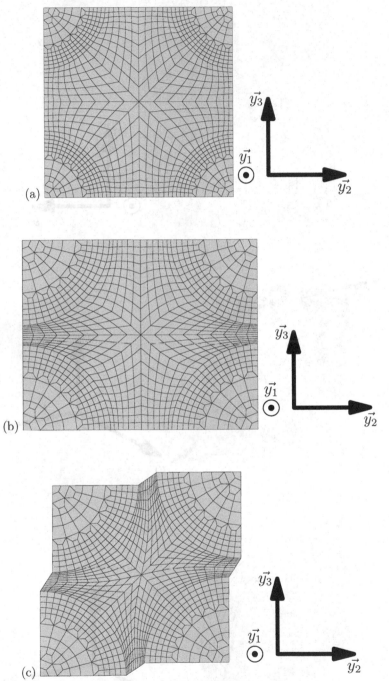

Figure 6.6: Periodic Method / $Y_{4QF}$ cell. Solutions of the cellular problem. (a) $\vec{\chi}^{11}$. (b) $\vec{\chi}^{22}$. (c) $\vec{\chi}^{23}$.

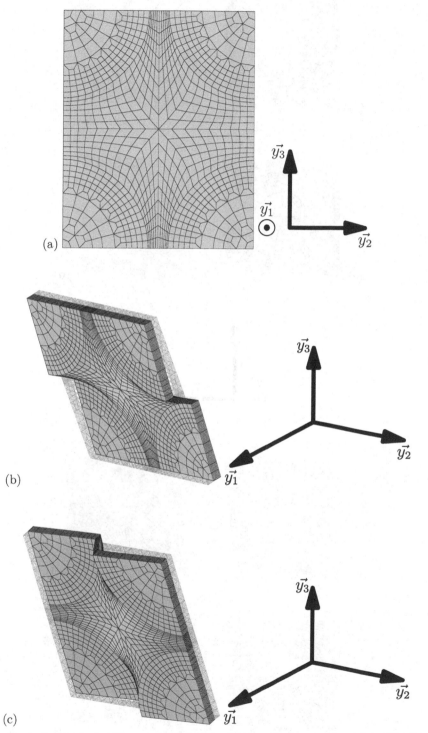

Figure 6.7: Periodic Method / $Y_{4QF}$ cell. Solutions of the cellular problem. (a) $\vec{\chi}^{33}$. (b) $\vec{\chi}^{13}$. (c) $\vec{\chi}^{12}$.

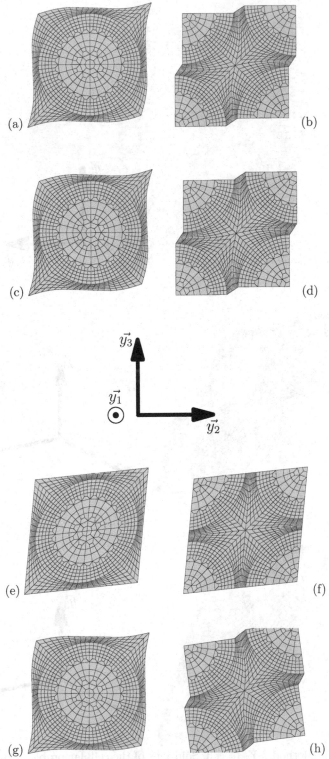

Figure 6.8: Deformed $Y$ cell for the cellular problem $kh = 23$ for two different geometries of the cell $(Y_{1F}, Y_{4QF})$. (abcd) Periodic homogenisation: (ab) strain approach, (cd) stress approach. (efgh) Effective moduli homogenisation: (ef) strain approach, (gh) stress approach.

| | PERIODIC HOMOGENISATION METHOD | | | | | |
|---|---|---|---|---|---|---|
| | Components in Voigt notation | | | | | |
| $a^H$ (MPa)<br>$\Delta_a$ (%,*)<br>$A^H$<br>$\Delta_A$ (%,**)<br>Usual constants<br>$\Delta_c$ (%,*) | $a^H_{11}$<br>(%)<br>$A^H_{11}$<br>(%)<br>$E^H_1$<br>(%) | $a^H_{12} = a^H_{13}$<br>(%)<br>$A^H_{12} = A^H_{13}$<br>(%)<br>$E^H_2 = E^H_3$<br>(%) | $a^H_{22} = a^H_{33}$<br>(%)<br>$A^H_{22} = A^H_{33}$<br>(%)<br>$\nu^H_{12} = \nu^H_{13}$<br>(%) | $a^H_{23}$<br>(%)<br>$A^H_{23}$<br>(%)<br>$\nu^H_{23}$<br>(%) | $a^H_{55} = a^H_{66}$<br>(%)<br>$A^H_{55} = A^H_{66}$<br>(%)<br>$G^H_{13} = G^H_{12}$<br>(%) | $a^H_{44}$<br>(%)<br>$A^H_{44}$<br>(%)<br>$G^H_{23}$<br>(%) |
| (*) Relative error compared to case $PH\varepsilon Y_{1F}$ / (*) Relative error compared to case $PH\sigma Y_{1F}$ | | | | | | |
| Reference case for strain approach | Case $PH\varepsilon Y_{1F}$ / Approach type: strain / Cell type: $Y_{1F}$ | | | | | |
| $a^H$ obtained<br>$10^3 \times A^H = (a^H)^{-1}$<br>$\Delta_A$<br>Usual constants | 149970.<br>0.006841<br>0.<br>146178. | 6322.<br>-0.002052<br>0.<br>16847. | 17676.<br>0.059358<br>0.<br>0.300 | 3396.<br>-0.010670<br>0.<br>0.181 | 5558.<br>0.179930<br>0.<br>5558. | 3639.<br>0.274800<br>0.<br>3639. |
| Reference case for stress approach | Case $PH\sigma Y_{1F}$ / Approach type: stress / Cell type: $Y_{1F}$ | | | | | |
| $10^3 \times A^H$ obtained<br>$a^H = (A^H)^{-1}$<br>$\Delta_a$<br>Usual constants<br>$\Delta_c$ | 0.006841<br>149970.<br>0.<br>146178.<br>0. | -0.002052<br>6322.<br>0.<br>16847.<br>0. | 0.059358<br>17676.<br>0.<br>0.300<br>0. | -0.010670<br>3396.<br>0.<br>0.181<br>0. | 0.179930<br>5558.<br>0.<br>5558.<br>0. | 0.274800<br>3639.<br>0.<br>3639.<br>0. |
| | Case $PH\varepsilon Y_{4QF}$ / Approach type: strain / Cell type: $Y_{4QF}$ | | | | | |
| $a^H$ obtained<br>$\Delta_a$<br>$10^3 \times A^H = (a^H)^{-1}$<br>$\Delta_A$<br>Usual constants<br>$\Delta_c$ | 149970.<br>0.<br>0.006841<br>0.<br>146178.<br>0. | 6322.<br>0.<br>-0.002052<br>0.<br>16847.<br>0. | 17676.<br>0.<br>0.059358<br>0.<br>0.300<br>0. | 3396.<br>0.<br>-0.010670<br>0.<br>0.181<br>0. | 5558.<br>0.<br>0.179930<br>0.<br>5558.<br>0. | 3639.<br>0.<br>0.274800<br>0.<br>3639.<br>0. |
| | Case $PH\sigma Y_{4QF}$ / Approach type: stress / Cell type: $Y_{4QF}$ | | | | | |
| $10^3 \times A^H$ obtained<br>$\Delta_A$<br>$a^H = (A^H)^{-1}$<br>$\Delta_a$<br>Usual constants<br>$\Delta_c$ | 0.006841<br>0.<br>149970.<br>0.<br>146178.<br>0. | -0.002052<br>0.<br>6322.<br>0.<br>16847.<br>0. | 0.059358<br>0.<br>17676.<br>0.<br>0.300<br>0. | -0.010670<br>0.<br>3396.<br>0.<br>0.181<br>0. | 0.179930<br>0.<br>5558.<br>0.<br>5558.<br>0. | 0.274800<br>0.<br>3639.<br>0.<br>3639.<br>0. |

**Table 6.1** Periodic Method applied to fibre/matrix unidirectional composite. Results for rigidity tensor $a^H$ and flexibility tensor $A^H$ for strain and stress approach, for both cells $Y_{1F}$ and $Y_{4QF}$. The fibre and the matrix are assumed to be homogeneous and isotropic and their behaviour is linear elastic. Their usual properties (Young modulus and Poisson coefficient) are: $E_{fibre} = 227000$ MPa, $\nu_{fibre} = 0.30$, $E_{matrix} = 3000$ MPa, $\nu_{matrix} = 0.30$. The fibre volume fraction is 0.64. It appears that the results are independent of the cell used and the approach used: the Periodic Method is then independent of the used cell, and the strain and stress approach are strictly equivalent.

Effective Moduli Method, which does not consider the periodicity of the RVE has been used, that says:

- the problems $(\mathcal{P}^{kh-\varepsilon})$ / $(\mathcal{P}^{m-\varepsilon})$ which give the approximations $a^{H-\varepsilon}$ and $A^{H-\varepsilon}$ of $a^H$ and $A^H$ have been solved;

- the problems $(\mathcal{P}^{kh-\sigma})$ / $(\mathcal{P}^{m-\sigma})$ which give the approximations $a^{H-\sigma}$ and $A^{H-\sigma}$ of $a^H$ and $A^H$ have been solved,

for both $Y_{1F}$ and $Y_{4QF}$ cells (Table 6.2).

The components of the rigidity and flexibility tensors and the usual mechanical constants have been calculated for each case: both homogenisation methods, both approaches (stress and

| EFFECTIVE MODULI HOMOGENISATION METHOD | | | | | | |
|---|---|---|---|---|---|---|
| Components in Voigt notation | | | | | | |
| $a^H$ (MPa)<br>$\Delta_a$ (%,*)<br>$A^H$<br>$\Delta_A$ (%,**)<br>Usual constants<br>$\Delta_c$ (%,*) | $a_{11}^H$<br>(%)<br>$A_{11}^H$<br>(%)<br>$E_1^H$<br>(%) | $a_{12}^H = a_{13}^H$<br>(%)<br>$A_{12}^H = A_{13}^H$<br>(%)<br>$E_2^H = E_3^H$<br>(%) | $a_{22}^H = a_{33}^H$<br>(%)<br>$A_{22}^H = A_{33}^H$<br>(%)<br>$\nu_{12}^H = \nu_{13}^H$<br>(%) | $a_{23}^H$<br>(%)<br>$A_{23}^H$<br>(%)<br>$\nu_{23}^H$<br>(%) | $a_{55}^H = a_{66}^H$<br>(%)<br>$A_{55}^H = A_{66}^H$<br>(%)<br>$G_{13}^H = G_{12}^H$<br>(%) | $a_{44}^H$<br>(%)<br>$A_{44}^H$<br>(%)<br>$G_{23}^H$<br>(%) |
| (*) Relative error compared to case $PH\varepsilon Y_{1F}$ / (*) Relative error compared to case $PH\sigma Y_{1F}$ | | | | | | |
| **Case $ME\varepsilon Y_{1F}$ / Approach type: strain / Cell type: $Y_{1F}$** | | | | | | |
| $a^H$ obtained | 192990. | 78016. | 183950. | 76101. | 55584. | 53925. |
| $\Delta_a$ | 29. | 1134. | 941. | 2141. | 900. | 1382. |
| $10^3 \times A^H = (a^H)^{-1}$ | 0.006841 | -0.002052 | 0.007174 | -0.002097 | 0.017991 | 0.018544 |
| $\Delta_A$ | 0. | 0. | 88. | 80. | 90. | 93. |
| Usual constants | 146178. | 139385. | 0.300 | 0.292 | 55584. | 53925. |
| $\Delta_c$ | 0. | 727. | 0. | 61. | 900. | 1382. |
| **Case $ME\sigma Y_{1F}$ / Approach type: strain / Cell type: $Y_{1F}$** | | | | | | |
| $10^3 \times A^H$ obtained | 0.109560 | -0.028298 | 0.106900 | -0.037523 | 0.303440 | 0.282950 |
| $\Delta_A$ | 1502. | 1279. | 80. | 252. | 69. | 3. |
| $a^H = (A^H)^{-1}$ | 11564. | 4717. | 12594. | 5669. | 3296. | 3534. |
| $\Delta_a$ | 92. | 25. | 29. | 67. | 41. | 3. |
| Usual constants | 9127. | 9354. | 0.258 | 0.351 | 3296. | 3534. |
| $\Delta_c$ | 93. | 44. | 14. | 93. | 41. | 3. |
| **Case $ME\varepsilon Y_{4QF}$ / Approach type: strain / Cell type: $Y_{4QF}$** | | | | | | |
| $a^H$ obtained | 193160. | 78297. | 184490. | 76495. | 55608. | 54127. |
| $\Delta_a$ | 29. | 1139. | 944. | 2153. | 901. | 1387. |
| $10^3 \times A^H = (a^H)^{-1}$ | 0.006841 | -0.002052 | 0.007161 | -0.002098 | 0.017983 | 0.018475 |
| $\Delta_A$ | 0. | 0. | 88. | 80. | 90. | 93. |
| Usual constants | 146178. | 139642. | 0.300 | 0.293 | 55608. | 54127. |
| $\Delta_c$ | 0. | 728. | 0. | 62. | 901. | 1387. |
| **Case $ME\sigma Y_{4QF}$ / Approach type: strain / Cell type: $Y_{4QF}$** | | | | | | |
| $10^3 \times A^H$ obtained | 0.105940 | -0.024080 | 0.098797 | -0.039254 | 0.303320 | 0.277590 |
| $\Delta_A$ | 1449. | 1073. | 66. | 268. | 69. | 1. |
| $a^H = (A^H)^{-1}$ | 11566. | 4677. | 13911. | 6667. | 3297. | 3602. |
| $\Delta_a$ | 92. | 26. | 21. | 96. | 41. | 1. |
| Usual constants | 9439. | 10121. | 0.227 | 0.397 | 3297. | 3602. |
| $\Delta_c$ | 93. | 40. | 24. | 120. | 41. | 1. |

**Table 6.2** Effective Moduli Method applied to fibre/matrix unidirectional composite. Results for rigidity tensor $a^H$ and flexibility tensor $A^H$ for strain and stress approach, for both cells $Y_{1F}$ and $Y_{4QF}$. The fibre and the matrix are assumed to be homogeneous and isotropic and their behaviour is linear elastic. Their usual properties (Young modulus and Poisson coefficient) are: $E_{fibre} = 227000$ MPa, $\nu_{fibre} = 0.30$, $E_{matrix} = 3000$ MPa, $\nu_{matrix} = 0.30$. The fibre volume fraction is 0.64. It appears that the results are dependent of the cell used and the approach used: the Effective Moduli Method is then dependent of the used cell, and the strain and stress approach are not equivalent.

strain) and both geometry cells (Table 6.1, Table 6.2). As the considered material is periodic the results obtained with the Periodic Method are considered as the reference compared the Effective Moduli Method, for which the differences have been given: the differences go from 0% to more than 1000%. These large differences can be explained by the fact that the boundary conditions used for the Effective Moduli Method are good from a mechanical point of view, but very bad for the considered material.

A graphic comparison for the same macroscopic load is given for all cases (Figure 6.8) that allows the difference between each method (Effective Moduli Method or Periodic Methods) to be understood, each approach (stress or strain approach) and each cells ($Y_{1F}$ and $Y_{4QF}$).

# 7 Graphic illustrations of the multi-scale, homogenisation and localisation concepts

e shall try here to illustrate how to solve the $(\mathcal{P}_1)$ problem (§ 6.3.3) for some typical cases of uctures with different internal constitutions requiring or not a multi-scale process:

- case $OHC$ (One Homogeneous Constituent): the structure consists of one homogeneous constituent (Figure 6.9 (a));
- case $BHH$ (Big Homogeneous Heterogeneities): the structure consists of several homogeneous constituents forming heterogeneities for which the characteristic size is comparable to the characteristic size of the structure (Figure 6.9 (b));
- case $SHHNP$ (Small Homogeneous Heterogeneities No Periodic): the structure consists of two or more homogeneous constituents forming heterogeneities for which the characteristic size is very far to the characteristic size of the structure (Figure 6.10);
- case $SHHP$ (Small Homogeneous Heterogeneities Periodic): the structure consists of one homogeneous constituent on which heterogeneities (made with a second homogeneous constituent) are placed in a regular network so that it can be assumed that the total medium making up the structure can be considered as a periodic medium (Figure 6.14).

cases $OHC$ and $BHH$, the access to the stress and strain fields does not need a very fine nite Element discretisation even to get a good precision. The size of the problem to solve, from Finite Element point of view (a linear system to solve), is schematically strongly dependent of e size of the structure and unless to study a gigantic structure, a computer with a normal size enough to solve the problem.

In cases $SHHNP$ (Figure 6.10) and $SHHP$ (Figure 6.14), it is easy to understand that if the ernal constitution of a structure consist of small heterogeneities for which it is important to aluate precisely the stress or strain fields around them, the Finite Element discretisation must made with elements the size of which is less than the size of the heterogeneities. Therefore, e size of the problem to solve (strongly dependent of the size of the heterogeneities relatively the size of the structure) from a Finite Element point of view, even with a small size structure, n be extremely huge. So that, even the most powerful computer cannot be sufficient. In this se, multi-scale processes are one of the single possibility to be used. The multi-scale process n be schematically divided in four parts:

- Part one – Definition of the scales of the process. Usually, two scales: the microscopic one and the macroscopic one (case $SHHNP$: Figure 6.11 / case $SHHP$: Figure 6.15). And then, two types of problem: the $(\mathcal{P}^M)$ macroscopic problem and the $(\mathcal{P}^m)$ microscopic problem;
- Part two – Homogenisaion step. Solve the $(\mathcal{P}^m)$ microscopic problem (or its derivatives) and get the homogenised material (case $SHHNP$: Figure 6.12 / case $SHHP$: Figure 6.16);
- Part three and four – The localisation step. Solve the $(\mathcal{P}^M)$ macroscopic problem and get the macroscopic stress and strain fields at each point $M$ of the macroscopic scale (case $SHHNP$: Figure 6.13 (a) / case $SHHP$: Figure 6.17 (a)). Then, solve at each point $M$, the $(\mathcal{P}^m)$ microscopic problem and get the microscopic stress and strain fields at each point $P$ of the microscopic scale under the point $M$ (case $SHHNP$: Figure 6.13 (b) / case $SHHP$: Figure 6.17 (b)).

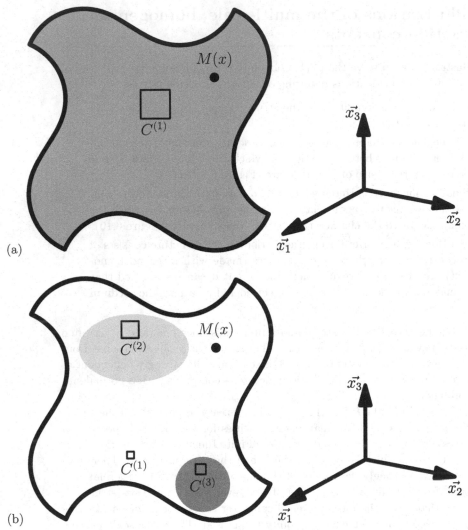

(a)

(b)

Figure 6.9: Case $OHC$ (a) and case $BHH$ (b). No need of a multi-scale process. Solving the $(\mathcal{P}_1)$ problem for the case when the structure consists of one (a) homogeneous constituent or several (b) homogeneous constituents forming heterogeneities for which the characteristic size is comparable to the characteristic size of the structure. Such situations are common and pose no particular technical problem. As no analytical resolution is possible a numerical resolution has to be used. The Finite Element Method is usually used. In these cases, the discretisation of the structure is usually made with elements for which the characteristic size is not far from the characteristic size of the structure. As a consequence, the number of degrees of freedom of the problem has a reasonable size and $(\mathcal{P}_1)$ can be solved. But the resolution gives no information concerning the inside of the RVE. (a) Structure consists of one homogeneous constituents $\mathcal{C}^{(1)}$ for which the properties of the RVE are coming from experimental results. (b) Structure consists of three homogeneous constituents $\mathcal{C}^{(I)}$ ($I = 1, 2, 3$). The constituents $\mathcal{C}^{(2)}$ and $\mathcal{C}^{(3)}$ can be see as big heterogeneities for which the characteristic size is comparable to the characteristic size of the structure. The structure cannot be seen as being made of only one homogeneous constituent. One RVE exists for each of the constituents, but no RVE can be defined covering every constituents. For both cases, the properties of the RVE of the constituents are obtained from experimental results. The square above the name of the constituent symbolises the RVE used for experimental tests.

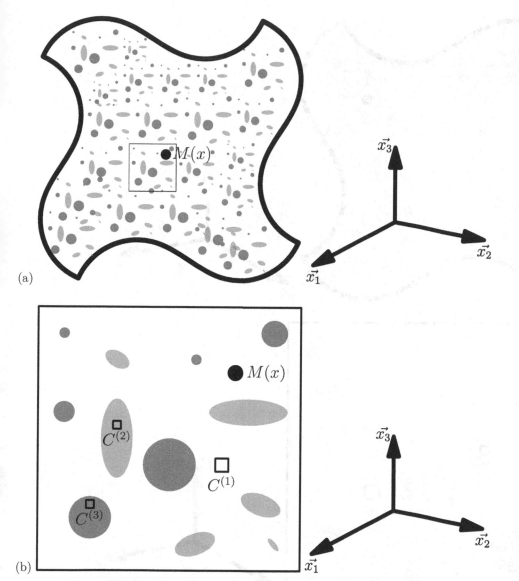

Case $SHHNP$. Need of a multi-scale process and no periodic medium. Solv-
g the ($\mathcal{P}_1$) problem for the case when the structure (a) consists of two or more homogeneous
nstituents (b) forming heterogeneities for which the characteristic size is very far to the charac-
ristic size of the structure. Such a situation is common (in particular with composite materials)
d poses technical problem. As no analytical resolution is possible a numerical resolution has
be used. The Finite Element Method is usually used. In this case, the discretization of the
ructure has to be made with elements for which the characteristic size is very far from the
aracteristic size of the structure. As a consequence, the number of degrees of freedom of the
oblem has a very large size and ($\mathcal{P}_1$) cannot be solved using the classical Finite Element
ethod. One solution is then to use a multi-scale process (homogenization step and localiza-
n step). (a) As illustrated here, the structure consists of three homogeneous constituents $\mathcal{C}^{(I)}$
$= 1, 2, 3$) for which the properties of the RVE are coming from experimental results. (b) The
nstituents $\mathcal{C}^{(2)}$ (ellipsoids, different sizes) and $\mathcal{C}^{(3)}$ (spheroids, different sizes) can be see as
ry small heterogeneities. The square above the name of the constituent symbolizes the RVE
ed for experimental tests.

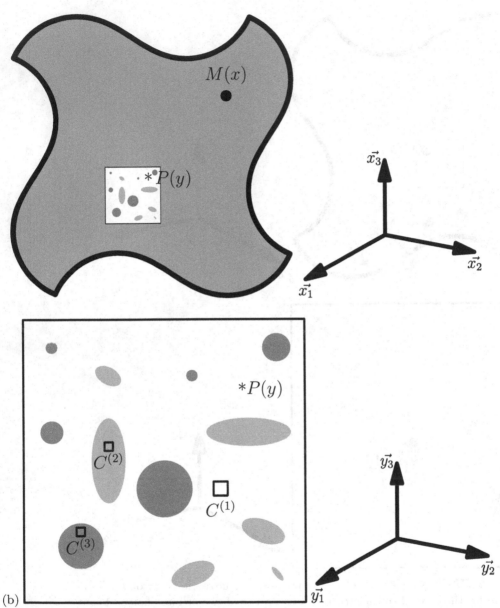

(b)

Figure 6.11: Using a multi-scale process to solve the $(\mathcal{P}_1)$ problem in case the structure (a) consists of two or more homogeneous constituents (b) forming heterogeneities for which the characteristic size is very far to the characteristic size of the structure. It is assumed that a RVE (b) exists. Multi-scale process / Part one – Define two scales: the microscopic scale (variable $y$) attached to the RVE for which the heterogenities can be seen, the macroscopic scale (variable $x$) attached to the structure for which the heterogenities have disappeared.

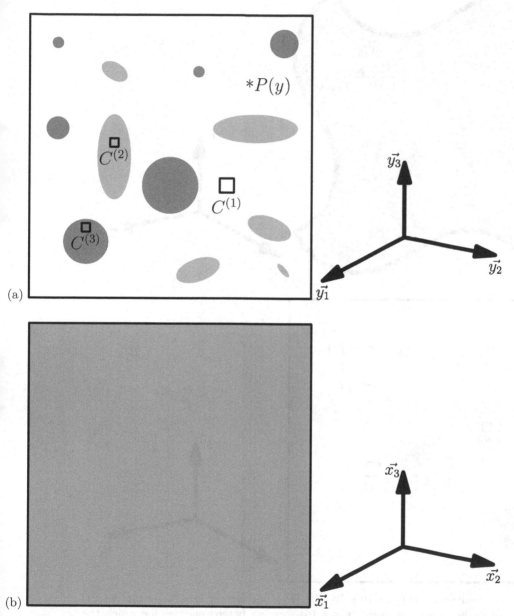

Figure 6.12: Using a multi-scale process to solve the $(\mathcal{P}_1)$ problem in case the structure (a) consists of two or more homogeneous constituents (b) forming heterogeneities for which the characteristic size is very far to the characteristic size of the structure. It is assumed that a RVE ) exists. Multi-scale process / Part two – Homogenisation step: solve the $(\mathcal{P}^m)$ problem (or derivatives) on the RVE (a) to obtain the (or an approximation of the) behaviour law of the omogenised material (b) characterised by $a^H$ and $A^H$.

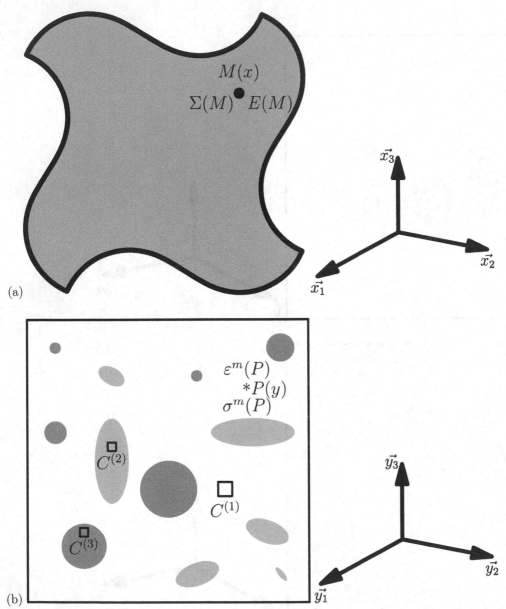

(a)

(b)

Figure 6.13:  Using a multi-scale process to solve the $(\mathcal{P}_1)$ problem in case the structure (a) consists of two or more homogeneous constituents (b) forming heterogeneities for which the characteristic size is very far to the characteristic size of the structure. It is assumed that a RVE (b) exists. (a) Multi-scale process / Part three - Solve the $(\mathcal{P}^M)$ problem to obtain the macroscopic stress and strain fields $\Sigma(M)$ and $E(M)$. (b) Multi-scale process / Part four – Localisation step for the point $M$ of the macroscopic scale: solve the $(\mathcal{P}^m)$ problem for which the RVE is loaded with the numerically obtained macroscopic stress and strain fields $\Sigma(M)$ and $E(M)$ and obtain the microscopic stress and strain fields $\sigma^m(P)$ and $\varepsilon^m(P)$ due to $\Sigma(M)$ and $E(M)$.

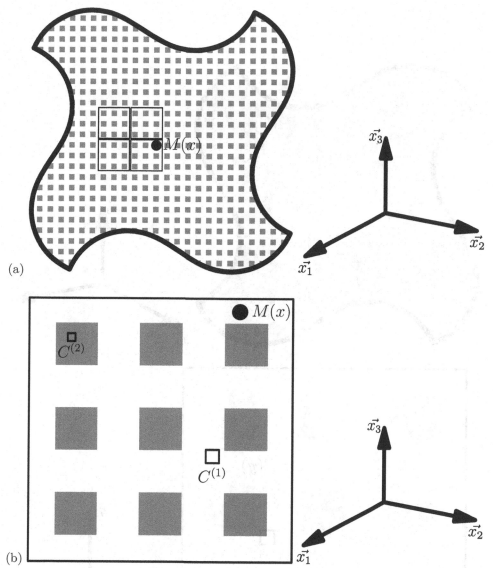

Figure 6.14: Case *SHHP*. Need of a multi-scale process and periodic medium. Solving the $(\mathcal{P}_1)$ problem in case the structure (a) consists of one homogeneous constituent on which heterogeneities (made with a second homogeneous constituent) are placed in a regular network so that it can be assumed that the total medium making up the structure can be considered as a periodic medium. The characteristic size of the heterogeneities is assumed to be very far to the characteristic size of the structure. Such a situation is common (in particular with composite materials) and poses technical problems. As analytical solutions do not exist, a numerical resolution has to be used. The Finite Element Method is usually chosen. In this case, the discretisation of the structure has to be made with elements for which the characteristic size is very far from the characteristic size of the structure. As a consequence, the number of degrees of freedom of the problem is very large size and $(\mathcal{P}_1)$ cannot be solved using the classical Finite Element Method. One solution is then to use a multi-scale process (homogenisation step and localisation step). (a) The square divided in four parts symbolises the RVE that can be defined for the medium making up the structure. (b) Each squared part is the periodic cell $Y$ of the RVE. The square above the name of the constituent symbolises the RVE used for experimental tests.

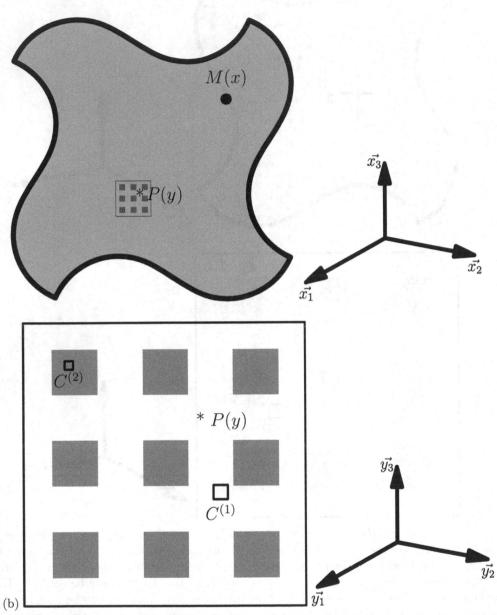

Figure 6.15: Using a multi-scale process to solve the $(\mathcal{P}_1)$ problem in case the structure (a) consists of a periodic medium formed by homogeneous heterogeneities for which their characteristic sizes are very far from the characteristic size of the structure. (b) The periodic cell $Y$. Multi-scale process / Part one – Define two scales: the microscopic scale (variable $y$) attached to the $Y$ cell for which the heterogenities can be seen, the macroscopic scale (variable $x$) attached to the structure for which the heterogenities have disappeared.

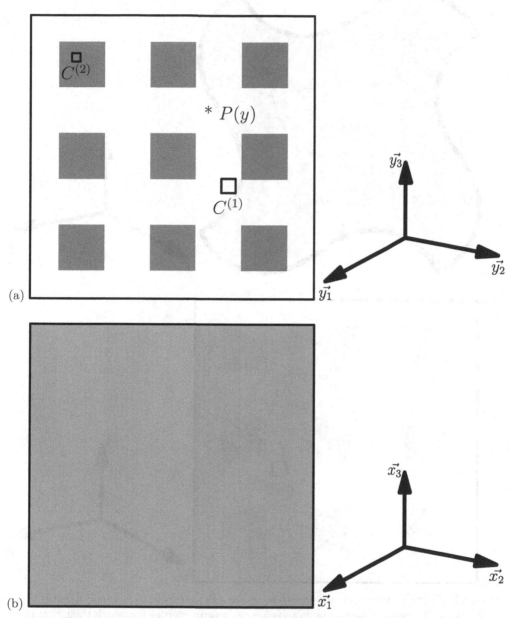

Figure 6.16: Using a multi-scale process to solve the $(\mathcal{P}_1)$ problem in case the structure (a) consists of a periodic medium formed by homogeneous heterogeneities for which their characteristic sizes are very far from the characteristic size of the structure. Multi-scale process / Part two – Homogenisation step: solve the $(\mathcal{P}^Y)$ problem on the cell $Y$ (a) to obtain the behaviour law of the homogenised material (b) characterised by $a^H$ and $A^H$.

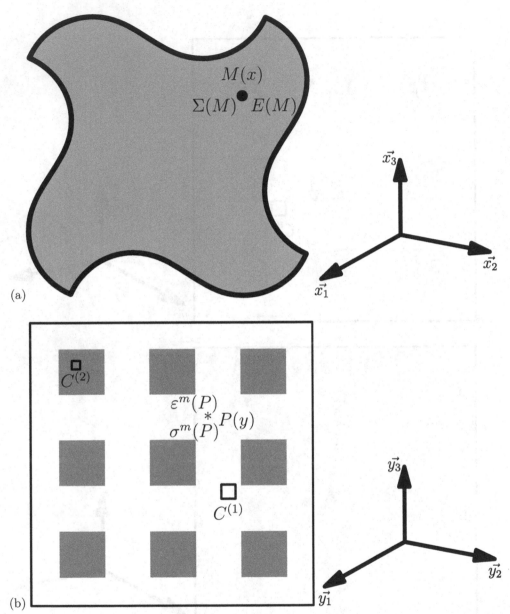

Figure 6.17: Using a multi-scale process to solve the $(\mathcal{P}_1)$ problem in case the structure (a) consists of a periodic medium formed by homogeneous heterogeneities for which their characteristic sizes are very far from the characteristic size of the structure. (a) Multi-scale process / Part three – Solve the $(\mathcal{P}^M)$ problem and obtain the macroscopic stress and strain fields $\Sigma(M)$ and $E(M)$. (b) Multi-scale process / Part four – Localisation step for the point $M$ of the macroscopic scale: solve the $(\mathcal{P}^Y)$ problem for which the cell $Y$ is loaded with the obtained macroscopic stress and strain fields $\Sigma(M)$ and $E(M)$ and get the microscopic stress and strain fields $\sigma^m(P)$ and $\varepsilon^m(P)$ due to $\Sigma(M)$ and $E(M)$. $\Sigma(M)$ and $E(M)$.

## 6.8 Conclusions

We have described here in general the concepts of homogenisation and localisation, forming the multi-scale processes but also in great detail the family of methods based on the averaging technique: the Effective Moduli Method and the Periodic Method. For the first, it can be applied without any particular hypothesis but gives results that can be debatable. For the second, the periodicity of the microstructure of the composite is necessary but it gives good and coherent results. It can be also mentionned that in case the studied material is made with only one isotropic constituent, all homogenisation methods considered in all approaches (stress or strain) give the same results.

# Revision exercises

### Exercise 6.1

What are the two major types of homogenisation methods? Can both methods be used for every type of material?

### Exercise 6.2

Does the Effective Moduli Method type always give a good approximation of the homogenised behaviour? Does it give the same equivalent behaviour in case of the stress and strain approach?

### Exercise 6.3

Does the periodic Method always give a good approximation of the homogenised behaviour? Is there an important hypothesis that should be verified for this method? Does it give the same equivalent behaviour in case of the stress and strain approach?

### Exercise 6.4

What are the Voigt and Reuss bounds? Are they used for both type of homogenisation methods?

### Exercise 6.5

In the case of the Periodic Method, why do the elementary problems have to be reformulated?

### Exercise 6.6

What is the role of the localization step of a multi-scale process? What is the role of the homogenisation step of a multi-scale process?

### Exercise 6.7

Do the results of the Periodic Method depend on the geometry of the elementary cell describing the RVE of the studied material? Do the results of the Effective Moduli Method depend on the geometry of the RVE of the studied material?

### Exercise 6.8

Periodic Method and Effective Moduli Method are based on the same general concept. Which one?

### Exercise 6.9

Does the macroscopic scale of a multi-scale process see the constituents that are present in the microscopic cell?

### Exercise 6.10

In which case does all homogenisation method types (Effective Moduli or Periodic) and all approaches (stress or strain) give the same results?

# Laminated composites

## 7.1   Introduction

The aim of this chapter is to study the modelling of laminates and in particular to see if it is possible to represent their macroscopic properties by an equivalent homogeneous material particularly using tensors of rigidity and flexibility. It is important to make the distinction between two cases: that where the total thickness of the laminate is small, so that then we call them thin laminates and those for which the overall thickness of the laminate is large, which we shall call thick laminates. In the first case classical plate theory is applicable and the analytical calculation of an equivalent behaviour of a homogeneous material is possible. In the second case classical plate theory is not applicable. Moreover, if the plies which make up the thick laminate have no order or are not repeated in an ordered manner it is not possible to associate an equivalent behaviour to that of a homogeneous material. However, if the plies show a periodicity in the thickness then the concepts of periodic homogenisation can be applied and the analytical calculation of an equivalent behaviour of a homogeneous plate becomes possible. We therefore will consider this second hypothesis. And then, this is an application of the homogenisation/localisation concepts in the case of a periodic structure, detailed in the previous chapter. In addition, over and above finding an equivalent behaviour we will see that, knowing the states of stress and strain existing in the equivalent medium, it is possible to calculate, analytically, the stresses and strains in each of the plies of the (thin or thick) laminate. The goal of calculating the stresses and strains in the plies is then to use them in determining the failure criteria governing the behaviour of the laminates. That will be the aim of the next chapter.

In particular we shall be treating the case of thick laminates composed of orientated unidirectional composite plies. Nevertheless, here we shall first give a more general description with the only restriction that the plies forming the laminate are considered to show linearly elastic orthotropic behaviour (within their anisotropic local framework).

Thus, the chapter initially consists of a section which recalls the essential points of linear applications and changes of base and the modified Voigt notation will be explained. Then this will be applied to the case of laws governing linear elastic behaviour for a large family of anisotropic materials, using the hypothesis, or not, of plane stress and the use of Voigt notation, which is considered classical and which is normally used, will be demonstrated. This will be followed by consideration of both thin and thick laminates showing the important points of both methods and giving demonstrations of the most important results.

## 7.2   Linear application and change of basis. Generalities

### 7.2.1   Definitions

Let's consider the vector space $E = \mathbb{R}^3$ on the field $K = \mathbb{R}$. It is said that an application from $\mathbb{R}^3$ in itself, which is to say a relation in which all vectors $\vec{u}$ are associated with one and only one element $f(\vec{u})$, is linear if and only if:

$$\forall \vec{A} \in E, \forall \vec{B} \in E, \forall \alpha \in K, \forall \beta \in K, \ f(\alpha \vec{A} + \beta \vec{B}) = \alpha f(\vec{A}) + \beta f(\vec{B})$$

We can then say that $f$ is a linear application of $\mathbb{R}^3$.

### 7.2.2   Matrix of a linear application

Consider the vector space $\mathbb{R}^3$ to which is associated the orthonormal basis $b = (\vec{x_1}, \vec{x_2}, \vec{x_3})$. Consider a linear application $l$ defined from $\mathbb{R}^3$ to $\mathbb{R}^3$:

$$\vec{v} = l(\vec{u}) \iff [\vec{v}]^b = [l]^b [\vec{u}]^b$$

$[\vec{v}]^b$ and $[\vec{u}]^b$ are the components in $b$, of the considered vectors and $[l]^b$ is the matrix of this linear application in $b$. One has:

$$[l]^b = \begin{pmatrix} L_{11} = \vec{x_1}.l(\vec{x_1}) & L_{12} = \vec{x_1}.l(\vec{x_2}) & L_{13} = \vec{x_1}.l(\vec{x_3}) \\ L_{21} = \vec{x_2}.l(\vec{x_1}) & L_{22} = \vec{x_2}.l(\vec{x_2}) & L_{23} = \vec{x_2}.l(\vec{x_3}) \\ L_{31} = \vec{x_3}.l(\vec{x_1}) & L_{32} = \vec{x_3}.l(\vec{x_2}) & L_{33} = \vec{x_3}.l(\vec{x_3}) \end{pmatrix}$$

### 7.2.3   Change of basis matrix

Consider the vector space $\mathbb{R}^3$ to which is associated two distinct orthonormal bases given by $b = (\vec{x_1}, \vec{x_2}, \vec{x_3})$ and $b' = (\vec{x_1}', \vec{x_2}', \vec{x_3}')$. It is supposed that one knows the coordinates of the vectors of $b'$ in $b$. They are written in the following manner:

$$\begin{cases} \vec{x_1}' = Q_{11}\vec{x_1} + Q_{21}\vec{x_2} + Q_{31}\vec{x_3} \\ \vec{x_2}' = Q_{12}\vec{x_1} + Q_{22}\vec{x_2} + Q_{32}\vec{x_3} \\ \vec{x_3}' = Q_{13}\vec{x_1} + Q_{23}\vec{x_2} + Q_{33}\vec{x_3} \end{cases}$$

Using these formulae a matrix can be constructed which is the change of basis matrix from $b$ to $b'$:

$$Q = \begin{pmatrix} Q_{11} & Q_{12} & Q_{13} \\ Q_{21} & Q_{22} & Q_{23} \\ Q_{31} & Q_{32} & Q_{33} \end{pmatrix}$$

**Property** – If the two bases $b$ and $b'$ are orthonormal then $Q^{-1} = Q^T$.

Let's consider a particular linear application $l_0$ which transforms the basis $b$ into the basis $b'$, which is to say that: $\vec{x_1}' = l_0(\vec{x_1})$, $\vec{x_2}' = l_0(\vec{x_2})$, $\vec{x_3}' = l_0(\vec{x_3})$. It can be seen that the matrix $Q$ is exactly the matrix in $b$ of the linear application $l_0$. Therefore:

$$[L_0]^b = Q = \begin{pmatrix} Q_{11} & Q_{12} & Q_{13} \\ Q_{21} & Q_{22} & Q_{23} \\ Q_{31} & Q_{32} & Q_{33} \end{pmatrix}$$

In other terms, $Q$ could be seen as the matrix (by its components in $b$) of the linear application (rotation) $q$ which transforms the basis $b$ into the basis $b'$.

### 7.2.4 Change of basis for a scalar, a vector, a second-order tensor

Let's consider the vector space $\mathbb{R}^3$ to which are associated two distinct orthonormal bases $b = (\vec{x_1}, \vec{x_2}, \vec{x_3})$ and $b' = (\vec{x_1}', \vec{x_2}', \vec{x_3}')$. Consider a scalar $s$, a vector $\vec{v}$ and a second-order tensor $t$. It can be noted respectively $[s]^b$, $[\vec{v}]^b$ and $[t]^b$ the matrix of their components in $b$ and $[s]^{b'}$, $[\vec{v}]^{b'}$ and $[t]^{b'}$ the matrix of their components in $b'$. The formulae of the change of basis are the following:

$$\begin{cases} [s]^{b'} = [s]^b \\ [\vec{v}]^{b'} = Q^T [\vec{v}]^b \\ [t]^{b'} = Q^T [t]^b Q \end{cases}$$

**Remark** – It is easy to demonstrate that the expression of the change of basis matrix $Q$ from $b$ to $b'$ is the same in bases $b$ and $b'$.

### 7.2.5 The modified Voigt notation. Change of basis of a second-order symmetrical tensor and of a fourth-order symmetrical tensor

Let's now consider that the second-order tensor $t$ is symmetrical so we note $(t_{ij} = t_{ji})_{i,j=1,2,3}$ the components in the basis $b$. The following notation is adopted, denoted as the modified Voigt notation: $t_i = t_{ii}$ for $i = 1, 2, 3$ and $t_{9-i-j} = t_{ij} = t_{ji}$ for $i \neq j$, $i, j = 1, 2, 3$. In this way the following conventional matricial notation is constructed: $t = (t_I)_{I=1,\ldots,6}$. In the basis $b'$, in a similar manner to that written for the basis $b$, it can be noted $t' = (t_I')_{I=1,\ldots,6}$. With this new notation it is possible to show that the transformation formulae can be written, with $P = (P_{IJ})_{I,J=1,\ldots,6}$ and $\bar{P} = (\bar{P}_{IJ})_{I,J=1,\ldots,6}$

$$\begin{cases} t = P t' \\ t' = \bar{P} t \end{cases} \Longleftrightarrow \begin{cases} t_I = P_{IJ} t_J' \\ t_I' = \bar{P}_{IJ} t_J \end{cases}$$

where:

$$P = \begin{pmatrix} Q_{11}^2 & Q_{12}^2 & Q_{13}^2 & 2Q_{13}Q_{12} & 2Q_{13}Q_{11} & 2Q_{11}Q_{12} \\ Q_{21}^2 & Q_{22}^2 & Q_{23}^2 & 2Q_{23}Q_{22} & 2Q_{23}Q_{21} & 2Q_{22}Q_{21} \\ Q_{31}^2 & Q_{32}^2 & Q_{33}^2 & 2Q_{33}Q_{32} & 2Q_{33}Q_{31} & 2Q_{32}Q_{31} \\ Q_{21}Q_{31} & Q_{22}Q_{32} & Q_{23}Q_{33} & Q_{23}Q_{32}+Q_{22}Q_{33} & Q_{23}Q_{31}+Q_{21}Q_{33} & Q_{22}Q_{31}+Q_{21}Q_{32} \\ Q_{11}Q_{31} & Q_{12}Q_{32} & Q_{13}Q_{33} & Q_{13}Q_{32}+Q_{12}Q_{33} & Q_{13}Q_{31}+Q_{11}Q_{33} & Q_{12}Q_{31}+Q_{11}Q_{32} \\ Q_{11}Q_{21} & Q_{12}Q_{22} & Q_{13}Q_{23} & Q_{13}Q_{22}+Q_{12}Q_{23} & Q_{13}Q_{21}+Q_{11}Q_{23} & Q_{12}Q_{21}+Q_{11}Q_{22} \end{pmatrix}$$

$$\bar{P} = \begin{pmatrix} Q_{11}^2 & Q_{31}^2 & Q_{31}^2 & 2Q_{31}Q_{21} & 2Q_{31}Q_{11} & 2Q_{11}Q_{21} \\ Q_{12}^2 & Q_{22}^2 & Q_{32}^2 & 2Q_{32}Q_{22} & 2Q_{32}Q_{12} & 2Q_{22}Q_{12} \\ Q_{13}^2 & Q_{23}^2 & Q_{33}^2 & 2Q_{33}Q_{23} & 2Q_{33}Q_{13} & 2Q_{23}Q_{13} \\ Q_{12}Q_{13} & Q_{22}Q_{23} & Q_{32}Q_{33} & Q_{32}Q_{23}+Q_{22}Q_{33} & Q_{32}Q_{13}+Q_{12}Q_{33} & Q_{22}Q_{13}+Q_{12}Q_{23} \\ Q_{11}Q_{13} & Q_{21}Q_{23} & Q_{31}Q_{33} & Q_{31}Q_{23}+Q_{21}Q_{33} & Q_{31}Q_{13}+Q_{11}Q_{33} & Q_{21}Q_{13}+Q_{11}Q_{23} \\ Q_{11}Q_{12} & Q_{21}Q_{22} & Q_{31}Q_{32} & Q_{31}Q_{22}+Q_{21}Q_{32} & Q_{31}Q_{12}+Q_{11}Q_{32} & Q_{21}Q_{12}+Q_{11}Q_{22} \end{pmatrix}$$

It can be easily shown that: $P\bar{P} = \bar{P}P = I_6$ where $I_6$ is the identity matrix of $\mathbb{R}^6$.

Let's consider a fourth-order tensor $c$ for which the components $(c_{ijkh})_{i,j,k,h=1,2,3}$ in the basis $b$ possesses the following properties (the fourth-order tensor is said to be symmetrical):

$$\begin{cases} c_{ijkh} = c_{jikh} \\ c_{ijkh} = c_{ijhk} \\ c_{ijkh} = c_{khij} \end{cases}$$

It is taken that this tensor is the linear link between two symmetrical second order tensors $\sigma$ and $\varepsilon$: $\sigma_{ij} = c_{ijkh}\varepsilon_{kh}$. For each of these two tensors the precedent notation is adopted. In this second convention, one writes in the basis $b$: $\sigma = C\varepsilon$ where:

$$C = \begin{pmatrix} C_{11}=c_{1111} & C_{12}=c_{1122} & C_{13}=c_{1133} & C_{14}=2c_{1123} & C_{15}=2c_{1113} & C_{16}=2c_{1112} \\ C_{21}=c_{2211} & C_{22}=c_{2222} & C_{23}=c_{2233} & C_{24}=2c_{2223} & C_{25}=2c_{2213} & C_{26}=2c_{2212} \\ C_{31}=c_{3311} & C_{32}=c_{3322} & C_{33}=c_{3333} & C_{34}=2c_{3323} & C_{35}=2c_{3313} & C_{36}=2c_{3312} \\ C_{41}=c_{2311} & C_{42}=c_{2322} & C_{43}=c_{2333} & C_{44}=2c_{2323} & C_{45}=2c_{2313} & C_{46}=2c_{2312} \\ C_{51}=c_{1311} & C_{52}=c_{1322} & C_{53}=c_{1333} & C_{54}=2c_{1323} & C_{55}=2c_{1313} & C_{56}=2c_{1312} \\ C_{61}=c_{1211} & C_{62}=c_{1222} & C_{63}=c_{1233} & C_{64}=2c_{1223} & C_{65}=2c_{1213} & C_{66}=2c_{1212} \end{pmatrix}$$

$$C = \begin{pmatrix} C_{11} & C_{12} & C_{13} & C_{41} & C_{51} & C_{61} \\ C_{12} & C_{22} & C_{23} & C_{42} & C_{52} & C_{26} \\ C_{13} & C_{23} & C_{33} & C_{43} & C_{53} & C_{63} \\ C_{41} & C_{42} & C_{43} & C_{44} & C_{45} & C_{46} \\ C_{51} & C_{52} & C_{53} & C_{45} & C_{55} & C_{56} \\ C_{61} & C_{62} & C_{63} & C_{46} & C_{56} & C_{66} \end{pmatrix}$$

In the basis $b'$, we write: $\sigma' = C'\varepsilon'$. It can be easily shown that:

$$\begin{cases} C' = \bar{P}CP \\ C = PC'\bar{P} \end{cases}$$

## 7.3    Application: linear elastic anisotropic behaviour law and change of basis. Voigt notation

### 7.3.1    Voigt notation for the linearly elastic anisotropic behaviour law

An anisotropic material is considered for which we note $b^{(loc)} = (\vec{x}_1^{(loc)}, \vec{x}_2^{(loc)}, \vec{x}_3^{(loc)})$ the local basis of anisotropy. As usual, $\varepsilon^{(loc)} = (\varepsilon_{ij}^{(loc)})_{i,j=1,2,3}$, respectively $\sigma^{(loc)} = (\sigma_{ij}^{(loc)})_{i,j=1,2,3}$, designate the second-order symmetrical tensor of the linearised strains, respectively the stresses in the basis $b^{(loc)}$. The linear elastic behaviour law, written in $b^{(loc)}$, gives the relation between $\varepsilon^{(loc)}$ and $\sigma^{(loc)}$:

$$\left\{ \begin{array}{l} \sigma^{(loc)} = a^{(loc)}\varepsilon^{(loc)} \iff \sigma^{(loc)}_{ij} = a^{(loc)}_{ijkh}\varepsilon^{(loc)}_{kh} \\ a^{(loc)}_{ijkh} = a^{(loc)}_{jikh} = a^{(loc)}_{ijhk} = a^{(loc)}_{khij} \end{array} \right.$$

$$\iff$$

$$\left\{ \begin{array}{l} \varepsilon^{(loc)} = A^{(loc)}\sigma^{(loc)} \iff \varepsilon^{(loc)}_{ij} = A^{(loc)}_{ijkh}\sigma^{(loc)}_{kh} \\ A^{(loc)}_{ijkh} = A^{(loc)}_{jikh} = A^{(loc)}_{ijhk} = A^{(loc)}_{khij} \end{array} \right.$$

where $a^{(loc)} = (a^{(loc)}_{ijkh})_{i,j,k,h=1,2,3}$ and $A^{(loc)} = (A^{(loc)}_{ijkh})_{i,j,k,h=1,2,3}$ designate respectively the fourth-order tensor of rigidity and flexibility (with their usual symmetries).

Taking account of all the symmetries of the indices as well as the stress and strain tensors and the flexibility and rigidity it becomes possible to write the above relations in the following matricial form:

$$\begin{pmatrix} \sigma^{(loc)}_{11} \\ \sigma^{(loc)}_{22} \\ \sigma^{(loc)}_{33} \\ \sigma^{(loc)}_{23} \\ \sigma^{(loc)}_{13} \\ \sigma^{(loc)}_{12} \end{pmatrix} = \begin{pmatrix} a^{(loc)}_{1111} & a^{(loc)}_{1122} & a^{(loc)}_{1133} & a^{(loc)}_{1123} & a^{(loc)}_{1113} & a^{(loc)}_{1112} \\ a^{(loc)}_{1122} & a^{(loc)}_{2222} & a^{(loc)}_{2233} & a^{(loc)}_{2223} & a^{(loc)}_{2213} & a^{(loc)}_{2212} \\ a^{(loc)}_{1133} & a^{(loc)}_{2233} & a^{(loc)}_{3333} & a^{(loc)}_{3323} & a^{(loc)}_{3313} & a^{(loc)}_{3312} \\ a^{(loc)}_{1123} & a^{(loc)}_{2223} & a^{(loc)}_{3323} & a^{(loc)}_{2323} & a^{(loc)}_{2313} & a^{(loc)}_{2312} \\ a^{(loc)}_{1113} & a^{(loc)}_{2213} & a^{(loc)}_{3313} & a^{(loc)}_{2313} & a^{(loc)}_{1313} & a^{(loc)}_{1312} \\ a^{(loc)}_{1112} & a^{(loc)}_{2212} & a^{(loc)}_{3312} & a^{(loc)}_{2312} & a^{(loc)}_{1312} & a^{(loc)}_{1212} \end{pmatrix} \begin{pmatrix} \varepsilon^{(loc)}_{11} \\ \varepsilon^{(loc)}_{22} \\ \varepsilon^{(loc)}_{33} \\ 2\varepsilon^{(loc)}_{23} \\ 2\varepsilon^{(loc)}_{13} \\ 2\varepsilon^{(loc)}_{12} \end{pmatrix}$$

$$\iff$$

$$\begin{pmatrix} \sigma^{(loc)}_{1} \\ \sigma^{(loc)}_{2} \\ \sigma^{(loc)}_{3} \\ \sigma^{(loc)}_{4} \\ \sigma^{(loc)}_{5} \\ \sigma^{(loc)}_{6} \end{pmatrix} = \begin{pmatrix} R^{(loc)}_{11} & R^{(loc)}_{12} & R^{(loc)}_{13} & R^{(loc)}_{14} & R^{(loc)}_{15} & R^{(loc)}_{16} \\ R^{(loc)}_{12} & R^{(loc)}_{22} & R^{(loc)}_{23} & R^{(loc)}_{24} & R^{(loc)}_{25} & R^{(loc)}_{26} \\ R^{(loc)}_{13} & R^{(loc)}_{23} & R^{(loc)}_{33} & R^{(loc)}_{34} & R^{(loc)}_{35} & R^{(loc)}_{36} \\ R^{(loc)}_{14} & R^{(loc)}_{24} & R^{(loc)}_{34} & R^{(loc)}_{44} & R^{(loc)}_{45} & R^{(loc)}_{46} \\ R^{(loc)}_{15} & R^{(loc)}_{25} & R^{(loc)}_{35} & R^{(loc)}_{45} & R^{(loc)}_{55} & R^{(loc)}_{56} \\ R^{(loc)}_{16} & R^{(loc)}_{26} & R^{(loc)}_{36} & R^{(loc)}_{46} & R^{(loc)}_{56} & R^{(loc)}_{66} \end{pmatrix} \begin{pmatrix} \varepsilon^{(loc)}_{1} \\ \varepsilon^{(loc)}_{2} \\ \varepsilon^{(loc)}_{3} \\ \varepsilon^{(loc)}_{4} \\ \varepsilon^{(loc)}_{5} \\ \varepsilon^{(loc)}_{6} \end{pmatrix}$$

$$\begin{pmatrix} \varepsilon^{(loc)}_{11} \\ \varepsilon^{(loc)}_{22} \\ \varepsilon^{(loc)}_{33} \\ 2\varepsilon^{(loc)}_{23} \\ 2\varepsilon^{(loc)}_{13} \\ 2\varepsilon^{(loc)}_{12} \end{pmatrix} = \begin{pmatrix} A^{(loc)}_{1111} & A^{(loc)}_{1122} & A^{(loc)}_{1133} & 2A^{(loc)}_{1123} & 2A^{(loc)}_{1113} & 2A^{(loc)}_{1112} \\ A^{(loc)}_{1122} & A^{(loc)}_{2222} & A^{(loc)}_{2233} & 2A^{(loc)}_{2223} & 2A^{(loc)}_{2213} & 2A^{(loc)}_{2212} \\ A^{(loc)}_{1133} & A^{(loc)}_{2233} & A^{(loc)}_{3333} & 2A^{(loc)}_{3323} & 2A^{(loc)}_{3313} & 2A^{(loc)}_{3312} \\ 2A^{(loc)}_{1123} & 2A^{(loc)}_{2223} & 2A^{(loc)}_{3323} & 4A^{(loc)}_{2323} & 4A^{(loc)}_{2313} & 4A^{(loc)}_{2312} \\ 2A^{(loc)}_{1113} & 2A^{(loc)}_{2213} & 2A^{(loc)}_{3313} & 4A^{(loc)}_{2313} & 4A^{(loc)}_{1313} & 4A^{(loc)}_{1312} \\ 2A^{(loc)}_{1112} & 2A^{(loc)}_{2212} & 2A^{(loc)}_{3312} & 4A^{(loc)}_{2312} & 4A^{(loc)}_{1312} & 4A^{(loc)}_{1212} \end{pmatrix} \begin{pmatrix} \sigma^{(loc)}_{11} \\ \sigma^{(loc)}_{22} \\ \sigma^{(loc)}_{33} \\ \sigma^{(loc)}_{23} \\ \sigma^{(loc)}_{13} \\ \sigma^{(loc)}_{12} \end{pmatrix}$$

$$\iff$$

$$\begin{pmatrix} \varepsilon^{(loc)}_{1} \\ \varepsilon^{(loc)}_{2} \\ \varepsilon^{(loc)}_{3} \\ \varepsilon^{(loc)}_{4} \\ \varepsilon^{(loc)}_{5} \\ \varepsilon^{(loc)}_{6} \end{pmatrix} = \begin{pmatrix} S^{(loc)}_{11} & S^{(loc)}_{12} & S^{(loc)}_{13} & S^{(loc)}_{14} & S^{(loc)}_{15} & S^{(loc)}_{16} \\ S^{(loc)}_{12} & S^{(loc)}_{22} & S^{(loc)}_{23} & S^{(loc)}_{24} & S^{(loc)}_{25} & S^{(loc)}_{26} \\ S^{(loc)}_{13} & S^{(loc)}_{23} & S^{(loc)}_{33} & S^{(loc)}_{34} & S^{(loc)}_{35} & S^{(loc)}_{36} \\ S^{(loc)}_{14} & S^{(loc)}_{24} & S^{(loc)}_{34} & S^{(loc)}_{44} & S^{(loc)}_{45} & S^{(loc)}_{46} \\ S^{(loc)}_{15} & S^{(loc)}_{25} & S^{(loc)}_{35} & S^{(loc)}_{45} & S^{(loc)}_{55} & S^{(loc)}_{56} \\ S^{(loc)}_{16} & S^{(loc)}_{26} & S^{(loc)}_{36} & S^{(loc)}_{46} & S^{(loc)}_{56} & S^{(loc)}_{66} \end{pmatrix} \begin{pmatrix} \sigma^{(loc)}_{1} \\ \sigma^{(loc)}_{2} \\ \sigma^{(loc)}_{3} \\ \sigma^{(loc)}_{4} \\ \sigma^{(loc)}_{5} \\ \sigma^{(loc)}_{6} \end{pmatrix}$$

In the (classical) Voigt notation (which is not the modified Voigt notation) the relations are written in the following manner where $I$ and $J$ belong to the ordinate list $\{1 := 11, 2 := 22, 3 := 33, 4 := 23, 5 := 13, 6 := 12\}$:

$$\begin{cases} \sigma^{(loc)} = R^{(loc)}\varepsilon^{(loc)} \iff \sigma_I^{(loc)} = R_{IJ}^{(loc)}\varepsilon_J^{(loc)} \\ R_{IJ}^{(loc)} = R_{JI}^{(loc)} \end{cases}$$

$$\iff$$

$$\begin{cases} \varepsilon^{(loc)} = S^{(loc)}\sigma^{(loc)} \iff \varepsilon_I^{(loc)} = S_{IJ}^{(loc)}\sigma_J^{(loc)} \\ S_{IJ}^{(loc)} = S_{JI}^{(loc)} \end{cases}$$

where $R^{(loc)} = (R_{IJ}^{(loc)})_{I,J=1,\ldots,6}$ et $S^{(loc)} = (S_{IJ}^{(loc)})_{I,J=1,\ldots,6}$. We have:

$$\begin{cases} (R^{(loc)})^{-1} = S^{(loc)} \\ (S^{(loc)})^{-1} = R^{(loc)} \end{cases}$$

## 7.3.2 Reminder of the most usual forms of the linear anisotropic behaviour law

**Triclinic behaviour**

$$\begin{pmatrix} \sigma_1^{(loc)} \\ \sigma_2^{(loc)} \\ \sigma_3^{(loc)} \\ \sigma_4^{(loc)} \\ \sigma_5^{(loc)} \\ \sigma_6^{(loc)} \end{pmatrix} = \begin{pmatrix} R_{11}^{(loc)} & R_{12}^{(loc)} & R_{13}^{(loc)} & R_{14}^{(loc)} & R_{15}^{(loc)} & R_{16}^{(loc)} \\ R_{12}^{(loc)} & R_{22}^{(loc)} & R_{23}^{(loc)} & R_{24}^{(loc)} & R_{25}^{(loc)} & R_{26}^{(loc)} \\ R_{13}^{(loc)} & R_{23}^{(loc)} & R_{33}^{(loc)} & R_{34}^{(loc)} & R_{35}^{(loc)} & R_{36}^{(loc)} \\ R_{14}^{(loc)} & R_{24}^{(loc)} & R_{34}^{(loc)} & R_{44}^{(loc)} & R_{45}^{(loc)} & R_{46}^{(loc)} \\ R_{15}^{(loc)} & R_{25}^{(loc)} & R_{35}^{(loc)} & R_{45}^{(loc)} & R_{55}^{(loc)} & R_{56}^{(loc)} \\ R_{16}^{(loc)} & R_{26}^{(loc)} & R_{36}^{(loc)} & R_{46}^{(loc)} & R_{56}^{(loc)} & R_{66}^{(loc)} \end{pmatrix} \begin{pmatrix} \varepsilon_1^{(loc)} \\ \varepsilon_2^{(loc)} \\ \varepsilon_3^{(loc)} \\ \varepsilon_4^{(loc)} \\ \varepsilon_5^{(loc)} \\ \varepsilon_6^{(loc)} \end{pmatrix}$$

$$\iff$$

$$\begin{pmatrix} \varepsilon_1^{(loc)} \\ \varepsilon_2^{(loc)} \\ \varepsilon_3^{(loc)} \\ \varepsilon_4^{(loc)} \\ \varepsilon_5^{(loc)} \\ \varepsilon_6^{(loc)} \end{pmatrix} = \begin{pmatrix} S_{11}^{(loc)} & S_{12}^{(loc)} & S_{13}^{(loc)} & S_{14}^{(loc)} & S_{15}^{(loc)} & S_{16}^{(loc)} \\ S_{12}^{(loc)} & S_{22}^{(loc)} & S_{23}^{(loc)} & S_{24}^{(loc)} & S_{25}^{(loc)} & S_{26}^{(loc)} \\ S_{13}^{(loc)} & S_{23}^{(loc)} & S_{33}^{(loc)} & S_{34}^{(loc)} & S_{35}^{(loc)} & S_{36}^{(loc)} \\ S_{14}^{(loc)} & S_{24}^{(loc)} & S_{34}^{(loc)} & S_{44}^{(loc)} & S_{45}^{(loc)} & S_{46}^{(loc)} \\ S_{15}^{(loc)} & S_{25}^{(loc)} & S_{35}^{(loc)} & S_{45}^{(loc)} & S_{55}^{(loc)} & S_{56}^{(loc)} \\ S_{16}^{(loc)} & S_{26}^{(loc)} & S_{36}^{(loc)} & S_{46}^{(loc)} & S_{56}^{(loc)} & S_{66}^{(loc)} \end{pmatrix} \begin{pmatrix} \sigma_1^{(loc)} \\ \sigma_2^{(loc)} \\ \sigma_3^{(loc)} \\ \sigma_4^{(loc)} \\ \sigma_5^{(loc)} \\ \sigma_6^{(loc)} \end{pmatrix}$$

$(\vec{x_1}^{(loc)})$ **monoclinic behaviour**

$$\begin{pmatrix} \sigma_1^{(loc)} \\ \sigma_2^{(loc)} \\ \sigma_3^{(loc)} \\ \sigma_4^{(loc)} \\ \sigma_5^{(loc)} \\ \sigma_6^{(loc)} \end{pmatrix} = \begin{pmatrix} R_{11}^{(loc)} & R_{12}^{(loc)} & R_{13}^{(loc)} & R_{14}^{(loc)} & 0 & 0 \\ R_{12}^{(loc)} & R_{22}^{(loc)} & R_{23}^{(loc)} & R_{24}^{(loc)} & 0 & 0 \\ R_{13}^{(loc)} & R_{23}^{(loc)} & R_{33}^{(loc)} & R_{34}^{(loc)} & 0 & 0 \\ R_{14}^{(loc)} & R_{24}^{(loc)} & R_{34}^{(loc)} & R_{44}^{(loc)} & 0 & 0 \\ 0 & 0 & 0 & 0 & R_{55}^{(loc)} & R_{56}^{(loc)} \\ 0 & 0 & 0 & 0 & R_{56}^{(loc)} & R_{66}^{(loc)} \end{pmatrix} \begin{pmatrix} \varepsilon_1^{(loc)} \\ \varepsilon_2^{(loc)} \\ \varepsilon_3^{(loc)} \\ \varepsilon_4^{(loc)} \\ \varepsilon_5^{(loc)} \\ \varepsilon_6^{(loc)} \end{pmatrix}$$

$$\iff$$

$$
\begin{pmatrix}
\varepsilon_1^{(loc)} \\
\varepsilon_2^{(loc)} \\
\varepsilon_3^{(loc)} \\
\varepsilon_4^{(loc)} \\
\varepsilon_5^{(loc)} \\
\varepsilon_6^{(loc)}
\end{pmatrix}
=
\begin{pmatrix}
S_{11}^{(loc)} & S_{12}^{(loc)} & S_{13}^{(loc)} & S_{14}^{(loc)} & 0 & 0 \\
S_{12}^{(loc)} & S_{22}^{(loc)} & S_{23}^{(loc)} & S_{24}^{(loc)} & 0 & 0 \\
S_{13}^{(loc)} & S_{23}^{(loc)} & S_{33}^{(loc)} & S_{34}^{(loc)} & 0 & 0 \\
S_{14}^{(loc)} & S_{24}^{(loc)} & S_{34}^{(loc)} & S_{44}^{(loc)} & 0 & 0 \\
0 & 0 & 0 & 0 & S_{55}^{(loc)} & S_{56}^{(loc)} \\
0 & 0 & 0 & 0 & S_{56}^{(loc)} & S_{66}^{(loc)}
\end{pmatrix}
\begin{pmatrix}
\sigma_1^{(loc)} \\
\sigma_2^{(loc)} \\
\sigma_3^{(loc)} \\
\sigma_4^{(loc)} \\
\sigma_5^{(loc)} \\
\sigma_6^{(loc)}
\end{pmatrix}
$$

$(\vec{x_3}^{(loc)})$ **monoclinic behaviour**

$$
\begin{pmatrix}
\sigma_1^{(loc)} \\
\sigma_2^{(loc)} \\
\sigma_3^{(loc)} \\
\sigma_4^{(loc)} \\
\sigma_5^{(loc)} \\
\sigma_6^{(loc)}
\end{pmatrix}
=
\begin{pmatrix}
R_{11}^{(loc)} & R_{12}^{(loc)} & R_{13}^{(loc)} & 0 & 0 & R_{16}^{(loc)} \\
R_{12}^{(loc)} & R_{22}^{(loc)} & R_{23}^{(loc)} & 0 & 0 & R_{26}^{(loc)} \\
R_{13}^{(loc)} & R_{23}^{(loc)} & R_{33}^{(loc)} & 0 & 0 & R_{36}^{(loc)} \\
0 & 0 & 0 & R_{44}^{(loc)} & R_{45}^{(loc)} & 0 \\
0 & 0 & 0 & R_{45}^{(loc)} & R_{55}^{(loc)} & 0 \\
R_{16}^{(loc)} & R_{26}^{(loc)} & R_{36}^{(loc)} & 0 & 0 & R_{66}^{(loc)}
\end{pmatrix}
\begin{pmatrix}
\varepsilon_1^{(loc)} \\
\varepsilon_2^{(loc)} \\
\varepsilon_3^{(loc)} \\
\varepsilon_4^{(loc)} \\
\varepsilon_5^{(loc)} \\
\varepsilon_6^{(loc)}
\end{pmatrix}
$$

$$\Longleftrightarrow$$

$$
\begin{pmatrix}
\varepsilon_1^{(loc)} \\
\varepsilon_2^{(loc)} \\
\varepsilon_3^{(loc)} \\
\varepsilon_4^{(loc)} \\
\varepsilon_5^{(loc)} \\
\varepsilon_6^{(loc)}
\end{pmatrix}
=
\begin{pmatrix}
S_{11}^{(loc)} & S_{12}^{(loc)} & S_{13}^{(loc)} & 0 & 0 & S_{16}^{(loc)} \\
S_{12}^{(loc)} & S_{22}^{(loc)} & S_{23}^{(loc)} & 0 & 0 & S_{26}^{(loc)} \\
S_{13}^{(loc)} & S_{23}^{(loc)} & S_{33}^{(loc)} & 0 & 0 & S_{36}^{(loc)} \\
0 & 0 & 0 & S_{44}^{(loc)} & S_{45}^{(loc)} & 0 \\
0 & 0 & 0 & S_{45}^{(loc)} & S_{55}^{(loc)} & 0 \\
S_{16}^{(loc)} & S_{26}^{(loc)} & S_{36}^{(loc)} & 0 & 0 & S_{66}^{(loc)}
\end{pmatrix}
\begin{pmatrix}
\sigma_1^{(loc)} \\
\sigma_2^{(loc)} \\
\sigma_3^{(loc)} \\
\sigma_4^{(loc)} \\
\sigma_5^{(loc)} \\
\sigma_6^{(loc)}
\end{pmatrix}
$$

**Orthotropic behaviour**

$$
\begin{pmatrix}
\sigma_1^{(loc)} \\
\sigma_2^{(loc)} \\
\sigma_3^{(loc)} \\
\sigma_4^{(loc)} \\
\sigma_5^{(loc)} \\
\sigma_6^{(loc)}
\end{pmatrix}
=
\begin{pmatrix}
R_{11}^{(loc)} & R_{12}^{(loc)} & R_{13}^{(loc)} & 0 & 0 & 0 \\
R_{12}^{(loc)} & R_{22}^{(loc)} & R_{23}^{(loc)} & 0 & 0 & 0 \\
R_{13}^{(loc)} & R_{23}^{(loc)} & R_{33}^{(loc)} & 0 & 0 & 0 \\
0 & 0 & 0 & R_{44}^{(loc)} & 0 & 0 \\
0 & 0 & 0 & 0 & R_{55}^{(loc)} & 0 \\
0 & 0 & 0 & 0 & 0 & R_{66}^{(loc)}
\end{pmatrix}
\begin{pmatrix}
\varepsilon_1^{(loc)} \\
\varepsilon_2^{(loc)} \\
\varepsilon_3^{(loc)} \\
\varepsilon_4^{(loc)} \\
\varepsilon_5^{(loc)} \\
\varepsilon_6^{(loc)}
\end{pmatrix}
$$

$$\Longleftrightarrow$$

$$
\begin{pmatrix}
\varepsilon_1^{(loc)} \\
\varepsilon_2^{(loc)} \\
\varepsilon_3^{(loc)} \\
\varepsilon_4^{(loc)} \\
\varepsilon_5^{(loc)} \\
\varepsilon_6^{(loc)}
\end{pmatrix}
=
\begin{pmatrix}
S_{11}^{(loc)} & S_{12}^{(loc)} & S_{13}^{(loc)} & 0 & 0 & 0 \\
S_{12}^{(loc)} & S_{22}^{(loc)} & S_{23}^{(loc)} & 0 & 0 & 0 \\
S_{13}^{(loc)} & S_{23}^{(loc)} & S_{33}^{(loc)} & 0 & 0 & 0 \\
0 & 0 & 0 & S_{44}^{(loc)} & 0 & 0 \\
0 & 0 & 0 & 0 & S_{55}^{(loc)} & 0 \\
0 & 0 & 0 & 0 & 0 & S_{66}^{(loc)}
\end{pmatrix}
\begin{pmatrix}
\sigma_1^{(loc)} \\
\sigma_2^{(loc)} \\
\sigma_3^{(loc)} \\
\sigma_4^{(loc)} \\
\sigma_5^{(loc)} \\
\sigma_6^{(loc)}
\end{pmatrix}
$$

Using the usual constants the following forms can be written:

$$
S^{(loc)} =
\begin{pmatrix}
\frac{1}{E_1} & \frac{-\nu_{21}}{E_2} & \frac{-\nu_{31}}{E_3} & 0 & 0 & 0 \\
\frac{-\nu_{12}}{E_1} & \frac{1}{E_2} & \frac{-\nu_{32}}{E_3} & 0 & 0 & 0 \\
\frac{-\nu_{13}}{E_1} & \frac{-\nu_{23}}{E_2} & \frac{1}{E_3} & 0 & 0 & 0 \\
0 & 0 & 0 & \frac{1}{G_{23}} & 0 & 0 \\
0 & 0 & 0 & 0 & \frac{1}{G_{13}} & 0 \\
0 & 0 & 0 & 0 & 0 & \frac{1}{G_{12}}
\end{pmatrix}
$$

$$
R^{(loc)} = \begin{pmatrix}
\frac{(1-\nu_{23}\nu_{32})E_1}{\delta} & \frac{(\nu_{23}\nu_{31}+\nu_{21})E_1}{\delta} & \frac{(\nu_{21}\nu_{32}+\nu_{31})E_1}{\delta} & 0 & 0 & 0 \\
\frac{(\nu_{13}\nu_{32}+\nu_{12})E_2}{\delta} & \frac{(1-\nu_{13}\nu_{31})E_2}{\delta} & \frac{(\nu_{12}\nu_{31}+\nu_{32})E_2}{\delta} & 0 & 0 & 0 \\
\frac{(\nu_{12}\nu_{23}+\nu_{13})E_3}{\delta} & \frac{(\nu_{13}\nu_{21}+\nu_{23})E_3}{\delta} & \frac{(1-\nu_{12}\nu_{21})E_3}{\delta} & 0 & 0 & 0 \\
0 & 0 & 0 & G_{23} & 0 & 0 \\
0 & 0 & 0 & 0 & G_{13} & 0 \\
0 & 0 & 0 & 0 & 0 & G_{12}
\end{pmatrix}
$$

with : $\delta = 1 - \nu_{12}\nu_{23}\nu_{31} - \nu_{13}\nu_{21}\nu_{32} - \nu_{12}\nu_{21} - \nu_{13}\nu_{31} - \nu_{23}\nu_{32}$.

$(\vec{x_1}^{(loc)})$ **transverse isotropic behaviour**

$$
\begin{pmatrix}
\sigma_1^{(loc)} \\
\sigma_2^{(loc)} \\
\sigma_3^{(loc)} \\
\sigma_4^{(loc)} \\
\sigma_5^{(loc)} \\
\sigma_6^{(loc)}
\end{pmatrix} = \begin{pmatrix}
R_{11}^{(loc)} & R_{12}^{(loc)} & R_{12}^{(loc)} & 0 & 0 & 0 \\
R_{12}^{(loc)} & R_{22}^{(loc)} & R_{23}^{(loc)} & 0 & 0 & 0 \\
R_{12}^{(loc)} & R_{23}^{(loc)} & R_{22}^{(loc)} & 0 & 0 & 0 \\
0 & 0 & 0 & \frac{R_{22}^{(loc)}-R_{23}^{(loc)}}{2} & 0 & 0 \\
0 & 0 & 0 & 0 & R_{66}^{(loc)} & 0 \\
0 & 0 & 0 & 0 & 0 & R_{66}^{(loc)}
\end{pmatrix} \begin{pmatrix}
\varepsilon_1^{(loc)} \\
\varepsilon_2^{(loc)} \\
\varepsilon_3^{(loc)} \\
\varepsilon_4^{(loc)} \\
\varepsilon_5^{(loc)} \\
\varepsilon_6^{(loc)}
\end{pmatrix}
$$

$$\Longleftrightarrow$$

$$
\begin{pmatrix}
\varepsilon_1^{(loc)} \\
\varepsilon_2^{(loc)} \\
\varepsilon_3^{(loc)} \\
\varepsilon_4^{(loc)} \\
\varepsilon_5^{(loc)} \\
\varepsilon_6^{(loc)}
\end{pmatrix} = \begin{pmatrix}
S_{11}^{(loc)} & S_{12}^{(loc)} & S_{12}^{(loc)} & 0 & 0 & 0 \\
S_{12}^{(loc)} & S_{22}^{(loc)} & S_{23}^{(loc)} & 0 & 0 & 0 \\
S_{12}^{(loc)} & S_{23}^{(loc)} & S_{22}^{(loc)} & 0 & 0 & 0 \\
0 & 0 & 0 & 2(S_{22}^{(loc)} - S_{23}^{(loc)}) & 0 & 0 \\
0 & 0 & 0 & 0 & S_{66}^{(loc)} & 0 \\
0 & 0 & 0 & 0 & 0 & S_{66}^{(loc)}
\end{pmatrix} \begin{pmatrix}
\sigma_1^{(loc)} \\
\sigma_2^{(loc)} \\
\sigma_3^{(loc)} \\
\sigma_4^{(loc)} \\
\sigma_5^{(loc)} \\
\sigma_6^{(loc)}
\end{pmatrix}
$$

$(\vec{x_3}^{(loc)})$ **transverse isotropic behaviour**

$$
\begin{pmatrix}
\sigma_1^{(loc)} \\
\sigma_2^{(loc)} \\
\sigma_3^{(loc)} \\
\sigma_4^{(loc)} \\
\sigma_5^{(loc)} \\
\sigma_6^{(loc)}
\end{pmatrix} = \begin{pmatrix}
R_{11}^{(loc)} & R_{12}^{(loc)} & R_{13}^{(loc)} & 0 & 0 & 0 \\
R_{12}^{(loc)} & R_{11}^{(loc)} & R_{13}^{(loc)} & 0 & 0 & 0 \\
R_{13}^{(loc)} & R_{13}^{(loc)} & R_{33}^{(loc)} & 0 & 0 & 0 \\
0 & 0 & 0 & R_{44}^{(loc)} & 0 & 0 \\
0 & 0 & 0 & 0 & R_{44}^{(loc)} & 0 \\
0 & 0 & 0 & 0 & 0 & \frac{R_{11}^{(loc)}-R_{12}^{(loc)}}{2}
\end{pmatrix} \begin{pmatrix}
\varepsilon_1^{(loc)} \\
\varepsilon_2^{(loc)} \\
\varepsilon_3^{(loc)} \\
\varepsilon_4^{(loc)} \\
\varepsilon_5^{(loc)} \\
\varepsilon_6^{(loc)}
\end{pmatrix}
$$

$$\Longleftrightarrow$$

$$
\begin{pmatrix}
\varepsilon_1^{(loc)} \\
\varepsilon_2^{(loc)} \\
\varepsilon_3^{(loc)} \\
\varepsilon_4^{(loc)} \\
\varepsilon_5^{(loc)} \\
\varepsilon_6^{(loc)}
\end{pmatrix} = \begin{pmatrix}
S_{11}^{(loc)} & S_{12}^{(loc)} & S_{13}^{(loc)} & 0 & 0 & 0 \\
S_{12}^{(loc)} & S_{11}^{(loc)} & S_{23}^{(loc)} & 0 & 0 & 0 \\
S_{13}^{(loc)} & S_{13}^{(loc)} & S_{33}^{(loc)} & 0 & 0 & 0 \\
0 & 0 & 0 & S_{44}^{(loc)} & 0 & 0 \\
0 & 0 & 0 & 0 & S_{44}^{(loc)} & 0 \\
0 & 0 & 0 & 0 & 0 & 2(S_{11}^{(loc)} - S_{12}^{(loc)})
\end{pmatrix} \begin{pmatrix}
\sigma_1^{(loc)} \\
\sigma_2^{(loc)} \\
\sigma_3^{(loc)} \\
\sigma_4^{(loc)} \\
\sigma_5^{(loc)} \\
\sigma_6^{(loc)}
\end{pmatrix}
$$

## Isotropic behaviour

Using the usual constant (Young modulus, Poisson coefficient, Lamé coefficients), we have (it can be pointed out that the $(loc)$ precision is not necessary because, for this type of behaviour, this law is the same whatever the orthonormal basis is):

$$
S^{(loc)} = \begin{pmatrix}
\frac{1}{E} & \frac{-\nu}{E} & \frac{-\nu}{E} & 0 & 0 & 0 \\
\frac{-\nu}{E} & \frac{1}{E} & \frac{-\nu}{E} & 0 & 0 & 0 \\
\frac{-\nu}{E} & \frac{-\nu}{E} & \frac{1}{E} & 0 & 0 & 0 \\
0 & 0 & 0 & \frac{2(1+\nu)}{E} & 0 & 0 \\
0 & 0 & 0 & 0 & \frac{2(1+\nu)}{E} & 0 \\
0 & 0 & 0 & 0 & 0 & \frac{2(1+\nu)}{E}
\end{pmatrix}
$$

$$
R^{(loc)} = \begin{pmatrix}
\frac{E(1-\nu)}{(1+\nu)(1-2\nu)} & \frac{\nu E}{(1+\nu)(1-2\nu)} & \frac{\nu E}{(1+\nu)(1-2\nu)} & 0 & 0 & 0 \\
\frac{\nu E}{(1+\nu)(1-2\nu)} & \frac{E(1-\nu)}{(1+\nu)(1-2\nu)} & \frac{\nu E}{(1+\nu)(1-2\nu)} & 0 & 0 & 0 \\
\frac{\nu E}{(1+\nu)(1-2\nu)} & \frac{\nu E}{(1+\nu)(1-2\nu)} & \frac{E(1-\nu)}{(1+\nu)(1-2\nu)} & 0 & 0 & 0 \\
0 & 0 & 0 & \frac{E}{2(1+\nu)} & 0 & 0 \\
0 & 0 & 0 & 0 & \frac{E}{2(1+\nu)} & 0 \\
0 & 0 & 0 & 0 & 0 & \frac{E}{2(1+\nu)}
\end{pmatrix}
$$

$$
S^{(loc)} = \begin{pmatrix}
\frac{\mu(\lambda+\mu)}{(3\lambda+2\mu)} & -\frac{\lambda\mu}{2(3\lambda+2\mu)} & -\frac{\lambda\mu}{2(3\lambda+2\mu)} & 0 & 0 & 0 \\
-\frac{\lambda\mu}{2(3\lambda+2\mu)} & \frac{\mu(\lambda+\mu)}{(3\lambda+2\mu)} & -\frac{\lambda\mu}{2(3\lambda+2\mu)} & 0 & 0 & 0 \\
-\frac{\lambda\mu}{2(3\lambda+2\mu)} & -\frac{\lambda\mu}{2(3\lambda+2\mu)} & \frac{\mu(\lambda+\mu)}{(3\lambda+2\mu)} & 0 & 0 & 0 \\
0 & 0 & 0 & 1/\mu & 0 & 0 \\
0 & 0 & 0 & 0 & 1/\mu & 0 \\
0 & 0 & 0 & 0 & 0 & 1/\mu
\end{pmatrix}
$$

$$
R^{(loc)} = \begin{pmatrix}
\lambda+2\mu & \lambda & \lambda & 0 & 0 & 0 \\
\lambda & \lambda+2\mu & \lambda & 0 & 0 & 0 \\
\lambda & \lambda & \lambda+2\mu & 0 & 0 & 0 \\
0 & 0 & 0 & \mu & 0 & 0 \\
0 & 0 & 0 & 0 & \mu & 0 \\
0 & 0 & 0 & 0 & 0 & \mu
\end{pmatrix}
$$

## Plane stress behaviour law

We can imagine that for certain geometries of a structure subjected to certain particular loads the existing stress states, outside of a given plane, will be zero or nearly so. This would be the case for thin structures loaded in their median plane. Let's suppose that that plane is generated by the vectors $(\vec{x}_1, \vec{x}_2)$. In this case the behaviour law can be taken in a simplified form obtained by using the restriction plane of the three-dimensional behaviour law written with the flexibility tensor:

$$
\begin{pmatrix}
\varepsilon_1^{(loc)} \\
\varepsilon_2^{(loc)} \\
\varepsilon_3^{(loc)} \\
\varepsilon_4^{(loc)} \\
\varepsilon_5^{(loc)} \\
\varepsilon_6^{(loc)}
\end{pmatrix}
=
\begin{pmatrix}
S_{11}^{(loc)} & S_{12}^{(loc)} & S_{13}^{(loc)} & S_{14}^{(loc)} & S_{15}^{(loc)} & S_{16}^{(loc)} \\
S_{12}^{(loc)} & S_{22}^{(loc)} & S_{23}^{(loc)} & S_{24}^{(loc)} & S_{25}^{(loc)} & S_{26}^{(loc)} \\
S_{13}^{(loc)} & S_{23}^{(loc)} & S_{33}^{(loc)} & S_{34}^{(loc)} & S_{35}^{(loc)} & S_{36}^{(loc)} \\
S_{14}^{(loc)} & S_{24}^{(loc)} & S_{34}^{(loc)} & S_{44}^{(loc)} & S_{45}^{(loc)} & S_{46}^{(loc)} \\
S_{15}^{(loc)} & S_{25}^{(loc)} & S_{35}^{(loc)} & S_{45}^{(loc)} & S_{55}^{(loc)} & S_{56}^{(loc)} \\
S_{16}^{(loc)} & S_{26}^{(loc)} & S_{36}^{(loc)} & S_{46}^{(loc)} & S_{56}^{(loc)} & S_{66}^{(loc)}
\end{pmatrix}
\begin{pmatrix}
\sigma_1^{(loc)} \\
\sigma_2^{(loc)} \\
0 \\
0 \\
0 \\
\sigma_6^{(loc)}
\end{pmatrix}
$$

$$\Longrightarrow$$

$$
\begin{pmatrix}
\varepsilon_1^{(loc)} \\
\varepsilon_2^{(loc)} \\
\varepsilon_6^{(loc)}
\end{pmatrix}
=
\begin{pmatrix}
S_{11}^{(loc)} & S_{12}^{(loc)} & S_{16}^{(loc)} \\
S_{12}^{(loc)} & S_{22}^{(loc)} & S_{26}^{(loc)} \\
S_{16}^{(loc)} & S_{26}^{(loc)} & S_{66}^{(loc)}
\end{pmatrix}
\begin{pmatrix}
\sigma_1^{(loc)} \\
\sigma_2^{(loc)} \\
\sigma_6^{(loc)}
\end{pmatrix}
$$

Then by inverting this relation we obtain $(\alpha, \beta = 1, 2, 6)$:

$$\begin{pmatrix} \sigma_1^{(loc)} \\ \sigma_2^{(loc)} \\ \sigma_6^{(loc)} \end{pmatrix} = \begin{pmatrix} R'^{(loc)}_{11} & R'^{(loc)}_{12} & R'^{(loc)}_{16} \\ R'^{(loc)}_{12} & R'^{(loc)}_{22} & R'^{(loc)}_{26} \\ R'^{(loc)}_{16} & R'^{(loc)}_{26} & R'^{(loc)}_{66} \end{pmatrix} \begin{pmatrix} \varepsilon_1^{(loc)} \\ \varepsilon_2^{(loc)} \\ \varepsilon_6^{(loc)} \end{pmatrix} \;,\; R'^{(loc)}_{\alpha\beta} = R^{(loc)}_{\alpha\beta} - \frac{R^{(loc)}_{\alpha3} R^{(loc)}_{\beta3}}{R^{(loc)}_{33}}$$

These relations are written in the following form $(\alpha, \beta = 1, 2, 6)$:

$$\begin{cases} \sigma^{(ploc)} = R^{(ploc)} \varepsilon^{(ploc)} \Longleftrightarrow \sigma_\alpha^{(ploc)} = R^{(ploc)}_{\alpha\beta} \varepsilon_\beta^{(ploc)} \\ R^{(ploc)}_{\alpha\beta} = R^{(ploc)}_{\beta\alpha} \end{cases}$$

$$\Longleftrightarrow$$

$$\begin{cases} \varepsilon^{(ploc)} = S^{(ploc)} \sigma^{(ploc)} \Longleftrightarrow \varepsilon_\alpha^{(ploc)} = S^{(ploc)}_{\alpha\beta} \sigma_\beta^{(ploc)} \\ S^{(ploc)}_{\alpha\beta} = S^{(ploc)}_{\beta\alpha} \end{cases}$$

where $R^{(ploc)} = (R^{(ploc)}_{\alpha\beta})_{\alpha,\beta=1,2,6}$ and $S^{(ploc)} = (S^{(ploc)}_{\alpha\beta})_{\alpha,\beta=1,2,6}$. We have:

$$\begin{cases} (R^{(ploc)})^{-1} = S^{(ploc)} \\ (S^{(ploc)})^{-1} = R^{(ploc)} \end{cases}$$

### 7.3.3    Change of basis. Case of general behaviour law

Consider the vector space $\mathbb{R}^3$ on the field $\mathbb{R}$ for which the orthonormal basis (defined as the reference basis) is $b^{(ref)} = (\vec{x}_1^{(ref)}, \vec{x}_2^{(ref)}, \vec{x}_3^{(ref)})$.

We consider an other basis $b^{(loc)}$. We suppose that the coordinates of the vectors of $b^{(loc)}$ are known in $b^{(ref)}$. They are written in the following manner:

$$\begin{cases} \vec{x}_1^{(loc)} = Q_{11}\vec{x}_1^{(ref)} + Q_{21}\vec{x}_2^{(ref)} + Q_{31}\vec{x}_3^{(ref)} \\ \vec{x}_2^{(loc)} = Q_{12}\vec{x}_1^{(ref)} + Q_{22}\vec{x}_2^{(ref)} + Q_{32}\vec{x}_3^{(ref)} \\ \vec{x}_3^{(loc)} = Q_{13}\vec{x}_1^{(ref)} + Q_{23}\vec{x}_2^{(ref)} + Q_{33}\vec{x}_3^{(ref)} \end{cases}$$

With the aid of these formulae it is possible to construct the $Q$, called the change of basis matrix from $b^{(ref)}$ to $b^{(loc)}$:

$$Q = \begin{pmatrix} Q_{11} & Q_{12} & Q_{13} \\ Q_{21} & Q_{22} & Q_{23} \\ Q_{31} & Q_{32} & Q_{33} \end{pmatrix}$$

The components of $Q$ are given in $b^{(ref)}$. As the two bases $b^{(ref)}$ and $b^{(loc)}$ are orthonormal then $Q^{-1} = Q^T$.

We define the following matrices:

$$P_\sigma = \begin{pmatrix} Q_{11}^2 & Q_{12}^2 & Q_{13}^2 & 2Q_{13}Q_{12} & 2Q_{13}Q_{11} & 2Q_{11}Q_{12} \\ Q_{21}^2 & Q_{22}^2 & Q_{23}^2 & 2Q_{23}Q_{22} & 2Q_{23}Q_{21} & 2Q_{22}Q_{21} \\ Q_{31}^2 & Q_{32}^2 & Q_{33}^2 & 2Q_{33}Q_{32} & 2Q_{33}Q_{31} & 2Q_{32}Q_{31} \\ Q_{21}Q_{31} & Q_{22}Q_{32} & Q_{23}Q_{33} & Q_{23}Q_{23}+Q_{22}Q_{33} & Q_{23}Q_{31}+Q_{21}Q_{33} & Q_{22}Q_{31}+Q_{21}Q_{32} \\ Q_{11}Q_{31} & Q_{12}Q_{32} & Q_{13}Q_{33} & Q_{13}Q_{32}+Q_{12}Q_{33} & Q_{13}Q_{31}+Q_{11}Q_{33} & Q_{12}Q_{31}+Q_{11}Q_{32} \\ Q_{11}Q_{21} & Q_{12}Q_{22} & Q_{13}Q_{23} & Q_{13}Q_{22}+Q_{12}Q_{23} & Q_{13}Q_{21}+Q_{11}Q_{23} & Q_{12}Q_{21}+Q_{11}Q_{22} \end{pmatrix}$$

$$\overline{P}_\sigma = \begin{pmatrix} Q_{11}^2 & Q_{21}^2 & Q_{31}^2 & 2Q_{31}Q_{21} & 2Q_{31}Q_{11} & 2Q_{11}Q_{21} \\ Q_{12}^2 & Q_{22}^2 & Q_{32}^2 & 2Q_{32}Q_{22} & 2Q_{32}Q_{12} & 2Q_{22}Q_{12} \\ Q_{13}^2 & Q_{23}^2 & Q_{33}^2 & 2Q_{33}Q_{23} & 2Q_{33}Q_{13} & 2Q_{23}Q_{13} \\ Q_{12}Q_{13} & Q_{22}Q_{23} & Q_{32}Q_{33} & Q_{32}Q_{23}+Q_{22}Q_{33} & Q_{32}Q_{13}+Q_{12}Q_{33} & Q_{22}Q_{13}+Q_{12}Q_{23} \\ Q_{11}Q_{13} & Q_{21}Q_{23} & Q_{31}Q_{33} & Q_{31}Q_{23}+Q_{21}Q_{33} & Q_{31}Q_{13}+Q_{11}Q_{33} & Q_{21}Q_{13}+Q_{11}Q_{23} \\ Q_{11}Q_{12} & Q_{21}Q_{22} & Q_{31}Q_{32} & Q_{31}Q_{22}+Q_{21}Q_{32} & Q_{31}Q_{12}+Q_{11}Q_{32} & Q_{21}Q_{12}+Q_{11}Q_{22} \end{pmatrix}$$

We construct the following matrices $P_\varepsilon$ and $\overline{P}_\varepsilon$:

$$\begin{cases} P_\varepsilon = (\overline{P}_\sigma)^T \\ \overline{P}_\varepsilon = (P_\sigma)^T \end{cases}$$

The matrix $\overline{P}_\sigma$ is constructed on the model of $P_\sigma$ where we use the components of the transposed matrix of $Q$. The matrix $\overline{P}_\varepsilon$ is constructed on the model of $P_\varepsilon$ where the components of the transposed matrix of $Q$. We have the following relations:

$$\begin{cases} \overline{P}_\sigma P_\sigma = I_6 \\ P_\sigma \overline{P}_\sigma = I_6 \end{cases} \iff \begin{cases} \overline{P}_\sigma = (P_\sigma)^{-1} \\ P_\sigma = (\overline{P}_\sigma)^{-1} \end{cases}$$

$$\begin{cases} \overline{P}_\varepsilon P_\varepsilon = I_6 \\ P_\varepsilon \overline{P}_\varepsilon = I_6 \end{cases} \iff \begin{cases} \overline{P}_\varepsilon = (P_\varepsilon)^{-1} \\ P_\varepsilon = (\overline{P}_\varepsilon)^{-1} \end{cases}$$

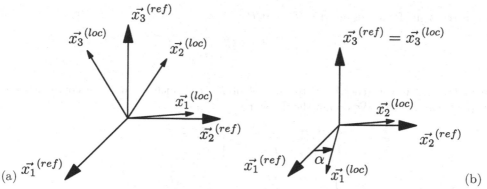

Figure 7.1: Change of basis. (a) General case. (b) Plane case, rotation around $\vec{x}_2^{(ref)}$ with angle $\alpha$.

### 7.3.4 Plane change of basis. Case of general behaviour law

Consider the vector space $\mathbb{R}^3$ on the field $\mathbb{R}$ for which the orthonormal basis (defined as the reference basis) is $b^{(ref)} = (\vec{x}_1^{(ref)}, \vec{x}_2^{(ref)}, \vec{x}_3^{(ref)})$.

We consider an other basis $b^{(loc)}$. We suppose that the coordinates of the vectors of $b^{(loc)}$ are known in $b^{(ref)}$. They are written in the following manner:

$$\begin{cases} \vec{x}_1^{(loc)} = Q_{11}\vec{x}_1^{(ref)} + Q_{21}\vec{x}_2^{(ref)} = cos\alpha\,\vec{x}_1^{(ref)} + sin\alpha\,\vec{x}_2^{(ref)} = c\,\vec{x}_1^{(ref)} + s\,\vec{x}_2^{(ref)} \\ \vec{x}_2^{(loc)} = Q_{12}\vec{x}_1^{(ref)} + Q_{22}\vec{x}_2^{(ref)} = -sin\alpha\,\vec{x}_1^{(ref)} + cos\alpha\,\vec{x}_2^{(ref)} = -s\,\vec{x}_1^{(ref)} + c\,\vec{x}_2^{(ref)} \\ \vec{x}_3^{(loc)} = \vec{x}_3^{(ref)} \end{cases}$$

With the aid of these formulae it is possible to construct the matrix $Q$, called the change of basis matrix from $b^{(ref)}$ to $b^{(loc)}$:

$$Q = \begin{pmatrix} Q_{11} = cos\alpha = c & Q_{12} = -sin\alpha = -s & 0 \\ Q_{21} = sin\alpha = s & Q_{22} = cos\alpha = c & 0 \\ 0 & 0 & 1 \end{pmatrix}$$

We define the following matrices:

$$P_\sigma = \begin{pmatrix} Q_{11}^2 = c^2 & Q_{12}^2 = s^2 & 0 & 0 & 0 & 2Q_{11}Q_{12} = -2sc \\ Q_{21}^2 = s^2 & Q_{22}^2 = c^2 & 0 & 0 & 0 & 2Q_{22}Q_{21} = 2sc \\ 0 & 0 & 1 & 0 & 0 & 0 \\ 0 & 0 & 0 & Q_{22} = c & Q_{21} = s & 0 \\ 0 & 0 & 0 & Q_{12} = -s & Q_{11} = c & 0 \\ Q_{11}Q_{21} = sc & Q_{12}Q_{22} = -sc & 0 & 0 & 0 & Q_{12}Q_{21} + Q_{11}Q_{22} = c^2 - s^2 \end{pmatrix}$$

$$\overline{P}_\sigma = \begin{pmatrix} Q_{11}^2 = c^2 & Q_{21}^2 = s^2 & 0 & 0 & 0 & 2Q_{11}Q_{21} = 2sc \\ Q_{12}^2 = s^2 & Q_{22}^2 = c^2 & 0 & 0 & 0 & 2Q_{22}Q_{12} = -2sc \\ 0 & 0 & 1 & 0 & 0 & 0 \\ 0 & 0 & 0 & Q_{22} = c & Q_{12} = -s & 0 \\ 0 & 0 & 0 & Q_{21} = s & Q_{11} = c & 0 \\ Q_{11}Q_{12} = -sc & Q_{21}Q_{22} = sc & 0 & 0 & 0 & Q_{21}Q_{12} + Q_{11}Q_{22} = c^2 - s^2 \end{pmatrix}$$

We construct the following matrices $P_\varepsilon$ and $\overline{P}_\varepsilon$:

$$\begin{cases} P_\varepsilon = (\overline{P}_\sigma)^T \\ \overline{P}_\varepsilon = (P_\sigma)^T \end{cases}$$

The matrix $\overline{P}_\varepsilon$ is constructed on the model of $P_\varepsilon$ for which we use the components of the transpose matrix of $Q$. We obtain the following relations:

$$\begin{cases} \overline{P}_\sigma P_\sigma = I_6 \\ P_\sigma \overline{P}_\sigma = I_6 \end{cases} \Longleftrightarrow \begin{cases} \overline{P}_\sigma = (P_\sigma)^{-1} \\ P_\sigma = (\overline{P}_\sigma)^{-1} \end{cases}$$

$$\begin{cases} \overline{P}_\varepsilon P_\varepsilon = I_6 \\ P_\varepsilon \overline{P}_\varepsilon = I_6 \end{cases} \Longleftrightarrow \begin{cases} \overline{P}_\varepsilon = (P_\varepsilon)^{-1} \\ P_\varepsilon = (\overline{P}_\varepsilon)^{-1} \end{cases}$$

In the bases $b^{(ref)}$ and $b^{(loc)}$, we note the behaviour laws in the Voigt form in the following manner:

$$\begin{cases} \sigma^{(ref)} = R^{(ref)} \varepsilon^{(ref)} \\ \sigma^{(loc)} = R^{(loc)} \varepsilon^{(loc)} \end{cases} \quad \begin{cases} \varepsilon^{(ref)} = S^{(ref)} \sigma^{(ref)} \\ \varepsilon^{(loc)} = S^{(loc)} \sigma^{(loc)} \end{cases} \quad \begin{cases} R^{(ref)} S^{(ref)} = I_6 \\ R^{(loc)} S^{(loc)} = I_6 \end{cases}$$

We have the following formulae for the change of basis:

$$\begin{cases} \sigma^{(ref)} = P_\sigma \sigma^{(loc)} \\ \sigma^{(loc)} = \overline{P}_\sigma \sigma^{(ref)} \end{cases} \quad \begin{cases} \varepsilon^{(ref)} = P_\varepsilon \varepsilon^{(loc)} \\ \varepsilon^{(loc)} = \overline{P}_\varepsilon \varepsilon^{(ref)} \end{cases}$$

$$\begin{cases} R^{(loc)} = \overline{P}_\sigma R^{(ref)} P_\varepsilon \\ R^{(ref)} = P_\sigma R^{(loc)} \overline{P}_\varepsilon \end{cases} \quad \begin{cases} S^{(loc)} = \overline{P}_\varepsilon S^{(ref)} P_\sigma \\ S^{(ref)} = P_\varepsilon S^{(loc)} \overline{P}_\sigma \end{cases}$$

## 7.3.5   Plane change of basis. Case of plane stress behaviour law

Consider the vector space $\mathbb{R}^3$ on the field $\mathbb{R}$ for which the orthonormal basis (defined as the reference basis) is $b^{(ref)} = (\vec{x}_1^{(ref)}, \vec{x}_2^{(ref)}, \vec{x}_3^{(ref)})$.

We consider another basis $b^{(loc)}$. We suppose that the coordinates of the vectors of $b^{(loc)}$ are known in $b^{(ref)}$. They are written in the following manner:

$$\begin{cases} \vec{x}_1^{(loc)} = Q_{11}\vec{x}_1^{(ref)} + Q_{21}\vec{x}_2^{(ref)} = cos\alpha\, \vec{x}_1^{(ref)} + sin\alpha\, \vec{x}_2^{(ref)} = c\,\vec{x}_1^{(ref)} + s\,\vec{x}_2^{(ref)} \\ \vec{x}_2^{(loc)} = Q_{12}\vec{x}_1^{(ref)} + Q_{22}\vec{x}_2^{(ref)} = -sin\alpha\, \vec{x}_1^{(ref)} + cos\alpha\, \vec{x}_2^{(ref)} = -s\,\vec{x}_1^{(ref)} + c\,\vec{x}_2^{(ref)} \\ \vec{x}_3^{(loc)} = \vec{x}_3^{(ref)} \end{cases}$$

With the aid of these formulae it is possible to construct the matrix $Q$, called the change of basis matrix from $b^{(ref)}$ to $b^{(loc)}$:

$$Q = \begin{pmatrix} Q_{11} = \cos\alpha = c & Q_{12} = -\sin\alpha = -s & 0 \\ Q_{21} = \sin\alpha = s & Q_{22} = \cos\alpha = c & 0 \\ 0 & 0 & 1 \end{pmatrix}$$

We define the following matrices:

$$P_\sigma^{(p)} = \begin{pmatrix} Q_{11}^2 = c^2 & Q_{12}^2 = s^2 & 2Q_{11}Q_{12} = -2sc \\ Q_{21}^2 = s^2 & Q_{22}^2 = c^2 & 2Q_{22}Q_{21} = 2sc \\ Q_{11}Q_{21} = sc & Q_{12}Q_{22} = -sc & Q_{12}Q_{21} + Q_{11}Q_{22} = c^2 - s^2 \end{pmatrix}$$

$$\overline{P}_\sigma^{(p)} = \begin{pmatrix} Q_{11}^2 = c^2 & Q_{21}^2 = s^2 & 2Q_{11}Q_{21} = 2sc \\ Q_{12}^2 = s^2 & Q_{22}^2 = c^2 & 2Q_{22}Q_{12} = -2sc \\ Q_{11}Q_{12} = -sc & Q_{21}Q_{22} = sc & Q_{21}Q_{12} + Q_{11}Q_{22} = c^2 - s^2 \end{pmatrix}$$

We construct the following matrices $P_\varepsilon^{(p)}$ and $\overline{P}_\varepsilon^{(p)}$:

$$\begin{cases} P_\varepsilon^{(p)} = (\overline{P}_\sigma^{(p)})^T \\ \overline{P}_\varepsilon^{(p)} = (P_\sigma^{(p)})^T \end{cases}$$

The matrix $\overline{P}_\sigma^{(p)}$ is constructed on the model of $P_\sigma^{(p)}$ for which we use the components of the transpose matrix of $Q$. The matrix $\overline{P}_\varepsilon^{(p)}$ is constructed on the model of $P_\varepsilon^{(p)}$ for which we use the components of the transpose matrix of $Q$. We obtain the following relations:

$$\begin{cases} \overline{P}_\sigma^{(p)} P_\sigma^{(p)} = I_3 \\ P_\sigma^{(p)} \overline{P}_\sigma^{(p)} = I_3 \end{cases} \iff \begin{cases} \overline{P}_\sigma^{(p)} = (P_\sigma^{(p)})^{-1} \\ P_\sigma^{(p)} = (\overline{P}_\sigma^{(p)})^{-1} \end{cases}$$

$$\begin{cases} \overline{P}_\varepsilon^{(p)} P_\varepsilon^{(p)} = I_3 \\ P_\varepsilon^{(p)} \overline{P}_\varepsilon^{(p)} = I_3 \end{cases} \iff \begin{cases} \overline{P}_\varepsilon^{(p)} = (P_\varepsilon^{(p)})^{-1} \\ P_\varepsilon^{(p)} = (\overline{P}_\varepsilon^{(p)})^{-1} \end{cases}$$

In the bases $b^{(ref)}$ and $b^{(loc)}$, we note the behaviour laws under the plane stress assumption using the Voigt form in the following manner:

$$\begin{cases} \sigma^{(pref)} = R^{(pref)} \varepsilon^{(pref)} \\ \sigma^{(ploc)} = R^{(ploc)} \varepsilon^{(ploc)} \end{cases} \quad \begin{cases} \varepsilon^{(pref)} = S^{(pref)} \sigma^{(pref)} \\ \varepsilon^{(ploc)} = S^{(ploc)} \sigma^{(ploc)} \end{cases} \quad \begin{cases} R^{(pref)} S^{(pref)} = I_3 \\ R^{(ploc)} S^{(ploc)} = I_3 \end{cases}$$

We have the following change-in-basis formulae:

$$\begin{cases} \sigma^{(pref)} = P_\sigma^{(p)} \sigma^{(ploc)} \\ \sigma^{(ploc)} = \overline{P}_\sigma^{(p)} \sigma^{(pref)} \end{cases} \quad \begin{cases} \varepsilon^{(pref)} = P_\varepsilon^{(p)} \varepsilon^{(ploc)} \\ \varepsilon^{(ploc)} = \overline{P}_\varepsilon^{(p)} \varepsilon^{(pref)} \end{cases}$$

$$\begin{cases} R^{(ploc)} = \overline{P}_\sigma^{(p)} R^{(pref)} P_\varepsilon^{(p)} \\ R^{(pref)} = P_\sigma^{(p)} R^{(ploc)} \overline{P}_\varepsilon^{(p)} \end{cases} \quad \begin{cases} S^{(ploc)} = \overline{P}_\varepsilon^{(p)} S^{(pref)} P_\sigma^{(p)} \\ S^{(pref)} = P_\varepsilon^{(p)} S^{(ploc)} \overline{P}_\sigma^{(p)} \end{cases}$$

## 7.4 Thin laminated plates of orthotropic plies

We consider the physical affine space $\varepsilon^3$, for which the associated vector space is $\mathbb{R}^3$, its reference frame is $R = (O, b)$, $b = (\vec{x}_1, \vec{x}_2, \vec{x}_3)$ designating its orthonormal basis and $O$ its origin.

We consider a structure such as a thin plate for which the median plane is defined by the vectors $(\vec{x}_1^{(ref)}, \vec{x}_2^{(ref)})$. The unit vector normal to this plane, which defines the thickness, is therefore the vector $\vec{x}_3^{(ref)}$. We denote $R^{(ref)}$ the reference frame of the plate for which $b^{(ref)} = (\vec{x}_1^{(ref)}, \vec{x}_2^{(ref)}, \vec{x}_3^{(ref)})$ defined its basis (so-called, the reference basis). We assume that $R$ and $R^{(ref)}$ coincide. Given the particular geometry of this structure we suppose that the stress field at each point of the plate is a state of plane stress within its plane.

We suppose that the plate is laminated and formed of a stack of $n$ homogeneous plies indicated as $k$ $(k = 1, \ldots, n)$. The numbering of the plies is as follows with increasing numbers describing the plies with the number of ply one being the lowest and the uppermost ply $n$. The total thickness of this thin laminate is $2e$ so that the plate is between the plane $x_3 = -e$ and the plane $x_3 = e$.

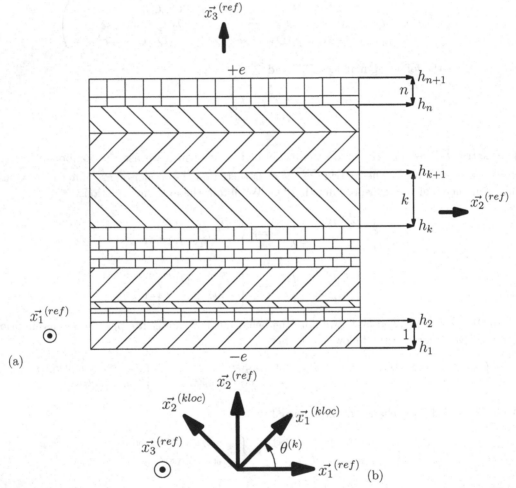

Figure 7.2: Thin laminated plate. (a) Description. (b) Reference basis of the laminate and local basis of the ply $k$.

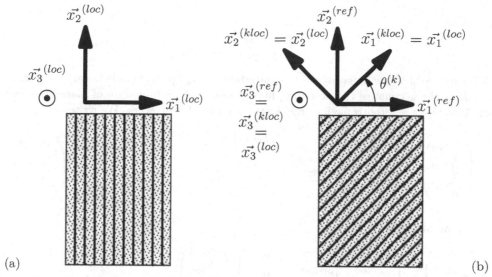

Figure 7.3: Schematic view of a fibre/resin unidirectional composite that could be the constitutive material of the ply $k$ of the laminate. (a) Anisotropic local basis of the unidirectional material. (b) Reference basis of the laminate and local basis of the ply $k$ of the laminate, orientated with angle $\theta^{(k)}$ relatively to the reference basis of the laminate.

## 7.4.1 Description of the plies

The ply $k$ of the laminate has a thickness of $e_k = h_{k+1} - h_k$. It is therefore situated between the planes given by the equations $x_3 = h_{k+1}$ and $x_3 = h_k$. We consider that it is homogeneous and made of an orthotropic material for which the basis of its anisotropic framework is $b^{(loc)} = (\vec{x}_1^{(loc)}, \vec{x}_2^{(loc)}, \vec{x}_3^{(loc)})$. We denote $R^{(kloc)}$ the local frame of the ply $k$ for which $b^{(kloc)}$ defined its basis (so-called, the local basis of the ply $k$): $b^{(kloc)} = (\vec{x}_1^{(kloc)}, \vec{x}_2^{(kloc)}, \vec{x}_3^{(kloc)})$. We suppose that $b^{(kloc)} = b^{(loc)}$ and $\vec{x}_3^{(kloc)} = \vec{x}_3^{(loc)} = \vec{x}_3^{(ref)}$ (Figure 7.2). We note $\theta^{(k)} = (\widehat{\vec{x}_1^{(ref)}, \vec{x}_1^{(kloc)}})$. For example the material in the ply $k$ could be a unidirectional fibre/resin composite of which the fibres are orientated in the direction of the vector $\vec{x}_1^{(kloc)}$ making an angle $\theta^{(k)}$ with the vector $\vec{x}_1^{(ref)}$ (Figure 7.3).

We give to each ply $k$ a change of basis matrix from the basis $b^{(ref)}$ to the basis $b^{(kloc)}$:

$$Q^{(k)} = \begin{pmatrix} cos\theta^{(k)} = c^{(k)} & -sin\theta^{(k)} = -s^{(k)} & 0 \\ sin\theta^{(k)} = s^{(k)} & cos\theta^{(k)} = c^{(k)} & 0 \\ 0 & 0 & 1 \end{pmatrix}$$

and the following matrices:

$$P_\sigma^{(pk)} = \begin{pmatrix} (c^{(k)})^2 & (s^{(k)})^2 & -2s^{(k)}c^{(k)} \\ (s^{(k)})^2 & (c^{(k)})^2 & 2s^{(k)}c^{(k)} \\ s^{(k)}c^{(k)} & -s^{(k)}c^{(k)} & (c^{(k)})^2 - (s^{(k)})^2 \end{pmatrix}$$

$$\overline{P}_\sigma^{(pk)} = \begin{pmatrix} (c^{(k)})^2 & (s^{(k)})^2 & 2s^{(k)}c^{(k)} \\ (s^{(k)})^2 & (c^{(k)})^2 & -2s^{(k)}c^{(k)} \\ -s^{(k)}c^{(k)} & s^{(k)}c^{(k)} & (c^{(k)})^2 - (s^{(k)})^2 \end{pmatrix}$$

$$P_\varepsilon^{(pk)} = \begin{pmatrix} (c^{(k)})^2 & (s^{(k)})^2 & -s^{(k)}c^{(k)} \\ (s^{(k)})^2 & (c^{(k)})^2 & s^{(k)}c^{(k)} \\ 2s^{(k)}c^{(k)} & -2s^{(k)}c^{(k)} & (c^{(k)})^2 - (s^{(k)})^2 \end{pmatrix}$$

$$\overline{P}_\varepsilon^{(pk)} = \begin{pmatrix} (c^{(k)})^2 & (s^{(k)})^2 & s^{(k)}c^{(k)} \\ (s^{(k)})^2 & (c^{(k)})^2 & -s^{(k)}c^{(k)} \\ -2s^{(k)}c^{(k)} & 2s^{(k)}c^{(k)} & (c^{(k)})^2 - (s^{(k)})^2 \end{pmatrix}$$

We suppose that the behaviour of the material making up the plies is linearly elastic. In this way, taking the account of the preceding hypotheses, a ply is in a state of plane stress and its elastic behaviour is orthotropic linearly elastic in its local framework. We write:

$$\begin{pmatrix} \varepsilon_1^{(kloc)} \\ \varepsilon_2^{(kloc)} \\ \varepsilon_6^{(kloc)} \end{pmatrix} = \begin{pmatrix} S_{11}^{(kloc)} = \frac{1}{E_1^{(kloc)}} & S_{12}^{(kloc)} = -\frac{\nu_{12}^{(kloc)}}{E_1^{(kloc)}} & 0 \\ S_{12}^{(kloc)} = -\frac{\nu_{12}^{(kloc)}}{E_1^{(kloc)}} & S_{22}^{(kloc)} = \frac{1}{E_2^{(kloc)}} & 0 \\ 0 & 0 & S_{66}^{(kloc)} = \frac{1}{G_{12}^{(kloc)}} \end{pmatrix} \begin{pmatrix} \sigma_1^{(kloc)} \\ \sigma_2^{(kloc)} \\ \sigma_6^{(kloc)} \end{pmatrix}$$

Inversing we obtain:

$$\begin{pmatrix} \sigma_1^{(kloc)} \\ \sigma_2^{(kloc)} \\ \sigma_6^{(kloc)} \end{pmatrix} = \begin{pmatrix} R'_{11}^{(kloc)} & R'_{12}^{(kloc)} & 0 \\ R'_{12}^{(kloc)} & R'_{11}^{(kloc)} & 0 \\ 0 & 0 & R'_{66}^{(kloc)} \end{pmatrix} \begin{pmatrix} \varepsilon_1^{(kloc)} \\ \varepsilon_2^{(kloc)} \\ \varepsilon_6^{(kloc)} \end{pmatrix}$$

$$= \begin{pmatrix} \frac{E_1^{(kloc)}}{1-(\nu_{12}^{(kloc)})^2 E_2^{(kloc)}/E_1^{(kloc)}} & \frac{\nu_{12}^{(kloc)} E_2^{(kloc)}}{1-(\nu_{12}^{(kloc)})^2 E_2^{(kloc)}/E_1^{(kloc)}} & 0 \\ \frac{\nu_{12}^{(kloc)} E_2^{(kloc)}}{1-(\nu_{12}^{(kloc)})^2 E_2^{(kloc)}/E_1^{(kloc)}} & \frac{E_2^{(kloc)}}{1-(\nu_{12}^{(kloc)})^2 E_2^{(kloc)}/E_1^{(kloc)}} & 0 \\ 0 & 0 & G_{12}^{(kloc)} \end{pmatrix} \begin{pmatrix} \varepsilon_1^{(kloc)} \\ \varepsilon_2^{(kloc)} \\ \varepsilon_6^{(kloc)} \end{pmatrix}$$

We write the relations in the following form $(\alpha, \beta = 1, 2, 6)$:

$$\begin{cases} \sigma^{(pkloc)} = R^{(pkloc)}\varepsilon^{(pkloc)} \iff \sigma_\alpha^{(pkloc)} = R_{\alpha\beta}^{(pkloc)}\varepsilon_\beta^{(pkloc)} \\ \varepsilon^{(pkloc)} = S^{(pkloc)}\sigma^{(pkloc)} \iff \varepsilon_\alpha^{(pkloc)} = S_{\alpha\beta}^{(pkloc)}\sigma_\beta^{(pkloc)} \end{cases}$$

where $R^{(pkloc)} = (R_{\alpha\beta}^{(pkloc)})_{\alpha,\beta=1,2,6}$ and $S^{(pkloc)} = (S_{\alpha\beta}^{(pkloc)})_{\alpha,\beta=1,2,6}$. One has: $(R^{(pkloc)})^{-1} = S^{(pkloc)}$ and $(S^{(pkloc)})^{-1} = R^{(pkloc)}$.

In the basis $b^{(ref)}$, the behaviour law in plane stress has the form $(\alpha, \beta = 1, 2, 6)$:

$$\begin{cases} \sigma^{(pkref)} = R^{(pkref)}\varepsilon^{(pkref)} \iff \sigma_\alpha^{(pkref)} = R_{\alpha\beta}^{(pkref)}\varepsilon_\beta^{(pkref)} \\ \varepsilon^{(pkref)} = S^{(pkref)}\sigma^{(pkref)} \iff \varepsilon_\alpha^{(pkref)} = S_{\alpha\beta}^{(pkref)}\sigma_\beta^{(pkref)} \end{cases}$$

where $R^{(pkref)} = (R_{\alpha\beta}^{(pkref)})_{\alpha,\beta=1,2,6}$ and $S^{(pkref)} = (S_{\alpha\beta}^{(pkref)})_{\alpha,\beta=1,2,6}$. We have: $(R^{(pkref)})^{-1} = S^{(pkref)}$ and $(S^{(pkref)})^{-1} = R^{(pkref)}$.

We have the following change of basis formulae:

$$\begin{cases} \sigma^{(pkref)} = P_\sigma^{(pk)} \sigma^{(pkloc)} \\ \sigma^{(pkloc)} = \overline{P}_\sigma^{(pk)} \sigma^{(pkref)} \end{cases} \qquad \begin{cases} \varepsilon^{(pkref)} = P_\varepsilon^{(pk)} \varepsilon^{(pkloc)} \\ \varepsilon^{(pkloc)} = \overline{P}_\varepsilon^{(pk)} \varepsilon^{(pkref)} \end{cases}$$

$$\begin{cases} R^{(pkloc)} = \overline{P}_\sigma^{(pk)} R^{(pkref)} P_\varepsilon^{(pk)} \\ R^{(pkref)} = P_\sigma^{(pk)} R^{(pkloc)} \overline{P}_\varepsilon^{(pk)} \end{cases} \qquad \begin{cases} S^{(pkloc)} = \overline{P}_\varepsilon^{(pk)} S^{(pkref)} P_\sigma^{(pk)} \\ S^{(pkref)} = P_\varepsilon^{(pk)} S^{(pkloc)} \overline{P}_\sigma^{(pk)} \end{cases}$$

Thus:

$$S^{(pkloc)} = \begin{pmatrix} S_{11}^{(pkloc)} & S_{12}^{(pkloc)} & 0 \\ S_{12}^{(pkloc)} & S_{22}^{(pkloc)} & 0 \\ 0 & 0 & S_{66}^{(pkloc)} \end{pmatrix}$$

$$R^{(pkloc)} = \begin{pmatrix} R_{11}^{(pkloc)} & R_{12}^{(pkloc)} & 0 \\ R_{12}^{(pkloc)} & R_{11}^{(pkloc)} & 0 \\ 0 & 0 & R_{66}^{(pkloc)} \end{pmatrix}$$

$$S^{(pkref)} = \begin{pmatrix} S_{11}^{(pkref)} & S_{12}^{(pkref)} & S_{16}^{(pkref)} \\ S_{12}^{(pkref)} & S_{22}^{(pkref)} & S_{26}^{(pkref)} \\ S_{16}^{(pkref)} & S_{26}^{(pkref)} & S_{66}^{(pkref)} \end{pmatrix}$$

$$R^{(pkref)} = \begin{pmatrix} R_{11}^{(pkref)} & R_{12}^{(pkref)} & R_{16}^{(pkref)} \\ R_{12}^{(pkref)} & R_{11}^{(pkref)} & R_{26}^{(pkref)} \\ R_{16}^{(pkref)} & R_{26}^{(pkref)} & R_{66}^{(pkref)} \end{pmatrix}$$

## 7.4.2 Kinematics of thick plates under the Kirshhoff-Love hypothesis. Membrane strain and bending strain of the plate

Given the particular geometry of the structure we estimate that its kinematics can be adapted and simplified compared to the general case. So we write, initially, that to reach the point $M$ in the plate, with coordinates $(x_i)_{i=1,2,3}$ in the framework $R$, we pass through the point $M_0$, which is the orthogonal projection of $M$ in the median plane:

$$\overrightarrow{OM} = x_1 \vec{x}_1^{(ref)} + x_2 \vec{x}_2^{(ref)} + x_3 \vec{x}_3^{(ref)} = \overrightarrow{OM_0} + x_3 \vec{n}(M_0)$$

where the vector $\vec{n}(M_0)$ designates the unit normal to the plate at $M_0$ (here it is the vector $\vec{x}_3^{(ref)}$). Then we estimate the displacement of $M$ can be calculated from the point $M_0$. Using the hypothesis of small perturbations it is easy to show that the vector $\vec{n}(M_0)$ becomes the vector $\vec{n}(M_0) + \vec{\Omega}(M_0) \wedge \vec{n}(M_0)$, where $\vec{\Omega}(M_0)$ defines the rotation of the normal in $M_0$. Finally, using the Kirshhoff-Love hypothesis (the normal in $M_0$ remains normal to the median plan of the plate, even after deformation), the following form of the displacement of $M$ is obtained:

$$\vec{u}(M) = \begin{pmatrix} u_1(x_1, x_2, x_3) \\ u_2(x_1, x_2, x_3) \\ u_3(x_1, x_2, x_3) \end{pmatrix}$$

$$= \begin{pmatrix} u_1(x_1, x_2, x_3 = 0) \\ u_2(x_1, x_2, x_3 = 0) \\ u_3(x_1, x_2, x_3 = 0) \end{pmatrix} + x_3 \begin{pmatrix} -\frac{\partial u_3(x_1, x_2, x_3=0)}{\partial x_1} \\ -\frac{\partial u_3(x_1, x_2, x_3=0)}{\partial x_2} \\ 0 \end{pmatrix} = \vec{u}(M_0) + x_3 \overrightarrow{\omega(M_0)}$$

If we calculate the plane components of the strain tensor (the only ones which interest us as we are considering plane stress) at the point $M$, in the basis $b^{(ref)}$, we obtain:

$$\varepsilon^{(pkref)}(M) = \begin{pmatrix} \varepsilon_1^{(kref)}(M) = \frac{\partial u_1(M)}{\partial x_1} \\ \varepsilon_2^{(kref)}(M) = \frac{\partial u_2(M)}{\partial x_2} \\ \varepsilon_6^{(kref)}(M) = \frac{1}{2}\left( \frac{\partial u_1(M)}{\partial x_2} + \frac{\partial u_2(M)}{\partial x_1} \right) \end{pmatrix}$$

$$= \begin{pmatrix} \frac{\partial u_1(M_0)}{\partial x_1} \\ \frac{\partial u_2(M_0)}{\partial x_2} \\ \frac{1}{2}\left( \frac{\partial u_1(M_0)}{\partial x_2} + \frac{\partial u_2(M_0)}{\partial x_1} \right) \end{pmatrix} + x_3 \begin{pmatrix} -\frac{\partial^2 u_3(M_0)}{\partial x_1^2} \\ -\frac{\partial^2 u_3(M_0)}{\partial x_2^2} \\ -\frac{\partial^2 u_3(M_0)}{\partial x_1 \partial x_2} \end{pmatrix}$$

We put:

$$\varepsilon^{0(pref)}(M_0) = \begin{pmatrix} \varepsilon_1^{0(ref)}(M_0) \\ \varepsilon_2^{0(ref)}(M_0) \\ \varepsilon_6^{0(ref)}(M_0) \end{pmatrix} = \begin{pmatrix} \frac{\partial u_1(M_0)}{\partial x_1} \\ \frac{\partial u_2(M_0)}{\partial x_2} \\ \frac{1}{2}\left( \frac{\partial u_1(M_0)}{\partial x_2} + \frac{\partial u_2(M_0)}{\partial x_1} \right) \end{pmatrix}$$

$$K^{(pref)}(M_0) = \begin{pmatrix} K_1^{(ref)}(M_0) \\ K_2^{(ref)}(M_0) \\ K_6^{(ref)}(M_0) \end{pmatrix} = \begin{pmatrix} -\frac{\partial^2 u_3(M_0)}{\partial x_1^2} \\ -\frac{\partial^2 u_3(M_0)}{\partial x_2^2} \\ -\frac{\partial^2 u_3(M_0)}{\partial x_1 \partial x_2} \end{pmatrix}$$

We finally obtain: $\varepsilon^{(pkref)}(M) = \varepsilon^{0(pref)}(M_0) + x_3 K^{(pref)}(M_0)$ with $h_{k+1} \leq x_3 \leq h_k$ for the ply $k$, $\varepsilon^{0(pref)}(M_0)$ is called the membrane deformation of the median plane of the laminate in $M_0$, $K^{(pref)}(M_0)$ is called the bending deformation of the median plane of the laminate, in $M_0$.

### 7.4.3 Membrane stress and bending stress of the plate

The membrane strain and the bending strain of the median plane of the laminate represent the global strain of the median plane of the laminate. In this way we can construct equivalent quantities in term of loading. In order to do that we define:

- membrane stress of the plate by:

$$N(M_0) = \begin{pmatrix} N_1(M_0) \\ N_2(M_0) \\ N_6(M_0) \end{pmatrix} = \begin{pmatrix} \int_{-e}^{+e} \sigma_1^{(pkref)}(M) dx_3 \\ \int_{-e}^{+e} \sigma_2^{(pkref)}(M) dx_3 \\ \int_{-e}^{+e} \sigma_6^{(pkref)}(M) dx_3 \end{pmatrix}$$

$$= \left( N_\alpha(M_0) = \int_{-e}^{+e} \sigma_\alpha^{(pkref)}(M) dx_3 \right)_{\alpha=1,2,6}$$

- bending stress of the plate by:

$$M(M_0) = \begin{pmatrix} M_1(M_0) \\ M_2(M_0) \\ M_6(M_0) \end{pmatrix} = \begin{pmatrix} \int_{-e}^{+e} x_3 \sigma_1^{(pkref)}(M) dx_3 \\ \int_{-e}^{+e} x_3 \sigma_2^{(pkref)}(M) dx_3 \\ \int_{-e}^{+e} x_3 \sigma_6^{(pkref)}(M) dx_3 \end{pmatrix}$$

$$= \left( M_\alpha(M_0) = \int_{-e}^{+e} x_3 \sigma_\alpha^{(pkref)}(M) dx_3 \right)_{\alpha=1,2,6}$$

### 7.4.4 Behaviour of a laminate

In the basis $b^{(ref)}$, at the point $M$, the behaviour in plane stress is written ($\alpha, \beta = 1, 2, 6$):

$$\sigma_\alpha^{(pkref)}(M) = R_{\alpha\beta}^{(pkref)}(M) \varepsilon_\beta^{(pkref)}(M)$$

In addition, we have: $\varepsilon_\beta^{(pkref)}(M) = \varepsilon_\beta^{0(pref)}(M_0) + x_3 K_\beta^{(pref)}(M_0)$. Finally:

$$\sigma_\alpha^{(pkref)}(M) = R_{\alpha\beta}^{(pkref)}(M) \left( \varepsilon_\beta^{0(pref)}(M_0) + x_3 K_\beta^{(pref)}(M_0) \right)$$

So we can deduce, by integration over the thickness:

$$N_\alpha(M_0) = \int_{-e}^{+e} \sigma_\alpha^{(pkref)}(M) dx_3$$

$$= \int_{-e}^{+e} R_{\alpha\beta}^{(pkref)}(M) \left( \varepsilon_\beta^{0(pref)}(M_0) + x_3 K_\beta^{(pref)}(M_0) \right) dx_3$$

$$= \left( \int_{-e}^{+e} R_{\alpha\beta}^{(pkref)}(M) dx_3 \right) \varepsilon_\beta^{0(pref)}(M_0) + \left( \int_{-e}^{+e} x_3 R_{\alpha\beta}^{(pkref)}(M) dx_3 \right) K_\beta^{(pref)}(M_0)$$

Then we can deduce, again by integrating over the thickness:

$$M_\alpha(M_0) = \int_{-e}^{+e} x_3 \sigma_\alpha^{(pkref)}(M)dx_3$$

$$= \int_{-e}^{+e} x_3 R_{\alpha\beta}^{(pkref)}(M) \left( \varepsilon_\beta^{0(pref)}(M_0) + x_3 K_\beta^{(pref)}(M_0) \right) dx_3$$

$$= \left( \int_{-e}^{+e} x_3 R_{\alpha\beta}^{(pkref)}(M)dx_3 \right) \varepsilon_\beta^{0(pref)}(M_0) + \left( \int_{-e}^{+e} x_3^2 R_{\alpha\beta}^{(pkref)}(M)dx_3 \right) K_\beta^{(pref)}(M_0)$$

Supposing that the plies are homogeneous we put:

$$A = \left( A_{\alpha\beta} = \int_{-e}^{+e} R_{\alpha\beta}^{(pkref)}(M)dx_3 = \sum_{k=1}^{n} R_{\alpha\beta}^{(pkref)} (h_{k+1} - h_k) \right)_{\alpha=1,2,6}$$

$$B = \left( B_{\alpha\beta} = \int_{-e}^{+e} x_3 R_{\alpha\beta}^{(pkref)}(M)dx_3 = \sum_{k=1}^{n} R_{\alpha\beta}^{(pkref)} \frac{(h_{k+1})^2 - (h_k)^2}{2} \right)_{\alpha=1,2,6}$$

$$D = \left( D_{\alpha\beta} = \int_{-e}^{+e} x_3^2 R_{\alpha\beta}^{(pkref)}(M)dx_3 = \sum_{k=1}^{n} R_{\alpha\beta}^{(pkref)} \frac{(h_{k+1})^3 - (h_k)^3}{3} \right)_{\alpha=1,2,6}$$

Taking account of these notations we obtain finally:

$$\begin{cases} N_\alpha(M_0) = A_{\alpha\beta}\varepsilon_\beta^{0(pref)}(M_0) + B_{\alpha\beta}K_\beta^{(pref)}(M_0) \\ M_\alpha(M_0) = B_{\alpha\beta}\varepsilon_\beta^{0(pref)}(M_0) + D_{\alpha\beta}K_\beta^{(pref)}(M_0) \end{cases}$$

The matrix $A$ is called the membrane rigidity matrix of the laminate. The matrix $D$ is called the bending rigidity matrix of the laminate. The matrix $B$ is called the matrix of membrane/bending coupling of the laminate. It is often written:

$$\begin{pmatrix} N_1(M_0) \\ N_2(M_0) \\ N_6(M_0) \\ M_1(M_0) \\ M_2(M_0) \\ M_6(M_0) \end{pmatrix} = \begin{pmatrix} A_{11} & A_{12} & A_{16} & B_{11} & B_{12} & B_{16} \\ A_{12} & A_{22} & A_{26} & B_{12} & B_{22} & B_{26} \\ A_{16} & A_{26} & A_{36} & B_{16} & B_{26} & B_{66} \\ B_{11} & B_{12} & B_{16} & D_{11} & D_{12} & D_{16} \\ B_{12} & B_{22} & B_{26} & D_{12} & D_{22} & D_{26} \\ B_{16} & B_{26} & B_{66} & D_{16} & D_{26} & D_{66} \end{pmatrix} \begin{pmatrix} \varepsilon_1^{0(pref)}(M_0) \\ \varepsilon_2^{0(pref)}(M_0) \\ \varepsilon_6^{0(pref)}(M_0) \\ K_1^{(pref)}(M_0) \\ K_2^{(pref)}(M_0) \\ K_6^{(pref)}(M_0) \end{pmatrix}$$

### 7.4.5 Calculation of the stresses and strains of plies in the reference basis and in the local basis of the plies

Solving the equations of laminate theory (deduced from those of the formulation of a problem defined in the framework of continuum mechanics) allows access to $\varepsilon^{0(pref)}(M_0)$, $K^{(pref)}(M_0)$, $N(M_0)$ and $M(M_0)$. These values are expressed in $b^{(ref)}$.

In this way it becomes possible to calculate, in $b^{(ref)}$, the strains $\varepsilon^{(pkref)}(M)$ in each ply $k$ and at each point $M$ by writing:

$$\varepsilon^{(pkref)}(M) = \varepsilon^{0(pref)}(M_0) + x_3 K^{(pref)}(M_0)$$

with $h_{k+1} \le x_3 \le h_k$ for the ply $k$. In addition, using the behaviour in plane stress we obtain the (plane) stresses, in each ply $k$ and at each point $M$ by writing:

$$\sigma_\alpha^{(pkref)}(M) = R_{\alpha\beta}^{(pkref)}(M)\varepsilon_\beta^{(pkref)}(M)$$

Furthermore, in each ply $k$ and at each point $M$, the change of basis formulae allow values to be obtained in the local basis $b^{(kloc)}$ of the ply $k$:

$$\left\{ \begin{array}{l} \varepsilon^{(kloc)}(M) = \overline{P}_\varepsilon^{(pk)}\, \varepsilon^{(pkref)}(M) \\ \sigma^{(pkloc)}(M) = \overline{P}_\sigma^{(pk)}\, \sigma^{(pkref)}(M) \end{array} \right.$$

We can then verify:

$$\left\{ \begin{array}{l} \sigma^{(pkloc)} = R^{(pkloc)}\varepsilon^{(pkloc)} \Longleftrightarrow \sigma_\alpha^{(pkloc)} = R_{\alpha\beta}^{(pkloc)}\varepsilon_\beta^{(pkloc)} \\ \varepsilon^{(pkloc)} = S^{(pkloc)}\sigma^{(pkloc)} \Longleftrightarrow \varepsilon_\alpha^{(pkloc)} = S_{\alpha\beta}^{(pkloc)}\sigma_\beta^{(pkloc)} \end{array} \right.$$

What is the interest in evaluating the plane stresses and strains in the local basis of each ply $k$? The most important interest is that the writing of a model (of damage for example) written at the level of a ply and in the local basis allows us to avoid the need to identify that which would depend on the orientation of the ply within the laminate. The model is identified once and for all in the local basis and is used in this basis, once the stresses and strains have been calculated. This will be illustrated in the chapter concerning the failure criteria.

## 7.5 Thick laminates of periodic stacking sequence of orientated orthotropic plies

### 7.5.1 Description of the laminate and its periodic cell

We consider the physical affine space $\varepsilon^3$, for which the associated vector space is $\mathbb{R}^3$, its reference frame is $R = (O, b)$, $b = (\vec{x}_1, \vec{x}_2, \vec{x}_3)$ designating its orthonormal basis and $O$ its origin.

We consider a structure such as a thick plate for which the median plane is defined by the vectors $(\vec{x}_1^{(ref)}, \vec{x}_2^{(ref)})$. The unit vector normal to this plane, which defines the thickness, is therefore the vector $\vec{x}_3^{(ref)}$. We denote $R^{(ref)}$ the reference frame of the plate for which $b^{(ref)} = (\vec{x}_1^{(ref)}, \vec{x}_2^{(ref)}, \vec{x}_3^{(ref)})$ defines its basis (so-called, the reference basis). We assume that $R$ and $R^{(ref)}$ coincide.

We suppose that the plate is periodic laminated formed with a periodic stack $Y$ ($N$ times) in the thickness direction, of $n$ homogeneous plies indicated as $k$ ($k = 1, \ldots, n$). The frame attached to the cell $Y$ is $R^{(Y)}$ and $b^{(Y)} = (\vec{y}_1, \vec{y}_2, \vec{y}_3)$ defines its basis. The numbering of the plies, in the $Y$ periodic cell, is as follows with increasing numbers describing the plies with the number of ply one being the lowest and the uppermost ply $n$. The total thickness of the periodic cell $Y$ is $h$ so that the cell is between the plane $y_3 = O$ and the plane $y_3 = h$. The total thickness of this thick laminate is $2e$ so that the plate is between the plane $x_3 = -e$ and the plane $x_3 = e$.

In the $Y$ cell, in $b^{(Y)}$, the fourth-order rigidity tensor is noted $a(y) = (a_{ijkh}(y))_{i,j,k,h=1,2,3}$ where the dependence with $y$ underlines that the constitutive material of the cell is inhomogeneous. But, here, because the plies are assumed to be homogeneous: $a(y_3) = (a_{ijkh}(y_3))_{i,j,k,h=1,2,3}$. At least, for sake of simplicity (but this does not affected the generality), we assume that $b^{(ref)} = b^{(Y)}$.

### 7.5.2 Description of the plies

The ply $k$ of the cell $Y$ has a thickness of $e_k = h_{k+1} - h_k$. It is therefore situated between the planes given by the equations $y_3 = h_{k+1}$ and $y_3 = h_k$. We consider that it is homogeneous and made of an orthotropic material for which the basis of its anisotropic framework is $b^{(loc)} = (\vec{x}_1^{(loc)}, \vec{x}_2^{(loc)}, \vec{x}_3^{(loc)})$. We denote $R^{(kloc)}$ the local frame of the ply $k$ for which $b^{(kloc)}$ defined its basis (so-called, the local basis of the ply $k$): $b^{(kloc)} = (\vec{x}_1^{(kloc)}, \vec{x}_2^{(kloc)}, \vec{x}_3^{(kloc)})$. We suppose that

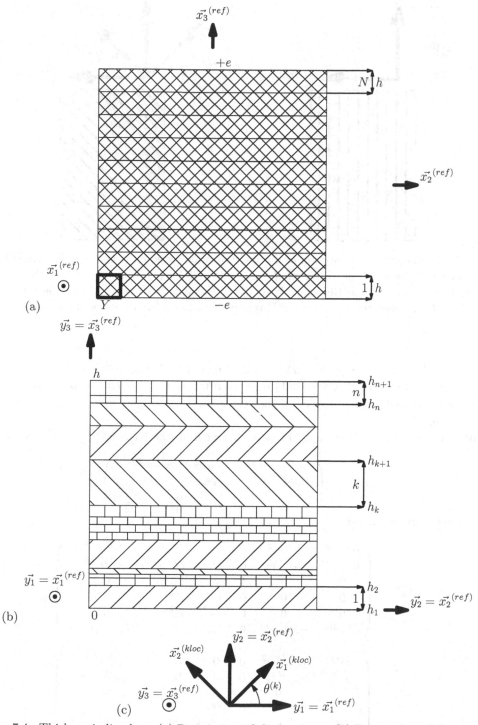

Figure 7.4: Thick periodic plate. (a) Description of the laminate. (b) Description of the periodic cell $Y$. (c) Reference frame of the laminate and local frame of the plies.

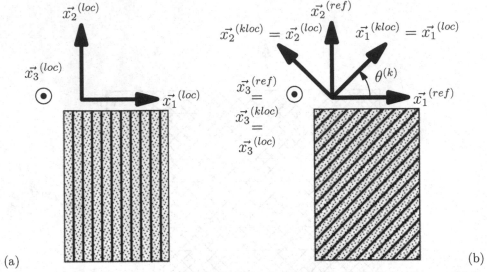

(a)                                                                                                 (b)

Figure 7.5: Schematic presentation of a fibre/resin unidirectional composite that could be the constitutive material of the ply $k$ of the periodic cell $Y$. (a) Anisotropic local basis of the unidirectional material. (b) Reference basis of the laminate and local basis of the ply $k$ of the periodic cell $Y$, orientated with angle $\theta^{(k)}$ relatively to the reference basis of the laminate.

$b^{(kloc)} = b^{(loc)}$ and $\vec{x}_3^{(kloc)} = \vec{x}_3^{(loc)} = \vec{x}_3^{(ref)} = \vec{y}_3$ (Figure 7.5). We note $\theta^{(k)} = (\widehat{\vec{x}_1^{(ref)}, \vec{x}_1^{(kloc)}})$. For example the material in the ply $k$ could be a unidirectional fibre/resin composite of which the fibres are orientated in the direction of the vector $\vec{x}_1^{(kloc)}$ making an angle $\theta^{(k)}$ with the vector $\vec{x}_1^{(ref)}$ (Figure 7.5).

We give to each ply $k$ a change of basis matrix from the basis $b^{(ref)} = b^{(Y)}$ to the basis $b^{(kloc)}$:

$$Q^{(k)} = \begin{pmatrix} \cos\theta^{(k)} = c^{(k)} & -\sin\theta^{(k)} = -s^{(k)} & 0 \\ \sin\theta^{(k)} = s^{(k)} & \cos\theta^{(k)} = c^{(k)} & 0 \\ 0 & 0 & 1 \end{pmatrix}$$

and the following matrices:

$$P_\sigma^{(k)} = \begin{pmatrix} (c^{(k)})^2 & (s^{(k)})^2 & 0 & 0 & 0 & -2s^{(k)}c^{(k)} \\ (s^{(k)})^2 & (c^{(k)})^2 & 0 & 0 & 0 & 2s^{(k)}c^{(k)} \\ 0 & 0 & 1 & 0 & 0 & 0 \\ 0 & 0 & 0 & c^{(k)} & s^{(k)} & 0 \\ 0 & 0 & 0 & -s^{(k)} & c^{(k)} & 0 \\ s^{(k)}c^{(k)} & -s^{(k)}c^{(k)} & 0 & 0 & 0 & (c^{(k)})^2 - (s^{(k)})^2 \end{pmatrix}$$

$$\overline{P}_\sigma^{(k)} = \begin{pmatrix} (c^{(k)})^2 & (s^{(k)})^2 & 0 & 0 & 0 & 2s^{(k)}c^{(k)} \\ (s^{(k)})^2 & (c^{(k)})^2 & 0 & 0 & 0 & -2s^{(k)}c^{(k)} \\ 0 & 0 & 1 & 0 & 0 & 0 \\ 0 & 0 & 0 & c^{(k)} & -s^{(k)} & 0 \\ 0 & 0 & 0 & s^{(k)} & c^{(k)} & 0 \\ -s^{(k)}c^{(k)} & s^{(k)}c^{(k)} & 0 & 0 & 0 & (c^{(k)})^2 - (s^{(k)})^2 \end{pmatrix}$$

$$
P_\varepsilon^{(k)} = \begin{pmatrix}
(c^{(k)})^2 & (s^{(k)})^2 & 0 & 0 & 0 & -s^{(k)}c^{(k)} \\
(s^{(k)})^2 & (c^{(k)})^2 & 0 & 0 & 0 & s^{(k)}c^{(k)} \\
0 & 0 & 1 & 0 & 0 & 0 \\
0 & 0 & 0 & c^{(k)} & s^{(k)} & 0 \\
0 & 0 & 0 & -s^{(k)} & c^{(k)} & 0 \\
2s^{(k)}c^{(k)} & -2s^{(k)}c^{(k)} & 0 & 0 & 0 & (c^{(k)})^2 - (s^{(k)})^2
\end{pmatrix}
$$

$$
\overline{P}_\varepsilon^{(k)} = \begin{pmatrix}
(c^{(k)})^2 & (s^{(k)})^2 & 0 & 0 & 0 & s^{(k)}c^{(k)} \\
(s^{(k)})^2 & (c^{(k)})^2 & 0 & 0 & 0 & -s^{(k)}c^{(k)} \\
0 & 0 & 1 & 0 & 0 & 0 \\
0 & 0 & 0 & c^{(k)} & -s^{(k)} & 0 \\
0 & 0 & 0 & s^{(k)} & c^{(k)} & 0 \\
-2s^{(k)}c^{(k)} & 2s^{(k)}c^{(k)} & 0 & 0 & 0 & (c^{(k)})^2 - (s^{(k)})^2
\end{pmatrix}
$$

We suppose that the behaviour of the material making up the plies is linearly elastic. In this way, taking account of the preceding hypotheses, a ply is orthotropic linearly elastic in its local framework. We write:

$$
\begin{pmatrix}
\varepsilon_1^{(kloc)} \\
\varepsilon_2^{(kloc)} \\
\varepsilon_3^{(kloc)} \\
\varepsilon_4^{(kloc)} \\
\varepsilon_5^{(kloc)} \\
\varepsilon_6^{(kloc)}
\end{pmatrix}
=
\begin{pmatrix}
S_{11}^{(kloc)} & S_{12}^{(kloc)} & S_{13}^{(kloc)} & 0 & 0 & 0 \\
S_{12}^{(kloc)} & S_{22}^{(kloc)} & S_{23}^{(kloc)} & 0 & 0 & 0 \\
S_{13}^{(kloc)} & S_{23}^{(kloc)} & S_{33}^{(kloc)} & 0 & 0 & 0 \\
0 & 0 & 0 & S_{44}^{(kloc)} & 0 & 0 \\
0 & 0 & 0 & 0 & S_{55}^{(kloc)} & 0 \\
0 & 0 & 0 & 0 & 0 & S_{66}^{(kloc)}
\end{pmatrix}
\begin{pmatrix}
\sigma_1^{(kloc)} \\
\sigma_2^{(kloc)} \\
\sigma_3^{(kloc)} \\
\sigma_4^{(kloc)} \\
\sigma_5^{(kloc)} \\
\sigma_6^{(kloc)}
\end{pmatrix}
$$

Inversing we obtain:

$$
\begin{pmatrix}
\sigma_1^{(kloc)} \\
\sigma_2^{(kloc)} \\
\sigma_3^{(kloc)} \\
\sigma_4^{(kloc)} \\
\sigma_5^{(kloc)} \\
\sigma_6^{(kloc)}
\end{pmatrix}
=
\begin{pmatrix}
R_{11}^{(kloc)} & R_{12}^{(kloc)} & R_{13}^{(kloc)} & 0 & 0 & 0 \\
R_{12}^{(kloc)} & R_{22}^{(kloc)} & R_{23}^{(kloc)} & 0 & 0 & 0 \\
R_{13}^{(kloc)} & R_{23}^{(kloc)} & R_{33}^{(kloc)} & 0 & 0 & 0 \\
0 & 0 & 0 & R_{44}^{(kloc)} & 0 & 0 \\
0 & 0 & 0 & 0 & R_{55}^{(kloc)} & 0 \\
0 & 0 & 0 & 0 & 0 & R_{66}^{(kloc)}
\end{pmatrix}
\begin{pmatrix}
\varepsilon_1^{(kloc)} \\
\varepsilon_2^{(kloc)} \\
\varepsilon_3^{(kloc)} \\
\varepsilon_4^{(kloc)} \\
\varepsilon_5^{(kloc)} \\
\varepsilon_6^{(kloc)}
\end{pmatrix}
$$

We write the relations in the following form $(i, j = 1, \ldots, 6)$:

$$
\begin{cases}
\sigma^{(kloc)} = R^{(kloc)}\varepsilon^{(kloc)} \iff \sigma_i^{(kloc)} = R_{ij}^{(kloc)}\varepsilon_j^{(kloc)} \\
\varepsilon^{(kloc)} = S^{(kloc)}\sigma^{(kloc)} \iff \varepsilon_i^{(kloc)} = S_{ij}^{(kloc)}\sigma_j^{(kloc)}
\end{cases}
$$

where $R^{(kloc)} = (R_{ij}^{(kloc)})_{i,j=1,\ldots,6}$ and $S^{(kloc)} = (S_{ij}^{(kloc)})_{i,j=1,\ldots,6}$. On a : $(R^{(kloc)})^{-1} = S^{(kloc)}$ and $(S^{(kloc)})^{-1} = R^{(kloc)}$.

In the basis $b^{(ref)}$, the behaviour law has the form $(i, j = 1, \ldots, 6)$ :

$$
\begin{cases}
\sigma^{(kref)} = R^{(kref)}\varepsilon^{(kref)} \iff \sigma_i^{(kref)} = R_{ij}^{(kref)}\varepsilon_j^{(kref)} \\
\varepsilon^{(kref)} = S^{(kref)}\sigma^{(kref)} \iff \varepsilon_i^{(kref)} = S_{ij}^{(kref)}\sigma_j^{(kref)}
\end{cases}
$$

where $R^{(kref)} = (R_{ij}^{(kref)})_{i,j=1,\ldots,6}$ and $S^{(kref)} = (S_{ij}^{(kref)})_{i,j=1,\ldots,6}$. On a : $(R^{(kref)})^{-1} = S^{(kref)}$ and $(S^{(kref)})^{-1} = R^{(kref)}$.

We have the following change of basis formulae:

$$
\begin{cases}
\sigma^{(kref)} = P_\sigma^{(k)} \sigma^{(kloc)} \\
\sigma^{(kloc)} = \overline{P}_\sigma^{(k)} \sigma^{(kref)}
\end{cases}
\qquad
\begin{cases}
\varepsilon^{(kref)} = P_\varepsilon^{(k)} \varepsilon^{(kloc)} \\
\varepsilon^{(kloc)} = \overline{P}_\varepsilon^{(k)} \varepsilon^{(kref)}
\end{cases}
$$

$$\begin{cases} R^{(kloc)} = \overline{P}_\sigma^{(k)} \, R^{(kref)} \, P_\varepsilon^{(k)} \\ R^{(kref)} = P_\sigma^{(k)} \, R^{(kloc)} \, \overline{P}_\varepsilon^{(k)} \end{cases} \qquad \begin{cases} S^{(kloc)} = \overline{P}_\varepsilon^{(k)} \, S^{(kref)} \, P_\sigma^{(k)} \\ S^{(kref)} = P_\varepsilon^{(k)} \, S^{(kloc)} \, \overline{P}_\sigma^{(k)} \end{cases}$$

Thus:

$$S^{(kloc)} = \begin{pmatrix} S_{11}^{(kloc)} & S_{12}^{(kloc)} & S_{13}^{(kloc)} & 0 & 0 & 0 \\ S_{12}^{(kloc)} & S_{22}^{(kloc)} & S_{23}^{(kloc)} & 0 & 0 & 0 \\ S_{13}^{(kloc)} & S_{23}^{(kloc)} & S_{33}^{(kloc)} & 0 & 0 & 0 \\ 0 & 0 & 0 & S_{44}^{(kloc)} & 0 & 0 \\ 0 & 0 & 0 & 0 & S_{55}^{(kloc)} & 0 \\ 0 & 0 & 0 & 0 & 0 & S_{66}^{(kloc)} \end{pmatrix}$$

$$R^{(kloc)} = \begin{pmatrix} R_{11}^{(kloc)} & R_{12}^{(kloc)} & R_{13}^{(kloc)} & 0 & 0 & 0 \\ R_{12}^{(kloc)} & R_{22}^{(kloc)} & R_{23}^{(kloc)} & 0 & 0 & 0 \\ R_{13}^{(kloc)} & R_{23}^{(kloc)} & R_{33}^{(kloc)} & 0 & 0 & 0 \\ 0 & 0 & 0 & R_{44}^{(kloc)} & 0 & 0 \\ 0 & 0 & 0 & 0 & R_{55}^{(kloc)} & 0 \\ 0 & 0 & 0 & 0 & 0 & R_{66}^{(kloc)} \end{pmatrix}$$

$$S^{(kref)} = \begin{pmatrix} S_{11}^{(kref)} & S_{12}^{(kref)} & S_{13}^{(kref)} & 0 & 0 & S_{16}^{(kref)} \\ S_{12}^{(kref)} & S_{22}^{(kref)} & S_{23}^{(kref)} & 0 & 0 & S_{26}^{(kref)} \\ S_{13}^{(kref)} & S_{23}^{(kref)} & S_{33}^{(kref)} & 0 & 0 & S_{36}^{(kref)} \\ 0 & 0 & 0 & S_{44}^{(kref)} & S_{45}^{(kref)} & 0 \\ 0 & 0 & 0 & S_{45}^{(kref)} & S_{55}^{(kref)} & 0 \\ S_{16}^{(kref)} & S_{26}^{(kref)} & S_{36}^{(kref)} & 0 & 0 & S_{66}^{(kref)} \end{pmatrix}$$

$$R^{(kref)} = \begin{pmatrix} R_{11}^{(kref)} & R_{12}^{(kref)} & R_{13}^{(kref)} & 0 & 0 & R_{16}^{(kref)} \\ R_{12}^{(kref)} & R_{22}^{(kref)} & R_{23}^{(kref)} & 0 & 0 & R_{26}^{(kref)} \\ R_{13}^{(kref)} & R_{23}^{(kref)} & R_{33}^{(kref)} & 0 & 0 & R_{36}^{(kref)} \\ 0 & 0 & 0 & R_{44}^{(kref)} & R_{45}^{(kref)} & 0 \\ 0 & 0 & 0 & R_{45}^{(kref)} & R_{55}^{(kref)} & 0 \\ R_{16}^{(kref)} & R_{26}^{(kref)} & R_{36}^{(kref)} & 0 & 0 & R_{66}^{(kref)} \end{pmatrix}$$

### 7.5.3    Formulations of the cellular problems for the case of a periodic laminate

Recall that in the framework of periodic homogenisation using the averaging technique, for the case of linear elasticity, known as the homogenisation behaviour, the relation which exists between the averages on a periodic cell $Y$ of the microscopic strain field and the microscopic stress field when the fields at the microscopic level are solutions of the following problem $(\mathcal{P}^Y)$ applied to the cell:

$$(\mathcal{P}^Y) \begin{cases} \text{Find } (\sigma^m, \vec{u}^m)/ \\ \vec{div}\, \sigma^m = \vec{0}, \, \forall M \in \Omega^Y \\ \sigma^m = a(y_3)\varepsilon^m(\vec{u}^m) \\ \sigma^m \text{ and } \varepsilon^m(\vec{u}^m) \text{ are Y-periodic} \\ < \varepsilon_{ij}^m(\vec{u}^m) > = E_{ij} \text{ where } E \text{ is a given 2nd-order symmetric tensor} \end{cases}$$

with $< q > = \frac{1}{|D|} \int_D q(y) dy$ is the average of $q$ over $D$ with a size $|D|$. Thus, according to the hypothesis of small perturbations, the homogenised behaviour is linearly elastic and defined by the relationship between the average values, called also macroscopic values:

$$\Sigma = a^H E \Longleftrightarrow \Sigma_{ij} = a^H_{ijkh} E_{kh}$$

where:

- $E = <\varepsilon^m(\vec{u}^m)>$ is the macroscopic strain ;
- $\Sigma = <\sigma^m>$ is the macroscopic stress.

So as to aid later calculations we consider the framework using the Voigt notation. The problem $(\mathcal{P}^Y)$ can therefore be written as :

$$(\mathcal{P}^Y) \begin{cases} \text{Find } (\sigma^m, \vec{u}^m)/ \\ \forall M \in \Omega^Y \begin{cases} \frac{\partial \sigma_1^m}{\partial y_1} + \frac{\partial \sigma_6^m}{\partial y_2} + \frac{\partial \sigma_5^m}{\partial y_3} = 0 \\ \frac{\partial \sigma_6^m}{\partial y_1} + \frac{\partial \sigma_2^m}{\partial y_2} + \frac{\partial \sigma_4^m}{\partial y_3} = 0 \\ \frac{\partial \sigma_5^m}{\partial y_1} + \frac{\partial \sigma_4^m}{\partial y_2} + \frac{\partial \sigma_3^m}{\partial y_3} = 0 \end{cases} \\ \sigma^m = R(y_3)\varepsilon^m(\vec{u}^m) \Longleftrightarrow \sigma_I^m = R_{IJ}(y_3)\varepsilon_J^m(\vec{u}^m) \\ \sigma^m \text{ and } \varepsilon^m(\vec{u}^m) \text{ are Y-periodic} \\ <\sigma_I^m> = \Sigma_I \\ <\varepsilon_I^m(\vec{u}^m)> = E_I \end{cases}$$

where $R(y_3) = R^{(kref)}$ si $h_{k+1} \geq y_3 \geq h_k$. Consequently, the homogenised behaviour is:

$$\Sigma = a^H E \Longleftrightarrow \Sigma_I = R^H_{IJ} E_J$$

The $(\mathcal{P}^Y)$ problem is linear. Its solutions are then linearly dependent of its datas. We can there write:

- $\sigma_I^m = \alpha_{IK} E_K$ ;
- $u_i^m = v_i^K E_K$.

The first relation gives:

$$<\sigma_I^m> = <\alpha_{IK} E_K> = <\alpha_{IK}> E_K = \Sigma_I \Longrightarrow R^H_{IK} = <\alpha_{IK}>$$

The second relation gives:

$$\sigma_I^m = R^{kref}_{IJ}(y_3)\varepsilon_J^m(\vec{v}^K E_K)$$

Consequently:

$$\sigma_I^m = R^{kref}_{IJ}(y_3)\varepsilon_J^m(\vec{v}^K) E_K = s_I^K E_K$$

where we note:

$$s_I^K = R^{kref}_{IJ}(y_3)\varepsilon_J^m(\vec{v}^K)$$

Then

$$<\sigma_I^m> = <s_I^K E_K> = <s_I^K> E_K \Longrightarrow R^H_{IK} = <s_I^K>$$

We can then found the tensor $a^H$ by solving the six following cellular problems:

$$(\mathcal{P}^{K-Y}) \begin{cases} \text{Find } (s^K, \vec{v}^K)/ \\ \forall M \in \Omega^Y \begin{cases} \frac{\partial s_1^K}{\partial y_1} + \frac{\partial s_6^K}{\partial y_2} + \frac{\partial s_5^K}{\partial y_3} = 0 \\ \frac{\partial s_6^K}{\partial y_1} + \frac{\partial s_2^K}{\partial y_2} + \frac{\partial s_4^K}{\partial y_3} = 0 \\ \frac{\partial s_5^K}{\partial y_1} + \frac{\partial s_4^K}{\partial y_2} + \frac{\partial s_3^K}{\partial y_3} = 0 \end{cases} \\ s^K = R(y_3)\varepsilon^m(\vec{v}^K) \Longleftrightarrow s_I^K = R_{IJ}(y_3)\varepsilon_J^m(\vec{v}^K) \\ s^K \text{ and } \varepsilon^m(\vec{v}^K) \text{ are Y-periodic} \\ <\varepsilon_I^m(\vec{v}^K)> = E_I^K = \delta_{IK} \end{cases}$$

As the imposition of the periodicity condition for the deformation is difficult to carry out the cellular problems $(\mathcal{P}^{K-Y})$ are reformulated by a change of variable on which the periodicity condition is directly written (rather than its derivatives). Thus we write:

$$\vec{\chi}^K = -\vec{v}^K + <grad(\vec{v}^K)>\vec{y} \iff \chi_i^K = -v_i^K + <\frac{\partial v_i^K}{\partial y_j}>y_j$$

We then deduce:

$$\varepsilon_{ij}^m(\vec{\chi}^K) = \frac{1}{2}\left(\frac{\partial \chi_i^K}{\partial y_j}+\frac{\partial \chi_j^K}{\partial y_i}\right) = -\frac{1}{2}\left(\frac{\partial v_i^K}{\partial y_j}+\frac{\partial v_j^K}{\partial y_i}\right) + \frac{1}{2}\left(<\frac{\partial v_i^K}{\partial y_j}>+<\frac{\partial v_j^K}{\partial y_i}>\right)$$

$$= -\frac{1}{2}\left(\frac{\partial v_i^K}{\partial y_j}+\frac{\partial v_j^K}{\partial y_i}\right) + <\frac{1}{2}\left(\frac{\partial v_i^K}{\partial y_j}+\frac{\partial v_j^K}{\partial y_i}\right)> = -\varepsilon_{ij}^m(\vec{v}^K) + <\varepsilon_{ij}^m(\vec{v}^K)>$$

$$\iff \varepsilon^m(\vec{v}^K) = -\varepsilon^m(\vec{\chi}^K) + <\varepsilon^m(\vec{v}^K)> = -\varepsilon^m(\vec{\chi}^K) + E^K \iff \varepsilon_I^m(\vec{v}^K) = -\varepsilon_I^m(\vec{\chi}^K) + E_I^K$$

Therefore: $s^K = R(y_3)\varepsilon^m(\vec{v}^K) = R(y_3)\left(-\varepsilon^m(\vec{\chi}^K) + E^K\right) = -R(y_3)\varepsilon^m(\vec{\chi}^K) + R(y_3)E^K$. We pose:

$$\sigma^K = R(y_3)\varepsilon^m(\vec{\chi}^K)$$

And then: $s^K = -\sigma^K + R(y_3)E^K \iff s_I^K = -\sigma_I^K + R_{IJ}(y_3)E_J^K = -\sigma_I^K + R_{IK}(y_3)$. With the new variables, the cellular problems $(\mathcal{P}^{K-Y})$ change their form and become the following $(\overline{P}_Y^{kh})$ cellular problems (in the sense of the distributions theory):

$$(\overline{\mathcal{P}}^{K-Y})\begin{cases} \text{Find } (\sigma^K, \vec{\chi}^K)/ \\ \forall M \in \Omega^Y \begin{cases} \frac{\partial \sigma_1^K}{\partial y_1}+\frac{\partial \sigma_6^K}{\partial y_2}+\frac{\partial \sigma_5^K}{\partial y_3} = \frac{\partial R_{1K}}{\partial y_1}+\frac{\partial R_{6K}}{\partial y_2}+\frac{\partial R_{5K}}{\partial y_3} \\ \frac{\partial \sigma_6^k}{\partial y_1}+\frac{\partial \sigma_2^k}{\partial y_2}+\frac{\partial \sigma_4^k}{\partial y_3} = \frac{\partial R_{6K}}{\partial y_1}+\frac{\partial R_{2K}}{\partial y_2}+\frac{\partial R_{4K}}{\partial y_3} \\ \frac{\partial \sigma_5^k}{\partial y_1}+\frac{\partial \sigma_4^k}{\partial y_2}+\frac{\partial \sigma_3^k}{\partial y_3} = \frac{\partial R_{5K}}{\partial y_1}+\frac{\partial R_{4K}}{\partial y_2}+\frac{\partial R_{3K}}{\partial y_3} \end{cases} \\ \sigma^K = R(y_3)\varepsilon^m(\vec{\chi}^K) \iff \sigma_I^K = R_{IJ}(y_3)\varepsilon_J^m(\vec{\chi}^K) \\ \sigma^K \text{ and } \vec{\chi}^K \text{ are Y-periodic} \end{cases}$$

With $R_{IK}^H = <s_I^K>$ and $s^K = -R(y_3)\varepsilon^m(\vec{\chi}^K) + R(y_3)E^K$, the homogeneised coefficients take therefore the following form:

$$R_{IK}^H = <R_{IK}(y)> - <R_{IJ}(y)\varepsilon_J^m(\vec{\chi}^K)>$$

It is possible, in the particular case treated here, to consider several points which will help in solving these problems.

Firstly, we should notice that where the plies of the periodically stratified laminate cell are each made of a homogeneous material, which is specific to each ply, only the variable which indicates the order of layers ($y_3$) allows the different materials to be distinguished.

For the same reason the dimensions of this cell in the directions indicated by the vectors $\vec{y}_1$ and $\vec{y}_2$ can be taken anyhow. Consequently:

- for the $(\mathcal{P}^Y)$ problem, $\sigma^m$ and $\varepsilon^m(\vec{u}^m)$ do only depend on the space variable $y_3$: $\sigma^m = \sigma^m(y_3)$ and $\varepsilon^m(\vec{u}^m) = \varepsilon^m(\vec{u}^m)(y_3)$ ;
- for the $(\overline{\mathcal{P}}^{K-Y})$ problem, $\sigma^K$, $\varepsilon^m(\vec{\chi}^K)$ and $\vec{\chi}^K$ do only depend on the space variable $y_3$.

We can then calculate, for the six $K$ cellular problems, the strain tensor:

$$\varepsilon^m(\vec{\chi}^K) = \begin{pmatrix} \varepsilon_1^m(\vec{\chi}^K) = 0 \\ \varepsilon_2^m(\vec{\chi}^K) = 0 \\ \varepsilon_3^m(\vec{\chi}^K) = \dfrac{d\chi_3^K(y_3)}{dy_3} \\ \varepsilon_4^m(\vec{\chi}^K) = \dfrac{d\chi_2^K(y_3)}{dy_3} \\ \varepsilon_5^m(\vec{\chi}^K) = \dfrac{d\chi_1^K(y_3)}{dy_3} \\ \varepsilon_6^m(\vec{\chi}^K) = 0 \end{pmatrix}$$

The behaviour law in the cell of the laminate is then, with the following form:

$$\begin{pmatrix} \sigma_1^K \\ \sigma_2^K \\ \sigma_3^K \\ \sigma_4^K \\ \sigma_5^K \\ \sigma_6^K \end{pmatrix} = \begin{pmatrix} R_{11}(y_3) & R_{12}(y_3) & R_{13}(y_3) & 0 & 0 & R_{16}(y_3) \\ R_{12}(y_3) & R_{22}(y_3) & R_{23}(y_3) & 0 & 0 & R_{26}(y_3) \\ R_{13}(y_3) & R_{23}(y_3) & R_{33}(y_3) & 0 & 0 & R_{36}(y_3) \\ 0 & 0 & 0 & R_{44}(y_3) & R_{45}(y_3) & 0 \\ 0 & 0 & 0 & R_{45}(y_3) & R_{55}(y_3) & 0 \\ R_{16}(y_3) & R_{26}(y_3) & R_{36}(y_3) & 0 & 0 & R_{66}(y_3) \end{pmatrix} \begin{pmatrix} 0 \\ 0 \\ \dfrac{d\chi_3^K(y_3)}{dy_3} \\ \dfrac{d\chi_2^K(y_3)}{dy_3} \\ \dfrac{d\chi_1^K(y_3)}{dy_3} \\ 0 \end{pmatrix}$$

So that:

$$\begin{cases} \sigma_1^K = R_{13}(y_3)\dfrac{d\chi_3^K(y_3)}{dy_3} \\ \sigma_2^K = R_{23}(y_3)\dfrac{d\chi_3^K(y_3)}{dy_3} \\ \sigma_3^K = R_{33}(y_3)\dfrac{d\chi_3^K(y_3)}{dy_3} \\ \sigma_4^K = R_{44}(y_3)\dfrac{d\chi_2^K(y_3)}{dy_3} + R_{45}(y_3)\dfrac{d\chi_1^K(y_3)}{dy_3} \\ \sigma_5^K = R_{45}(y_3)\dfrac{d\chi_2^K(y_3)}{dy_3} + R_{55}(y_3)\dfrac{d\chi_1^K(y_3)}{dy_3} \\ \sigma_6^K = R_{36}(y_3)\dfrac{d\chi_3^K(y_3)}{dy_3} \end{cases}$$

### 7.5.4 Getting the homogenised behaviour without solving the cellular problems

Using the previous remarks, the problem $(\mathcal{P}^Y)$ gives ($l_3$, $l_4$ and $l_5$ are constantes) :

$$(\mathcal{P}^Y) \begin{cases} \text{Find } (\sigma^m, \vec{u}^m)/ \\ \forall M \in \Omega^Y \begin{cases} \dfrac{d\sigma_5^m(y_3)}{dy_3} = 0 \Longrightarrow \sigma_5^m(y_3) = l_5 \Longrightarrow <\sigma_5^m> = \Sigma_5 = constante \\ \dfrac{d\sigma_4^m(y_3)}{dy_3} = 0 \Longrightarrow \sigma_4^m(y_3) = l_4 \Longrightarrow <\sigma_4^m> = \Sigma_4 = constante \\ \dfrac{d\sigma_3^m(y_3)}{dy_3} = 0 \Longrightarrow \sigma_3^m(y_3) = l_3 \Longrightarrow <\sigma_3^m> = \Sigma_3 = constante \end{cases} \\ \sigma^m = a(y_3)\varepsilon^m(\vec{u}^m) \Longleftrightarrow \sigma_I^m = R_{IJ}^{kref}(y_3)\varepsilon_J^m(\vec{u}^m) \\ \sigma^m \text{ and } \varepsilon^m(\vec{u}^m) \text{ are Y-periodic} \\ <\sigma_I^m> = \Sigma_I \\ <\varepsilon_I^m(\vec{u}^m)> = E_I \end{cases}$$

The compatibility equations, for a second-order symmetric tensor $e$, in the framework of the Voigt notations, are:

$$\begin{cases} \dfrac{\partial^2 e_1(y)}{\partial y_2^2} - \dfrac{\partial^2 e_6(y)}{\partial y_2 \partial y_1} + \dfrac{\partial^2 e_2(y)}{\partial y_1^2} = 0 \\ \dfrac{\partial^2 e_2(y)}{\partial y_3^2} - \dfrac{\partial^2 e_4(y)}{\partial y_3 \partial y_2} + \dfrac{\partial^2 e_3(y)}{\partial y_2^2} = 0 \\ \dfrac{\partial^2 e_1(y)}{\partial y_3^2} - \dfrac{\partial^2 e_5(y)}{\partial y_3 \partial y_1} + \dfrac{\partial^2 e_3(y)}{\partial y_1^2} = 0 \\ 2\dfrac{\partial^2 e_1(y)}{\partial y_3 \partial y_2} + \dfrac{\partial^2 e_4(y)}{\partial y_1^2} - \dfrac{\partial^2 e_5(y)}{\partial y_1 \partial y_2} - \dfrac{\partial^2 e_6(y)}{\partial y_3 \partial y_1} = 0 \\ 2\dfrac{\partial^2 e_2(y)}{\partial y_3 \partial y_1} - \dfrac{\partial^2 e_4(y)}{\partial y_2 \partial y_1} + \dfrac{\partial^2 e_5(y)}{\partial y_2^2} - \dfrac{\partial^2 e_6(y)}{\partial y_3 \partial y_2} = 0 \\ 2\dfrac{\partial^2 e_3(y)}{\partial y_2 \partial y_1} - \dfrac{\partial^2 e_4(y)}{\partial y_3 \partial y_1} - \dfrac{\partial^2 e_5(y)}{\partial y_2 \partial y_3} + \dfrac{\partial^2 e_6(y)}{\partial y_3^2} = 0 \end{cases}$$

Then, for $(\mathcal{P}^Y)$, the compatibility equations give:

$$\begin{cases} \frac{d^2\varepsilon_2^m(\vec{u}^m)(y_3)}{dy_3^2} = 0 \Longrightarrow \varepsilon_2^m(\vec{u}^m)(y_3) = k_2y_3 + l_2 \\ \frac{d^2\varepsilon_1^m(\vec{u}^m)(y_3)}{dy_3^2} = 0 \Longrightarrow \varepsilon_1^m(\vec{u}^m)(y_3) = k_1y_3 + l_1 \\ \frac{d^2\varepsilon_6^m(\vec{u}^m)(y_3)}{dy_3^2} = 0 \Longrightarrow \varepsilon_6^m(\vec{u}^m)(y_3) = k_6y_3 + l_6 \end{cases}$$

where $k_1$, $k_2$, $k_6$, $l_1$, $l_2$ and $l_6$ are constantes. But, invoking that $\varepsilon^m(\vec{u}^m)$ is periodic, we deduce hat $k_1$, $k_2$ are $k_6$ equal to zero. It comes then:

$$\begin{cases} \varepsilon_2^m(\vec{u}^m)(y_3) = l_2 \Longrightarrow< \varepsilon_2^m(\vec{u}^m) >= \varepsilon_2^m(\vec{u}^m) := \varepsilon_2^m = E_2 = constante \\ \varepsilon_1^m(\vec{u}^m)(y_3) = l_1 \Longrightarrow< \varepsilon_1^m(\vec{u}^m) >= \varepsilon_1^m(\vec{u}^m) := \varepsilon_1^m = E_1 = constante \\ \varepsilon_6^m(\vec{u}^m)(y_3) = l_6 \Longrightarrow< \varepsilon_6^m(\vec{u}^m) >= \varepsilon_6^m(\vec{u}^m) := \varepsilon_6^m = E_6 = constante \end{cases}$$

We can then, built the following constant vector $\vec{A}$ :

$$\vec{A} = \begin{pmatrix} \varepsilon_1^m \\ \varepsilon_2^m \\ \varepsilon_6^m \\ \sigma_5^m \\ \sigma_4^m \\ \sigma_3^m \end{pmatrix} = \begin{pmatrix} E_1 \\ E_2 \\ E_6 \\ \Sigma_5 \\ \Sigma_4 \\ \Sigma_3 \end{pmatrix}$$

And, we built also the vector $\vec{B}$ only depending on the $y_3$ variable, and we calcute its average:

$$\vec{B}(y_3) = \begin{pmatrix} \varepsilon_5^m(\vec{u}^m)(y_3) := \varepsilon_5^m(y_3) \\ \varepsilon_4^m(\vec{u}^m)(y_3) := \varepsilon_4^m(y_3) \\ \varepsilon_3^m(\vec{u}^m)(y_3) := \varepsilon_3^m(y_3) \\ \sigma_1^m(y_3) \\ \sigma_2^m(y_3) \\ \sigma_6^m(y_3) \end{pmatrix} \Longrightarrow< \vec{B}(y_3) >= \begin{pmatrix} < \varepsilon_5^m(y_3) >= E_5 \\ < \varepsilon_4^m(y_3) >= E_4 \\ < \varepsilon_3^m(y_3) >= E_3 \\ < \sigma_1^m(y_3) >= \Sigma_1 \\ < \sigma_2^m(y_3) >= \Sigma_2 \\ < \sigma_6^m(y_3) >= \Sigma_6 \end{pmatrix}$$

The behaviour law, for $(\mathcal{P}^Y)$ is:

$$\sigma_I^m = R_{IJ}(y_3)\varepsilon_J^m$$

$$\Longleftrightarrow$$

$$\begin{cases} (a)\, \sigma_1^m(y_3) = R_{11}(y_3)\varepsilon_1^m + R_{12}(y_3)\varepsilon_2^m + R_{13}(y_3)\varepsilon_3^m(y_3) + R_{16}(y_3)\varepsilon_6^m \\ (b)\, \sigma_2^m(y_3) = R_{12}(y_3)\varepsilon_1^m + R_{22}(y_3)\varepsilon_2^m + R_{23}(y_3)\varepsilon_3^m(y_3) + R_{26}(y_3)\varepsilon_6^m \\ (c)\, \sigma_3^m = R_{31}(y_3)\varepsilon_1^m + R_{32}(y_3)\varepsilon_2^m + R_{33}(y_3)\varepsilon_3^m(y_3) + R_{36}(y_3)\varepsilon_6^m \\ (d)\, \sigma_4^m = R_{44}(y_3)\varepsilon_4^m(y_3) + R_{45}(y_3)\varepsilon_5^m(y_3) \\ (e)\, \sigma_5^m = R_{45}(y_3)\varepsilon_4^m(y_3) + R_{55}(y_3)\varepsilon_5^m(y_3) \\ (f)\, \sigma_6^m(y_3) = R_{16}(y_3)\varepsilon_1^m + R_{26}(y_3)\varepsilon_2^m + R_{36}(y_3)\varepsilon_3^m + R_{66}(y_3)\varepsilon_6^m \end{cases}$$

The idea now is to reorganise the constitutive law so as to make appear the vectors $\vec{A}$ and $\vec{B}(y_3)$. Then:

$$(c) \Longrightarrow \varepsilon_3^m(y_3) = -\frac{R_{13}(y_3)}{R_{33}(y_3)}\varepsilon_1^m - \frac{R_{23}(y_3)}{R_{33}(y_3)}\varepsilon_2^m - \frac{R_{36}(y_3)}{R_{33}(y_3)}\varepsilon_6^m + \frac{1}{R_{33}(y_3)}\sigma_3^m$$

$$(a) \Longrightarrow \sigma_1^m(y_3) = \left(R_{11}(y_3) - \frac{R_{13}(y_3)^2}{R_{33}(y_3)}\right)\varepsilon_1^m + \left(R_{12}(y_3) - \frac{R_{13}(y_3)R_{23}(y_3)}{R_{33}(y_3)}\right)\varepsilon_2^m + \dots$$

$$\dots \left(R_{16}(y_3) - \frac{R_{13}(y_3)R_{36}(y_3)}{R_{33}(y_3)}\right)\varepsilon_6^m + \frac{R_{13}(y_3)}{R_{33}(y_3)}\sigma_3^m$$

$$(b) \implies \sigma_2^m(y_3) = \left( R_{12}(y_3) - \frac{R_{23}(y_3)R_{13}(y_3)}{R_{33}(y_3)} \right) \varepsilon_1^m + \left( R_{22}(y_3) - \frac{R_{23}(y_3)^2}{R_{33}(y_3)} \right) \varepsilon_2^m + \dots$$

$$\dots \left( R_{26}(y_3) - \frac{R_{23}(y_3)R_{36}(y_3)}{R_{33}(y_3)} \right) \varepsilon_6^m + \frac{R_{23}(y_3)}{R_{33}(y_3)} \sigma_3^m$$

$$(f) \implies \sigma_6^m(y_3) = \left( R_{16}(y_3) - \frac{R_{36}(y_3)R_{13}(y_3)}{R_{33}(y_3)} \right) \varepsilon_1^m + \left( R_{26}(y_3) - \frac{R_{36}(y_3)R_{23}(y_3)}{R_{33}(y_3)} \right) \varepsilon_2^m + \dots$$

$$\dots \left( R_{66}(y_3) - \frac{R_{36}(y_3)^2}{R_{33}(y_3)} \right) \varepsilon_6^m + \frac{R_{36}(y_3)}{R_{33}(y_3)} \sigma_3^m$$

$$(d)(e) \implies \varepsilon_4^m(y_3) = \frac{R_{55}(y_3)}{R_{44}(y_3)R_{55}(y_3) - R_{45}(y_3)^2} \sigma_4^m - \frac{R_{45}(y_3)}{R_{44}(y_3)R_{55}(y_3) - R_{45}(y_3)^2} \sigma_5^m$$

$$(d)(e) \implies \varepsilon_5^m(y_3) = -\frac{R_{45}(y_3)}{R_{44}(y_3)R_{55}(y_3) - R_{45}(y_3)^2} \sigma_4^m + \frac{R_{44}(y_3)}{R_{44}(y_3)R_{55}(y_3) - R_{45}(y_3)^2} \sigma_5^m$$

Finally, we obtain:

$$\begin{pmatrix} \varepsilon_5^m(y_3) \\ \varepsilon_4^m(y_3) \\ \varepsilon_3^m(y_3) \\ \sigma_1^m(y_3) \\ \sigma_2^m(y_3) \\ \sigma_6^m(y_3) \end{pmatrix} = \begin{pmatrix} 0 & 0 & 0 & C_{14}(y_3) & C_{15}(y_3) & 0 \\ 0 & 0 & 0 & C_{24}(y_3) & C_{25}(y_3) & 0 \\ C_{31}(y_3) & C_{32}(y_3) & C_{33}(y_3) & 0 & 0 & C_{36}(y_3) \\ C_{41}(y_3) & C_{42}(y_3) & C_{43}(y_3) & 0 & 0 & C_{46}(y_3) \\ C_{51}(y_3) & C_{52}(y_3) & C_{53}(y_3) & 0 & 0 & C_{56}(y_3) \\ C_{61}(y_3) & C_{62}(y_3) & C_{63}(y_3) & 0 & 0 & C_{66}(y_3) \end{pmatrix} \begin{pmatrix} \varepsilon_1^m \\ \varepsilon_2^m \\ \varepsilon_6^m \\ \sigma_5^m \\ \sigma_4^m \\ \sigma_3^m \end{pmatrix}$$

$$\Longleftrightarrow$$

$$\vec{B}(y_3) = C(y_3)\vec{A}$$

with:

$$\begin{cases} C_{14}(y_3) = \frac{R_{44}(y_3)}{R_{44}(y_3)R_{55}(y_3) - R_{45}(y_3)^2} \\ C_{15}(y_3) = -\frac{R_{45}(y_3)}{R_{44}(y_3)R_{55}(y_3) - R_{45}(y_3)^2} \end{cases}$$

$$\begin{cases} C_{24}(y_3) = -\frac{R_{45}(y_3)}{R_{44}(y_3)R_{55}(y_3) - R_{45}(y_3)^2} \\ C_{25}(y_3) = \frac{R_{55}(y_3)}{R_{44}(y_3)R_{55}(y_3) - R_{45}(y_3)^2} \end{cases}$$

$$\begin{cases} C_{31}(y_3) = -\frac{R_{13}(y_3)}{R_{33}(y_3)} \\ C_{32}(y_3) = -\frac{R_{23}(y_3)}{R_{33}(y_3)} \\ C_{33}(y_3) = -\frac{R_{36}(y_3)}{R_{33}(y_3)} \\ C_{36}(y_3) = \frac{1}{R_{33}(y_3)} \end{cases}$$

$$\begin{cases} C_{41}(y_3) = \left( R_{11}(y_3) - \frac{R_{13}(y_3)^2}{R_{33}(y_3)} \right) \\ C_{42}(y_3) = \left( R_{12}(y_3) - \frac{R_{13}(y_3)R_{23}(y_3)}{R_{33}(y_3)} \right) \\ C_{43}(y_3) = \left( R_{16}(y_3) - \frac{R_{13}(y_3)R_{36}(y_3)}{R_{33}(y_3)} \right) \\ C_{46}(y_3) = \frac{R_{13}(y_3)}{R_{33}(y_3)} \end{cases}$$

$$
\begin{cases}
C_{51}(y_3) = \left( R_{12}(y_3) - \dfrac{R_{23}(y_3)R_{13}(y_3)}{R_{33}(y_3)} \right) \\[2mm]
C_{52}(y_3) = \left( R_{22}(y_3) - \dfrac{R_{23}(y_3)^2}{R_{33}(y_3)} \right) \\[2mm]
C_{53}(y_3) = \left( R_{26}(y_3) - \dfrac{R_{23}(y_3)R_{36}(y_3)}{R_{33}(y_3)} \right) \\[2mm]
C_{56}(y_3) = \dfrac{R_{23}(y_3)}{R_{33}(y_3)}
\end{cases}
$$

$$
\begin{cases}
C_{61}(y_3) = \left( R_{16}(y_3) - \dfrac{R_{36}(y_3)R_{13}(y_3)}{R_{33}(y_3)} \right) \\[2mm]
C_{62}(y_3) = \left( R_{26}(y_3) - \dfrac{R_{36}(y_3)R_{23}(y_3)}{R_{33}(y_3)} \right) \\[2mm]
C_{63}(y_3) = \left( R_{66}(y_3) - \dfrac{R_{36}(y_3)^2}{R_{33}(y_3)} \right) \\[2mm]
C_{66}(y_3) = \dfrac{R_{36}(y_3)}{R_{33}(y_3)}
\end{cases}
$$

By taking the average of this expression, we obtain:

$$
< \vec{B}(y_3) > = < C(y_3)\vec{A} > = < C(y_3) > \vec{A}
$$

Therefore by putting $< C(y_3) > = C^H = (C^H_{IJ})_{I,J=1,\dots,6}$

$$
\begin{pmatrix} E_5 \\ E_4 \\ E_3 \\ \Sigma_1 \\ \Sigma_2 \\ \Sigma_6 \end{pmatrix}
=
\begin{pmatrix}
0 & 0 & 0 & C^H_{14} & C^H_{15} & 0 \\
0 & 0 & 0 & C^H_{24} & C^H_{25} & 0 \\
C^H_{31} & C^H_{32} & C^H_{33} & 0 & 0 & C^H_{36} \\
C^H_{41} & C^H_{42} & C^H_{43} & 0 & 0 & C^H_{46} \\
C^H_{51} & C^H_{52} & C^H_{53} & 0 & 0 & C^H_{56} \\
C^H_{61} & C^H_{62} & C^H_{63} & 0 & 0 & C^H_{66}
\end{pmatrix}
\begin{pmatrix} E_1 \\ E_2 \\ E_6 \\ \Sigma_5 \\ \Sigma_4 \\ \Sigma_3 \end{pmatrix}
$$

with:

$$
\begin{cases}
C^H_{14} = < \dfrac{R_{44}(y_3)}{R_{44}(y_3)R_{55}(y_3)-R_{45}(y_3)^2} > \\[3mm]
C^H_{15} = - < \dfrac{R_{45}(y_3)}{R_{44}(y_3)R_{55}(y_3)-R_{45}(y_3)^2} >
\end{cases}
$$

$$
\begin{cases}
C^H_{24} = - < \dfrac{R_{45}(y_3)}{R_{44}(y_3)R_{55}(y_3)-R_{45}(y_3)^2} > \\[3mm]
C^H_{25} = < \dfrac{R_{55}(y_3)}{R_{44}(y_3)R_{55}(y_3)-R_{45}(y_3)^2} >
\end{cases}
$$

$$
\begin{cases}
C^H_{31} = - < \dfrac{R_{13}(y_3)}{R_{33}(y_3)} > \\[2mm]
C^H_{32} = - < \dfrac{R_{23}(y_3)}{R_{33}(y_3)} > \\[2mm]
C^H_{33} = - < \dfrac{R_{36}(y_3)}{R_{33}(y_3)} > \\[2mm]
C^H_{36} = < \dfrac{1}{R_{33}(y_3)} >
\end{cases}
$$

$$
\begin{cases}
C^H_{41} = < R_{11}(y_3) - \dfrac{R_{13}(y_3)^2}{R_{33}(y_3)} > \\[2mm]
C^H_{42} = < R_{12}(y_3) - \dfrac{R_{13}(y_3)R_{23}(y_3)}{R_{33}(y_3)} > \\[2mm]
C^H_{43} = < R_{16}(y_3) - \dfrac{R_{13}(y_3)R_{36}(y_3)}{R_{33}(y_3)} > \\[2mm]
C^H_{46} = < \dfrac{R_{13}(y_3)}{R_{33}(y_3)} >
\end{cases}
$$

$$
\begin{cases}
C^H_{51} = < R_{12}(y_3) - \dfrac{R_{23}(y_3)R_{13}(y_3)}{R_{33}(y_3)} > \\[2mm]
C^H_{52} = < R_{22}(y_3) - \dfrac{R_{23}(y_3)^2}{R_{33}(y_3)} > \\[2mm]
C^H_{53} = < R_{26}(y_3) - \dfrac{R_{23}(y_3)R_{36}(y_3)}{R_{33}(y_3)} > \\[2mm]
C^H_{56} = < \dfrac{R_{23}(y_3)}{R_{33}(y_3)} >
\end{cases}
$$

$$\begin{cases} C_{61}^H =< R_{16}(y_3) - \frac{R_{36}(y_3)R_{13}(y_3)}{R_{33}(y_3)} > \\ C_{62}^H =< R_{26}(y_3) - \frac{R_{36}(y_3)R_{23}(y_3)}{R_{33}(y_3)} > \\ C_{63}^H =< R_{66}(y_3) - \frac{R_{36}(y_3)^2}{R_{33}(y_3)} > \\ C_{66}^H =< \frac{R_{36}(y_3)}{R_{33}(y_3)} > \end{cases}$$

By reorganising the previous expression according to macroscopic stresses and strains, we obtain:

$$\begin{pmatrix} \Sigma_1 \\ \Sigma_2 \\ \Sigma_3 \\ \Sigma_4 \\ \Sigma_5 \\ \Sigma_6 \end{pmatrix} = \begin{pmatrix} R_{11}^H & R_{12}^H & R_{13}^H & 0 & 0 & R_{16}^H \\ R_{12}^H & R_{22}^H & R_{23}^H & 0 & 0 & R_{26}^H \\ R_{13}^H & R_{23}^H & R_{33}^H & 0 & 0 & R_{36}^H \\ 0 & 0 & 0 & R_{44}^H & R_{45}^H & 0 \\ 0 & 0 & 0 & R_{45}^H & R_{55}^H & 0 \\ R_{16}^H & R_{26}^H & R_{36}^H & 0 & 0 & R_{66}^H \end{pmatrix} \begin{pmatrix} E_1 \\ E_2 \\ E_3 \\ E_4 \\ E_5 \\ E_6 \end{pmatrix}$$

with:

$$\begin{cases} R_{11}^H = C_{41}^H - \frac{C_{46}^H C_{31}^H}{C_{36}^H} \\ R_{12}^H = C_{42}^H - \frac{C_{46}^H C_{32}^H}{C_{36}^H} = C_{51}^H - \frac{C_{56}^H C_{31}^H}{C_{36}^H} \\ R_{13}^H = \frac{C_{46}^H}{C_{36}^H} = -\frac{C_{31}^H}{C_{36}^H} \\ R_{16}^H = C_{43}^H - \frac{C_{46}^H C_{33}^H}{C_{36}^H} = C_{61}^H - \frac{C_{66}^H C_{31}^H}{C_{36}^H} \end{cases}$$

$$\begin{cases} R_{22}^H = C_{52}^H - \frac{C_{56}^H C_{32}^H}{C_{36}^H} \\ R_{23}^H = \frac{C_{56}^H}{C_{36}^H} = -\frac{C_{32}^H}{C_{36}^H} \\ R_{26}^H = C_{53}^H - \frac{C_{56}^H C_{33}^H}{C_{36}^H} = C_{62}^H - \frac{C_{66}^H C_{32}^H}{C_{36}^H} \end{cases}$$

$$\begin{cases} R_{33}^H = \frac{1}{C_{36}^H} \\ R_{36}^H = -\frac{C_{33}^H}{C_{36}^H} = \frac{C_{66}^H}{C_{36}^H} \end{cases}$$

$$\begin{cases} R_{44}^H = \frac{C_{14}^H}{C_{14}^H C_{25}^H - C_{24}^H C_{15}^H} \\ R_{45}^H = -\frac{C_{24}^H}{C_{14}^H C_{25}^H - C_{24}^H C_{15}^H} = -\frac{C_{15}^H}{C_{14}^H C_{25}^H - C_{24}^H C_{15}^H} \\ R_{55}^H = \frac{C_{25}^H}{C_{14}^H C_{25}^H - C_{24}^H C_{15}^H} \\ R_{66}^H = C_{63}^H - \frac{C_{66}^H C_{33}^H}{C_{36}^H} \end{cases}$$

Returning to the notations explicitly showing the rigidity of the layers $(k)$ expressed in the reference frame of the laminate and therefore in the basis $b^{(ref)}$, $R^{(kref)} = (R_{IJ}^{(kref)})_{I,J=1,2,3,4,5,6}$, we finally obtain:

$$
\left\{
\begin{aligned}
R_{11}^H &= < R_{11}^{(kref)} - \frac{(R_{13}^{(kref)})^2}{R_{33}^{(kref)}} > + \frac{< \frac{R_{13}^{(kref)}}{R_{33}^{(kref)}} >^2}{< \frac{1}{R_{33}^{(kref)}} >} \\[2mm]
R_{12}^H &= < R_{12}^{(kref)} - \frac{R_{13}^{(kref)} R_{23}^{(kref)}}{R_{33}^{(kref)}} > + \frac{< \frac{R_{13}^{(kref)}}{R_{33}^{(kref)}} > < \frac{R_{23}^{(kref)}}{R_{33}^{(kref)}} >}{< \frac{1}{R_{33}^{(kref)}} >} \\[2mm]
R_{22}^H &= < R_{22}^{(kref)} - \frac{(R_{23}^{(kref)})^2}{R_{33}^{(kref)}} > + \frac{< \frac{R_{23}^{(kref)}}{R_{33}^{(kref)}} >^2}{< \frac{1}{R_{33}^{(kref)}} >} \\[2mm]
R_{13}^H &= \frac{< \frac{R_{13}^{(kref)}}{R_{33}^{(kref)}} >}{< \frac{1}{R_{33}^{(kref)}} >} \\[2mm]
R_{23}^H &= \frac{< \frac{R_{23}^{(kref)}}{R_{33}^{(kref)}} >}{< \frac{1}{R_{33}^{(kref)}} >} \\[2mm]
R_{33}^H &= \frac{1}{< \frac{1}{R_{33}^{(kref)}} >}
\end{aligned}
\right.
$$

$$
\left\{
\begin{aligned}
R_{44}^H &= \frac{< \frac{R_{44}^{(kref)}}{R_{44}^{(kref)} R_{55}^{(kref)} - (R_{45}^{(kref)})^2} >}{< \frac{R_{44}^{(kref)}}{R_{44}^{(kref)} R_{55}^{(kref)} - (R_{45}^{(kref)})^2} > < \frac{R_{55}^{(kref)}}{R_{44}^{(kref)} R_{55}^{(kref)} - (R_{45}^{(kref)})^2} > - (< \frac{R_{45}^{(kref)}}{R_{44}^{(kref)} R_{55}^{(kref)} - (R_{45}^{(kref)})^2} >)^2} \\[2mm]
R_{45}^H &= \frac{< \frac{R_{45}^{(kref)}}{R_{44}^{(kref)} R_{55}^{(kref)} - (R_{45}^{(kref)})^2} >}{< \frac{R_{44}^{(kref)}}{R_{44}^{(kref)} R_{55}^{(kref)} - (R_{45}^{(kref)})^2} > < \frac{R_{55}^{(kref)}}{R_{44}^{(kref)} R_{55}^{(kref)} - (R_{45}^{(kref)})^2} > - (< \frac{R_{45}^{(kref)}}{R_{44}^{(kref)} R_{55}^{(kref)} - (R_{45}^{(kref)})^2} >)^2} \\[2mm]
R_{55}^H &= \frac{< \frac{R_{55}^{(kref)}}{R_{44}^{(kref)} R_{55}^{(kref)} - (R_{45}^{(kref)})^2} >}{< \frac{R_{44}^{(kref)}}{R_{44}^{(kref)} R_{55}^{(kref)} - (R_{45}^{(kref)})^2} > < \frac{R_{55}^{(kref)}}{R_{44}^{(kref)} R_{55}^{(kref)} - (R_{45}^{(kref)})^2} > - (< \frac{R_{45}^{(kref)}}{R_{44}^{(kref)} R_{55}^{(kref)} - (R_{45}^{(kref)})^2} >)^2}
\end{aligned}
\right.
$$

$$
\left\{
\begin{aligned}
R_{16}^H &= < R_{16}^{(kref)} - \frac{R_{13}^{(kref)} R_{36}^{(kref)}}{R_{33}^{(kref)}} > + \frac{< \frac{R_{13}^{(kref)}}{R_{33}^{(kref)}} > < \frac{R_{36}^{(kref)}}{R_{33}^{(kref)}} >}{< \frac{1}{R_{33}^{(kref)}} >} \\[2mm]
R_{26}^H &= < R_{26}^{(kref)} - \frac{R_{23}^{(kref)} R_{36}^{(kref)}}{R_{33}^{(kref)}} > + \frac{< \frac{R_{23}^{(kref)}}{R_{33}^{(kref)}} > < \frac{R_{36}^{(kref)}}{R_{33}^{(kref)}} >}{< \frac{1}{R_{33}^{(kref)}} >} \\[2mm]
R_{36}^H &= \frac{< \frac{R_{36}^{(kref)}}{R_{33}^{(kref)}} >}{< \frac{1}{R_{33}^{(kref)}} >} \\[2mm]
R_{66}^H &= < R_{66}^{(kref)} - \frac{(R_{36}^{(kref)})^2}{R_{33}^{(kref)}} > + \frac{< \frac{R_{36}^{(kref)}}{R_{33}^{(kref)}} >^2}{< \frac{1}{R_{33}^{(kref)}} >}
\end{aligned}
\right.
$$

## 7.5.5   Calculation of the stresses and strains in the plies making up the periodic cell. Solving the cellular problems

### Formulation of the cellular problems

To find the stresses and strains in the plies of the periodic cell during the macroscopic deformation $E$ requires solving the cellular problems. It should be noted that this solution will also supply the homogenised rigidity tensor (obtained in the preceding section by another method).

The formulation of these cellular problems, given the only relationship of $y_3$ with different values, is as follows:

$$(\overline{\mathcal{P}}^{K-Y})\begin{cases} \text{Find } (\sigma^K(y_3), \vec{\chi}^K(y_3))/ \\ \forall M \in \Omega^Y \begin{cases} \frac{d\sigma_5^K(y_3)}{dy_3} = \frac{dR_{5K}(y_3)}{dy_3} \\ \frac{d\sigma_4^K(y_3)}{dy_3} = \frac{dR_{4K}(y_3)}{dy_3} \\ \frac{d\sigma_3^K(y_3)}{dy_3} = \frac{dR_{3K}(y_3)}{dy_3} \end{cases} \\ \begin{cases} \sigma_1^K = R_{13}(y_3)\frac{d\chi_3^K(y_3)}{dy_3} \\ \sigma_2^K = R_{23}(y_3)\frac{d\chi_3^K(y_3)}{dy_3} \\ \sigma_3^K = R_{33}(y_3)\frac{d\chi_3^K(y_3)}{dy_3} \\ \sigma_4^K = R_{44}(y_3)\frac{d\chi_2^K(y_3)}{dy_3} + R_{45}(y_3)\frac{d\chi_1^K(y_3)}{dy_3} \\ \sigma_5^K = R_{45}(y_3)\frac{d\chi_2^K(y_3)}{dy_3} + R_{55}(y_3)\frac{d\chi_1^K(y_3)}{dy_3} \\ \sigma_6^K = R_{36}(y_3)\frac{d\chi_3^K(y_3)}{dy_3} \end{cases} \\ \sigma^K \text{ and } \vec{\chi}^K \text{ are Y-periodic} \end{cases}$$

**Calculation of the homogenised rigidity tensor and the stresses and strains within the plies**

Solving the cellular problem $K$ gives $\vec{\chi}^K(y_3)$ and subsequently the tensor of localisation of the stress $s^K$ which is associated with it:

$$s_I^K = -R_{IJ}(y_3)\varepsilon_J^m(\vec{\chi}^K) + R_{IK}(y_3) = -\sigma_I^K + R_{IK}(y_3)$$

When the six cellular problems are solved we obtain:

- the tensor of homogenised rigidity $R^H = (R_{IK}^H)_{I,J=1,...,6}$ by:

$$R_{IK}^H = < s_I^K >$$

- the stresses $\sigma^m$ and the strains $\varepsilon^m(\vec{u}^m)$ in each ply $k$ $(h_{k+1} \geq y_3 \geq h_k)$ in the reference framee, then by change of basis, in the local basis, during the macroscopic strain $E = (E_K)_{I=1,...,6}$:

$$\sigma_I^m(h_{k+1} \geq y_3 \geq h_k) = s_I^K(h_{k+1} \geq y_3 \geq h_k)E_K$$

$$\implies \varepsilon_I^m(\vec{u}^m)(h_{k+1} \geq y_3 \geq h_k) = S_{IJ}(h_{k+1} \geq h_{k+1} \geq y_3 \geq h_k)\sigma_J^m(h_{k+1} \geq h_{k+1} \geq y_3 \geq h_k)$$

$$\implies \varepsilon_I^m(\vec{u}^m)(h_{k+1} \geq y_3 \geq h_k) = S_{IJ}^{(kref)}\sigma_J^m(h_{k+1} \geq h_{k+1} \geq y_3 \geq h_k)$$

The change of basis formulae give:

$$\begin{cases} \sigma^{(kloc)}(h_{k+1} \geq y_3 \geq h_k) = \overline{P}_\sigma^{(k)}\sigma^{(kref)}(h_{k+1} \geq y_3 \geq h_k) = \overline{P}_\sigma^{(k)}\sigma^m(h_{k+1} \geq h_{k+1} \geq y_3 \geq h_k) \\ \varepsilon^{(kloc)}(h_{k+1} \geq y_3 \geq h_k) = \overline{P}_\varepsilon^{(k)}\varepsilon^{(kref)}(h_{k+1} \geq y_3 \geq h_k) = \overline{P}_\varepsilon^{(k)}\varepsilon^m(\vec{u}^m)(h_{k+1} \geq y_3 \geq h_k) \end{cases}$$

**Solving the cellular problem $K = 1$**

$$(\overline{\mathcal{P}}^{1-Y}) \begin{cases} \text{Find } (\sigma^{(1)}(y_3), \vec{\chi}^{(1)}(y_3))/ \\ \forall M \in \Omega^Y \begin{cases} \dfrac{d\sigma_5^{(1)}(y_3)}{dy_3} = 0 \\ \dfrac{d\sigma_4^{(1)}(y_3)}{dy_3} = 0 \\ \dfrac{d\sigma_3^{(1)}(y_3)}{dy_3} = \dfrac{dR_{31}(y_3)}{dy_3} \end{cases} \\ \begin{cases} \sigma_1^{(1)} = R_{13}(y_3)\dfrac{d\chi_3^{(1)}(y_3)}{dy_3} \\ \sigma_2^{(1)} = R_{23}(y_3)\dfrac{d\chi_3^{(1)}(y_3)}{dy_3} \\ \sigma_3^{(1)} = R_{33}(y_3)\dfrac{d\chi_3^{(1)}(y_3)}{dy_3} \\ \sigma_4^{(1)} = R_{44}(y_3)\dfrac{d\chi_2^{(1)}(y_3)}{dy_3} + R_{45}(y_3)\dfrac{d\chi_1^{(1)}(y_3)}{dy_3} \\ \sigma_5^{(1)} = R_{45}(y_3)\dfrac{d\chi_2^{(1)}(y_3)}{dy_3} + R_{55}(y_3)\dfrac{d\chi_1^{(1)}(y_3)}{dy_3} \\ \sigma_6^{(1)} = R_{36}(y_3)\dfrac{d\chi_3^{(1)}(y_3)}{dy_3} \end{cases} \\ \sigma^{(1)} \text{ and } \vec{\chi}^{(1)} \text{ are Y-periodic} \end{cases}$$

$$\Longrightarrow \begin{cases} \dfrac{d\sigma_5^{(1)}(y_3)}{dy_3} = 0 \Longrightarrow \sigma_5^{(1)}(y_3) = C_5 \Longrightarrow C_5 = R_{45}(y_3)\dfrac{d\chi_2^{(1)}(y_3)}{dy_3} + R_{55}(y_3)\dfrac{d\chi_1^{(1)}(y_3)}{dy_3} \\ \dfrac{d\sigma_4^{(1)}(y_3)}{dy_3} = 0 \Longrightarrow \sigma_4^{(1)}(y_3) = C_4 \Longrightarrow C_4 = R_{44}(y_3)\dfrac{d\chi_2^{(1)}(y_3)}{dy_3} + R_{45}(y_3)\dfrac{d\chi_1^{(1)}(y_3)}{dy_3} \\ \dfrac{d\sigma_3^{(1)}(y_3)}{dy_3} = \dfrac{dR_{31}(y_3)}{dy_3} \Longrightarrow \sigma_3^{(1)}(y_3) = R_{31}(y_3) + C_3 \Longrightarrow R_{31}(y_3) + C_3 = R_{33}(y_3)\dfrac{d\chi_3^{(1)}(y_3)}{dy_3} \end{cases}$$

$$\Longrightarrow \begin{cases} \dfrac{d\chi_1^{(1)}(y_3)}{dy_3} = \dfrac{R_{44}(y_3)C_5 - R_{45}(y_3)C_4}{R_{44}(y_3)R_{55}(y_3) - (R_{45}(y_3))^2} \\ \dfrac{d\chi_2^{(1)}(y_3)}{dy_3} = \dfrac{-R_{45}(y_3)C_5 + R_{55}(y_3)C_4}{R_{44}(y_3)R_{55}(y_3) - (R_{45}(y_3))^2} \\ \dfrac{d\chi_3^{(1)}(y_3)}{dy_3} = \dfrac{R_{31}(y_3)}{R_{33}(y_3)} + \dfrac{C_3}{R_{33}(y_3)} \end{cases}$$

$$\Longrightarrow \begin{cases} \displaystyle\int_{\chi_1^{(1)}(0)}^{\chi_1^{(1)}(y_3)} d\chi_1^{(1)}(y_3) = \chi_1^{(1)}(y_3) - \chi_1^{(1)}(0) \\ \qquad\qquad = C_5 \displaystyle\int_0^{y_3} \dfrac{R_{44}(y_3)}{R_{44}(y_3)R_{55}(y_3) - (R_{45}(y_3))^2} dy_3 \cdots \\ -C_4 \displaystyle\int_0^{y_3} \dfrac{R_{45}(y_3)}{R_{44}(y_3)R_{55}(y_3) - (R_{45}(y_3))^2} dy_3 \\ \displaystyle\int_{\chi_2^{(1)}(0)}^{\chi_2^{(1)}(y_3)} d\chi_2^{(1)}(y_3) = \chi_2^{(1)}(y_3) - \chi_2^{(1)}(0) \\ \qquad\qquad = -C_5 \displaystyle\int_0^{y_3} \dfrac{R_{45}(y_3)}{R_{44}(y_3)R_{55}(y_3) - (R_{45}(y_3))^2} dy_3 \cdots \\ +C_4 \displaystyle\int_0^{y_3} \dfrac{R_{55}(y_3)}{R_{44}(y_3)R_{55}(y_3) - (R_{45}(y_3))^2} dy_3 \\ \displaystyle\int_{\chi_3^{(1)}(0)}^{\chi_3^{(1)}(y_3)} d\chi_3^{(1)}(y_3) = \chi_3^{(1)}(y_3) - \chi_3^{(1)}(0) \\ \qquad\qquad = \displaystyle\int_0^{y_3} \dfrac{R_{31}(y_3)}{R_{33}(y_3)} dy_3 + C_3 \displaystyle\int_0^{y_3} \dfrac{1}{R_{33}(y_3)} dy_3 \end{cases}$$

We calculate the integration constants $C_3$, $C_4$ and $C_5$ by making the preceeding integration operation from 0 to $y_3 = h$ and by invoking that the fonctions $\chi_i^{(1)}(y_3)$ $(i = 1, 2, 3)$ are periodic.

Then: $\chi_i^{(1)}(h) = \chi_i^{(1)}(0)$. We deduce:

$$
\begin{cases}
C_4 = 0 \\
C_5 = 0 \\
C_3 = -\dfrac{\int_0^h \frac{R_{31}(y_3)}{R_{33}(y_3)}\,dy_3}{\int_0^h \frac{1}{R_{33}(y_3)}\,dy_3} = -\dfrac{<\frac{R_{31}(y_3)}{R_{33}(y_3)}>}{<\frac{1}{R_{33}(y_3)}>}
\end{cases}
$$

$$
\Longrightarrow
\begin{cases}
\chi_1^{(1)}(y_3) = \chi_1^{(1)}(0) \\
\chi_2^{(1)}(y_3) = \chi_2^{(1)}(0) \\
\chi_3^{(1)}(y_3) = \chi_3^{(1)}(0) + \displaystyle\int_0^{y_3} \frac{R_{31}(y_3)}{R_{33}(y_3)}\,dy_3 - \frac{<\frac{R_{31}(y_3)}{R_{33}(y_3)}>}{<\frac{1}{R_{33}(y_3)}>}\int_0^{y_3}\frac{1}{R_{33}(y_3)}\,dy_3
\end{cases}
$$

Finally:

$$
\varepsilon^m(\vec{\chi}^{(1)}) =
\begin{pmatrix}
\varepsilon_1^m(\vec{\chi}^{(1)}) = 0 \\
\varepsilon_2^m(\vec{\chi}^{(1)}) = 0 \\
\varepsilon_3^m(\vec{\chi}^{(1)}) = \dfrac{R_{31}(y_3)}{R_{33}(y_3)} - \dfrac{<\frac{R_{31}(y_3)}{R_{33}(y_3)}>}{<\frac{1}{R_{33}(y_3)}>}\dfrac{1}{R_{33}(y_3)} \\
\varepsilon_4^m(\vec{\chi}^{(1)}) = 0 \\
\varepsilon_5^m(\vec{\chi}^{(1)}) = 0 \\
\varepsilon_6^m(\vec{\chi}^{(1)}) = 0
\end{pmatrix}
$$

$$
\sigma^{(1)} =
\begin{pmatrix}
\sigma_1^{(1)} = \dfrac{R_{31}^2(y_3)}{R_{33}(y_3)} - \dfrac{<\frac{R_{31}(y_3)}{R_{33}(y_3)}>}{<\frac{1}{R_{33}(y_3)}>}\dfrac{R_{13}(y_3)}{R_{33}(y_3)} \\[2mm]
\sigma_2^{(1)} = \dfrac{R_{23}(y_3)R_{31}(y_3)}{R_{33}(y_3)} - \dfrac{<\frac{R_{31}(y_3)}{R_{33}(y_3)}>}{<\frac{1}{R_{33}(y_3)}>}\dfrac{R_{23}(y_3)}{R_{33}(y_3)} \\[2mm]
\sigma_3^{(1)} = \dfrac{R_{33}(y_3)R_{31}(y_3)}{R_{33}(y_3)} - \dfrac{<\frac{R_{31}(y_3)}{R_{33}(y_3)}>}{<\frac{1}{R_{33}(y_3)}>} \\[2mm]
\sigma_4^{(1)} = 0 \\
\sigma_5^{(1)} = 0 \\
\sigma_6^{(1)} = \dfrac{R_{36}(y_3)R_{31}(y_3)}{R_{33}(y_3)} - \dfrac{<\frac{R_{31}(y_3)}{R_{33}(y_3)}>}{<\frac{1}{R_{33}(y_3)}>}\dfrac{R_{36}(y_3)}{R_{33}(y_3)}
\end{pmatrix}
$$

And we obtain the expression of the localisation tensor associated to this cellular problem:

$$
s^{(1)} =
\begin{pmatrix}
s_1^{(1)} = -\sigma_1^{(1)} + R_{11}(y_3) \\
s_2^{(1)} = -\sigma_2^{(1)} + R_{21}(y_3) \\
s_3^{(1)} = -\sigma_3^{(1)} + R_{31}(y_3) \\
s_4^{(1)} = -\sigma_4^{(1)} + 0 \\
s_5^{(1)} = -\sigma_5^{(1)} + 0 \\
s_6^{(1)} = -\sigma_6^{(1)} + R_{61}(y_3)
\end{pmatrix}
=
\begin{pmatrix}
R_{11}(y_3) - \dfrac{R_{31}(y_3)R_{31}(y_3)}{R_{33}(y_3)} + \dfrac{<\frac{R_{31}(y_3)}{R_{33}(y_3)}>}{<\frac{1}{R_{33}(y_3)}>}\dfrac{R_{13}(y_3)}{R_{33}(y_3)} \\[2mm]
R_{21}(y_3) - \dfrac{R_{23}(y_3)R_{31}(y_3)}{R_{33}(y_3)} + \dfrac{<\frac{R_{31}(y_3)}{R_{33}(y_3)}>}{<\frac{1}{R_{33}(y_3)}>}\dfrac{R_{23}(y_3)}{R_{33}(y_3)} \\[2mm]
\dfrac{<\frac{R_{31}(y_3)}{R_{33}(y_3)}>}{<\frac{1}{R_{33}(y_3)}>} \\[2mm]
0 \\
0 \\
R_{61}(y_3) - \dfrac{R_{36}(y_3)R_{31}(y_3)}{R_{33}(y_3)} + \dfrac{<\frac{R_{31}(y_3)}{R_{33}(y_3)}>}{<\frac{1}{R_{33}(y_3)}>}\dfrac{R_{36}(y_3)}{R_{33}(y_3)}
\end{pmatrix}
$$

**Solving the cellular problem $K = 2$**

$$(\overline{\mathcal{P}}^{2-Y}) \begin{cases} \text{Find } (\sigma^{(2)}(y_3), \vec{\chi}^{(2)}(y_3))/ \\ \forall M \in \Omega^Y \begin{cases} \dfrac{d\sigma_5^{(2)}(y_3)}{dy_3} = 0 \\ \dfrac{d\sigma_4^{(2)}(y_3)}{dy_3} = 0 \\ \dfrac{d\sigma_3^{(2)}(y_3)}{dy_3} = \dfrac{dR_{32}(y_3)}{dy_3} \end{cases} \\ \begin{cases} \sigma_1^{(2)} = R_{13}(y_3)\dfrac{d\chi_3^{(2)}(y_3)}{dy_3} \\ \sigma_2^{(2)} = R_{23}(y_3)\dfrac{d\chi_3^{(2)}(y_3)}{dy_3} \\ \sigma_3^{(2)} = R_{33}(y_3)\dfrac{d\chi_3^{(2)}(y_3)}{dy_3} \\ \sigma_4^{(2)} = R_{44}(y_3)\dfrac{d\chi_2^{(2)}(y_3)}{dy_3} + R_{45}(y_3)\dfrac{d\chi_1^{(2)}(y_3)}{dy_3} \\ \sigma_5^{(2)} = R_{45}(y_3)\dfrac{d\chi_2^{(2)}(y_3)}{dy_3} + R_{55}(y_3)\dfrac{d\chi_1^{(2)}(y_3)}{dy_3} \\ \sigma_6^{(2)} = R_{36}(y_3)\dfrac{d\chi_3^{(2)}(y_3)}{dy_3} \end{cases} \\ \sigma^{(2)} \text{ and } \vec{\chi}^{(2)} \text{ are Y-periodic} \end{cases}$$

$$\Longrightarrow \begin{cases} \dfrac{d\sigma_5^{(2)}(y_3)}{dy_3} = 0 \Longrightarrow \sigma_5^{(2)}(y_3) = C_5 \Longrightarrow C_5 = R_{45}(y_3)\dfrac{d\chi_2^{(2)}(y_3)}{dy_3} + R_{55}(y_3)\dfrac{d\chi_1^{(2)}(y_3)}{dy_3} \\ \dfrac{d\sigma_4^{(2)}(y_3)}{dy_3} = 0 \Longrightarrow \sigma_4^{(2)}(y_3) = C_4 \Longrightarrow C_4 = R_{44}(y_3)\dfrac{d\chi_2^{(2)}(y_3)}{dy_3} + R_{45}(y_3)\dfrac{d\chi_1^{(2)}(y_3)}{dy_3} \\ \dfrac{d\sigma_3^{(2)}(y_3)}{dy_3} = \dfrac{dR_{32}(y_3)}{dy_3} \Longrightarrow \sigma_3^{(2)}(y_3) = R_{32}(y_3) + C_3 \Longrightarrow R_{32}(y_3) + C_3 = R_{33}(y_3)\dfrac{d\chi_3^{(2)}(y_3)}{dy_3} \end{cases}$$

$$\Longrightarrow \begin{cases} \dfrac{d\chi_1^{(2)}(y_3)}{dy_3} = \dfrac{R_{44}(y_3)C_5 - R_{45}(y_3)C_4}{R_{44}(y_3)R_{55}(y_3) - (R_{45}(y_3))^2} \\ \dfrac{d\chi_2^{(2)}(y_3)}{dy_3} = \dfrac{-R_{45}(y_3)C_5 + R_{55}(y_3)C_4}{R_{44}(y_3)R_{55}(y_3) - (R_{45}(y_3))^2} \\ \dfrac{d\chi_3^{(2)}(y_3)}{dy_3} = \dfrac{R_{32}(y_3)}{R_{33}(y_3)} + \dfrac{C_3}{R_{33}(y_3)} \end{cases}$$

$$\Longrightarrow \begin{cases} \displaystyle\int_{\chi_1^{(2)}(0)}^{\chi_1^{(2)}(y_3)} d\chi_1^{(2)}(y_3) = \chi_1^{(2)}(y_3) - \chi_1^{(2)}(0) \\ \qquad\qquad = C_5 \displaystyle\int_0^{y_3} \dfrac{R_{44}(y_3)}{R_{44}(y_3)R_{55}(y_3) - (R_{45}(y_3))^2} dy_3 \cdots \\ -C_4 \displaystyle\int_0^{y_3} \dfrac{R_{45}(y_3)}{R_{44}(y_3)R_{55}(y_3) - (R_{45}(y_3))^2} dy_3 \\ \displaystyle\int_{\chi_2^{(2)}(0)}^{\chi_2^{(2)}(y_3)} d\chi_2^{(2)}(y_3) = \chi_2^{(2)}(y_3) - \chi_2^{(2)}(0) \\ \qquad\qquad = -C_5 \displaystyle\int_0^{y_3} \dfrac{R_{45}(y_3)}{R_{44}(y_3)R_{55}(y_3) - (R_{45}(y_3))^2} dy_3 \cdots \\ +C_4 \displaystyle\int_0^{y_3} \dfrac{R_{55}(y_3)}{R_{44}(y_3)R_{55}(y_3) - (R_{45}(y_3))^2} dy_3 \\ \displaystyle\int_{\chi_3^{(2)}(0)}^{\chi_3^{(2)}(y_3)} d\chi_3^{(2)}(y_3) = \chi_3^{(2)}(y_3) - \chi_3^{(2)}(0) \\ \qquad\qquad = \displaystyle\int_0^{y_3} \dfrac{R_{32}(y_3)}{R_{33}(y_3)} dy_3 + C_3 \displaystyle\int_0^{y_3} \dfrac{1}{R_{33}(y_3)} dy_3 \end{cases}$$

We calculate the integration constants $C_3$, $C_4$ and $C_5$ by making the preceeding integration operation from 0 to $y_3 = h$ and by invoking that the fonctions $\chi_i^{(2)}(y_3)$ $(i = 1, 2, 3)$ are periodic.

Then: $\chi_i^{(2)}(h) = \chi_i^{(2)}(0)$. We deduce:

$$\begin{cases} C_4 = 0 \\ C_5 = 0 \\ C_3 = -\dfrac{\int_0^h \frac{R_{32}(y_3)}{R_{33}(y_3)} dy_3}{\int_0^h \frac{1}{R_{33}(y_3)} dy_3} = -\dfrac{< \frac{R_{32}(y_3)}{R_{33}(y_3)} >}{< \frac{1}{R_{33}(y_3)} >} \end{cases}$$

$$\implies \begin{cases} \chi_1^{(2)}(y_3) = \chi_1^{(2)}(0) \\ \chi_2^{(2)}(y_3) = \chi_2^{(2)}(0) \\ \chi_3^{(2)}(y_3) = \chi_3^{(2)}(0) + \displaystyle\int_0^{y_3} \frac{R_{32}(y_3)}{R_{33}(y_3)} dy_3 - \frac{< \frac{R_{32}(y_3)}{R_{33}(y_3)} >}{< \frac{1}{R_{33}(y_3)} >} \int_0^{y_3} \frac{1}{R_{33}(y_3)} dy_3 \end{cases}$$

Finally

$$\varepsilon^m(\vec{\chi}^{(2)}) = \begin{pmatrix} \varepsilon_1^m(\vec{\chi}^{(2)}) = 0 \\ \varepsilon_2^m(\vec{\chi}^{(2)}) = 0 \\ \varepsilon_3^m(\vec{\chi}^{(2)}) = \dfrac{R_{32}(y_3)}{R_{33}(y_3)} - \dfrac{< \frac{R_{32}(y_3)}{R_{33}(y_3)} >}{< \frac{1}{R_{33}(y_3)} >} \dfrac{1}{R_{33}(y_3)} \\ \varepsilon_4^m(\vec{\chi}^{(2)}) = 0 \\ \varepsilon_5^m(\vec{\chi}^{(2)}) = 0 \\ \varepsilon_6^m(\vec{\chi}^{(2)}) = 0 \end{pmatrix}$$

$$\sigma^{(2)} = \begin{pmatrix} \sigma_1^{(2)} = \dfrac{R_{13}(y_3)R_{32}(y_3)}{R_{33}(y_3)} - \dfrac{< \frac{R_{32}(y_3)}{R_{33}(y_3)} >}{< \frac{1}{R_{33}(y_3)} >} \dfrac{R_{13}(y_3)}{R_{33}(y_3)} \\ \sigma_2^{(2)} = \dfrac{(R_{32}(y_3))^2}{R_{33}(y_3)} - \dfrac{< \frac{R_{32}(y_3)}{R_{33}(y_3)} >}{< \frac{1}{R_{33}(y_3)} >} \dfrac{R_{23}(y_3)}{R_{33}(y_3)} \\ \sigma_3^{(2)} = \dfrac{R_{33}(y_3)R_{32}(y_3)}{R_{33}(y_3)} - \dfrac{< \frac{R_{32}(y_3)}{R_{33}(y_3)} >}{< \frac{1}{R_{33}(y_3)} >} \\ \sigma_4^{(2)} = 0 \\ \sigma_5^{(2)} = 0 \\ \sigma_6^{(2)} = \dfrac{R_{36}(y_3)R_{32}(y_3)}{R_{33}(y_3)} - \dfrac{< \frac{R_{32}(y_3)}{R_{33}(y_3)} >}{< \frac{1}{R_{33}(y_3)} >} \dfrac{R_{36}(y_3)}{R_{33}(y_3)} \end{pmatrix}$$

And we obtain the expression of the localisation tensor associated to this cellular problem:

$$s^{(2)} = \begin{pmatrix} s_1^{(2)} = -\sigma_1^{(2)} + R_{12}(y_3) \\ s_2^{(2)} = -\sigma_2^{(2)} + R_{22}(y_3) \\ s_3^{(2)} = -\sigma_3^{(2)} + R_{32}(y_3) \\ s_4^{(2)} = -\sigma_4^{(2)} + 0 \\ s_5^{(2)} = -\sigma_5^{(2)} + 0 \\ s_6^{(2)} = -\sigma_6^{(2)} + R_{62}(y_3) \end{pmatrix} = \begin{pmatrix} R_{12}(y_3) - \dfrac{R_{13}(y_3)R_{32}(y_3)}{R_{33}(y_3)} + \dfrac{< \frac{R_{32}(y_3)}{R_{33}(y_3)} >}{< \frac{1}{R_{33}(y_3)} >} \dfrac{R_{13}(y_3)}{R_{33}(y_3)} \\ R_{22}(y_3) - \dfrac{(R_{32}(y_3))^2}{R_{33}(y_3)} + \dfrac{< \frac{R_{32}(y_3)}{R_{33}(y_3)} >}{< \frac{1}{R_{33}(y_3)} >} \dfrac{R_{23}(y_3)}{R_{33}(y_3)} \\ \dfrac{< \frac{R_{32}(y_3)}{R_{33}(y_3)} >}{< \frac{1}{R_{33}(y_3)} >} \\ 0 \\ 0 \\ R_{62}(y_3) - \dfrac{R_{36}(y_3)R_{32}(y_3)}{R_{33}(y_3)} + \dfrac{< \frac{R_{32}(y_3)}{R_{33}(y_3)} >}{< \frac{1}{R_{33}(y_3)} >} \dfrac{R_{36}(y_3)}{R_{33}(y_3)} \end{pmatrix}$$

**Solving the cellular problem $K = 3$**

$$
(\overline{\mathcal{P}}^{3-Y})
\begin{cases}
\text{Find } (\sigma^{(3)}(y_3), \vec{\chi}^{(3)}(y_3))/ \\[4pt]
\forall M \in \Omega^Y
\begin{cases}
\frac{d\sigma_5^{(3)}(y_3)}{dy_3} = 0 \\[6pt]
\frac{d\sigma_4^{(3)}(y_3)}{dy_3} = 0 \\[6pt]
\frac{d\sigma_3^{(3)}(y_3)}{dy_3} = \frac{dR_{33}(y_3)}{dy_3}
\end{cases} \\[20pt]
\begin{cases}
\sigma_1^{(3)} = R_{13}(y_3)\frac{d\chi_3^{(3)}(y_3)}{dy_3} \\[6pt]
\sigma_2^{(3)} = R_{23}(y_3)\frac{d\chi_3^{(3)}(y_3)}{dy_3} \\[6pt]
\sigma_3^{(3)} = R_{33}(y_3)\frac{d\chi_3^{(3)}(y_3)}{dy_3} \\[6pt]
\sigma_4^{(3)} = R_{44}(y_3)\frac{d\chi_2^{(3)}(y_3)}{dy_3} + R_{45}(y_3)\frac{d\chi_1^{(3)}(y_3)}{dy_3} \\[6pt]
\sigma_5^{(3)} = R_{45}(y_3)\frac{d\chi_2^{(3)}(y_3)}{dy_3} + R_{55}(y_3)\frac{d\chi_1^{(3)}(y_3)}{dy_3} \\[6pt]
\sigma_6^{(3)} = R_{36}(y_3)\frac{d\chi_3^{(3)}(y_3)}{dy_3}
\end{cases} \\[20pt]
\sigma^{(3)} \text{ and } \vec{\chi}^{(3)} \text{ are Y-periodic}
\end{cases}
$$

$$
\Longrightarrow
\begin{cases}
\frac{d\sigma_5^{(3)}(y_3)}{dy_3} = 0 \Longrightarrow \sigma_5^{(3)}(y_3) = C_5 \Longrightarrow C_5 = R_{45}(y_3)\frac{d\chi_2^{(3)}(y_3)}{dy_3} + R_{55}(y_3)\frac{d\chi_1^{(3)}(y_3)}{dy_3} \\[8pt]
\frac{d\sigma_4^{(3)}(y_3)}{dy_3} = 0 \Longrightarrow \sigma_4^{(3)}(y_3) = C_4 \Longrightarrow C_4 = R_{44}(y_3)\frac{d\chi_2^{(3)}(y_3)}{dy_3} + R_{45}(y_3)\frac{d\chi_1^{(3)}(y_3)}{dy_3} \\[8pt]
\frac{d\sigma_3^{(3)}(y_3)}{dy_3} = \frac{dR_{33}(y_3)}{dy_3} \Longrightarrow \sigma_3^{(3)}(y_3) = R_{33}(y_3) + C_3 \Longrightarrow R_{33}(y_3) + C_3 = R_{33}(y_3)\frac{d\chi_3^{(3)}(y_3)}{dy_3}
\end{cases}
$$

$$
\Longrightarrow
\begin{cases}
\frac{d\chi_1^{(3)}(y_3)}{dy_3} = \frac{R_{44}(y_3)C_5 - R_{45}(y_3)C_4}{R_{44}(y_3)R_{55}(y_3) - (R_{45}(y_3))^2} \\[8pt]
\frac{d\chi_2^{(3)}(y_3)}{dy_3} = \frac{-R_{45}(y_3)C_5 + R_{55}(y_3)C_4}{R_{44}(y_3)R_{55}(y_3) - (R_{45}(y_3))^2} \\[8pt]
\frac{d\chi_3^{(3)}(y_3)}{dy_3} = 1 + \frac{C_3}{R_{33}(y_3)}
\end{cases}
$$

$$
\Longrightarrow
\begin{cases}
\displaystyle\int_{\chi_1^{(3)}(0)}^{\chi_1^{(3)}(y_3)} d\chi_1^{(3)}(y_3) = \chi_1^{(3)}(y_3) - \chi_1^{(3)}(0) \\[10pt]
\qquad\qquad = C_5 \displaystyle\int_0^{y_3} \frac{R_{44}(y_3)}{R_{44}(y_3)R_{55}(y_3) - (R_{45}(y_3))^2} dy_3 \dots \\[12pt]
-C_4 \displaystyle\int_0^{y_3} \frac{R_{45}(y_3)}{R_{44}(y_3)R_{55}(y_3) - (R_{45}(y_3))^2} dy_3 \\[12pt]
\displaystyle\int_{\chi_2^{(3)}(0)}^{\chi_2^{(3)}(y_3)} d\chi_2^{(3)}(y_3) = \chi_2^{(3)}(y_3) - \chi_2^{(3)}(0) \\[10pt]
\qquad\qquad = -C_5 \displaystyle\int_0^{y_3} \frac{R_{45}(y_3)}{R_{44}(y_3)R_{55}(y_3) - (R_{45}(y_3))^2} dy_3 \dots \\[12pt]
+C_4 \displaystyle\int_0^{y_3} \frac{R_{55}(y_3)}{R_{44}(y_3)R_{55}(y_3) - (R_{45}(y_3))^2} dy_3 \\[12pt]
\displaystyle\int_{\chi_3^{(3)}(0)}^{\chi_3^{(3)}(y_3)} d\chi_3^{(3)}(y_3) = \chi_3^{(3)}(y_3) - \chi_3^{(3)}(0) \\[10pt]
\qquad\qquad = y_3 + C_3 \displaystyle\int_0^{y_3} \frac{1}{R_{33}(y_3)} dy_3
\end{cases}
$$

We calculate the integration constants $C_3$, $C_4$ and $C_5$ by making the preceeding integration operation from 0 to $y_3 = h$ and by invoking that the fonctions $\chi_i^{(3)}(y_3)$ $(i = 1, 2, 3)$ are periodic.

Then: $\chi_i^{(3)}(h) = \chi_i^{(3)}(0)$. We deduce:

$$\begin{cases} C_4 = 0 \\ C_5 = 0 \\ C_3 = -\dfrac{h}{\int_0^h \frac{1}{R_{33}(y_3)} dy_3} = -\dfrac{1}{< \frac{1}{R_{33}(y_3)} >} \end{cases}$$

$$\implies \begin{cases} \chi_1^{(3)}(y_3) = \chi_1^{(3)}(0) \\ \chi_2^{(3)}(y_3) = \chi_2^{(3)}(0) \\ \chi_3^{(3)}(y_3) = \chi_3^{(3)}(0) + y_3 - \dfrac{1}{< \frac{1}{R_{33}(y_3)} >} \int_0^{y_3} \dfrac{1}{R_{33}(y_3)} dy_3 \end{cases}$$

Finally

$$\varepsilon^m(\vec{\chi}^{(3)}) = \begin{pmatrix} \varepsilon_1^m(\vec{\chi}^{(3)}) = 0 \\ \varepsilon_2^m(\vec{\chi}^{(3)}) = 0 \\ \varepsilon_3^m(\vec{\chi}^{(3)}) = 1 - \dfrac{1}{< \frac{1}{R_{33}(y_3)} >} \dfrac{1}{R_{33}(y_3)} \\ \varepsilon_4^m(\vec{\chi}^{(3)}) = 0 \\ \varepsilon_5^m(\vec{\chi}^{(3)}) = 0 \\ \varepsilon_6^m(\vec{\chi}^{(3)}) = 0 \end{pmatrix}$$

$$\sigma^{(3)} = \begin{pmatrix} \sigma_1^{(3)} = R_{31}(y_3) - \dfrac{1}{< \frac{1}{R_{33}(y_3)} >} \dfrac{R_{31}(y_3)}{R_{33}(y_3)} \\ \sigma_2^{(3)} = R_{23}(y_3) - \dfrac{1}{< \frac{1}{R_{33}(y_3)} >} \dfrac{R_{23}(y_3)}{R_{33}(y_3)} \\ \sigma_3^{(3)} = R_{33}(y_3) - \dfrac{1}{< \frac{1}{R_{33}(y_3)} >} \\ \sigma_4^{(3)} = 0 \\ \sigma_5^{(3)} = 0 \\ \sigma_6^{(3)} = R_{36}(y_3) - \dfrac{1}{< \frac{1}{R_{33}(y_3)} >} \dfrac{R_{36}(y_3)}{R_{33}(y_3)} \end{pmatrix}$$

And we obtain the expression of the localisation tensor associated to this cellular problem:

$$s^{(3)} = \begin{pmatrix} s_1^{(3)} = -\sigma_1^{(3)} + R_{13}(y_3) \\ s_2^{(3)} = -\sigma_2^{(3)} + R_{23}(y_3) \\ s_3^{(3)} = -\sigma_3^{(3)} + R_{33}(y_3) \\ s_4^{(3)} = -\sigma_4^{(3)} + 0 \\ s_5^{(3)} = -\sigma_5^{(3)} + 0 \\ s_6^{(3)} = -\sigma_6^{(3)} + R_{63}(y_3) \end{pmatrix} = \begin{pmatrix} \dfrac{1}{< \frac{1}{R_{33}(y_3)} >} \dfrac{R_{31}(y_3)}{R_{33}(y_3)} \\ \dfrac{1}{< \frac{1}{R_{33}(y_3)} >} \dfrac{R_{23}(y_3)}{R_{33}(y_3)} \\ \dfrac{1}{< \frac{1}{R_{33}(y_3)} >} \\ 0 \\ 0 \\ \dfrac{1}{< \frac{1}{R_{33}(y_3)} >} \dfrac{R_{36}(y_3)}{R_{33}(y_3)} \end{pmatrix}$$

**Solving the cellular problem $K = 4$**

$$
(\overline{\mathcal{P}}^{4-Y})
\begin{cases}
\text{Find } (\sigma^{(4)}(y_3), \vec{\chi}^{(4)}(y_3))/ \\[4pt]
\forall M \in \Omega^Y
\begin{cases}
\frac{d\sigma_5^{(4)}(y_3)}{dy_3} = \frac{dR_{54}(y_3)}{dy_3} \\[4pt]
\frac{d\sigma_4^{(4)}(y_3)}{dy_3} = \frac{dR_{44}(y_3)}{dy_3} \\[4pt]
\frac{d\sigma_3^{(4)}(y_3)}{dy_3} = 0
\end{cases} \\[20pt]
\begin{cases}
\sigma_1^{(4)} = R_{13}(y_3)\frac{d\chi_3^{(4)}(y_3)}{dy_3} \\[6pt]
\sigma_2^{(4)} = R_{23}(y_3)\frac{d\chi_3^{(4)}(y_3)}{dy_3} \\[6pt]
\sigma_3^{(4)} = R_{33}(y_3)\frac{d\chi_3^{(4)}(y_3)}{dy_3} \\[6pt]
\sigma_4^{(4)} = R_{44}(y_3)\frac{d\chi_2^{(4)}(y_3)}{dy_3} + R_{45}(y_3)\frac{d\chi_1^{(4)}(y_3)}{dy_3} \\[6pt]
\sigma_5^{(4)} = R_{45}(y_3)\frac{d\chi_2^{(4)}(y_3)}{dy_3} + R_{55}(y_3)\frac{d\chi_1^{(4)}(y_3)}{dy_3} \\[6pt]
\sigma_6^{(4)} = R_{36}(y_3)\frac{d\chi_3^{(4)}(y_3)}{dy_3}
\end{cases} \\[20pt]
\sigma^{(4)} \text{ and } \vec{\chi}^{(4)} \text{ are } Y\text{-periodic}
\end{cases}
$$

$$
\Longrightarrow
\begin{cases}
\frac{d\sigma_5^{(4)}(y_3)}{dy_3} = \frac{dR_{54}(y_3)}{dy_3} \Longrightarrow \sigma_5^{(4)}(y_3) = R_{54}(y_3) + C_5 \\[8pt]
\qquad\qquad \Longrightarrow R_{54}(y_3) + C_5 = R_{45}(y_3)\frac{d\chi_2^{(4)}(y_3)}{dy_3} + R_{55}(y_3)\frac{d\chi_1^{(4)}(y_3)}{dy_3} \\[8pt]
\frac{d\sigma_4^{(4)}(y_3)}{dy_3} = \frac{dR_{44}(y_3)}{dy_3} \Longrightarrow \sigma_4^{(4)}(y_3) = R_{44}(y_3) + C_4 \\[8pt]
\qquad\qquad \Longrightarrow R_{44}(y_3) + C_4 = R_{44}(y_3)\frac{d\chi_2^{(4)}(y_3)}{dy_3} + R_{45}(y_3)\frac{d\chi_1^{(4)}(y_3)}{dy_3} \\[8pt]
\frac{d\sigma_3^{(4)}(y_3)}{dy_3} = 0 \Longrightarrow \sigma_3^{(4)}(y_3) = C_3 \Longrightarrow C_3 = R_{33}(y_3)\frac{d\chi_3^{(4)}(y_3)}{dy_3}
\end{cases}
$$

$$
\Longrightarrow
\begin{cases}
\frac{d\chi_1^{(4)}(y_3)}{dy_3} = \frac{R_{44}(y_3)C_5 - R_{45}(y_3)C_4}{R_{44}(y_3)R_{55}(y_3) - (R_{45}(y_3))^2} \\[8pt]
\frac{d\chi_2^{(4)}(y_3)}{dy_3} = 1 + \frac{-R_{45}(y_3)C_5 + R_{55}(y_3)C_4}{R_{44}(y_3)R_{55}(y_3) - (R_{45}(y_3))^2} \\[8pt]
\frac{d\chi_3^{(4)}(y_3)}{dy_3} = \frac{C_3}{R_{33}(y_3)}
\end{cases}
$$

$$\Longrightarrow \begin{cases} \int_{\chi_1^{(4)}(0)}^{\chi_1^{(4)}(y_3)} d\chi_1^{(4)}(y_3) = \chi_1^{(4)}(y_3) - \chi_1^{(4)}(0) \\ \qquad = C_5 \int_0^{y_3} \dfrac{R_{44}(y_3)}{R_{44}(y_3)R_{55}(y_3) - (R_{45}(y_3))^2} dy_3 \cdots \\ -C_4 \int_0^{y_3} \dfrac{R_{45}(y_3)}{R_{44}(y_3)R_{55}(y_3) - (R_{45}(y_3))^2} dy_3 \\ \int_{\chi_2^{(4)}(0)}^{\chi_2^{(4)}(y_3)} d\chi_2^{(4)}(y_3) = \chi_2^{(4)}(y_3) - \chi_2^{(4)}(0) \\ \qquad = y_3 - C_5 \int_0^{y_3} \dfrac{R_{45}(y_3)}{R_{44}(y_3)R_{55}(y_3) - (R_{45}(y_3))^2} dy_3 \cdots \\ +C_4 \int_0^{y_3} \dfrac{R_{55}(y_3)}{R_{44}(y_3)R_{55}(y_3) - (R_{45}(y_3))^2} dy_3 \\ \int_{\chi_3^{(4)}(0)}^{\chi_3^{(4)}(y_3)} d\chi_3^{(4)}(y_3) = \chi_3^{(4)}(y_3) - \chi_3^{(4)}(0) \\ \qquad = C_3 \int_0^{y_3} \dfrac{1}{R_{33}(y_3)} dy_3 \end{cases}$$

We calculate the integration constants $C_3$, $C_4$ and $C_5$ by making the preceeding integration operation from 0 to $y_3 = h$ and by invoking that the fonctions $\chi_i^{(4)}(y_3)$ $(i = 1, 2, 3)$ are periodic. Then: $\chi_i^{(4)}(h) = \chi_i^{(4)}(0)$. We deduce:

$$\begin{cases} C_4 = -\dfrac{< \frac{R_{44}(y_3)}{R_{44}(y_3)R_{55}(y_3) - (R_{45}(y_3))^2} >}{< \frac{R_{44}(y_3)}{R_{44}(y_3)R_{55}(y_3) - (R_{45}(y_3))^2} > < \frac{R_{55}(y_3)}{R_{44}(y_3)R_{55}(y_3) - (R_{45}(y_3))^2} > - (< \frac{R_{45}(y_3)}{R_{44}(y_3)R_{55}(y_3) - (R_{45}(y_3))^2} >)^2} \\ C_5 = -\dfrac{< \frac{R_{45}(y_3)}{R_{44}(y_3)R_{55}(y_3) - (R_{45}(y_3))^2} >}{< \frac{R_{44}(y_3)}{R_{44}(y_3)R_{55}(y_3) - (R_{45}(y_3))^2} > < \frac{R_{55}(y_3)}{R_{44}(y_3)R_{55}(y_3) - (R_{45}(y_3))^2} > - (< \frac{R_{45}(y_3)}{R_{44}(y_3)R_{55}(y_3) - (R_{45}(y_3))^2} >)^2} \\ C_3 = 0 \end{cases}$$

$$\Longrightarrow \begin{cases} \chi_1^{(4)}(y_3) = \chi_1^{(4)}(0) + C_5 \int_0^{y_3} \dfrac{R_{44}(y_3)}{R_{44}(y_3)R_{55}(y_3) - (R_{45}(y_3))^2} dy_3 \cdots \\ -C_4 \int_0^{y_3} \dfrac{R_{45}(y_3)}{R_{44}(y_3)R_{55}(y_3) - (R_{45}(y_3))^2} dy_3 \\ \chi_2^{(4)}(y_3) = \chi_2^{(4)}(0) + y_3 - C_5 \int_0^{y_3} \dfrac{R_{45}(y_3)}{R_{44}(y_3)R_{55}(y_3) - (R_{45}(y_3))^2} dy_3 \cdots \\ +C_4 \int_0^{y_3} \dfrac{R_{55}(y_3)}{R_{44}(y_3)R_{55}(y_3) - (R_{45}(y_3))^2} dy_3 \\ \chi_3^{(4)}(y_3) = \chi_3^{(4)}(0) \end{cases}$$

Finally

$$\varepsilon^m(\vec{\chi}^{(4)}) = \begin{pmatrix} \varepsilon_1^m(\vec{\chi}^{(4)}) = 0 \\ \varepsilon_2^m(\vec{\chi}^{(4)}) = 0 \\ \varepsilon_3^m(\vec{\chi}^{(4)}) = 0 \\ \varepsilon_4^m(\vec{\chi}^{(4)}) = 1 + \dfrac{-R_{45}(y_3)C_5 + R_{55}(y_3)C_4}{R_{44}(y_3)R_{55}(y_3) - (R_{45}(y_3))^2} \\ \varepsilon_5^m(\vec{\chi}^{(4)}) = \dfrac{R_{44}(y_3)C_5 - R_{45}(y_3)C_4}{R_{44}(y_3)R_{55}(y_3) - (R_{45}(y_3))^2} \\ \varepsilon_6^m(\vec{\chi}^{(4)}) = 0 \end{pmatrix}$$

$$\sigma^{(4)} = \begin{pmatrix} \sigma_1^{(4)} = 0 \\ \sigma_2^{(4)} = 0 \\ \sigma_3^{(4)} = 0 \\ \sigma_4^{(4)} = \dfrac{R_{44}(y_3)}{R_{44}(y_3)R_{55}(y_3)-(R_{45}(y_3))^2} \\ \sigma_5^{(4)} = \dfrac{R_{45}(y_3)}{R_{44}(y_3)R_{55}(y_3)-(R_{45}(y_3))^2} \\ \sigma_6^{(4)} = 0 \end{pmatrix}$$

$$- \begin{pmatrix} 0 \\ 0 \\ 0 \\ \dfrac{<\frac{R_{44}(y_3)}{R_{44}(y_3)R_{55}(y_3)-(R_{45}(y_3))^2}>}{<\frac{R_{44}(y_3)}{R_{44}(y_3)R_{55}(y_3)-(R_{45}(y_3))^2}><\frac{R_{55}(y_3)}{R_{44}(y_3)R_{55}(y_3)-(R_{45}(y_3))^2}>-(<\frac{R_{45}(y_3)}{R_{44}(y_3)R_{55}(y_3)-(R_{45}(y_3))^2}>)^2} \\ \dfrac{<\frac{R_{45}(y_3)}{R_{44}(y_3)R_{55}(y_3)-(R_{45}(y_3))^2}>}{<\frac{R_{44}(y_3)}{R_{44}(y_3)R_{55}(y_3)-(R_{45}(y_3))^2}><\frac{R_{55}(y_3)}{R_{44}(y_3)R_{55}(y_3)-(R_{45}(y_3))^2}>-(<\frac{R_{45}(y_3)}{R_{44}(y_3)R_{55}(y_3)-(R_{45}(y_3))^2}>)^2} \\ 0 \end{pmatrix}$$

And we obtain the expression of the localisation tensor associated to this cellular problem:

$$s^{(4)} = \begin{pmatrix} s_1^{(4)} = -\sigma_1^{(4)} + 0 \\ s_2^{(4)} = -\sigma_2^{(4)} + 0 \\ s_3^{(4)} = -\sigma_3^{(4)} + 0 \\ s_4^{(4)} = -\sigma_4^{(4)} + 0 \\ s_5^{(4)} = -\sigma_5^{(4)} + 0 \\ s_6^{(4)} = -\sigma_6^{(4)} + 0 \end{pmatrix}$$

$$= \begin{pmatrix} 0 \\ 0 \\ 0 \\ \dfrac{<\frac{R_{44}(y_3)}{R_{44}(y_3)R_{55}(y_3)-(R_{45}(y_3))^2}>}{<\frac{R_{44}(y_3)}{R_{44}(y_3)R_{55}(y_3)-(R_{45}(y_3))^2}><\frac{R_{55}(y_3)}{R_{44}(y_3)R_{55}(y_3)-(R_{45}(y_3))^2}>-(<\frac{R_{45}(y_3)}{R_{44}(y_3)R_{55}(y_3)-(R_{45}(y_3))^2}>)^2} \\ \dfrac{<\frac{R_{45}(y_3)}{R_{44}(y_3)R_{55}(y_3)-(R_{45}(y_3))^2}>}{<\frac{R_{44}(y_3)}{R_{44}(y_3)R_{55}(y_3)-(R_{45}(y_3))^2}><\frac{R_{55}(y_3)}{R_{44}(y_3)R_{55}(y_3)-(R_{45}(y_3))^2}>-(<\frac{R_{45}(y_3)}{R_{44}(y_3)R_{55}(y_3)-(R_{45}(y_3))^2}>)^2} \\ 0 \end{pmatrix}$$

**Solving the cellular problem $K = 5$**

$$(\overline{\mathcal{P}}^{5-Y}) \begin{cases} \text{Find } (\sigma^{(5)}(y_3), \vec{\chi}^{(5)}(y_3))/ \\ \forall M \in \Omega^Y \begin{cases} \dfrac{d\sigma_5^{(5)}(y_3)}{dy_3} = \dfrac{dR_{55}(y_3)}{dy_3} \\ \dfrac{d\sigma_4^{(5)}(y_3)}{dy_3} = \dfrac{dR_{45}(y_3)}{dy_3} \\ \dfrac{d\sigma_3^{(5)}(y_3)}{dy_3} = 0 \end{cases} \\ \begin{cases} \sigma_1^{(5)} = R_{13}(y_3)\dfrac{d\chi_3^{(5)}(y_3)}{dy_3} \\ \sigma_2^{(5)} = R_{23}(y_3)\dfrac{d\chi_3^{(5)}(y_3)}{dy_3} \\ \sigma_3^{(5)} = R_{33}(y_3)\dfrac{d\chi_3^{(5)}(y_3)}{dy_3} \\ \sigma_4^{(5)} = R_{44}(y_3)\dfrac{d\chi_2^{(5)}(y_3)}{dy_3} + R_{45}(y_3)\dfrac{d\chi_1^{(5)}(y_3)}{dy_3} \\ \sigma_5^{(5)} = R_{45}(y_3)\dfrac{d\chi_2^{(5)}(y_3)}{dy_3} + R_{55}(y_3)\dfrac{d\chi_1^{(5)}(y_3)}{dy_3} \\ \sigma_6^{(5)} = R_{36}(y_3)\dfrac{d\chi_3^{(5)}(y_3)}{dy_3} \end{cases} \\ \sigma^{(5)} \text{ and } \vec{\chi}^{(5)} \text{ are Y-periodic} \end{cases}$$

$$\implies \begin{cases} \dfrac{d\sigma_5^{(5)}(y_3)}{dy_3} = \dfrac{dR_{55}(y_3)}{dy_3} \implies \sigma_5^{(5)}(y_3) = R_{55}(y_3) + C_5 \\ \qquad\qquad\qquad\qquad \implies R_{55}(y_3) + C_5 = R_{45}(y_3)\dfrac{d\chi_2^{(5)}(y_3)}{dy_3} + R_{55}(y_3)\dfrac{d\chi_1^{(5)}(y_3)}{dy_3} \\ \dfrac{d\sigma_4^{(5)}(y_3)}{dy_3} = \dfrac{dR_{45}(y_3)}{dy_3} \implies \sigma_4^{(5)}(y_3) = R_{45}(y_3) + C_4 \\ \qquad\qquad\qquad\qquad \implies R_{45}(y_3) + C_4 = R_{44}(y_3)\dfrac{d\chi_2^{(5)}(y_3)}{dy_3} + R_{45}(y_3)\dfrac{d\chi_1^{(5)}(y_3)}{dy_3} \\ \dfrac{d\sigma_3^{(5)}(y_3)}{dy_3} = 0 \implies \sigma_3^{(5)}(y_3) = C_3 \implies C_3 = R_{33}(y_3)\dfrac{d\chi_3^{(5)}(y_3)}{dy_3} \end{cases}$$

$$\implies \begin{cases} \dfrac{d\chi_1^{(5)}(y_3)}{dy_3} = 1 + \dfrac{R_{44}(y_3)C_5 - R_{45}(y_3)C_4}{R_{44}(y_3)R_{55}(y_3) - (R_{45}(y_3))^2} \\ \dfrac{d\chi_2^{(5)}(y_3)}{dy_3} = \dfrac{-R_{45}(y_3)C_5 + R_{55}(y_3)C_4}{R_{44}(y_3)R_{55}(y_3) - (R_{45}(y_3))^2} \\ \dfrac{d\chi_3^{(5)}(y_3)}{dy_3} = \dfrac{C_3}{R_{33}(y_3)} \end{cases}$$

$$\implies \begin{cases} \displaystyle\int_{\chi_1^{(5)}(0)}^{\chi_1^{(5)}(y_3)} d\chi_1^{(5)}(y_3) = \chi_1^{(5)}(y_3) - \chi_1^{(5)}(0) \\ \qquad\qquad = y_3 + C_5 \displaystyle\int_0^{y_3} \dfrac{R_{44}(y_3)}{R_{44}(y_3)R_{55}(y_3) - (R_{45}(y_3))^2} dy_3 \cdots \\ -C_4 \displaystyle\int_0^{y_3} \dfrac{R_{45}(y_3)}{R_{44}(y_3)R_{55}(y_3) - (R_{45}(y_3))^2} dy_3 \\ \displaystyle\int_{\chi_2^{(5)}(0)}^{\chi_2^{(5)}(y_3)} d\chi_2^{(5)}(y_3) = \chi_2^{(5)}(y_3) - \chi_2^{(5)}(0) \\ \qquad\qquad = -C_5 \displaystyle\int_0^{y_3} \dfrac{R_{45}(y_3)}{R_{44}(y_3)R_{55}(y_3) - (R_{45}(y_3))^2} dy_3 \cdots \\ +C_4 \displaystyle\int_0^{y_3} \dfrac{R_{55}(y_3)}{R_{44}(y_3)R_{55}(y_3) - (R_{45}(y_3))^2} dy_3 \\ \displaystyle\int_{\chi_3^{(5)}(0)}^{\chi_3^{(5)}(y_3)} d\chi_3^{(5)}(y_3) = \chi_3^{(5)}(y_3) - \chi_3^{(5)}(0) \\ \qquad\qquad = C_3 \displaystyle\int_0^{y_3} \dfrac{1}{R_{33}(y_3)} dy_3 \end{cases}$$

We calculate the integration constants $C_3$, $C_4$ and $C_5$ by making the preceeding integration operation from 0 to $y_3 = h$ and by invoking that the fonctions $\chi_i^{(5)}(y_3)$ ($i = 1, 2, 3$) are periodic. Then: $\chi_i^{(5)}(h) = \chi_i^{(5)}(0)$. We deduce:

$$\begin{cases} C_4 = -\dfrac{< \frac{R_{45}(y_3)}{R_{44}(y_3)R_{55}(y_3)-(R_{45}(y_3))^2} >}{< \frac{R_{44}(y_3)}{R_{44}(y_3)R_{55}(y_3)-(R_{45}(y_3))^2} > < \frac{R_{55}(y_3)}{R_{44}(y_3)R_{55}(y_3)-(R_{45}(y_3))^2} > - (< \frac{R_{45}(y_3)}{R_{44}(y_3)R_{55}(y_3)-(R_{45}(y_3))^2} >)^2} \\ C_5 = -\dfrac{< \frac{R_{55}(y_3)}{R_{44}(y_3)R_{55}(y_3)-(R_{45}(y_3))^2} >}{< \frac{R_{44}(y_3)}{R_{44}(y_3)R_{55}(y_3)-(R_{45}(y_3))^2} > < \frac{R_{55}(y_3)}{R_{44}(y_3)R_{55}(y_3)-(R_{45}(y_3))^2} > - (< \frac{R_{45}(y_3)}{R_{44}(y_3)R_{55}(y_3)-(R_{45}(y_3))^2} >)^2} \\ C_3 = 0 \end{cases}$$

$$\Longrightarrow \begin{cases} \chi_1^{(5)}(y_3) = \chi_1^{(5)}(0) + y_3 + C_5 \int_0^{y_3} \frac{R_{44}(y_3)}{R_{44}(y_3)R_{55}(y_3) - (R_{45}(y_3))^2} dy_3 \cdots \\ - C_4 \int_0^{y_3} \frac{R_{45}(y_3)}{R_{44}(y_3)R_{55}(y_3) - (R_{45}(y_3))^2} dy_3 \\ \chi_2^{(5)}(y_3) = \chi_2^{(5)}(0) - C_5 \int_0^{y_3} \frac{R_{45}(y_3)}{R_{44}(y_3)R_{55}(y_3) - (R_{45}(y_3))^2} dy_3 \cdots \\ + C_4 \int_0^{y_3} \frac{R_{55}(y_3)}{R_{44}(y_3)R_{55}(y_3) - (R_{45}(y_3))^2} dy_3 \\ \chi_3^{(5)}(y_3) = \chi_3^{(5)}(0) \end{cases}$$

Finally

$$\varepsilon^m(\vec{\chi}^{(5)}) = \begin{pmatrix} \varepsilon_1^m(\vec{\chi}^{(5)}) = 0 \\ \varepsilon_2^m(\vec{\chi}^{(5)}) = 0 \\ \varepsilon_3^m(\vec{\chi}^{(5)}) = 0 \\ \varepsilon_4^m(\vec{\chi}^{(5)}) = \frac{-R_{45}(y_3)C_5 + R_{55}(y_3)C_4}{R_{44}(y_3)R_{55}(y_3) - (R_{45}(y_3))^2} \\ \varepsilon_5^m(\vec{\chi}^{(5)}) = 1 + \frac{R_{44}(y_3)C_5 - R_{45}(y_3)C_4}{R_{44}(y_3)R_{55}(y_3) - (R_{45}(y_3))^2} \\ \varepsilon_6^m(\vec{\chi}^{(5)}) = 0 \end{pmatrix}$$

$$\sigma^{(5)} = \begin{pmatrix} \sigma_1^{(5)} = 0 \\ \sigma_2^{(5)} = 0 \\ \sigma_3^{(5)} = 0 \\ \sigma_4^{(5)} = \frac{R_{45}(y_3)}{R_{44}(y_3)R_{55}(y_3) - (R_{45}(y_3))^2} \\ \sigma_5^{(5)} = \frac{R_{55}(y_3)}{R_{44}(y_3)R_{55}(y_3) - (R_{45}(y_3))^2} \\ \sigma_6^{(5)} = 0 \end{pmatrix}$$

$$- \begin{pmatrix} 0 \\ 0 \\ 0 \\ \frac{<\frac{R_{45}(y_3)}{R_{44}(y_3)R_{55}(y_3)-(R_{45}(y_3))^2}>}{<\frac{R_{44}(y_3)}{R_{44}(y_3)R_{55}(y_3)-(R_{45}(y_3))^2}><\frac{R_{55}(y_3)}{R_{44}(y_3)R_{55}(y_3)-(R_{45}(y_3))^2}> - (<\frac{R_{45}(y_3)}{R_{44}(y_3)R_{55}(y_3)-(R_{45}(y_3))^2}>)^2} \\ \frac{<\frac{R_{55}(y_3)}{R_{44}(y_3)R_{55}(y_3)-(R_{45}(y_3))^2}>}{<\frac{R_{44}(y_3)}{R_{44}(y_3)R_{55}(y_3)-(R_{45}(y_3))^2}><\frac{R_{55}(y_3)}{R_{44}(y_3)R_{55}(y_3)-(R_{45}(y_3))^2}> - (<\frac{R_{45}(y_3)}{R_{44}(y_3)R_{55}(y_3)-(R_{45}(y_3))^2}>)^2} \\ 0 \end{pmatrix}$$

And we obtain the expression of the localisation tensor associated to this cellular problem:

$$s^{(5)} = \begin{pmatrix} s_1^{(5)} = -\sigma_1^{(5)} + 0 \\ s_2^{(5)} = -\sigma_2^{(5)} + 0 \\ s_3^{(5)} = -\sigma_3^{(5)} + 0 \\ s_4^{(5)} = -\sigma_4^{(5)} + 0 \\ s_5^{(5)} = -\sigma_5^{(5)} + 0 \\ s_6^{(5)} = -\sigma_6^{(5)} + 0 \end{pmatrix}$$

$$= \begin{pmatrix} 0 \\ 0 \\ 0 \\ \frac{<\frac{R_{45}(y_3)}{R_{44}(y_3)R_{55}(y_3)-(R_{45}(y_3))^2}>}{<\frac{R_{44}(y_3)}{R_{44}(y_3)R_{55}(y_3)-(R_{45}(y_3))^2}><\frac{R_{55}(y_3)}{R_{44}(y_3)R_{55}(y_3)-(R_{45}(y_3))^2}> - (<\frac{R_{45}(y_3)}{R_{44}(y_3)R_{55}(y_3)-(R_{45}(y_3))^2}>)^2} \\ \frac{<\frac{R_{55}(y_3)}{R_{44}(y_3)R_{55}(y_3)-(R_{45}(y_3))^2}>}{<\frac{R_{44}(y_3)}{R_{44}(y_3)R_{55}(y_3)-(R_{45}(y_3))^2}><\frac{R_{55}(y_3)}{R_{44}(y_3)R_{55}(y_3)-(R_{45}(y_3))^2}> - (<\frac{R_{45}(y_3)}{R_{44}(y_3)R_{55}(y_3)-(R_{45}(y_3))^2}>)^2} \\ 0 \end{pmatrix}$$

**Solving the cellular problem $K = 6$**

$$(\overline{\mathcal{P}}^{6-Y}) \begin{cases} \text{Find } (\sigma^{(6)}(y_3), \vec{\chi}^{(6)}(y_3))/ \\ \forall M \in \Omega^Y \begin{cases} \frac{d\sigma_5^{(6)}(y_3)}{dy_3} = 0 \\ \frac{d\sigma_4^{(6)}(y_3)}{dy_3} = 0 \\ \frac{d\sigma_3^{(6)}(y_3)}{dy_3} = \frac{dR_{36}(y_3)}{dy_3} \end{cases} \\ \begin{cases} \sigma_1^{(6)} = R_{13}(y_3)\frac{d\chi_3^{(6)}(y_3)}{dy_3} \\ \sigma_2^{(6)} = R_{23}(y_3)\frac{d\chi_3^{(6)}(y_3)}{dy_3} \\ \sigma_3^{(6)} = R_{33}(y_3)\frac{d\chi_3^{(6)}(y_3)}{dy_3} \\ \sigma_4^{(6)} = R_{44}(y_3)\frac{d\chi_2^{(6)}(y_3)}{dy_3} + R_{45}(y_3)\frac{d\chi_1^{(6)}(y_3)}{dy_3} \\ \sigma_5^{(6)} = R_{45}(y_3)\frac{d\chi_2^{(6)}(y_3)}{dy_3} + R_{55}(y_3)\frac{d\chi_1^{(6)}(y_3)}{dy_3} \\ \sigma_6^{(6)} = R_{36}(y_3)\frac{d\chi_3^{(6)}(y_3)}{dy_3} \end{cases} \\ \sigma^{(6)} \text{ and } \vec{\chi}^{(6)} \text{ are Y-periodic} \end{cases}$$

$$\implies \begin{cases} \frac{d\sigma_5^{(6)}(y_3)}{dy_3} = 0 \implies \sigma_5^{(6)}(y_3) = C_5 \implies C_5 = R_{45}(y_3)\frac{d\chi_2^{(6)}(y_3)}{dy_3} + R_{55}(y_3)\frac{d\chi_1^{(6)}(y_3)}{dy_3} \\ \frac{d\sigma_4^{(6)}(y_3)}{dy_3} = 0 \implies \sigma_4^{(6)}(y_3) = C_4 \implies C_4 = R_{44}(y_3)\frac{d\chi_2^{(6)}(y_3)}{dy_3} + R_{45}(y_3)\frac{d\chi_1^{(6)}(y_3)}{dy_3} \\ \frac{d\sigma_3^{(6)}(y_3)}{dy_3} = \frac{dR_{36}(y_3)}{dy_3} \implies \sigma_3^{(6)}(y_3) = R_{36}(y_3) + C_3 \implies R_{36}(y_3) + C_3 = R_{33}(y_3)\frac{d\chi_3^{(6)}(y_3)}{dy_3} \end{cases}$$

$$\implies \begin{cases} \frac{d\chi_1^{(6)}(y_3)}{dy_3} = \frac{R_{44}(y_3)C_5 - R_{45}(y_3)C_4}{R_{44}(y_3)R_{55}(y_3) - (R_{45}(y_3))^2} \\ \frac{d\chi_2^{(6)}(y_3)}{dy_3} = \frac{-R_{45}(y_3)C_5 + R_{55}(y_3)C_4}{R_{44}(y_3)R_{55}(y_3) - (R_{45}(y_3))^2} \\ \frac{d\chi_3^{(6)}(y_3)}{dy_3} = \frac{R_{36}(y_3)}{R_{33}(y_3)} + \frac{C_3}{R_{33}(y_3)} \end{cases}$$

$$\implies \begin{cases} \displaystyle\int_{\chi_1^{(6)}(0)}^{\chi_1^{(6)}(y_3)} d\chi_1^{(6)}(y_3) = \chi_1^{(6)}(y_3) - \chi_1^{(6)}(0) \\ \qquad\qquad = C_5 \int_0^{y_3} \frac{R_{44}(y_3)}{R_{44}(y_3)R_{55}(y_3) - (R_{45}(y_3))^2} dy_3 \cdots \\ -C_4 \int_0^{y_3} \frac{R_{45}(y_3)}{R_{44}(y_3)R_{55}(y_3) - (R_{45}(y_3))^2} dy_3 \\ \displaystyle\int_{\chi_2^{(6)}(0)}^{\chi_2^{(6)}(y_3)} d\chi_2^{(6)}(y_3) = \chi_2^{(6)}(y_3) - \chi_2^{(6)}(0) \\ \qquad\qquad = -C_5 \int_0^{y_3} \frac{R_{45}(y_3)}{R_{44}(y_3)R_{55}(y_3) - (R_{45}(y_3))^2} dy_3 \cdots \\ +C_4 \int_0^{y_3} \frac{R_{55}(y_3)}{R_{44}(y_3)R_{55}(y_3) - (R_{45}(y_3))^2} dy_3 \\ \displaystyle\int_{\chi_3^{(6)}(0)}^{\chi_3^{(6)}(y_3)} d\chi_3^{(6)}(y_3) = \chi_3^{(6)}(y_3) - \chi_3^{(6)}(0) \\ \qquad\qquad = \int_0^{y_3} \frac{R_{36}(y_3)}{R_{33}(y_3)} dy_3 + C_3 \int_0^{y_3} \frac{1}{R_{33}(y_3)} dy_3 \end{cases}$$

We calculate the integration constants $C_3$, $C_4$ and $C_5$ by making the preceeding integration operation from 0 to $y_3 = h$ and by invoking that the fonctions $\chi_i^{(6)}(y_3)$ $(i = 1, 2, 3)$ are periodic.

Then: $\chi_i^{(6)}(h) = \chi_i^{(6)}(0)$. We deduce:

$$
\begin{cases}
C_4 = 0 \\
C_5 = 0 \\
C_3 = -\dfrac{\int_0^h \frac{R_{36}(y_3)}{R_{33}(y_3)} dy_3}{\int_0^h \frac{1}{R_{33}(y_3)} dy_3} = -\dfrac{< \frac{R_{36}(y_3)}{R_{33}(y_3)} >}{< \frac{1}{R_{33}(y_3)} >}
\end{cases}
$$

$$
\implies
\begin{cases}
\chi_1^{(6)}(y_3) = \chi_1^{(6)}(0) \\
\chi_2^{(6)}(y_3) = \chi_2^{(6)}(0) \\
\chi_3^{(6)}(y_3) = \chi_3^{(6)}(0) + \int_0^{y_3} \frac{R_{36}(y_3)}{R_{33}(y_3)} dy_3 - \dfrac{< \frac{R_{36}(y_3)}{R_{33}(y_3)} >}{< \frac{1}{R_{33}(y_3)} >} \int_0^{y_3} \frac{1}{R_{33}(y_3)} dy_3
\end{cases}
$$

Finally

$$
\varepsilon^m(\vec{\chi}^{(6)}) =
\begin{pmatrix}
\varepsilon_1^m(\vec{\chi}^{(6)}) = 0 \\
\varepsilon_2^m(\vec{\chi}^{(6)}) = 0 \\
\varepsilon_3^m(\vec{\chi}^{(6)}) = \frac{R_{36}(y_3)}{R_{33}(y_3)} - \frac{< \frac{R_{36}(y_3)}{R_{33}(y_3)} >}{< \frac{1}{R_{33}(y_3)} >} \frac{1}{R_{33}(y_3)} \\
\varepsilon_4^m(\vec{\chi}^{(6)}) = 0 \\
\varepsilon_5^m(\vec{\chi}^{(6)}) = 0 \\
\varepsilon_6^m(\vec{\chi}^{(6)}) = 0
\end{pmatrix}
$$

$$
\sigma^{(6)} =
\begin{pmatrix}
\sigma_1^{(6)} = \frac{R_{13}(y_3)R_{36}(y_3)}{R_{33}(y_3)} - \frac{< \frac{R_{36}(y_3)}{R_{33}(y_3)} >}{< \frac{1}{R_{33}(y_3)} >} \frac{R_{13}(y_3)}{R_{33}(y_3)} \\
\sigma_2^{(6)} = \frac{R_{23}(y_3)R_{36}(y_3)}{R_{33}(y_3)} - \frac{< \frac{R_{36}(y_3)}{R_{33}(y_3)} >}{< \frac{1}{R_{33}(y_3)} >} \frac{R_{23}(y_3)}{R_{33}(y_3)} \\
\sigma_3^{(6)} = \frac{R_{33}(y_3)R_{36}(y_3)}{R_{33}(y_3)} - \frac{< \frac{R_{36}(y_3)}{R_{33}(y_3)} >}{< \frac{1}{R_{33}(y_3)} >} \\
\sigma_4^{(6)} = 0 \\
\sigma_5^{(6)} = 0 \\
\sigma_6^{(6)} = \frac{(R_{36}(y_3))^2}{R_{33}(y_3)} - \frac{< \frac{R_{36}(y_3)}{R_{33}(y_3)} >}{< \frac{1}{R_{33}(y_3)} >} \frac{R_{36}(y_3)}{R_{33}(y_3)}
\end{pmatrix}
$$

And we obtain the expression of the localisation tensor associated to this cellular problem:

$$
s^{(6)} =
\begin{pmatrix}
s_1^{(6)} = -\sigma_1^{(6)} + R_{16}(y_3) \\
s_2^{(6)} = -\sigma_2^{(6)} + R_{26}(y_3) \\
s_3^{(6)} = -\sigma_3^{(6)} + R_{36}(y_3) \\
s_4^{(6)} = -\sigma_4^{(6)} + 0 \\
s_5^{(6)} = -\sigma_5^{(6)} + 0 \\
s_6^{(6)} = -\sigma_6^{(6)} + R_{66}(y_3)
\end{pmatrix}
=
\begin{pmatrix}
R_{16}(y_3) - \frac{R_{13}(y_3)R_{36}(y_3)}{R_{33}(y_3)} + \frac{< \frac{R_{36}(y_3)}{R_{33}(y_3)} >}{< \frac{1}{R_{33}(y_3)} >} \frac{R_{13}(y_3)}{R_{33}(y_3)} \\
R_{26}(y_3) - \frac{R_{23}(y_3)R_{36}(y_3)}{R_{33}(y_3)} + \frac{< \frac{R_{36}(y_3)}{R_{33}(y_3)} >}{< \frac{1}{R_{33}(y_3)} >} \frac{R_{23}(y_3)}{R_{33}(y_3)} \\
\frac{< \frac{R_{36}(y_3)}{R_{33}(y_3)} >}{< \frac{1}{R_{33}(y_3)} >} \\
0 \\
0 \\
R_{66}(y_3) - \frac{(R_{36}(y_3))^2}{R_{33}(y_3)} + \frac{< \frac{R_{36}(y_3)}{R_{33}(y_3)} >}{< \frac{1}{R_{33}(y_3)} >} \frac{R_{36}(y_3)}{R_{33}(y_3)}
\end{pmatrix}
$$

## 7.6   Conclusion

This chapter presents analytical solutions to find equivalent behaviour of continua to that of of laminated composites consisting of thin and periodically thick stacking sequences of oriented unidirectional plies. In particular, for the latter case, this is a direct application of the periodic homogenisation method using the averaging technique. This allows the possibility of defining a multi-scale process at three different scales for the case in an application of a structure made as a laminate consisting of oriented unidirectional plies. The first defined scale is the microscopic level which can see the fibres and the matrix in the RVE of the unidirectional material. The second, defined as the mesoscopic scale, sees the plies of the laminate as homogeneous layers. The third is the macroscopic scale which sees the laminate as an homogeneous material.

The microscopic/mesoscopic homogenisation link allows the homogenised behaviour of the plies to be described. The mesoscopic/macroscopic homogenisation link allows the homogenised behaviour of the laminate to be described. After that, the mechanical problem of the failure of a structure can be solved at the macroscopic scale. The macroscopic stress and strain are obtained for each point of the macroscopic scale, everywhere in the structure. The macroscopic stress and strain fields and the macroscopic/mesoscopic localisation link allow the mesoscopic stress and strain at each point (of the mesoscopic scale) of each ply of the laminate to be obtained. Finally, the mesoscopic stress and strain fields and the mesoscopic/microscopic localisation link allow the microscopic stress and strain in each point (of the microscopic scale) of each fibre of the unidirectional material to be determined.

Such calculations are now very common for all hightech industrial technology applications. The following chapters give failure criteria that can be applied at the mesoscopic scale and modelling for the most important damages that can exist in laminates: matrix cracking and fibre break phenomenon. Finally the case of pressure vessels is considered.

# Revision exercises

### Exercise 7.1

For which types of laminates can the periodic homogenisation concepts be used?

### Exercise 7.2

For which types of laminates can the usual Kirshhoff-Love plate theory concepts be used?

### Exercise 7.3

Can the periodic homogenisation concepts always be used for thick laminates?

### Exercise 7.4

Describe the difference between the Voigt notation and the modified Voigt notation

### Exercise 7.5

In the plane stress state, how is the restricted plane behaviour law obtained?

### Exercise 7.6

When periodic homogenisation is used to homogenise a periodic laminate how are the macroscopic stress and strain defined compared to the microscopic stress and strain?

### Exercise 7.7

When periodic homogenisation is used to homogenise a periodic laminate is the homogenised behaviour obtained with a stress approach the same as the homogenised behaviour obtained with a strain approachl?

### Exercise 7.8

How many times does the change of basis matrix appear in the change of basis formulae for a scalar, a vector, a second-order tensor and a fourth-order tensor (like the rigidity or flexibility tensor)?

### Exercise 7.9

What is the Kirshhoff-Love hypothesis for laminates? Is it realistic for thick laminates?

### Exercise 7.10

Why is there the necessity to define local and reference frames for the thin or thick laminate theory? In which type of ply should this not be necessary?

8

# Failure criteria

## 8.1 Introduction

In this chapter, some failure models, which are widely used in designing composite material laminates for which the layers are orientated unidirectional fibre/resin composite, will be described. These models are generally developed based on the concept of strength. After a description of the most traditional criteria, we shall underline the limitations inherent in these models.

## 8.2 Typical failure mechanisms existing in the orientated unidirectional laminate

As unidirectional fibre/matrix composites were among the first composite materials to be used in high performance technologies all the failure criteria described here were originally developed for composite laminates made up of stacks of orientated unidirectional composite. It is this reference that we shall use to describe these criteria (Figure 8.1). The most common damage processes found in these laminates are (Figure 8.2) fibre failure (Figure 8.3) and intralaminar cracking (Figure 8.4, Figure 8.5). Intralaminar cracking is the result of the coalescence of smaller defects which appear in the resin matrix: microcracks (Figure 8.6), porosity (Figure 8.7), fibre-matrix debonding (Figure 8.8). The criteria written here attempt to explain the phenomena of fibre failure and intra-laminar cracking, which appear within plies.

It is important to note that the criteria which will be developed here are written in the framework of Damage Mechanics, which means describing a continuum which sees its properties degraded. There exists therefore a RVE in which all the damage processes considered can be included. The damage processes considered here, the fibre failure and interlaminar cracking, can be included in this framework and then be integrated in the criteria which will be written. However delamination (Figure 8.2, Figure 8.9) cannot be readily treated in this way as it is a damage process which has to be modelled by Fracture Mechanics rather than Damage Mechanics as its large size prevents it to be included in an RVE.

## 8.3 Framework of the writing of failure criteria. The ply strength constants

Let's consider an orthotropic material for which we note $b^{(loc)} = (\vec{x_1}^{(loc)}, \vec{x_2}^{(loc)}, \vec{x_3}^{(loc)})$ as the basis (called local of anisotropy) of the local frame of anisotropy. This material makes up the ply $k$, in which a plane stress state is assumed to exist, of a laminate of thin plate type or of the periodic cell of a thick periodic (in its thickness) plate type. We designate as $b^{(kloc)} = (\vec{x_1}^{(kloc)}, \vec{x_2}^{(kloc)}, \vec{x_3}^{(kloc)})$ the basis of the local framework of the ply $k$, oriented with respect to the reference base $b^{(ref)} = (\vec{x_1}^{(ref)}, \vec{x_2}^{(ref)}, \vec{x_3}^{(ref)})$ of the reference frame of the laminate. It is considered that the bases $b^{(loc)}$ and $b^{(kloc)}$ coincide. Here the material considered is a unidirectional fibre-matrix composite for which the direction of the fibres is given by the vector $\vec{x_1}^{(loc)}$.

When a material is anisotropic it is advisable, so as to overcome identifications depending on a reference frame, to write a model, particularly a failure model, in the local framework of anisotropy of the material using the expression, in this framework, of the stress and strain fields $\sigma^{(loc)}$, $\varepsilon^{(loc)}$ and of all the necessary properties used to establish the model.

Hence, for the ply $k$ of the considered laminate it is the stress and strain fields $\sigma^{(kloc)}$ and $\varepsilon^{(kloc)}$ which give the values of the $\sigma^{(loc)}$ and $\varepsilon^{(loc)}$ fields and the model can then be used independently of the base $b^{(ref)}$.

We are using the Voigt notation. Thus, under the hypotheses proposed the stress and strain fields are noted as $\sigma^{(loc)} = (\sigma_I^{(loc)})_{I=1,2,6}$ and $\varepsilon^{(loc)} = (\varepsilon_I^{(loc)})_{I=1,2,6}$.

In the laminate theories discussed in the previous chapter, the basic ply is homogenised which means that the details of the microstructure are not considered. Both the two- and three-dimensional laminate theory models are then used to calculate the stresses in each ply. Based on the material strength concept, the ply, which is anisotropic (usually, orthotropic), is then assumed to possess a set of strength constants. In this way a comparison with the stress state of the ply can be used to predict failure. These constants can be determined either by a model based on the mechanics of the processes involved and developed at the fibre/matrix level or by testing the ply material in different directions to reveal the anisotropy of the mechanical properties. The first approach, similar in terms of strength to models described in the previous chapter, is rarely used. These models based on the strength of the components are found to be poor in estimating ply strength. For this reason, ply strength constants are measured directly by testing the ply material.

For the most common failure criteria, five independent ply failure modes are assumed: tensile and compressive failures in the $\vec{x_1}^{(loc)}$ direction (the fibre direction), tensile and compressive failures in the $\vec{x_2}^{(loc)}$ direction (perpendicular to the fibre direction) and in-plane shear failure. Thus, at the most, five strength constants have to be determined by testing the basic ply of the laminate composite structure:

- $X_t$ is the tensile breaking stress in the fibre direction;
- $X_c$ is the compressive breaking stress in the fibre direction;
- $Y_t$ is the tensile breaking stress transverse to the fibre direction;
- $Y_c$ is the compressive breaking stress transverse to the fibre direction;

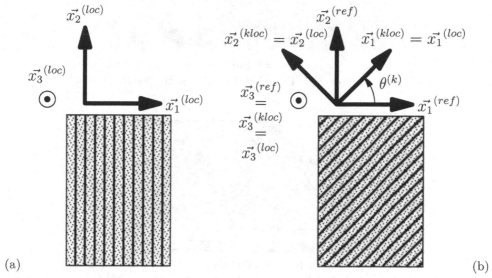

Figure 8.1: Schematic view of a fibre/resin unidirectional composite. (a) Anisotropic local frame of the unidirectional material. (b) Reference frame of the laminate and local frame of the $k$th layer of a laminate, orientated with angle $\theta^{(k)}$ relatively to the reference frame of the laminate.

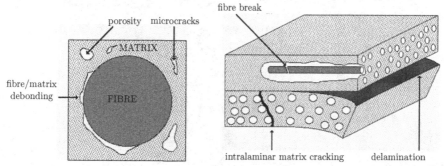

Figure 8.2: Schematic view of the most common damage mechanisms appearing in laminate for which the layers are orientated unidirectional fibre/resin composite (Thionnet, 1991).

I⟸ ≈ 0.150 mm ⟹I
Figure 8.3:  Fibre breaks (Kim et al., 2003).

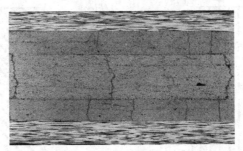

I⟸ ≈ 4.000 mm ⟹I
Figure 8.4:  Intralaminar matrix cracking of layers of laminate (Thionnet et al., 2002).

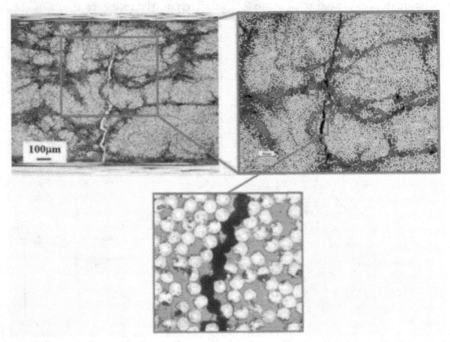

I⟸ ≈ 0.050 mm ⟹I
Figure 8.5:  Intralaminar matrix cracking of layers of laminate. Coalescence of small defects existing in the resin matrix (Revest, 2011).

I⟸ ≈ 1.000 mm ⟹I

Figure 8.6: Cracks in resin (Purslow, 1981, 1986).

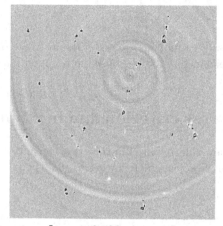

I⟸ ≈ 0.500 mm ⟹I

Figure 8.7: Porosities in resin (Rojek, 2020).

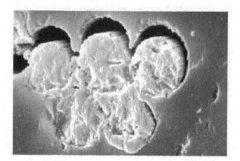

I⟸ ≈ 0.020 mm ⟹I

Figure 8.8: Fibre/matrix debonding (Revest, 2011).

- $S$ is the in-plane breaking stress;
- if it is assumed that the tensile and compressive strengths are equal, we have $X = X_t = X_c$ and $Y = Y_t = Y_c$.

Experimental determination of these constants is sometimes difficult. Results often exhibit wide scatter and it is not always possible to determine these constants independently. The state of stress is not always homogeneous depending on the loading mode and the size and the geometry of the tested specimen. Analysis of the state of stress has to be carefully made to be sure that the correct stress and the corresponding strength are determined. An example is the shear test

I$\Longleftarrow$ ≈ 0.500 mm $\Longrightarrow$I

Figure 8.9:  Delamination at the tip of an intralaminar matrix cracking (Okabe et al., 2008).

where a mixture of all stresses cannot be avoided. Despite these difficulties, the above five ply constants are routinely determined and widely used in common failure criteria for unidirectional composite materials.

## 8.4   Maximum stress criterion and maximum strain criterion

The maximum stress criterion is formulated in this way: the material of the ply at a given point is supposed to fail if one (or more) of the following five independant conditions is reached:

- condition 1 - $\sigma_1^{(loc)} \leq X_t$;
- condition 2 - $|\sigma_1^{(loc)}| \leq X_c$;
- condition 3 - $\sigma_2^{(loc)} \leq Y_t$;
- condition 4 - $|\sigma_2^{(loc)}| \leq Y_c$;
- condition 5 - $|\sigma_6^{(loc)}| \leq S$.

This criterion can be expressed in terms of strain as following:

- condition 1 - $\varepsilon_1^{(loc)} \leq X_t/E_1^{(loc)}$;
- condition 2 - $|\varepsilon_1^{(loc)}| \leq X_c/E_1^{(loc)}$;
- condition 3 - $\varepsilon_2^{(loc)} \leq Y_t/E_2^{(loc)}$;
- condition 4 - $|\varepsilon_2^{(loc)}| \leq Y_c/E_2^{(loc)}$;
- condition 5 - $|\varepsilon_6^{(loc)}| \leq S/G_{12}^{(loc)}$.

The conditions 1 and 2 reflect fibre break, the conditions 3, 4 and 5 reflect matrix break.

## 8.5   Quadratic criterion in stress space or in strain space

These criteria are often qualified as 'interaction criteria' because they assume that the ply material at a point can fail even if none of the five preceding conditions are satisfied. The premise of this assumption was postulated in the stress interaction effect in the yield criterion of Von Mises which is based on the stored distorsional energy. After a short reminder of this criterion, we are going to analyse the main quadratic criteria which are used in the design and sizing of composite material structures.

## 8.5.1 The Von Mises criterion

Isotropy requires that the boundary of the domain of the criterion be invariant under a change of axes. Therefore, a criterion is to be only expressed with invariant quantities. The Von Mises criterion corresponds to the isotropic-hardening state of metals. As metals generally exhibit plastic incompressibility and yield-independence with respect to hydrostatic stress, it is sufficient to use the deviatoric stress tensor defined by $s = \sigma - \frac{1}{3}Tr(\sigma)I_3$ where $I_3$ is the unit second-order tensor in $\mathbb{R}^3$ and $Tr(\sigma)$ is the trace of the stress tensor $\sigma$. The Von Mises criterion is expressed as a function of the deviatoric stress tensor by the threshold function $f(s_{II}, \sigma_s)$, where $s_{II} = \frac{1}{2}Tr(s^2)$ corresponds to the second invariant of the deviatoric stress tensor and $\sigma_s$ characterises the yield stress in simple tension. The Von Mises criterion postulates that the threshold is reached when $s_{II}$, which characterises a three-dimensional state of stress, is equivalent to the pure tensile one-dimensional state, the threshold of which is $\sigma_s$. As in the one-dimensional case, a pure tensile stress state (for the value $\sigma_s$) and the corresponding deviatoric stress are expressed, in Voigt notation, by:

$$\sigma = \begin{pmatrix} \sigma_1 = \sigma_s \\ \sigma_2 = 0 \\ \sigma_3 = 0 \\ \sigma_4 = 0 \\ \sigma_5 = 0 \\ \sigma_6 = 0 \end{pmatrix} \qquad s = \begin{pmatrix} s_1 = \frac{2}{3}\sigma_s \\ s_2 = -\frac{1}{3}\sigma_s \\ s_3 = -\frac{1}{3}\sigma_s \\ s_4 = 0 \\ s_5 = 0 \\ s_6 = 0 \end{pmatrix}$$

The Von Mises criterion is therefore written as:

$$f(s_{II}, \sigma_s) = s_{II} - \frac{1}{3}\sigma_s^2 = 0$$

If we develop this expression for a three-dimensional state of stress, we obtain a convex envelope in stress space:

$$\frac{1}{2}\left[(\sigma_1 - \sigma_2)^2 + (\sigma_2 - \sigma_3)^2 + (\sigma_3 - \sigma_1)^2 + 6(\sigma_4^2 + \sigma_5^2 + \sigma_5^2) - \sigma_s^2\right] = 0$$

This criterion has been established for isotropic materials. The criteria we are now going to present are extensions of the Von Mises criterion to materials with anisotropic properties.

## 8.5.2 The Hill criterion

The extension of the Von Mises criterion to anisotropic materials was first presented by Hill (1948). In this case the convex envelope which characterises the threshold in the local frame of the material is written as:

$$F(\sigma_2^{(loc)} - \sigma_3^{(loc)})^2 + G(\sigma_3^{(loc)} - \sigma_1^{(loc)})^2 + H(\sigma_1^{(loc)} - \sigma_2^{(loc)})^2$$

$$+2L\sigma_4^{(loc)2} + 2M\sigma_5^{(loc)2} + 2N\sigma_5^{(loc)2} = 1$$

where $F$, $G$, $H$, $L$, $M$ and $N$ are strength properties, in a similar way as $X_t$, $X_c$, $Y_t$, $Y_c$ and $S$. We can develop this expression and obtain the following:

$$(G+H)\sigma_1^{(loc)2} + (F+H)\sigma_2^{(loc)2} + (F+G)\sigma_3^{(loc)2}$$

$$-2H\sigma_1^{(loc)}\sigma_2^{(loc)} - 2G\sigma_1^{(loc)}\sigma_3^{(loc)} - 2F\sigma_2^{(loc)}\sigma_3^{(loc)}$$

$$+2L\sigma_4^{(loc)2} + 2M\sigma_5^{(loc)2} + 2N\sigma_6^{(loc)2} = 1$$

It can be then observed that this criterion does not distinguish between tensile and compressive loads.

### 8.5.3 The Tsai–Hill criterion

The simplified plane stress form of the Hill criterion has been proposed by (Azzi and Tsai, 1965), defining the Tsai–Hill criterion:

$$a\sigma_1^{(loc)2} + b\sigma_2^{(loc)2} - c\sigma_1^{(loc)}\sigma_2^{(loc)} + d\sigma_6^{(loc)2} = 1$$

where $a$, $b$, $c$ and $d$ are strength properties depending on $X$, $Y$ and $S$. It easy to find:

$$\left(\frac{\sigma_1^{(loc)}}{X}\right)^2 + \left(\frac{\sigma_2^{(loc)}}{Y}\right)^2 - \left(\frac{1}{X^2} + \frac{1}{Y^2}\right)\sigma_1^{(loc)}\sigma_2^{(loc)} + \left(\frac{\sigma_6^{(loc)}}{S}\right)^2 = 1$$

### 8.5.4 The Hoffman criterion

A possible generalisation of the Hill criterion, which takes into account possible differences between tensile and compressive behaviours has been proposed by (Hoffman, 1967):

$$F(\sigma_2^{(loc)} - \sigma_3^{(loc)})^2 + G(\sigma_3^{(loc)} - \sigma_1^{(loc)})^2 + H(\sigma_1^{(loc)} - \sigma_2^{(loc)})^2$$

$$+P\sigma_1^{(loc)} + Q\sigma_2^{(loc)} + R\sigma_3^{(loc)}$$

$$+2L\sigma_4^{(loc)2} + 2M\sigma_5^{(loc)2} + 2N\sigma_6^{(loc)2} = 1$$

where $F$, $G$, $H$, $P$, $Q$, $R$, $L$, $M$ and $N$ are strength properties, in a similar way as $X_t$, $X_c$, $Y_t$, $Y_c$ and $S$. We can develop this expression and obtain the following:

$$(G+H)\sigma_1^{(loc)2} + (F+H)\sigma_2^{(loc)2} + (F+G)\sigma_3^{(loc)2}$$

$$-2H\sigma_1^{(loc)}\sigma_2^{(loc)} - 2G\sigma_1^{(loc)}\sigma_3^{(loc)} - 2F\sigma_2^{(loc)}\sigma_3^{(loc)}$$

$$+P\sigma_1^{(loc)} + Q\sigma_2^{(loc)} + R\sigma_3^{(loc)}$$

$$+2L\sigma_4^{(loc)2} + 2M\sigma_5^{(loc)2} + 2N\sigma_6^{(loc)2} = 1$$

It can be then observed that this criterion can distinguish between tensile and compressive loads. The simplified plane stress form of the Hoffman criterion is the following:

$$a\sigma_1^{(loc)2} + b\sigma_2^{(loc)2} + c\sigma_1^{(loc)} + d\sigma_2^{(loc)} - e\sigma_1^{(loc)}\sigma_2^{(loc)} + f\sigma_6^{(loc)2} = 1$$

where $a$, $b$, $c$, $d$, $e$ and $f$ are strength properties depending on $X_t$, $X_c$, $Y_t$, $Y_c$ and $S$. It easy to find:

$$\frac{\sigma_1^{(loc)2}}{X_tX_c} + \frac{\sigma_2^{(loc)2}}{Y_tY_c} - \frac{\sigma_1^{(loc)}\sigma_2^{(loc)}}{X_tX_c}$$

$$+\frac{X_c - X_t}{X_tX_c}\sigma_1^{(loc)} + \frac{Y_c - Y_t}{Y_tY_c}\sigma_2^{(loc)} + \left(\frac{\sigma_6^{(loc)}}{S}\right)^2 = 1$$

### 8.5.5 The Tsai-Wu criterion

The general form of the Tsai-Wu criterion, working for a tridimensionnal stress state, has been proposed by Tsai and Wu (1971). The plane stress form of this criterion takes the following form:

$$F_{11}\sigma_1^{(loc)2} + 2F_{12}\sigma_1^{(loc)}\sigma_2^{(loc)} + F_{22}\sigma_2^{(loc)2} + F_{66}\sigma_6^{(loc)2} + F_1\sigma_1^{(loc)} + F_2\sigma_2^{(loc)} +$$

$$F_6\sigma_6^{(loc)} + 2F_{16}\sigma_1^{(loc)}\sigma_6^{(loc)} + 2F_{26}\sigma_2^{(loc)}\sigma_6^{(loc)} = 1$$

Invoking that this criterion is insensitive to the sign of the shear stress, it appears that: $F_6 = F_{16} = F_{26} = 0$. And then, the Tsai-Wu criterion, under plane stress assumption, has the following form:

$$F_{11}\sigma_1^{(loc)2} + 2F_{12}\sigma_1^{(loc)}\sigma_2^{(loc)} + F_{22}\sigma_2^{(loc)2} + F_{66}\sigma_6^{(loc)2} + F_1\sigma_1^{(loc)} + F_2\sigma_2^{(loc)} = 1$$

where $F_{11}$, $F_{12}$, $F_{22}$, $F_{66}$, $F_1$ and $F_2$ are strength properties depending on $X_t$, $X_c$, $Y_t$, $Y_c$ and $S$. As previously, by taking some simple cases of the stress state, five among the six constants can be easily determined:

$$F_{11} = \frac{1}{X_t X_c} \quad F_1 = \frac{X_c - X_t}{X_t X_c} \quad F_{22} = \frac{1}{Y_t Y_c} \quad F_2 = \frac{Y_c - Y_t}{X_t X_c} \quad F_{66} = \frac{1}{S^2}$$

Concerning $F_{12}$ by invoking that the stress state cannot be infinite, it gives:

$$-1 < \frac{F_{12}}{\sqrt{F_{11} F_{22}}} < 1$$

The value usually taken is $\frac{F_{12}}{\sqrt{F_{11} F_{22}}} = -\frac{1}{2}$ (Tsai, 1980). However, this value can be obtained with more precision, depending on the considered material, by making multiaxial experimental tests.

## 8.6 Progressive ply failures in laminates

The various ply failure criteria discussed above are used in the failure analysis of laminates. Failure of the laminate may thus be a process in which the weakest ply fails first whilst the rest of the plies in the laminate still carry the load. The progressive ply failure model assumes that when one ply fails, it no longer can carry the stress in it and then its contribution to laminate strength should be removed from the laminate. A new iterative analysis of the laminate then has to be carried out to determine if any of the remaining plies will fail under the previously applied loads. If one ply does fail due to the redistribution of the load, the failed ply is again removed from consideration. If no ply fails after load redistribution, the applied load is progressively increased until another ply fails. This procedure is repeated until the last ply of the laminate fails.

This fail-or-no-fail procedure is very conservative, because, in reality, the failed ply, instead of failing in a brittle manner, has yielded in a non-linear sense such as showing perfect-plasticity behaviour or strain hardening plasticity. In this case, the concerned ply remains inside the laminate and continues to carry the stress as a yielded ply. In general, the evolution of the fibre breaks and intralaminar matrix cracking starts from no damage to the maximum damage passing through all possible values. The two following chapters will illustrate a more detailed modelling of intralaminar cracking and the fibre break phenomenon based on the concept of Damage Mechanics.

## 8.7 Extensions of these criteria for materials other than uni-directional

The criteria used here consider orientated unidirectional laminates as this framework is that which corresponds to their original development. However now these same criteria are used for other types of laminates and especially laminates composed of plies of woven fibres embedded in resin (Figure 8.10). One of the reasons is that within the unidirectional laminates, as in the woven laminates (at least in their simplest forms) certain damage processes are similar: fibre failure, intralaminar cracking, fibre-matrix debonding (Figure 8.3 to Figure 8.8, Figure 8.11). We can point out that these similarities in damage processes are due to another similarity:

generally these materials are very anisotropic and consequently damages processes are guided by the microstructure of the material rather than the loading, as is the case for isotropic materials. It therefore seems natural, nevertheless constrained by the identification of processes, to use these criteria for both families of materials.

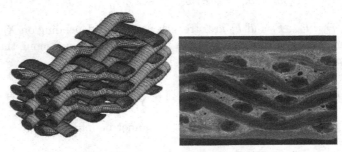

Figure 8.10:  Woven composite (Trabelsi, 2013).

I⟸ ≈ 10.000 mm ⟹I

Figure 8.11:  Woven composites and their most common damage phenomena (fibre breaks, intralaminar matrix crackings) (Trabelsi, 2013).

## 8.8   Conclusions

The two inner ply failure modes previously described (fibre break, intralaminar cracking) do not have the same consequences for the lifetime of the laminates as the delamination, which is an outer ply damage. Delamination growth is more damaging for the laminates than intralaminar cracking. Delamination can grow in size under an increasing load and render the laminate structurally useless. If a delamination crack became unstable, this damage could provoke the failure of the structure. On the contrary, this is generally not the case with intralaminar cracking or fibre breaks where neighbouring plies can constrain the effect of these damages and maintain the residual load-carrying capacity of the structure. As a failure mode, delamination can occur near any stress risers in the laminate, such as holes, cut-outs. . ., in addition to free edges. Even the formation of intralaminar cracks can induce localised delamination at the tips of intralaminar cracks. On the other hand, if a delamination crack occurs first, stress concentrations can develop at the crack tip area, which in turn can precipitate intralaminar cracking and fibre breaks. Continuous formation of these competing damages processes means that the microstructure changes continuously when under load. This dependence on the loading history of the laminate in service becomes crucial when long-term behaviour has to be analysed. As has already been pointed out, to have a more accurate prediction of the failure of a laminate, it is necessary to have a complete understanding of all the most important damage processes occurring when the composite is under load, from the microstructure to the laminated structure.

# Revision exercises

## Exercise 8.1

If the five constants $X_t$, $X_c$, $Y_t$, $Y_c$ and $S$ are referred to the local frame of anisotropy of a unidirectional composite material $b^{(loc)} = (\vec{x_1}^{(loc)}, \vec{x_2}^{(loc)}, \vec{x_3}^{(loc)})$, draw the failure surface in the $(\sigma_1^{(loc)}, \sigma_2^{(loc)})$ plane ($\sigma_6^{(loc)}$ is assumed to be equal to zero) according to the maximum stress criterion and the Tsai-Wu criterion. Numerical values: $X_t = 2280$ MPa, $X_c = 335$ MPa, $Y_t = 57$ MPa, $Y_c = 158$ MPa and $S = 71$ MPa.

## Exercise 8.2

Is it a good idea to use the Von Mises criterion in case of a unidirectional composite material ?

## Exercise 8.3

The local frame of anisotropy of a unidirectional composite material is defined as $b^{(loc)} = (\vec{x_1}^{(loc)}, \vec{x_2}^{(loc)}, \vec{x_3}^{(loc)})$. In the objective to use the Tsai-Wu criterion, the five constants $X_t$, $X_c$, $Y_t$, $Y_c$ and $S$ are measured in this frame. Now, a new local frame of anisotropy is defined: $b^{(loc)'} = (\vec{x_1}^{(loc)'}, \vec{x_2}^{(loc)'}, \vec{x_3}^{(loc)'})$ for which the Tsai-Wu criterion will be also used with the constants $X_t'$, $X_c'$, $Y_t'$, $Y_c'$ and $S'$. It is assumed that: $\vec{x_1}^{(loc)'} = \vec{x_2}^{(loc)}$ and $\vec{x_2}^{(loc)'} = -\vec{x_1}^{(loc)}$. Is there a link between $(X_t, X_c, Y_t, Y_c$ and $S)$ and $(X_t', X_c', Y_t', Y_c'$ and $S')$ ?

# 9

# Multiscale modelling of the intralaminar cracking phenomenon. From Fracture to Damage Mechanics: modelling using the concept of Crack Opening Mode

## 9.1   Introduction

The objective here is to model one of the most common damage processes that exist in composite laminates, known commonly as transverse cracking, but which should really be described as intralaminar cracking (Figure 9.1).

The model will be included in a general framework, those of the Damage Mechanics, by using the description of the phenomenon in the concepts of Fracture Mechanics. First, this general framework will be described and then, the following very important property of the unidirectional composite will be used: the geometry of the damage is not governed by the stress field but by the arrangement of the constituent of the material.

## 9.2   Aim and framework of the modelling of intralaminar cracking existing in composite laminates

Damage present in a material in the form of plane cracks induces the decrease of some of its mechanical properties. For example, in the case of a tensile loading, and according to the plane in which the cracks are located, a decrease of the axial modulus can be observed. The same observation can be made for the shear modulus in the case of shear loading. If the sign of the loading changes, the axial modulus is restored. It is not the case for the shear modulus. This effect is usually called the unilateral effect of damage or damage activation/deactivation.

The modelling that can be made of the damageable elastic behaviour of a microcracked body needs to be written in a way that takes into account:

- $(c1)$: any discontinuity in the stress/strain relation;
- $(c2)$: anisotropy induced by damage;
- $(c3)$: decreases of the appropriate modulus;
- $(c4)$: damage activation/deactivation.

The Principles of Physics and Mechanics also have to be satisfied. Hence, the modelling should be:

- $(c5)$: S-invariant (where S denotes the material symmetry group);
- $(c6)$: objective;
- $(c7)$: in agreement with the Second Principle of Thermodynamics.

The development of a behaviour law for the degradation of a material can be considered on different scales. The first, that of the pioneers of Continuum Damage Mechanics (CDM), such as Kachanov (1958), was at the level of the structure or macroscopic level. At this scale, the microstructural phenomena are generally not considered. Originally, the models which were constructed used scalar macroscopic variables and the notion of effective stress. These models rarely reflect the reality of damage at the macroscopic level as for example the unilateral effect of damage or damage-induced anisotropy. The preliminary work of Kachanov was limited to creep damage and particular loading conditions. These early approaches were later improved, essentially by the use of more powerful mathematical techniques than those involving scalar variables. Initially; for models dealing with the behaviour of metals, n-tensorial ($n = 1, 2, 4$) variables were used to describe damage at the macroscopic level (Cordebois and Sidoroff, 1979; Leckie and Onat, 1981; Krajcinovic and Fonseka, 1981; Murakami and Ohno, 1981; Litewka, 1985; Chaboche, 1993). The idea of 2-tensorial variables was firstly proposed by Vakulenko and Kachanov (1971). Several other properties of damage tensors have been outlined by Betten (1981, 1986). Cauvin and Testa (1999) used eight-rank tensors and showed that this damage variable could be reduced into a fourth-rank tensor within the general theory of anisotropic elasticity. They have also shown that the 4-tensor damage variable is sufficient to describe anisotropic

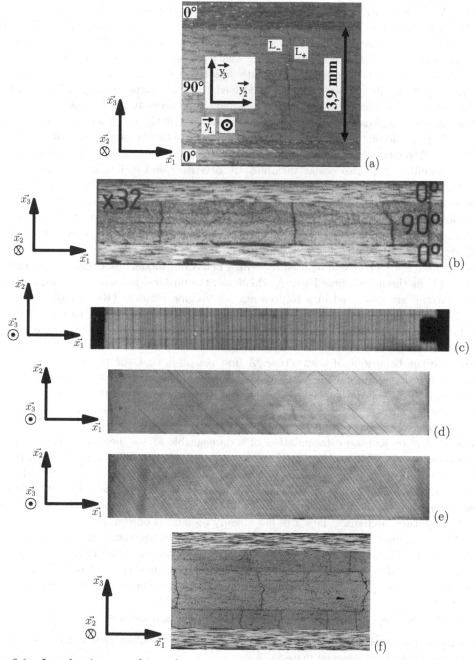

Figure 9.1: Intralaminar cracking phenomenon. The basis of the reference frame of the specimen is $(\vec{x_2}, \vec{x_1}, \vec{x_3})$. The basis of the local anisotropic frame of the unidirectional material is $(\vec{y_1}, \vec{y_2}, \vec{y_3})$. (a) $(0°_2, 90°_4)_s$ specimen. Fatigue loading. Optical microscopy observation. Indication of the local anisotropy basis $(\vec{y_1}, \vec{y_2}, \vec{y_3})$ of the material of the 90° ply. (b) $(0°_2, 90°_2)_s$ specimen. Monotonic quasi-static loading. Optical microscopy observation, magnification ×32. (c) $(0°_2, 90°_2)_s$ specimen. Monotonic quasi-static loading. Saturation intralaminar cracking state. X-ray observation. (d) $(0°_2, +45°_2, -45°_2)_s$ specimen. Fatigue loading. X-ray observation. (e) $(0°_2, +55°_2, -55°_2)_s$ specimen. Fatigue loading. X-ray observation. Saturation intralaminar cracking state. (f) $(0°_2, +55°_2, -55°_2)_s$ specimen. Fatigue loading. Optical microscopy observation, magnification ×45. Intralaminar cracking in the 90° ply of a $(0°, 90°, 0°)$ specimen (Thionnet et al., 2002).

damage. More recently, Voyiadjis and Kattan (2007a,b) applied the concept of fabric tensor introduced by Kanatani (1984). For composites, a vectorial approach was proposed by Talreja (1985). We may also mention the work of Ladeveze and LeDantec (1992) who derived a laminate ply failure model where micro-cracking and fibre debonding were described using CDM. Matzenmiller et al. (1995) derived a model with anisotropic damage to describe the elastic-brittle behaviour of composites. These models are derived based on phenomenological approaches.

Most of the time, none of these models satisfies all the previous conditions. The main reason for these shortcomings can be that the models are written on the macroscopic scale without correctly taking into account the microscopic phenomena. Another reason is, as shown by Leguillon and Sanchez-Palencia (1982) in their microscopic approach, that in the case of a plane crack and 2D-modelling, only two scalar variables are necessary and sufficient to obtain a coherent behaviour law. However, these models use tensorial variables to describe the damage at the macroscopic level and the induced superabundance of variables is often the (direct or indirect) reason for the inconsistencies. Nevertheless, these models are widely used for finite element calculations of structures. They provide results which are often satisfactory, with reasonable calculation times.

The second scale which can be used to write a behaviour model describing the degradation of a material is at the microscopic level. At this level, the physical mechanisms governing damage in the material are described in a Representative Volume Element (RVE) (Allen et al., 1987; Allen, 2001; Andrieux, 1983; Lee et al., 1989; Mauge and Kachanov, 1994; Hoenig, 1979; Hashin, 1985; Leguillon and Sanchez-Palencia, 1982; Talreja, 1991). These models have often been used to justify earlier macroscopic models. Up until recently these models have very rarely been used to calculate the behaviour of a structure. A first reason is that the parameters used in these models require experimental investigations which are difficult to achieve. A second reason is that the approaches based on the notion of a RVE may show some difficulty in going from the microscopic level to the macroscopic level: for often purely conceptual reasons, the transition from one scale to another cannot be solved analytically.

The goal is to propose a formulation of a damageable elastic behaviour law verifying the conditions $(c1 - c7)$. This work is based on the vectorial description of damage made by Talreja (1985, 1991), on the microscopic descriptions of the crack made by Leguillon and Sanchez-Palencia (1982); Allen et al. (1987); Allen (2001); Lee et al. (1989); Andrieux (1983) and on the following concept: the definition of the Crack Opening Mode for Damage Mechanics is given as it exists for Fracture Mechanics. This idea has already been mentioned in previous works (Aussedat et al., 1995; Thionnet and Renard, 1999; Thionnet, 2001) conducted in the framework of plane stress. However, the increasing use of thick composite materials in industrial structures allows this assumption to be ignored. Therefore, this assumption is discarded here. It is important to mention that our previous studies are not the 2D-restriction of the three-dimensional case presented here.

The different steps of the approach answer the following questions:

- what is the number and the type of the macroscopic variables that can be used in a model of the microcracked material ?
- how can these variables be used to build the concept of Crack Opening Mode for Damage Mechanics as it exists for Fracture Mechanics ?
- how can we build up a coherent model of a damageable elastic behaviour law according to the conditions $(c1 - c7)$ ?

A damaged elastic composite (or material) is studied. The damage is a network of identical plane microcracks. We assume that the geometry of damage spreads along preferential directions that are indicated by the constituents of the material whatever the loading. Contrary to what happens in the case of isotropic materials (metals for instance) the damage is not guided by the applied loading. The microstructure of the material is assumed to be periodic and no friction occurs in case cracks are closed. The appearance of the microcracks generates no residual stress field. Crack interaction, well described by Kachanov (1958) and Gorbatikh and Kachanov (2000), is not taken into account, except that we suppose that the microcracks are periodic. Similarly, the size effect (as illustrated for instance by Camanho et al. (2007)) is not studied in the present work. The damage state of the material is given and constant, denoted by the scalar variable $\alpha$. For instance, $\alpha$ can be representative of the crack density existing in the RVE. The Small Perturbations Hypothesis is made. The undamaged material is assumed to be orthotropic.

The Cartesian frame $R = (O, \vec{x_1}, \vec{x_2}, \vec{x_3})$ is related to the macroscopic scale where $x = (x_i)_{i=1,2,3}$ describes the position of the point $M$ and $\vec{U} = (U_i)_{i=1,2,3}$ defines the displacement of $M$. At this scale, $\sigma$ defines the stress tensor and $\varepsilon$ defines the strain tensor. The Cartesian frame $R_y = (O_y, \vec{y_1}, \vec{y_2}, \vec{y_3})$ is related to the microscopic scale where $y = (y_i)_{i=1,2,3}$ describes the position of the point $M$ and $\vec{u} = (u_i)_{i=1,2,3}$ defines the displacement of $M$. At this scale, $\mathbf{s}$ defines the stress tensor and $\mathbf{e}$ defines the strain tensor. $R_y$ is the orthotropic frame of the undamaged material. The axes of $R$ and $R_y$ are assumed to be identical. The material symmetry group of the undamaged material is then $S_8 = \{T_i, i = 0, \ldots, 7\}$ with: $T_0 = diag(1,1,1)$, $T_1 = diag(-1,1,1)$, $T_2 = diag(1,-1,1)$, $T_3 = diag(1,1,-1)$, $T_4 = diag(1,-1,-1)$, $T_5 = diag(-1,1,-1)$, $T_6 = diag(-1,-1,1)$, $T_7 = -T_0$.

The geometry of the elementary cell is a parallelepiped containing a single crack $F$ (Figure 9.2), symmetrically placed with respect to the axes of the frame. Its dimensions are: $2a \times 2b \times 2c$. Then, the material symmetry group of the unloaded damaged material is $S_8$. The external boundary surface is denoted by $\partial\Omega$. The surfaces $L_+$ and $L_-$ define the microcrack surfaces (or lips). Without loading they are in the $(\vec{y_1}, \vec{y_3})$ plane. Their dimensions are assumed to be known. We have: $\partial\overline{\Omega} = \partial\Omega \cup L_+ \cup L_-$. The domain delimited by $\partial\Omega$ (the volume of the material plus the volume of the crack) is denoted by $\Omega$. The domain delimited by $\partial\overline{\Omega}$ (the volume of the material) is denoted by $\overline{\Omega}$. The outer unit normal vector at each point of the boundary is denoted by $\vec{n} = (n_i)_{i=1,2,3}$.

A point $M$ of the crack is defined by two opposite points: the first one, called $M_+$, belongs to $L_+$, the other one, $M_-$, belongs to $L_-$. It follows that: $-\vec{n}(M_+) = \vec{n}(M_-) = \vec{y_2} = \vec{n}(M)$. The jump of the quantity $Q$ will be denoted by $[Q(M)] = Q(M_+) - Q(M_-)$. We define: $[Q]_F = \int_{L_+} Q(M_+)dS - \int_{L_-} Q(M_-)dS = \int_F [Q(M)]dS$. The components of the vector $[\vec{u}(M)]$ are called *the microscopic displacement jumps at the point $M$*. The components of the vector $[\vec{u}]_F$ are called *the macroscopic displacement jumps*.

The modelling is based on a realistic microscopic approach of the damage phenomenon and especially of the unilateral condition. Therefore the homogenisation of a body containing microcracks is analysed at the microscopic level. The shape of the unloaded microcracks is assumed to be planar. The microscopic analysis of the microcracks is carried out using this given geometry. Obviously, if the geometry is different, the result statements are no longer valid. However, the methodology used to build the modelling is general and could probably be applied to any geometry. Finally, it is important to indicate that no information is given about the size of cracks though using the framework of Damage Mechanics implicitly infers that they are small compared to those of the RVE. The considered microcracks are typically intralaminar microcracks that can be found in the plies of classic laminated composites.

## 9.3 Micro-macro transition for the unilateral effect of damage and the displacement jumps

Elasticity problems for a microcracked body have been studied for a long time and by many authors. Under the assumption of a continuum, exact analytical solutions of these problems exist. Without hypotheses, even for elementary geometries, exact analytical solutions do not often exist and numerical calculations have been usually used to obtain an estimation of the

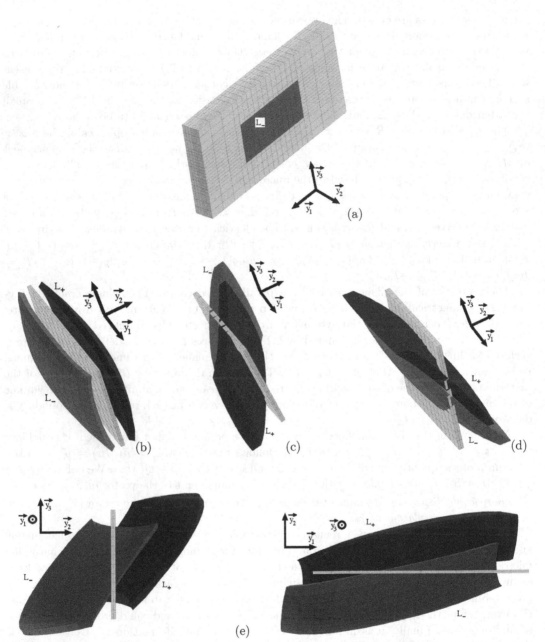

Figure 9.2: Illustration of the Crack Opening Mode at the microscopic scale. (a) Half of the elementary cell of the microcracked body (in red: the $L_-$ side of the crack). (b) $m_1 = 1$, $m_{23} = 2$: $[u_1]_F = 0$, $[u_2]_F > 0$ and $[u_3]_F = 0$ (this case corresponds to Mode I in Fracture Mechanics). (c) $m_1 = 2$, $m_{23} = 2$: $[u_1]_F \neq 0$, $[u_2]_F = 0$ and $[u_3]_F = 0$ (this case corresponds to Mode II in Fracture Mechanics). (d) $m_1 = 2$, $m_{23} = 3$: $[u_1]_F = 0$, $[u_2]_F = 0$ and $[u_3]_F \neq 0$ (this case corresponds to Mode III in Fracture Mechanics). (b-d) The macroscopic loadings which give these configurations are indicated by (b), (c), (d) in Table 9.3 and Table 9.4. (e) $m_1 = 1$, $m_{23} = 2.5$: $[u_1]_F \neq 0$, $[u_2]_F \neq 0$ and $[u_3]_F \neq 0$ obtained with the macroscopic loading ($\varepsilon_{11} = 1$, $\varepsilon_{23} = 1$, $\varepsilon_{12} = 1$) (this case corresponds to a mixed mode Mode I/Mode II/Mode III in Fracture Mechanics). (b-e) In grey, the plane of the unloaded crack.

solutions (Andrieux, 1983; Mauge and Kachanov, 1994; Hoenig, 1979; Hashin, 1985; Leguillon and Sanchez-Palencia, 1982; Eshelby, 1957; Muskhelishvili, 1953; Taya and Chou, 1981).

## 9.3.1 Definitions and hypotheses

To obtain the behaviour of the material equivalent to a periodic microcracked body, the problem to solve is written as a set of equations (Eqs. (9.31) to (9.35), § 9.8). Without friction, for a given damage state, Leguillon and Sanchez-Palencia (1982) demonstrate the hyperelastic character of the equivalent material behaviour and the uniqueness of the displacement solution of the problem.

The macroscopic strain is: $\varepsilon = \frac{1}{|\Omega|} \int_{\partial \Omega} \frac{1}{2}(u_i(M)n_j(M) + u_j(M)n_i(M))dS$. The average of the microscopic strain is: $\mathbf{E} = \frac{1}{|\Omega|} \int_{\overline{\Omega}} \mathbf{e}(M)dy$. We have:

$$\varepsilon = \mathbf{E} + \int_F [\vec{u}(M)] \otimes_s \vec{n}(M)dS = \mathbf{E} + \begin{pmatrix} 0 & [u_1]_F & 0 \\ [u_1]_F & [u_2]_F & [u_3]_F \\ 0 & [u_3]_F & 0 \end{pmatrix} \tag{9.1}$$

Then, the macroscopic description of a microcracked body requires the macroscopic strain and three other variables ($[u_i]_F$, $i = 1, 2, 3$). However, for a given damage state, the equivalent material behaviour law is hyperelastic. Thus:

$$\begin{cases} [\vec{u}]_F = \eta_\varepsilon(\varepsilon, \alpha) \text{ (strain formulation)} \\ [\vec{u}]_F = \eta_\sigma(\sigma, \alpha) \text{ (stress formulation)} \end{cases} \tag{9.2}$$

The next step of the modelling is to write an explicit form for these relations. Two definitions should be given beforehand. The first one ($D1$) defines a micro-macro transition for the unilateral condition. The second one ($D2$) defines the role of the functions $\eta_\varepsilon$ and $\eta_\sigma$.

**Definition ($D1$)** - The unilateral contact is checked at the macroscopic level if it is checked at the first order at the microscopic level.

**Definition ($D2$)** - The geometrical state of the microcrack network is known at the macroscopic level if the quantities $([u_i]_F)_{i=1,2,3}$ are known. More precisely: as for a given damage state the macroscopic behaviour is hyperelastic, $([u_i]_F)_{i=1,2,3}$ is a function of the macroscopic strain or stress. So the geometrical state of the microcrack network is known at the macroscopic level if the functions $\eta_\varepsilon$ or $\eta_\sigma$ (Eq. 9.2) are known.

## 9.3.2 Macroscopic formulation of the unilateral condition

To find the expressions of $\eta_\varepsilon(\varepsilon, \alpha)$ and $\eta_\sigma(\sigma, \alpha)$ (Eq. 9.2) (for a given elementary cell and a given material), the results of the preceding authors are combined with a numerical resolution of equations (Eqs. (9.31) to (9.35)). It can be concluded that in the strain (respectively, stress) space, the boundary between the configurations for which the microcracks lips are open ($[u_2]_F > 0$) or closed ($[u_2]_F = 0$) is a plane. The equation of this plane is: $F_\varepsilon(\varepsilon_{11}, \varepsilon_{22}, \varepsilon_{33}) = k_a \varepsilon_{22} + k_b \varepsilon_{11} + k_c \varepsilon_{33} = 0$, (respectively, $F_\sigma(\sigma_{11}, \sigma_{22}, \sigma_{33}) = k_p \sigma_{22} + k_q \sigma_{11} + k_r \sigma_{33} = 0$). For the strain (respectively, stress) formulation:

- the crack surfaces are open if: $F_\varepsilon(\varepsilon_{11}, \varepsilon_{22}, \varepsilon_{33}) > 0$ (respectively, $F_\sigma(\sigma_{11}, \sigma_{22}, \sigma_{33}) > 0$);
- the crack surfaces are closed if: $F_\varepsilon(\varepsilon_{11}, \varepsilon_{22}, \varepsilon_{33}) \leq 0$ (respectively, $F_\sigma(\sigma_{11}, \sigma_{22}, \sigma_{33}) \leq 0$).

$k_a$, $k_b$ and $k_c$ (respectively, $k_p$, $k_q$ and $k_r$) are constants depending on the geometrical characteristics of the cell and on the material (§ 9.8).

### 9.3.3   Macroscopic formulation of the displacement jumps

The uniqueness of the displacement solution (up to an additive constant vector) and the linearity of the equations Equation (9.31) to Equation (9.35) give $[\vec{u}]_F$ (Equation (9.3) and Equation (9.4)).

$$[\vec{u}]_F = \eta_\varepsilon(\varepsilon, \alpha) = \left( \begin{array}{l} [u_1]_F = k_1^\varepsilon \varepsilon_{12} \\ [u_2]_F = \left\{ \begin{array}{l} k_2^\varepsilon F_\varepsilon(\varepsilon_{11}, \varepsilon_{22}, \varepsilon_{33}) \text{ if } F_\varepsilon(\varepsilon_{11}, \varepsilon_{22}, \varepsilon_{33}) > 0 \\ 0 \text{ if } F_\varepsilon(\varepsilon_{11}, \varepsilon_{22}, \varepsilon_{33}) \leq 0 \end{array} \right. \\ [u_3]_F = k_3^\varepsilon \varepsilon_{23} \end{array} \right) \qquad (9.3)$$

$$[\vec{u}]_F = \eta_\sigma(\sigma, \alpha) = \left( \begin{array}{l} [u_1]_F = k_1^\sigma \sigma_{12} \\ [u_2]_F = \left\{ \begin{array}{l} k_2^\sigma F_\sigma(\sigma_{11}, \sigma_{22}, \sigma_{33}) \text{ if } F_\sigma(\sigma_{11}, \sigma_{22}, \sigma_{33}) > 0 \\ 0 \text{ if } F_\sigma(\sigma_{11}, \sigma_{22}, \sigma_{33}) \leq 0 \end{array} \right. \\ [u_3]_F = k_3^\sigma \sigma_{23} \end{array} \right) \qquad (9.4)$$

The quantities $(k_i^\varepsilon)_{i=1,2,3}$ (respectively, $(k_i^\sigma)_{i=1,2,3}$) are constants depending on the geometrical characteristics of the cell and on the material.

## 9.4   Equivalent material symmetry group, induced anisotropy and consequences

The material symmetry group of the unloaded microcracked material is $S_8$. Results found in the literature (Eshelby, 1957; Muskhelishvili, 1953; Taya and Chou, 1981) and the numerical resolution of equations (Eqs. (9.31) to (9.35)) (§ 9.8, Table 9.3, Table 9.4) show that the deformed geometry of the cell, induced by axial or $(\vec{y_1}, \vec{y_3})$ shear loading, keep the same material symmetry group. In the case of $(\vec{y_2}, \vec{y_3})$ (respectively, $(\vec{y_1}, \vec{y_2})$) shear loading, the material symmetry group of the equivalent material is $S_4^{23} = \{T_0, T_7, T_1, T_4\}$ (respectively, $S_4^{12} = \{T_0, T_7, T_3, T_6\}$).

The Theory of Invariants gives invariants under a given symmetry group for a given set of tensorial variables (Thionnet et al., 2003; Thionnet and Martin, 2006). Then, the material symmetry group which will be used in the modelling needs to be able to take into account all types of anisotropy induced by all possible changes in the cracks geometry. For this reason, the intersection between the material symmetry groups existing for the different loadings needs to be settled in order to construct the most general model. Thus, the material symmetry group used in the following is $S_2 = \{T_0, T_7\}$. The material symmetry group is thus determined once and for all.

This group expresses that the modelling must be unaffected by the change of sign in the vectorial quantities that can be used. Thus, Definition $(D2)$ can be expressed in the form $(D2')$.

**Definition** $(D2')$ - The geometrical state of the network of microcracks is known at the macroscopic level if the quantities $(\|[u_i]_F\|)_{i=1,2,3}$ are known.

## 9.5   Definition of the Crack Opening Mode of a microcracked body in the framework of Damage Mechanics

### 9.5.1   Extension of the Crack Opening Mode of Fracture Mechanics

In the framework of Fracture Mechanics, the geometry of a plane crack, at the point $M$, is defined by using a concept which can take three different values. In order to explain this concept, the two fronts of the crack should be identified. Here (Figure 9.2), they are chosen as the segments of the crack which are parallel to the $(\vec{y_3})$ axis. Then:

- if the crack faces move in opposite directions and perpendicularly to the plane of the crack, the crack is in Mode I (opening mode, Figure 9.2). Here, this case corresponds to: $[u_1(M)] = 0$, $[u_2(M)] > 0$, $[u_3(M)] = 0$;
- if the crack faces move on the same plane and perpendicularly to the fronts of the crack, the crack is in Mode II (sliding mode, Figure 9.2). Here, this case corresponds to: $[u_1(M)] \neq 0$, $[u_2(M)] = 0$, $[u_3(M)] = 0$;
- if the crack faces move on the same plane and parallel to the fronts of the crack, the crack is in Mode III (out-of-plane shearing or tearing mode, Figure 9.2). Here, this case corresponds to: $[u_1(M)] = 0$, $[u_2(M)] = 0$, $[u_3(M)] \neq 0$.

This concept, defined for Fracture Mechanics, is now extended to Damage Mechanics. Scalar state variables will be used as indicators of the opening mode of a network of microcracks. More precisely, these variables will describe the geometrical state of the microcrack network (i.e., the geometrical state of the crack in the periodic cell) at the macroscopic level. In particular, one of these variables will evolve from 2 to 3 to describe all possible geometrical states of a crack between Mode II and Mode III. Furthermore (§ 9.6) these variables will be included in a model of a hyperelastic behaviour law for a microcracked composite.

Two of these variables are denoted by $m_1$ and $m_{23}$ and called *mode 1* and *mode 2-3*. They represent, for Damage Mechanics, the notion of Crack Opening Mode as it exists in Fracture Mechanics. In reference to the usual language, they are defined in the following manner:

- whatever the displacement jumps along the $(\vec{y_1})$ and $(\vec{y_3})$ axes, $m_1$ is equal to 1 if the displacement jump along the $(\vec{y_2})$ axis is not zero, and equal to 2 if it is zero. This definition is the extension of the definition of Mode I;
- whatever the displacement jumps along the $(\vec{y_2})$ axis:

  . $m_{23}$ is equal to 2 if the only non-zero displacement jump is parallel to the $(\vec{y_1})$ axis;

  . $m_{23}$ is equal to 3 if the only non-zero displacement jump is parallel to the $(\vec{y_3})$ axis;

  . $m_{23}$ is equal to any intermediate value in mixed cases.

  This definition is the extension of the definition of Mode II and Mode III.

Two other variables are defined. They are noted $r_1$ and $r_{23}$ and are called *radius 1* and *radius 2-3*. They describe the space between the lips of the microcracks.

The set $(m_1, m_{23}, r_1, r_{23})$ defines the Crack Opening Mode of the microcracks network.

In the following, possible expressions for these variables as functions of $([u_i]_F)_{i=1,2,3}$ are identified. Then by using previous studies (Aussedat et al., 1995; Thionnet and Renard, 1999; Thionnet, 2001), expressions of these variables as functions of the macroscopic strain or stress will be obtained.

## 9.5.2 Expression of the mode and radius variables using macroscopic displacement jumps

Using the macroscopic displacement jumps, $m_1$, $m_{23}$, $r_1$ and $r_{23}$ are written in the following way:

$$m_1 = m_1^u([\vec{u}]_F) = \begin{cases} 1 \text{ if } [u_2]_F > 0 \\ 2 \text{ if } [u_2]_F = 0 \end{cases} \tag{9.5}$$

$$m_{23} = m_{23}^u([\vec{u}]_F) = \begin{cases} \frac{2[u_1]_F^2 + 3[u_3]_F^2}{[u_1]_F^2 + [u_3]_F^2} \text{ if } [u_3]_F \neq 0 \\ 2 \text{ if } [u_3]_F = 0 \end{cases} \tag{9.6}$$

$$r_1 = r_1^u([\vec{u}]_F) = \begin{cases} \sqrt{[u_2]_F^2} \text{ if } [u_2]_F > 0 \\ 0 \text{ if } [u_2]_F = 0 \end{cases} \tag{9.7}$$

$$r_{23} = r_{23}^u([\vec{u}]_F) = \sqrt{[u_1]_F^2 + [u_3]_F^2} \tag{9.8}$$

### 9.5.3   Expression of the mode and radius variables using macroscopic strain or stress

Taking into account the expressions of the displacement jumps as a function of the macroscopic strain (§ 9.3.3), we have:

$$m_1 = m_1^\varepsilon(\varepsilon, \alpha) = \begin{cases} 1 \text{ if } \varepsilon_{22} + k_b\varepsilon_{11} + k_c\varepsilon_{33} > 0 \\ 2 \text{ if } \varepsilon_{22} + k_b\varepsilon_{11} + k_c\varepsilon_{33} \leq 0 \end{cases} \tag{9.9}$$

$$m_{23} = m_{23}^\varepsilon(\varepsilon, \alpha) = \begin{cases} \dfrac{2\frac{\varepsilon_{12}^2}{\varepsilon_{12}^c(\alpha)^2} + 3\frac{\varepsilon_{23}^2}{\varepsilon_{23}^c(\alpha)^2}}{\frac{\varepsilon_{12}^2}{\varepsilon_{12}^c(\alpha)^2} + \frac{\varepsilon_{23}^2}{\varepsilon_{23}^c(\alpha)^2}} \text{ if } \varepsilon_{23} \neq 0 \\ 2 \text{ if } \varepsilon_{23} = 0 \end{cases} \tag{9.10}$$

$$r_1 = r_1^\varepsilon(\varepsilon, \alpha) = \begin{cases} \sqrt{\dfrac{(\varepsilon_{22} + k_b\varepsilon_{11} + k_c\varepsilon_{33})^2}{\varepsilon_{22}^c(\alpha)^2}} \text{ if } \varepsilon_{22} + k_b\varepsilon_{11} + k_c\varepsilon_{33} > 0 \\ 0 \text{ if } \varepsilon_{22} + k_b\varepsilon_{11} + k_c\varepsilon_{33} \leq 0 \end{cases} \tag{9.11}$$

$$r_{23} = r_{23}^\varepsilon(\varepsilon, \alpha) = \sqrt{\dfrac{\varepsilon_{12}^2}{\varepsilon_{12}^c(\alpha)^2} + \dfrac{\varepsilon_{23}^2}{\varepsilon_{23}^c(\alpha)^2}} \tag{9.12}$$

$\varepsilon_{22}^c(\alpha)$, $\varepsilon_{12}^c(\alpha)$ and $\varepsilon_{23}^c(\alpha)$ are the failure strains of the equivalent material depending on the damage variable.

Taking into account the expressions of the displacement jumps as a function of the macroscopic stress (§ 9.3.3), we have:

$$m_1 = m_1^\sigma(\sigma, \alpha) = \begin{cases} 1 \text{ if } \sigma_{22} > 0 \\ 2 \text{ if } \sigma_{22} \leq 0 \end{cases} \tag{9.13}$$

$$m_{23} = m_{23}^\sigma(\sigma, \alpha) = \begin{cases} \dfrac{2\frac{\sigma_{12}^2}{\sigma_{12}^c(\alpha)^2} + 3\frac{\sigma_{23}^2}{\sigma_{23}^c(\alpha)^2}}{\frac{\sigma_{12}^2}{\sigma_{12}^c(\alpha)^2} + \frac{\sigma_{23}^2}{\sigma_{23}^c(\alpha)^2}} \text{ if } \sigma_{23} \neq 0 \\ 2 \text{ if } \sigma_{23} = 0 \end{cases} \tag{9.14}$$

$$r_1 = r_1^\sigma(\sigma, \alpha) = \begin{cases} \sqrt{\dfrac{\sigma_{22}^2}{\sigma_{22}^c(\alpha)^2}} \text{ if } \sigma_{22} > 0 \\ 0 \text{ if } \sigma_{22} \leq 0 \end{cases} \tag{9.15}$$

$$r_{23} = r_{23}^\sigma(\sigma, \alpha) = \sqrt{\dfrac{\sigma_{12}^2}{\sigma_{12}^c(\alpha)^2} + \dfrac{\sigma_{23}^2}{\sigma_{23}^c(\alpha)^2}} \tag{9.16}$$

$\sigma_{22}^c(\alpha)$, $\sigma_{12}^c(\alpha)$ and $\sigma_{23}^c(\alpha)$ are the failure stresses of the equivalent material depending on the damage variable.

Table 9.2 presents the characteristic cases of the Crack Opening Mode for Damage Mechanics and the correspondence with Fracture Mechanics. Each of these cases defines an infinity of possible states. For instance, Case 6 defines all intermediate states between Mode II and Mode III.

## 9.6　Application: a general hyperelastic behaviour law for a microcracked composite

The concept of Crack Opening Mode for Damage Mechanics is applied here with the objective of building a general hyperelastic behaviour law for a microcracked composite. It will be pointed out that this concept allows the conditions $(c1 - c7)$ to be easily enforced.

### 9.6.1　Framework of the model

The hypotheses used in this model are the following:

- at the microscopic level, the considered material is damaged by a network of identical plane microcracks which are guided by the material constituents and not by the loading. The microstructure of the material is assumed to be periodic. The undamaged material is orthotropic. When the material is unloaded, the geometry of the microcracks respects this orthotropic symmetry;

- the microcracks appear instantaneously and then, the damage evolves by multiplication of microcracks;

- there is no friction for closed microcracks surfaces;

- the appearance of the microcracks does not induce a residual stress field;

- the dissipation created by the appearance of the microcracks is small. Thus, the temperature elevation during any transformation is assumed to be insignificant. The temperature is then assumed uniform and constant with time;

- we work in the framework of Damage Mechanics. Thus, the damaged material will be replaced by an equivalent one. $R$ is assumed to be the principal orthotropic frame (§ 9.2) of this material;

- Small Perturbations Hypothesis;

- the model is built up in the framework of the Thermodynamics of Continuous Media using the Local State Hypothesis;

- the damage phenomenon is time-independent;

- contrary to previous works, the model is carried out without limitations to strain or stress tensor components. More precisely, the plane strain or stress hypothesis is not used;

- the material symmetry group of the equivalent material is $S_2 = \{T_0, T_7\}$.

The strain tensor $\varepsilon$ is taken as a state variable. Its associated variable is the stress tensor $\sigma$. Four scalar variables are used to model the damage: $\alpha$ defines quantitatively the phenomenon, $m_{23}$, $r_1$ and $r_{23}$ define the geometrical state of the microcracks. Their associated variables are respectively denoted by $A$, $M_{23}$, $R_1$ and $R_{23}$.

### 9.6.2　Vectorial description of the damage phenomenon

By reference to previous works (Talreja, 1985; Thionnet, 2001; Thionnet and Renard, 1999), the damage phenomenon will be modelled using a vector. This vector has three non-zero components that can evolve with the damage state. Even though $[\vec{u}]_F$ was not chosen, it could have been. Instead, the vector used has the following form: $\vec{V} = (V_i(\alpha, m_{23}, r_1, r_{23}))_{i=1,2,3}$. In order to obtain a model coherent with the definition of $m_{23}$, $r_1$ and $r_{23}$, the direction of $\vec{V}$ does not implicitly depend on the number of cracks. There follows: $\vec{V} = f(\alpha)\vec{U}(m_{23}, r_1, r_{23})$. To obtain a coherent model with the definition of $m_{23}$, $r_1$ and $r_{23}$, the vector $\vec{U}$ should be built such that:

$$\vec{U} = \begin{pmatrix} U_1(m_{23}, r_1, r_{23}) = U_{T1}(m_{23}, r_1, r_{23}) \begin{cases} = 0 \text{ if } r_{23} = 0 \,(\Leftrightarrow m_{23} = 2), \forall r_1 \ (a) \\ = 0 \text{ if } m_{23} = 3, \forall r_{23} \neq 0, \forall r_1 \ (b) \\ \neq 0 \text{ if } m_{23} = 2, \forall r_{23} \neq 0, \forall r_1 \ (c) \\ \neq 0 \text{ if } 2 < m_{23} < 3, \forall r_{23} \neq 0, \forall r_1 \ (d) \end{cases} \\ U_2(m_{23}, r_1, r_{23}) = U_N(m_{23}, r_1, r_{23}) \begin{cases} = 0 \text{ if } r_1 = 0, \forall m_{23}, \forall r_{23} \ (e) \\ \neq 0 \text{ if } r_1 \neq 0, \forall m_{23}, \forall r_{23} \ (f) \end{cases} \\ U_3(m_{23}, r_1, r_{23}) = U_{T3}(m_{23}, r_1, r_{23}) \begin{cases} = 0 \text{ if } r_{23} = 0 \,(\Leftrightarrow m_{23} = 2), \forall r_1 \ (g) \\ = 0 \text{ if } m_{23} = 2, \forall r_{23} \neq 0, \forall r_1 \ (h) \\ \neq 0 \text{ if } m_{23} = 3, \forall r_{23} \neq 0, \forall r_1 \ (i) \\ \neq 0 \text{ if } 2 < m_{23} < 3, \forall r_{23} \neq 0, \forall r_1 \ (j) \end{cases} \end{pmatrix}$$

(9.17)

The justification of the conditions $(a - j)$ can be made in the following way:

- $(a)$ and $(g)$: $r_{23} = 0$ and Equation (9.8) indicate that tangential displacement jumps should be zero and $m_{23} = 2$ (Eq. 9.6). Then, in this case, $\vec{U}$ should be built so that: $U_1 = 0$ and $U_3 = 0$;
- $(b)$ and $(i)$: Equation (9.6) indicates that $m_{23} = 3$ can be obtained if and only if $[u_3]_F \neq 0$ and $[u_1]_F = 0$ (and then $r_{23} \neq 0$). Then, in this case, $\vec{U}$ should be built such that: $U_1 = 0$ and $U_3 \neq 0$;
- $(h)$ and $(c)$: Equation (9.6) indicates that $m_{23} = 2$ can be obtained if and only if $[u_3]_F = 0$. In addition, if one assumes $r_{23} \neq 0$, Equation (9.8) shows that $[u_1]_F \neq 0$. Then, in this case, $\vec{U}$ should be built such that: $U_1 \neq 0$ and $U_3 = 0$;
- $(d)$ and $(j)$: these cases are in-between the two preceding cases;
- $(e)$ and $(f)$: $r_1 = 0$ and Equation (9.7) indicate that the normal displacement jump $[u_2]_F$ should be zero. Then, in this case, $\vec{U}$ should be built such that: $U_2 = 0$ if $r_1 = 0$ and $U_2 \neq 0$ if $r_1 \neq 0$.

These conditions induce $(c8 - c12)$:

- $U_{T1}(m_{23} = 2, r_1, r_{23} = 0) = 0, \forall r_1 \ (c8)$;
- $U_{T3}(m_{23} = 2, r_1, r_{23} = 0) = 0, \forall r_1 \ (c9)$;
- $U_N(m_{23}, r_1 = 0, r_{23}) = 0, \forall r_{23}, \forall m_{23} \ (c10)$;
- $U_{T1}(m_{23} = 3, r_1, r_{23}) = 0, \forall r_{23} \neq 0, \forall r_1 \ (c11)$;
- $U_{T3}(m_{23} = 2, r_1, r_{23}) = 0, \forall r_{23} \neq 0, \forall r_1 \ (c12)$.

If the conditions $(c8 - c12)$ are satisfied, the condition $(c4)$ is checked. We will see in the following that these conditions will be also used to verify the continuity of the stress/strain relation.

### 9.6.3 Building of the objective state function

For convenience of notation, we state: $m = (m_{23}, r_1, r_{23})$. The state function is written $\psi = \psi(\varepsilon, \alpha, m) = \varphi(\varepsilon, \vec{V}(\alpha, m))$ in a way to obtain a damageable elastic behaviour law. Then: $\varphi(\varepsilon, \vec{V}) = \varphi_{20}(\varepsilon) + \varphi_{22}(\varepsilon, \vec{V}) + \overline{\Phi}(\vec{V})$ where $\varphi_{20}$ and $\varphi_{22}$ are polynomial functions. The subscripts indicate the partial degree respectively associated with $\varepsilon$ and $\vec{V}$. The function $\overline{\Phi}$ is a function which should be $S_2$-invariant, objective and zero when the material is unloaded. Here, the explicit form of $\overline{\Phi}$ is not necessary: thus, it does not need to be given.

Applying the Noether Theorem (Thionnet et al., 2003; Thionnet and Martin, 2006) leads to expressing an integrity basis of the $S_2$-invariant polynomial functions of $\varepsilon$ and $\vec{V}$:

$$F_2 = \{\varepsilon_{11}, \varepsilon_{22}, \varepsilon_{33}, \varepsilon_{23}, \varepsilon_{13}, \varepsilon_{12}, V_1^2, V_2^2, V_3^2, V_2 V_3, V_1 V_3, V_1 V_2\}$$

The most general form of $\varphi$ is then:

$$\varphi(\varepsilon, \vec{V}) = \frac{1}{2} \sum_{\substack{i,j,k,h=1 \\ i \leq j, k \leq h}}^{3} C_{ijkh}^{0} \varepsilon_{ij} \varepsilon_{kh} + \frac{1}{2} \sum_{\substack{i,j,k,h,p,q=1 \\ i \leq j, k \leq h, p \leq q}}^{3} D_{ijkhpq} \varepsilon_{ij} \varepsilon_{kh} V_p V_q + \overline{\Phi}(\vec{V}) \qquad (9.18)$$

The polynomial part of $\varphi$ is $S_2$-invariant but, at this time, not defined as an objective function. Then we assume that $C^0 = (C_{ijkh}^0)_{i,j,k,h=1,2,3}$ and $D = (D_{ijkhpq})_{i,j,k,h,p,q=1,2,3}$ should be respectively tensors of fourth and sixth order. The duality between these tensors and the quantity they are associated with, leads $\varphi$ to be an objective function (Thionnet et al., 2003; Thionnet and Martin, 2006). By using the Voigt Notations and the state variables, $\psi(\varepsilon, \alpha, m)$ takes the following form:

$$\psi(\varepsilon, \alpha, m) = \psi^0(\varepsilon) + \sum_{\substack{p,q=1 \\ p \leq q}}^{3} \psi^{(pq)}(\varepsilon, V_p(\alpha, m), V_q(\alpha, m)) + \overline{\Phi}(\vec{V}(\alpha, m)) \qquad (9.19)$$

$$\begin{cases} \psi^{(0)}(\varepsilon) = F^{(0)}(\varepsilon) = \frac{1}{2} C_{ij}^0 \varepsilon_i \varepsilon_j \\ \psi^{(pq)}(\varepsilon, V_p(\alpha, m), V_q(\alpha, m)) = F^{(pq)}(\varepsilon) f^2(\alpha) U_p(m) U_q(m) \\ F^{(pq)}(\varepsilon) = \frac{1}{2} D_{ij}^{(pq)} \varepsilon_i \varepsilon_j \end{cases} \qquad (9.20)$$

This general method to obtain the state function satisfies the conditions $(c2, c5, c6)$.

## 9.6.4  Laws of state

We have:

$$d\psi(\varepsilon, \alpha, m) = \sigma(\varepsilon, \alpha, m) d\varepsilon + A(\varepsilon, \alpha, m) d\alpha + M_{23}(\varepsilon, \alpha, m) dm_{23} \ldots$$

$$\ldots + R_1(\varepsilon, \alpha, m) dr_1 + R_{23}(\varepsilon, \alpha, m) dr_{23} = \frac{\partial \psi(\varepsilon, \alpha, m)}{\partial \varepsilon} d\varepsilon + \frac{\partial \psi(\varepsilon, \alpha, m)}{\partial \alpha} d\alpha + \frac{\partial \psi(\varepsilon, \alpha, m)}{\partial m_{23}} dm_{23} \ldots$$

$$\ldots + \frac{\partial \psi(\varepsilon, \alpha, m)}{\partial r_1} dr_1 + \frac{\partial \psi(\varepsilon, \alpha, m)}{\partial r_{23}} dr_{23} \qquad (9.21)$$

Thus, the laws of state are:

$$\sigma_i = \left( C_{ij}^0 + f^2(\alpha) \sum_{\substack{p,q=1 \\ p \leq q}}^{3} D_{ij}^{(pq)} U_p(m) U_q(m) \right) \varepsilon_j = C(\alpha, m) \varepsilon_j \qquad (9.22)$$

$$A(\varepsilon, \alpha, m) = f(\alpha) f'(\alpha) \left( \sum_{\substack{p,q=1 \\ p<q}}^{3} D_{ij}^{(pq)} U_p(m) U_q(m) \right) \varepsilon_i \varepsilon_j + \frac{\partial \Phi}{\partial \alpha}(\alpha, m) \qquad (9.23)$$

$$M_{23}(\varepsilon, \alpha, m) = \frac{\partial \Phi}{\partial m_{23}}(\alpha, m) + \frac{1}{2} f^2(\alpha) \sum_{\substack{p,q=1 \\ p<q}}^{3} D_{ij}^{(pq)} \left( \frac{\partial U_p}{\partial m_{23}}(m) U_q(m) + U_p(m) \frac{\partial U_q}{\partial m_{23}}(m) \right) \varepsilon_i \varepsilon_j \qquad (9.24)$$

$$R_1(\varepsilon, \alpha, m) = \frac{\partial \Phi}{\partial r_1}(\alpha, m) + \frac{1}{2} f^2(\alpha) \sum_{\substack{p,q=1 \\ p<q}}^{3} D_{ij}^{(pq)} \left( \frac{\partial U_p}{\partial r_1}(m) U_q(m) + U_p(m) \frac{\partial U_q}{\partial r_1}(m) \right) \varepsilon_i \varepsilon_j \qquad (9.25)$$

$$R_{23}(\varepsilon, \alpha, m) = \frac{\partial \Phi}{\partial r_{23}}(\alpha, m) + \frac{1}{2} f^2(\alpha) \sum_{\substack{p,q=1 \\ p<q}}^{3} D_{ij}^{(pq)} \left( \frac{\partial U_p}{\partial r_{23}}(m) U_q(m) + U_p(m) \frac{\partial U_q}{\partial r_{23}}(m) \right) \varepsilon_i \varepsilon_j$$

$$(9.26)$$

### 9.6.5 Damage evolution law

The variables $m_{23}$, $r_1$ and $r_{23}$ are not dissipative. The Clausius-Duhem Inequality becomes: $-A(\varepsilon, \alpha, m)\dot{\alpha} \geq 0, \forall \dot{\alpha} \geq 0$. A classical way to obtain the evolution law of $\alpha$ is to use a scalar criterion with the following form: $c(\varepsilon, \alpha, m) = A_c(\alpha, m) - A(\varepsilon, \alpha, m)$. Then, the consistency hypothesis:

$$\begin{cases} c(\varepsilon, \alpha, m) = 0 \\ dc(\varepsilon, \alpha, m) = 0 \end{cases} \qquad (9.27)$$

which can be written during the damage process leads to the evolution law:

$$\begin{cases} d\alpha = \frac{N}{\frac{\partial A_c(\alpha, m)}{\partial \alpha} - \frac{\partial^2 \psi(\varepsilon, \alpha, m)}{\partial \alpha^2}} \\ N = \frac{\partial^2 \psi(\varepsilon, \alpha, m)}{\partial \alpha \partial \varepsilon} d\varepsilon + \left( \frac{\partial^2 \psi(\varepsilon, \alpha, m)}{\partial \alpha \partial m_{23}} - \frac{\partial A_c(\alpha, m)}{\partial m_{23}} \right) dm_{23} + \cdots \\ \cdots \left( \frac{\partial^2 \psi(\varepsilon, \alpha, m)}{\partial \alpha \partial r_1} - \frac{\partial A_c(\alpha, m)}{\partial r_1} \right) dr_1 + \left( \frac{\partial^2 \psi(\varepsilon, \alpha, m)}{\partial \alpha \partial r_{23}} - \frac{\partial A_c(\alpha, m)}{\partial r_{23}} \right) dr_{23} \end{cases} \qquad (9.28)$$

$A_c(\alpha, m)$ is the damage threshold. There is no evolution of damage if $c(\varepsilon, \alpha, m) < 0$.

### 9.6.6 Application to the microcracking phenomenon appearing in carbon/epoxy laminates

The proposed modelling (in the case of the stress formulation, Eqs. (9.13) to (9.16) is applied for the microcracking phenomenon which appears in carbon/epoxy laminates. The concept of stacking disoriented plies to form a laminated structure leads to the development of a model for the unidirectional ply. A multiscale procedure between the laminate and the ply allows the mechanical response of any laminate subjected to any multiaxial loading to be calculated (Thionnet and Renard, 1999). The local frame of the unidirectional material is defined by $(\vec{y_1}, \vec{y_2}, \vec{y_3})$ for which the $(\vec{y_1})$ axis is aligned with fibres and the $(\vec{y_3})$ axis defines the thickness of the ply. The plane of the ply (and of the laminate) is then the $(\vec{y_1}, \vec{y_2})$ plane. The loadings are assumed to be applied on the plane of the laminate. Typical observations of intralaminar cracks in the 90° ply of a $(0°, 90°, 0°)$ laminate, made under $\vec{y_2}$-traction loading (50% of the failure stress of the specimen), show (Figure 9.1):

- the microcracks are located in the plane $(\vec{y_1}, \vec{y_3})$;
- the microcracks occur all along the height of the cracked ply, and all along the width of the specimen;
- the normal displacement jump is weak compared to sizes of the crack.

It can also be observed that in the case of more complex loadings, these characteristics are preserved. This last property mainly characterises the materials showing strong anisotropy. Contrary to isotropic materials their damage is guided by the applied loading. So for this particular material, we assume:

- the anisotropy induced by the microcracks geometry changes can be neglected (Hypothesis ($H1$));

- the influence of the set $(m_{23}, r_1, r_{23})$ can be neglected in the expression of $C(\alpha, m_{23}, r_1, r_{23})$ except in the vicinity of the deactivation states (Hypothesis $(H2)$). This last remark is very important: it allows the continuity of the behaviour law to be obtained (conditions $(c8 - c12)$).

The damage variable $\alpha$ is defined by: $\alpha = e \times d$ where $e$ is the thickness of the damaged ply and $d$ the microcracks density $(nb/mm)$ (Thionnet and Renard, 1993).

The material considered here is a typical carbon/epoxy composite the properties of which are assumed to be known (but they are not given because of industrial confidentiality). The numerical values obtained in the following identification process are calculated with these properties.

As a consequence of the hypotheses $(H1 - H2)$, it can be assumed that the tensor $C(\alpha, m_{23}, r_1, r_{23})$ has the same form as the tensor $C^0$:

$$C(\alpha, m_{23}, r_1, r_{23}) = C^0 + f^2(\alpha)U_N^2(r_1)\begin{pmatrix} D_{11}^{(22)} & D_{12}^{(22)} & D_{13}^{(22)} & 0 & 0 & 0 \\ D_{12}^{(22)} & D_{22}^{(22)} & D_{23}^{(22)} & 0 & 0 & 0 \\ D_{13}^{(22)} & D_{23}^{(22)} & D_{33}^{(22)} & 0 & 0 & 0 \\ 0 & 0 & 0 & 0 & 0 & 0 \\ 0 & 0 & 0 & 0 & 0 & 0 \\ 0 & 0 & 0 & 0 & 0 & 0 \end{pmatrix} + \ldots$$

$$\ldots f^2(\alpha)\begin{pmatrix} 0 & 0 & 0 & 0 & 0 & 0 \\ 0 & 0 & 0 & 0 & 0 & 0 \\ 0 & 0 & 0 & 0 & 0 & 0 \\ 0 & 0 & 0 & D_{44}^{(33)}U_{T3}^2(m_{23}, r_{23}) & 0 & 0 \\ 0 & 0 & 0 & 0 & 0 & 0 \\ 0 & 0 & 0 & 0 & 0 & D_{66}^{(11)}U_{T1}^2(m_{23}, r_{23}) \end{pmatrix} \tag{9.29}$$

$$\vec{U}(m_{23}, r_1, r_{23}) = \begin{pmatrix} U_{T1}(m_{23}, r_{23}) = (1 - e^{-X_T r_{23}})(1 - e^{-X_{T1}(3-m_{23})}) \\ U_N(r_1) = 1 - e^{-X_N r_1} \\ U_{T3}(m_{23}, r_{23}) = (1 - e^{-X_T r_{23}})(1 - e^{-X_{T3}(m_{23}-2)}) \end{pmatrix} \tag{9.30}$$

where $X_T$, $X_N$, $X_{T1}$ and $X_{T3}$ are large positive real values (for instance, $X_T = X_N = X_{T1} = X_{T3} = 100$). By solving the equations Eqs. (9.31) to (9.35) for different values of $\alpha$ (e.g., for different sizes of the elementary cell), we obtain the curves of stiffness reduction (Thionnet and Renard, 1993). By smoothing all of these curves, the function $f(\alpha)$ and the coefficients are identified: $f(\alpha) = \sqrt{\frac{\alpha}{1+\alpha}}$ and, for instance, $D_{11}^{(22)} = 0$, $D_{22}^{(22)} \approx -1.9C_{22}^0$.

General and detailed indications about the way to identify the tensor $C$ and the damage evolution law can be found in Thionnet (2010).

Some typical cases can be used to illustrate the modelling (assuming $\alpha \neq 0$):

- the damaged material is unloaded. This case corresponds to: $r_1 = 0$, $r_{23} = 0$, $m_1 = 2$, $m_{23} = 2$. We get: $C(\alpha, m_{23} = 2, r_1 = 0, r_{23} = 0) = C^0$;
- the microcrack faces are opened but not sheared. This case corresponds to: $[u_1]_F = 0$, $[u_2]_F > 0$, $[u_3]_F = 0$, e.g., $r_1 \neq 0$, $r_{23} = 0$, $m_{23} = 2$. The behaviour law takes the following form:

$$C(\alpha, m_{23} = 2, r_1, r_{23} = 0) = C^0 + f^2(\alpha)U_N^2(r_1)\begin{pmatrix} D_{11}^{(22)} & D_{12}^{(22)} & D_{13}^{(22)} & 0 & 0 & 0 \\ D_{12}^{(22)} & D_{22}^{(22)} & D_{23}^{(22)} & 0 & 0 & 0 \\ D_{13}^{(22)} & D_{23}^{(22)} & D_{33}^{(22)} & 0 & 0 & 0 \\ 0 & 0 & 0 & 0 & 0 & 0 \\ 0 & 0 & 0 & 0 & 0 & 0 \\ 0 & 0 & 0 & 0 & 0 & 0 \end{pmatrix}$$

- the microcrack faces are closed and sheared. This case corresponds to the setting of the unilateral effect of damage for which $[u_1]_F \neq 0$, $[u_2]_F = 0$, $[u_3]_F \neq 0$, e.g., $r_1 = 0$, $r_{23} \neq 0$, $2 \leq m_{23} \leq 3$. The behaviour law takes the following form:

$$C(\alpha, m_{23}, r_1 = 0, r_{23}) = C^0 + f^2(\alpha) \begin{pmatrix} 0 & 0 & 0 & 0 & 0 & 0 \\ 0 & 0 & 0 & 0 & 0 & 0 \\ 0 & 0 & 0 & 0 & 0 & 0 \\ 0 & 0 & 0 & D_{44}^{(33)} U_{T3}^2(m_{23}, r_{23}) & 0 & 0 \\ 0 & 0 & 0 & 0 & 0 & 0 \\ 0 & 0 & 0 & 0 & 0 & D_{66}^{(11)} U_{T1}^2(m_{23}, r_{23}) \end{pmatrix}$$

The unilateral effect of damage is then correctly taken into account.

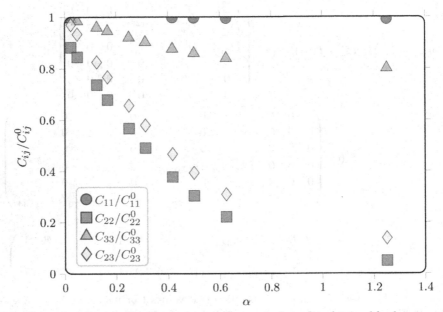

Figure 9.3: Several stiffness reductions versus $\alpha$ (numerical results obtained by homogenisation). (1) $\frac{C_{11}}{C_{11}^0}$. (2) $\frac{C_{22}}{C_{22}^0}$. (3) $\frac{C_{33}}{C_{33}^0}$. (4) $\frac{C_{23}}{C_{23}^0}$.

**Characteristic functioning and stability of the model**

To show the main characteristics of the model we study the responses of two stacking sequences (which one assumes never break): the first one is a $(90°)$ specimen (Figure 9.4 a), the second one is a $(45°)$ (Figure 9.4 b). The applied loading is first tensile and positive (Figure 9.4, O/A/B/C/B/D/O)) and secondly decreasing down to a negative value (Figure 9.4, O/E). Some comments can be found in Table 9.1 and the unilateral effect of damage is illustrated in the following way:

- in the case of the $(90°)$ specimen, the elastic slope of the undamaged material at (O) is equal to the restored slope (O/E) when the microcracks are closed again;
- in the case of the $(45°)$ specimen, the elastic slope of the undamaged material at (O) is not equal to the restored slope (O/E) when the microcracks closed again: the global stiffness of the specimen takes into account the damaged shear modulus which

is different from its initial value because the unilateral condition of damage does not affect this modulus.

We also observe that when damage reaches its maximum value (saturation, Figure 9.4 ab, A → B) in both cases, the model is not unstable. This property results from the fact that maximum decreases of the rigidity tensor components – as a function of damage – are not very important.

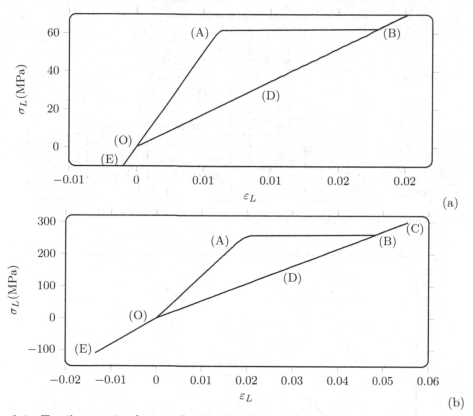

Figure 9.4: Tensile test simulations. Longitudinal stress ($\sigma_L$, MPa) versus longitudinal strain ($\varepsilon_L$). (a) (90°) specimen. (b) (45°) specimen.

## Comparison between experience and simulation

A tensile loading is applied on a $(40°_2, 90°_2, 40°_2)$ specimen (Figure 9.5) (Thionnet and Renard, 1993). The values of the Crack Opening Mode on the different plies are:

- $(m_1 = 1, m_{23} = 2, r_1 = 0.046, r_{23} = 0.272 \times 10^{-03})$ for the 40° plies;
- $(m_1 = 1, m_{23} = 2, r_1 = 0.053, r_{23} = 0.139 \times 10^{-04})$ for the 90° ply.

The evolution of the damage in each ply follows the damage threshold $A_c$ which corresponds to its own Crack Opening Mode. The difference between numerical and experimental curves comes from the fact that the only non-linearity introduced in the behaviour of the material is the non-linearity due to damage. The viscoelastic (or viscoplastic) character of the behaviour is not taken into account.

| Case of the (90°) specimen | | | | | | |
|---|---|---|---|---|---|---|
| *Points* | $r_1$ | $m_1$ | $r_{23}$ | $m_{23}$ | correspondence with Fracture Mechanics | Damage evolution |
| (O) | 0 | 2 | 0 | 2 | Mode II | 0 |
| (O/A/B) | $\neq 0$ | 1 | 0 | 2 | Mode I | ↗ |
| (B) | $\neq 0$ | 1 | 0 | 2 | Mode I | saturation |
| (B/C) | $\neq 0$ | 1 | 0 | 2 | Mode I | → |
| (C/B/D/O) | $\neq 0$ | 1 | 0 | 2 | Mode I | → |
| (O/E) | 0 | 2 | 0 | 2 | Mode II | → |
| Case of the (45°) specimen | | | | | | |
| *Points* | $r_1$ | $m_1$ | $r_{23}$ | $m_{23}$ | correspondence with Fracture Mechanics | Damage evolution |
| (O) | 0 | 2 | 0 | 2 | Mode II | 0 |
| (O/A/B) | $\neq 0$ | 1 | $\neq 0$ | 2 | mixed Mode I / Mode II | ↗ |
| (B) | $\neq 0$ | 1 | $\neq 0$ | 2 | mixed Mode I / Mode II | saturation |
| (B/C) | $\neq 0$ | 1 | $\neq 0$ | 2 | mixed Mode I / Mode II | → |
| (C/B/D/O) | $\neq 0$ | 1 | $\neq 0$ | 2 | mixed Mode I / Mode II | → |
| (O/E) | 0 | 2 | $\neq 0$ | 2 | Mode II | → |

**Table 9.1**   Comments for the two simulations.

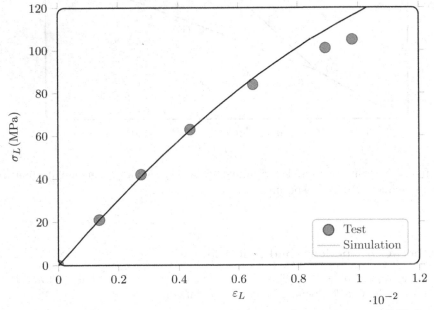

Figure 9.5:  Tensile loading applied to a $(40°_2, 90°_2)_s$ specimen. $\sigma_L$ (MPa) versus $\varepsilon_L$.

## 9.7   Conclusions

Many studies have been carried out in order to build a coherent macroscopic behaviour law for a composite containing microcracks. All of them are only partially coherent and none of them are complete. The shortcomings are generally due to the fact that these models do not take into account microscopic descriptions of the damage phenomena, especially the microscopic displacement jumps on the surfaces of the microcracks. The studies of Leguillon and Sanchez-Palencia (1982), at the microscopic level, give the most important results that can be used to solve the problems.

By using their results and those of some others taken from literature, which often discuss the homogenisation of microcracked bodies, the model has proposed a new concept for Damage Mechanics: the Crack Opening Mode as it exists for Fracture Mechanics. This concept uses 3 scalar variables to describe the geometrical state of the microcracks. Then, it becomes very simple to write a hyperelastic behaviour law for a microcracked body that will be coherent with the Principles of Continuum Mechanics and Thermodynamics as well as the Objectivity Principle and the damage activation/deactivation.

The notion of Crack Opening Mode has been already mentioned (Aussedat et al., 1995; Thionnet and Renard, 1999; Thionnet, 2001) made under the plane stress or strain hypothesis. However, here, a more general three-dimensional framework has been studied. The concept of Crack Opening Mode as it exists in the previous studies is not the 2D-restriction of the Crack Opening Mode that is defined here: it seems that by starting from this idea, it is not possible to solve simultaneously the continuity of the stress/strain relation at any point and the use of 3 variables to describe the damage phenomenon. A solution could exist but it was not found.

The generalisation of the model can easily be made for a finite number of microcrack networks. Thus, this formalism can be used for composite materials, for which the known geometry of the reinforcement is a favored place for the damage phenomena, independently from the loading directions.

The framework chosen herein assumes that the number and the direction of the damage phenomena that can appear are predetermined (guided by the constituents of the material). Within this framework, it has been possible to build the model of the hyperelastic microcracked body. However, for isotropic materials, the damage phenomena that can appear depend on the intensity and the direction of the loading. In this case, the number of vectors and their directions that must be taken into account in the modelling are unknown and it is not possible to use the Local State Hypothesis to build the model. Thus, before writing any model on this problem, it is necessary to find the mathematical concept that can be used to describe an infinite number of damage phenomena which can take into account the unilateral condition for each of them.

## 9.8 Appendix: crack surface displacement results calculated numerically on the unit cell of the microcracked body

To obtain the behaviour of the material equivalent of a periodic microcracked body, the problem to solve is written as a set of equations (Eqs. (9.31) to (9.35)):

- local equilibrium:

$$\vec{div}\,\mathbf{s}(M) = \vec{0},\ \forall M \in \overline{\Omega} \tag{9.31}$$

- behaviour law (orthotropic material):

$$s_{ij}(M) = C_{ijkh}e_{kh}(M) \tag{9.32}$$

- unilateral condition without friction on the microcracks surfaces:

$$\forall M \in F \begin{cases} [\vec{u}(M).\vec{n}(M)] \geq 0 \\ [\mathbf{s}(M)\vec{n}(M)] = \vec{0} \\ \mathbf{s}(M_-)\vec{n}(M).\vec{n}(M) < 0 \text{ if } [\vec{u}(M).\vec{n}(M)] = 0 \\ \mathbf{s}(M_-)\vec{n}(M).\vec{n}(M) = 0 \text{ if } [\vec{u}(M).\vec{n}(M)] > 0 \end{cases} \tag{9.33}$$

- in the case of a strain formulation:

$$\begin{cases} \mathbf{e}(M) \text{ periodic} \\ \frac{1}{|\Omega|} \int_{\Omega} \mathbf{e}(M) dy \text{ given} \end{cases} \qquad (9.34)$$

- in the case of a stress formulation:

$$\begin{cases} \mathbf{s}(M) \text{ periodic} \\ \frac{1}{|\Omega|} \int_{\Omega} \mathbf{s}(M) dy \text{ given} \end{cases} \qquad (9.35)$$

The boundary conditions are some kind of *Signorini's problem*. The uniqueness of the solutions $\mathbf{s}(M)$ and $\vec{u}(M)$ (up to an additive translation) can be shown (Duvaut and Lions, 1972; Leguillon and Sanchez-Palencia, 1982). Without friction, for a given damage state, Leguillon and Sanchez-Palencia (1982) demonstrate the hyperelastic character of the equivalent material behaviour.

For instance, one solves equations (Eqs. 9.31–9.35) for $a = b = c = 0.5$ mm, for an isotropic material (Young's modulus equal to 1000 Mpa, Poisson's ratio equal to 0.25) and the microcracks surfaces are square (0.215 mm × 0.215 mm). The detailed results are given in (Tables 9.20 and 9.3) in the case of the strain approach and in (Table 9.4) in the case of the stress approach. The values obtained (Table 9.5) are: $k_a = 1, k_b = k_c = \frac{1}{3}$ and $k_p = 1, k_q = k_r = 0$. The definition of the different columns of the two tables (Table 9.3, Table 9.4) are the followings:

- column 1 indicates which non-zero macroscopic component is applied on the unit cell. The letters in parentheses refer to (Fig. 9.2).
- columns 2, 3, 4 indicate if the microscopic displacement jump components are zero (0), negligible ($\simeq 0^n$, infinitely small terms of order n) or non zero ($>, \neq$);
- columns 5, 6, 7 indicate – by application of Definition ($D1$) – if the macroscopic displacement jump components are zero (0), negligible ($\simeq 0^n$, infinitely small terms of order n) or non zero ($>, \neq$).

| | $r_1$ | $m_1$ | $r_{23}$ | $m_{23}$ | Correspondence with Fracture Mechanics | Illustration |
|---|---|---|---|---|---|---|
| Case 1 | $\neq 0$ | 1 | 0 | 2 | Mode I | Fig. 9.2(b) |
| Case 2 | $= 0$ | 2 | $\neq 0$ | 2 | Mode II | Fig. 9.2(c) |
| Case 3 | $= 0$ | 2 | $\neq 0$ | 3 | Mode III | Fig. 9.2(d) |
| Case 4 | $\neq 0$ | 1 | $\neq 0$ | 2 | mixed Mode I / Mode II | |
| Case 5 | $\neq 0$ | 1 | $\neq 0$ | 3 | mixed Mode I / Mode III | |
| Case 6 | $= 0$ | 2 | $\neq 0$ | $2 < m_{23} < 3$ | mixed Mode II / Mode III | |
| Case 7 | $\neq 0$ | 1 | $\neq 0$ | $2 < m_{23} < 3$ | mixed Mode I / Mode II / Mode III | Fig. 9.2(e) |

**Table 9.2**   Characteristic cases of the Crack Opening Mode

| $\varepsilon_{ij} \neq 0$ | $[u_1(\forall M \in F)]$ | $[u_2(\forall M \in F)]$ | $[u_3(\forall M \in F)]$ | $[u_1]_F$ | $[u_2]_F$ | $[u_3]_F$ |
|---|---|---|---|---|---|---|
| $\varepsilon_{11}$ (b) | $=0$ | $> 0$ si $\varepsilon_{11} > 0$ <br> $= 0$ si $\varepsilon_{11} \leq 0$ | $=0$ | $=0$ | $> 0$ if $\varepsilon_{11} > 0$ <br> $= 0$ if $\varepsilon_{11} \leq 0$ | $=0$ |
| $\varepsilon_{22}$ (b) | $=0$ | $> 0$ si $\varepsilon_{22} > 0$ <br> $= 0$ si $\varepsilon_{22} \leq 0$ | $=0$ | $=0$ | $> 0$ if $\varepsilon_{22} > 0$ <br> $= 0$ if $\varepsilon_{22} \leq 0$ | $=0$ |
| $\varepsilon_{33}$ (b) | $=0$ | $> 0$ si $\varepsilon_{33} > 0$ <br> $= 0$ si $\varepsilon_{33} \leq 0$ | $=0$ | $=0$ | $> 0$ if $\varepsilon_{33} > 0$ <br> $= 0$ if $\varepsilon_{33} \leq 0$ | $=0$ |
| $\varepsilon_{23}$ (d) | $\simeq 0^1$ | $\simeq 0^2$ | $\neq 0$ | $=0$ | $=0$ | $\neq 0$ |
| $\varepsilon_{13}$ | $\simeq 0^2$ | $\simeq 0^2$ | $\simeq 0^2$ | $=0$ | $=0$ | $=0$ |
| $\varepsilon_{12}$ (c) | $\neq 0$ | $\simeq 0^2$ | $\simeq 0^1$ | $\neq 0$ | $=0$ | $=0$ |

**Table 9.3**  Conclusions on the microscopic and macroscopic displacement jumps in case macroscopic strains are applied to the unit cell.

| $\sigma_{ij} \neq 0$ | $[u_1(\forall M \in F)]$ | $[u_2(\forall M \in F)]$ | $[u_3(\forall M \in F)]$ | $[u_1]_F$ | $[u_2]_F$ | $[u_3]_F$ |
|---|---|---|---|---|---|---|
| $\sigma_{11}$ | $=0$ | $=0$ | $=0$ | $=0$ | $=0$ | $=0$ |
| $\sigma_{22}$ (b) | $=0$ | $> 0$ si $\sigma_{22} > 0$ <br> $= 0$ si $\sigma_{22} \leq 0$ | $=0$ | $=0$ | $> 0$ si $\sigma_{22} > 0$ <br> $= 0$ si $\sigma_{22} \leq 0$ | $=0$ |
| $\sigma_{33}$ | $=0$ | $=0$ | $=0$ | $=0$ | $=0$ | $=0$ |
| $\sigma_{23}$ (d) | $\simeq 0^1$ | $\simeq 0^2$ | $\neq 0$ | $=0$ | $=0$ | $\neq 0$ |
| $\sigma_{13}$ | $\simeq 0^2$ | $\simeq 0^2$ | $\simeq 0^2$ | $=0$ | $=0$ | $=0$ |
| $\sigma_{12}$ (c) | $\neq 0$ | $\simeq 0^2$ | $\simeq 0^1$ | $\neq 0$ | $=0$ | $=0$ |

**Table 9.4**  Conclusions on the microscopic and macroscopic displacement jumps in case macroscopic stress are applied to the unit cell.

| $\varepsilon_{11}$ | $\varepsilon_{22}$ | $\varepsilon_{33}$ | $\sigma_{11}$ | $\sigma_{22}$ | $\sigma_{33}$ |
|---|---|---|---|---|---|
| 0 | 0 | 0 | 0 | 0 | 0 |
| 0.001 | -0.00025 | -0.00025 | 1 | 0 | 0 |
| -0.00025 | -0.00025 | 0.001 | 0 | 0 | 1 |
| 0.001 | -0.00033 | 0 | 1 | 0.00358 | 0.26787 |
| 0 | -0.00033 | 0.001 | 0.26787 | 0.00358 | 1 |

**Table 9.5**  Macroscopic strain and stress values

# Revision exercises

### Exercise 9.1

How many internal variables (number, mathematical type and physical meaning) are necessary to exactly describe a network of plane microcracks in a 2D description in the framework of Damage Mechanics?

### Exercise 9.2

How many internal variables (number, mathematical type and physical meaning) are necessary to exactly describe a network of plane microcracks in a 3D description in the framework of Damage Mechanics?

### Exercise 9.3

Consider that a network of identical planar microcracks develops in a material which is initially intact and isotropic does that mean it becomes anisotropic?

### Exercise 9.4

What are the four essential conditions necessary to verify a damage model (using Damage Mechanics) to model plane microcracks, in addition to the conditions associated with the verification of the Main Principles of the Physics and Mechanics?

### Exercise 9.5

What is called the unilateral effect of damage?

### Exercise 9.6

What is the type of behaviour if it is assumed that a given and no evolutive damage state (a network of identical plane microcracks) exists in a material, assuming that there is neither friction between the microcrack surfaces nor any other dissipative phenomenon?

### Exercise 9.7

What is the effect of the opening of a planar microcrack network on the behaviour of a material? Is there an influence of the type of the loading applied?

### Exercise 9.8

Why is it not possible to use only one scalar internal model to model the behaviour of a plane microcrack network in the Damage Mechanics framework?

### Exercise 9.9

In a general way explain why each dissipative internal variable used in a model made in the framework of Damage Mechanics should have an evolution law.

**Exercise 9.10**

Which principle of Thermodynamics has to be verified to obtain a thermodynamically admissible model?

# 10

# Multiscale modelling of the fibre break phenomenon. Application to the design of a high-performance pressure vessel

# 10.1   Introduction

The purpose of this chapter is to show how a scientific understanding of the physical processes involved in the degradation of composite materials can lead to improvements in design and safety of structures. In particular the application of this approach to the optimisation of carbon fibre pressure vessels will be discussed. These structures represent not only the biggest market for carbon fibres but by their nature require high reliability so as to avoid dangerous accidents. This requires a detailed understanding of the failure processes with the use of computer modelling relating the behaviour of the whole structure to the microstructure on the scale of the fibres. This must be achieved with simulations based on physical processes but also by using an efficient modelling approach so as to limit calculation times. The result is a much improved understanding of the failure processes and the means to optimise the design of composite pressure vessels whilst retaining acceptable safety margins and reducing their cost of manufacture.

For clarity all the steps involved in this study will be discussed outlining the structured scientific approach adopted.

## 10.2 Nomenclature

- $R_y$: framework at the microscopic scale;
- $R$: framework at the macroscopic scale, the framework of the structure;
- $\vec{v}$, $s$, $e$: displacement, stress and linearised strain fields at the microscopic scale;
- $\vec{u}$, $\sigma$, $\varepsilon$: displacement, stress and linearised strain fields at the macroscopic scale;
- $\mathcal{V}$: scalar field defining the local fibre volume fraction $V_f$;
- $\mathcal{D}$: scalar field defining the damage state;
- $\mathcal{P}_i^C$ ($i = 1, ..., p$): $p$ behavioural properties regrouped in a set denoted $\mathbb{P}^C$;
- $\mathcal{P}^R$: failure properties;
- RVE($M$): Representative Volume Element at point $M$;
- $R_{loc} = (O_{loc}, \vec{x_1}^{loc}, \vec{x_2}^{loc}, \vec{x_3}^{loc})$: anisotropic framework of the material;
- $R_{loc}(M)$: local anisotropic framework of the material at point $M$;
- $c$: behaviour 4th-order tensor of the linear elasticity (rigidity or flexibility);
- $c^{loc}$: behaviour 4th-order tensor of the linear elasticity (rigidity and flexibility) in the anisotropic framework of the material, $C^{loc}$ rigidity, $S^{loc}$ flexibility;
- $c^{ref}$ behaviour 4th-order tensor of the linear elasticity (rigidity or flexibility) in the observational framework, $C^{ref}$ rigidity, $S^{ref}$ flexibility;
- most of these quantities are defined at point $M$ and time $t$.

## 10.3 Goal of the study

The goal of this study is to establish how a calculation procedure can be developed which has calculation times and costs acceptable to industrial norms so as to optimise the design of composite pressure vessels whilst ensuring quantifiable levels of confidences in their performance. This will be achieved by Finite Element numerical simulation.

Composites owe their properties primarily to the reinforcing fibres which should be arranged so as to support applied loads. Unidirectional carbon fibre composites are the most widely used constitutive elements in industrial high-performance composite structures even if they are arranged to create stratified structures with the fibres in different plies arranged in a variety of directions. Degradations which can develop in these plies are prejudicial for the mechanical integrity of the structure. The most important degradation mechanism is the breakage of fibres, which if they occur in sufficiently large numbers will lead to the total failure of the structure. It is therefore necessary to understand the phenomenon of fibre failure on the scale of the fibres. This means that the failure process must be simulated at a scale for which a characteristic length is of the order of the fibre characteristic dimension: its radius, for instance, which is of order of microns. This scale is defined as the microscopic scale. This scale sees the constituents of the composite: the fibres, the matrix and their interfaces.

The results at this level will then be included in the calculation at the scale of the complete industrial structure which is the defined goal. However, such a calculation is immediately confronted with the problem of passing from the microscopic to the macroscopic. It is not possible to envisage basing the Finite Element calculation of the whole structure by simulating the behaviour of all the fibres in the composite. This would entail impossibly long and expensive calculation times because of the very large number of fibres making up the structure and the very large numbers of degrees of freedom involved. What to do?

A multi-scale process is the solution to this problem as it is capable of both addressing the finest parts of the structure, together with their physical characteristics, and describing the whole structure with reasonable calculation times. Such a two-scale approach has been used in tackling this problem and the approach is described as $FE^2$ and has been used in the works of Feyel (2003), Thionnet and Renard (1998) and Souza et al. (2008). Despite the increasing power of calculators it is clear that without taking into account the specificities of the mechanisms studied or other details of the processes involved the $FE^2$ multi-scale process would not be powerful enough to result in calculation times acceptable for practical use, so the process has had to be simplified.

The multi-scale process, even simplified, will allow the relations between the microscopic scale and the scale of the structure, described as the macroscopic scale, to be understood. For that, at the macroscopic scale, it is necessary to define the minimum volume of the composite which is statistically representative of this material at the macroscopic scale: this volume is called the Representative Volume Element (RVE). Basically and in the most general manner a multi-scale process consists of applying the evaluation of the macroscopic fields (usually stresses and strains) at each point in the macroscopic structure experienced by the RVE, described at the microscopic scale and existing under every macroscopic point in the structure.

As a first step, after the description of the physical processes implicit at the microscopic scale, it is the precise description of the RVE and its microstructure which must be made. It is this which will be included in the model as well as the analyses of the physical phenomena which will be considered.

The model which was used here has been developed based on the understanding that fibre breaks are the primary failure mechanism in composite structures by Blassiau (2005); Blassiau et al. (2006a,b, 2008). It is solidly based on earlier studies by many researchers into the failure of unidirectional composites which have included analytical, statistical and less often numerical approaches with occasional considerations of the viscoelastic nature of the matrix. The following, non-exhaustive list of studies merit being cited: Rosen (1964), Cox (1951), Zweben (1968), Hedgepeth (1961), Ochiai et al. (1991), Curtin (2000); Curtin and Takeda (1998); Curtin and Ibnabdeljalil (1997), Goree and Gross (1980), Harlow and Phoenix (1978a,b), Scop and Argon (1967, 1969), Kong (1979), Batdorf (1982a,b), Nedele and Wisnom (1994b,a), Hedgepeth and Dyke (1967), Baxevanakis (1994), Landis and McMeeking (1999); Landis et al. (2000), Phoenix (1997); Phoenix and Beyerlein (2000); Phoenix and Newman (2009); Mahesh and Phoenix (2004), Wisnom and Green (1995), Van Den Heuvel et al. (2004, 1998), Lifschitz and Rotem (1970), Lagoudas et al. (1989), Beyerlein et al. (1998). The above studies have led to an increasing understanding of failure processes in composites and inspired the model used here. The present model is written at the scale of the fibre (microscopic scale) and included in a simplified (Camara et al., 2011) multiscale $FE^2$ approach (Feyel, 2003; Thionnet and Renard, 1998) which allows composite structure failure to be calculated. It has been validated at the microscopic scale (Scott et al., 2012) and at the macroscopic (structural) scale (Chou et al., 2013).

## 10.4 Description of the physical phenomena which will be considered

### 10.4.1 Inventory of the phenomena

A unidirectional composite material reinforced with long fibres having diameters of a few microns will first be considered. These fibres are considered to show linear isotropic behaviour. It is accepted that the fibres are not really isotropic but it will be seen later that only their tensile properties are important in assessing the behaviour of a unidirectional composite so that cross sectional isotropy can be assumed although in more complex situations the fibres could be considered as anisotropic as their radial properties differ markedly from their longitudinal properties. The fibres are considered to be homogeneous and can fail in a cross-section normal to the main fibre axis when the average longitudinal stress exceeds a critical value. This critical value depends on the cross-section at the point of failure and reflects the inhomogeneous and random nature of critical defects in the fibre.

The stress field in a particular fibre depends on its proximity with any nearby broken fibre which may exist. Such fibre breaks may be the result of manipulation of the fibre bundle during manufacture of the composite or because the fibre was particularly weak and broke at low stresses on initial loading. Intact fibres neighbouring a fibre break will have to take up the load which the broken fibre had been supporting before its break. This phenomenon results in the redistribution of loads from the broken fibre to, particularly, neighbouring intact fibres. It can also be amplified by the existence of debonding between the broken fibre and the matrix. The effect can be further amplified over time by the viscous nature of the matrix allowing the matrix to relax locally around a fibre break with the result of increasing loads on neighbouring intact fibres. For simplification the behaviour of the matrix will be considered to be isotropic and linearly viscoelastic.

Finally all which has been described above depends on the local fibre volume fraction which is of a random nature due to the impossibility of placing such fine fibres in a definite format.

### 10.4.2 Illustration of stochastic nature of the fibre strength

It has long been known that, similar to other fibres, the strengths of carbon fibres, $\sigma_R$, show a wide scatter and this has to be described in a statistical manner (see Chapter 2). Tests made by Blassiau (2005) on carbon fibres with different gauge lengths can be mentioned to illustrate this phenomenon (Figure 10.1). A two-parameters Weibull function $P_R(\sigma_R)$ can be used to describe this statistical characteristic of fibre failure:

$$P_R(\sigma_R) = 1 - e^{-(\frac{L}{L_0})(\frac{\sigma_R}{\sigma_R^0})^{m_{\sigma_R}}} \tag{10.1}$$

where $\sigma_R^0$ and $m_{\sigma_R}$ are respectively the scale and the shape parameters of the Weibull function, $L$ and $L_0$ are respectively the tested and the reference gauge lengths of the fibres. Although the Weibull function is often defined by its scale and shape parameters, the mean $\mu_{\sigma_R}$ and the standard deviation $s_{\sigma_R}$ can also be used.

Within a structure, the value of the failure stress of a given fibre is then dependant on the considered point along the axis of the fibre, so that in the domain defining this structure, the function, denoted as $\sigma_R(M)$, giving a value of $\sigma_R$ at the given point $M$ defines a field of local fibre strength. Therefore, $\sigma_R(M)$ is implicitly a local quantity and in the following, the fibre strength should be understood as the local fibre strength. It should be also recalled that the fibre strength is the value of the force applied to the fibre when the fibre breaks divided by the cross section area at the point of failure of the fibre.

The model which is described here takes into account the random nature of the failure of fibres along their axes by using a Monte-Carlo process which will describe the failure stress field $\sigma_R(M)$.

### 10.4.3  Illustration of the stochastic nature of the local fibre volume fraction

The fineness of the carbon fibres means that their organisation within a composite cannot be precisely controlled and as a consequence the overall distribution of the fibres in a $(0°)$ carbon/epoxy laminate is never regular. This means that the local fibre volume fraction can vary widely. Microscopic image analysis (Figure 10.2) was performed on three $(0°)$ carbon/epoxy specimens to highlight and quantify the variation of local fibre volume fractions. The three specimens were cut from the same plate of $(0°)$ carbon/epoxy material and therefore were nominally identical. For each specimen, approximately 1700 observations were examined at many randomly selected locations and the size of each inspection area was 0.0025 mm$^2$ corresponding to the approximate cross-sectional area of the RVE, as defined by Blassiau (2005); Blassiau et al. (2006a,b, 2008) and Baxevanakis (1994) of a material with an average local fibre volume fraction $V_f$ of 0.64, which was the overall value given by the manufacturer. The statistical characteristic of $V_f$ is clearly shown (Figure 10.2) as a cumulative probability distribution $P_V(V_f)$ defined with a two-parameter Weibull function:

$$P_V(V_f) = 1 - e^{-(\frac{V_f}{V_f^0})^{m_{V_f}}} \tag{10.2}$$

where $V_f^0$ and $m_{V_f}$ are respectively the scale and the shape parameters of the Weibull function.

It should be clear that within the composite structure the manufacturing process is incapable of producing a regular array of fibres. Even though the overall fibre volume fraction can be controlled during manufacture the local fibre arrangement cannot be controlled and by nature is random. This has to be considered in the model and is taken into account at each point by a Monte-Carlo process.

### 10.4.4  Illustration of the viscous nature of the matrix

Tensile tests conducted on unidirectional specimens by Fuwa et al. (1976) showed that the acoustic emission activity, for the case of increasing monotonic loading, increased until the final failure of the specimen. It has been shown that this acoustic activity could be directly related to the breakage of fibres for this type of specimen: the global load applied was supported by the fibres that were increasingly loaded during the test. Then, they broke during the increasing of the load. Similar tests were conducted by Blassiau, but for when the load was held at a given constant value. In this case, it could be expected that, at the end of the increasing part of the load, the acoustic activity would stop. However this was not the case because fibres continued to break. The overall load remained constant but clearly the local load on some fibres increased causing their failure (Figure 10.3). The explanation is that the load on each fibre near fibre breaks continued to increase even if the global loading was held constant because the surrounding matrix relaxed due to its viscosity, resulting in an increase in the (local) load supported by the fibres. Another result found was that during this type of test, the evolution of the longitudinal strain of the specimen was not measurable or, in other terms, this evolution of the longitudinal strain was so small that it could be neglected. This means that in the framework of the model the viscoelastic character of the matrix must be taken into account at the level of the constituents but not at the scale of the structure.

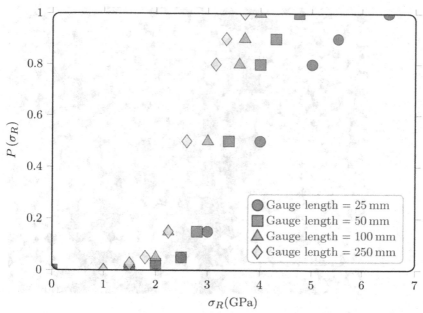

Figure 10.1: Typical experimental cumulative probability $P(\sigma_R)$ for carbon fibre strength $\sigma_R$ (Blassiau, 2005).

## 10.5 Description of the RVE of a unidirectional composite reinforced with continuous fibres. Definition of the framework associated with the microscopic scale and random local fibre volume fraction

The studies by Blassiau (2005); Blassiau et al. (2006a,b, 2008) and that of Baxevanakis (1994) have allowed the geometry and the microstructure of the RVE to be identified for the virgin material, which is considered undamaged and not to contain any broken fibres. The geometry is that of a parallelepiped with a length $L$ and a square section with sides of $h$ (Figure 10.4). Inside this geometry there are $N$ fibres, noted as $F_i$ ($i = 1, \ldots, N$). All the fibres are considered to be identical, arranged in parallel and are perfectly cylindrical with circular cross-sections. All fibres have therefore the same circular cross-section and diameter, and same cross-sectional area $S_F$. It is also supposed that the fibres are regularly spaced in a periodic square format. The research of Baxevanakis (1994), Blassiau (2005); Blassiau et al. (2006a,b, 2008) and confirm by Rojek (2020) have shown that whatever the fibre volume fraction considered the RVE has a length of $L$ equal to 4 mm and contains $N = 32$ fibres. The value of $h$ depends on the fibre volume fraction and the fibre diameters. For fibres with diameters of 8 microns the value of $h$ is then (Figure 10.5):

- $h = 0.050$ mm for a fibre volume fraction of $V_f = 0.64$;
- $h = 0.064$ mm for a fibre volume fraction of $V_f = 0.39$;
- $h = 0.092$ mm for a fibre volume fraction of $V_f = 0.19$.

For this geometry defined at the microscopic scale the fibres and the matrix are clearly represented and associated with the framework $R_y = (O_y, b_y)$ for which the base, which is taken to be orthonormal, is defined by $b_y = (\vec{y}_1, \vec{y}_2, \vec{y}_3)$ and $O_y$ the origin (Figure 10.4). The longitudinal axis of the fibres is aligned with the vector $\vec{y}_1$ and defines the length of the parallelepiped. $O_y$ is

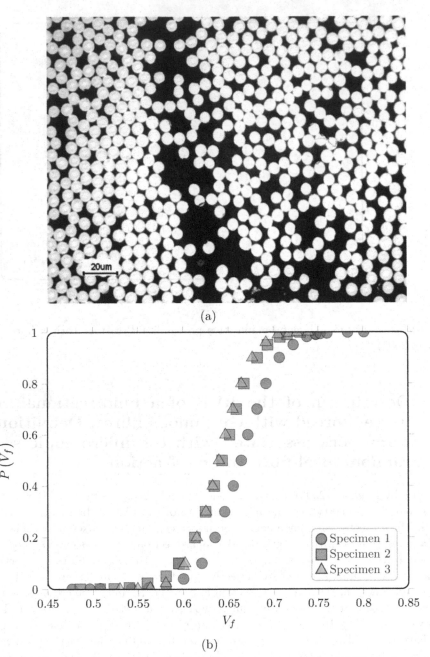

(a)

(b)

Figure 10.2: (a) Typical micrography of (0°) carbon/epoxy specimen. (b) Typical experimental cumulative probability of fibre volume fraction for three (0°) carbon/epoxy specimens coming from the same plate.

placed at the geometrical centre of one of the two square sections of the parallelepiped so that $\vec{y_1}$ is the normal vector and is directed towards the interior. The two other vectors $\vec{y_2}$ and $\vec{y_3}$ are such that they are orthogonal to the planes which define the other faces of the parallelepiped.

It is appropriate at this point to discuss the hypothesis of the periodic arrangement of the fibres within the RVE. It is clear that this ideal configuration does not reflect what really occurs

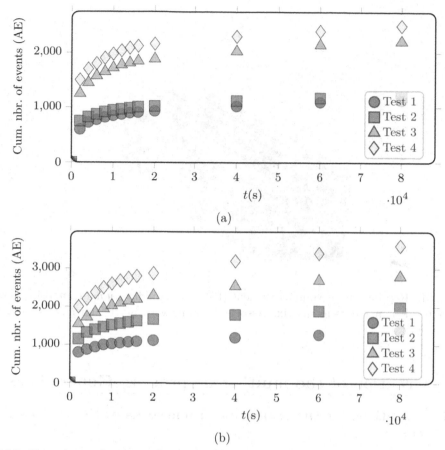

Figure 10.3: Experimental scatter of cumulative number of events obtained by Acoustic Emission for (0°) carbon/epoxy specimens under sustained loading at $X$ % of the failure strength of the specimen (Blassiau, 2005). (a) $X = 75\%$. (b) $X = 80\%$. One event is assumed to be one fibre break.

in a composite (Figure 10.2). In fact it is never found in a carbon fibre composite. So why consider this configuration when there is practically no chance of it being found in reality? The reason is that this configuration reflects the average fibre volume fraction around which all local values must oscillate. The results obtained (properties or other data) for the real configurations only vary as a second order with respect to the results obtained with the assumed periodicity.

The choice taken here to include the random nature of the local fibre distribution (§ 10.4.3) has the effect that:

- a given fibre volume fraction is represented by a RVE which is exactly similar to that which is described above but that the dimensions of the RVE are adapted to the volume fraction considered;

- during the application of the process of multi-scale simulation a Monte-Carlo function is used to describe the randomness of the local fibre volume fraction at each point in the structure.

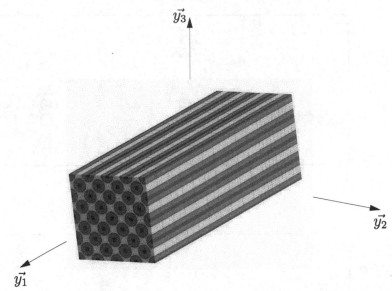

Figure 10.4: Representative Volume Element (RVE) of a unidirectional composite showing the fibres embedded in the resin matrix. Associated framework at the microscopic scale: $R_y = (O_y, \vec{y_1}, \vec{y_2}, \vec{y_3})$.

## 10.6   Creation of the multi-scale process. Generalities

### 10.6.1   Notations for the scales and frameworks of the multi-scale process

Generally a multi-scale process links several scales to one another. Here two distinct scales are considered: that which sees the constituents (fibres and matrix) inside the RVE as distinct and separate continua and secondly that which does not see them but considers the material as a single continuum. The first refers to the microscopic scale and the other the macroscopic scale. It is necessary to define a framework for each of these scales.

The geometry of the RVE has been previously identified as being a regular parallelepiped with a square cross-section in which $N = 32$ parallel fibres $F_i$ $(i = 1, \ldots, N)$ are embedded. To this geometry at the microscopic scale is associated a direct framework $R_y = (O_y, \vec{y_1}, \vec{y_2}, \vec{y_3})$ where the vector $\vec{y_1}$ defines the axis of the fibres and $y = (y_1, y_2, y_3)$ designates the coordinates of a point $P$ relative to this framework. The fields of displacement, stresses (supposed symmetrical) and deformation (linearised) are respectively noted as $\vec{v} = (v_i)_{i=1,2,3}$, $s = (s_{ij})_{i,j=1,2,3}$ and $e = (e_{ij})_{i,j=1,2,3}$. The values of these fields at a point $P$ at time $t$ are then respectively $\vec{v}(P,t)$, $s(P,t)$ and $e(P,t)$ of which the expression as a function of $y$ is written in the same manner so as to avoid multiplying the notations: $\vec{v}(y,t) = (v_i(y,t))_{i=1,2,3}$, $s(y,t) = (s_{ij}(y,t))_{i,j=1,2,3}$ and $e(y,t) = e(\vec{v}(y,t)) = (e_{ij}(y,t) = \frac{1}{2}(\frac{\partial(v_i(y,t))}{\partial y_j} + \frac{\partial(v_j(y,t))}{\partial y_i}))_{i,j=1,2,3}$.

The framework for the macroscopic scale which is that of the structure is $R = (O, b)$ for which the orthonormal base is $b = (\vec{x_1}, \vec{x_2}, \vec{x_3})$ and $O$ the origin. The triplet $x = (x_1, x_2, x_3)$ designates the coordinates of a point $M$ relative to this framework. The fields of displacements, stresses (supposed symmetrical) and strains (linearised) are respectively noted as $\vec{u} = (u_i)_{i=1,2,3}$, $\sigma = (\sigma_{ij})_{i,j=1,2,3}$ and $\varepsilon = (\varepsilon_{ij})_{i,j=1,2,3}$. The values of these fields at a point $M$ at time $t$ are then respectively $\vec{u}(M,t)$, $\sigma(M,t)$ and $\varepsilon(M,t)$ the expression can be equally noted as a function of $x$ in the same way so to reduce the number of notations: $\vec{u}(x,t) = (u_i(x,t))_{i=1,2,3}$, $\sigma(x,t) = (\sigma_{ij}(x,t))_{i,j=1,2,3}$ and $\varepsilon(x,t) = \varepsilon(\vec{u}(x,t)) = (\varepsilon_{ij}(x,t) = \frac{1}{2}(\frac{\partial(u_i(x,t))}{\partial x_j} + \frac{\partial(u_j(x,t))}{\partial x_i}))_{i,j=1,2,3}$.

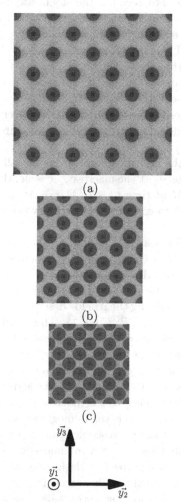

Figure 10.5: Representative Volume Elements (RVE) of a unidirectional fibre/resin composite. The fibres are shown in dark circles. (a) Fibre volume fraction of 19%. (b) Fibre volume fraction of de 39%. (c) Fibre volume fraction of 64%.

## 10.6.2    Organisation

### Process

In order to set up the multi-scale process it is useful to precisely describe the organisation of the phenomena which will lead to the failure of the structure. It is on this organisation of the physical phenomena (which have been revealed by earlier investigations and experimental observations) that it will be possible to construct the organisation of the multi-scale process. It should be recalled that the goal is to take into account: the stochastic nature of fibre failure, the random nature of the local fibre volume fraction and the viscoelastic nature of the resin matrix.

From the start it is considered that the composite structure is made up of fibres which are not arranged in a regular manner. This random irregularity results in an inhomogeneous distribution of the fibre volume fraction within the structure (§ 10.4.3, § 10.5), of which the associated RVE is itself made up of a regular array of fibres. This random characteristic is taken into account by attributing values to the local fibre volume fractions at each point $M$, through the use of a Monte-Carlo simulation. These values define the inhomogeneous character of the local fibre volume fraction, which is considered not to vary with time, through a scalar field $\mathcal{V}$ of which the value at point $M$ is $\mathcal{V}(M) = V_f(M)$.

It is then considered that the unloaded structure is initially (at the beginning of loading at time $t_0$) in a given state of damage. At the macroscopic scale, this state is represented by a scalar field $\mathcal{D}$ of which the value in $M$ at time $t_0$ is $\mathcal{D}(M, t_0)$. At $t_0$, it is usual (but not obligatory if the physical reality is different) to consider that at every point in the structure the material is undamaged.

The calculation can then be started however at the macroscopic scale it is necessary to know at $t_0$ and at every point $M$ of the structure the macroscopic behavioural properties of the material making up the structure. All these properties of varying mathematical natures (scalar, vectorial, second order tensorial) the number of which is given by $p$, are grouped in the following set $\mathbb{P}^C(M, t_0)$ containing and describing the $p$ fields described as $\mathcal{P}_i^C(M, t_0)$ $(i = 1, ..., p)$. This set is noted: $\mathbb{P}^C(M, t_0) = \left\{ \mathcal{P}_1^C(M, t_0), ..., \mathcal{P}_p^C(M, t_0) \right\}$. These properties are calculated at time $t_0$, by the solution, at the microscopic scale, of the problems of homogenisation applied to the RVE at point $M$, noted RVE($M$), and the average operations on all of the solutions of these problems. The dependence with respect to $M$ is due the local volume fraction being inhomogeneous and therefore the RVE is not the same at every point of the structure (Figure 10.5). To obtain the quantities (properties or other variables) at the macroscopic scale from quantities (properties or other variables) at the microscopic scale is the homogenisation phase of the multi-scale process. This means that the properties at the microscopic scale, as well as the geometry of the RVE($M$) are data. If the behavioural properties are really inhomogeneous over all the structure in its initial state it is also the case for the failure properties, designated as $\mathcal{P}^R(M, t_0)$. In effect, our wish is to take into account the random nature of the fibre strength along their axes. Thus, physically, that means, for example, that two neighbouring fibres, ostensively geometrically and behaviourally very similar and subjected to the same loads in the structure, do not break simultaneously or in the same place. This random nature is achieved in the simulation by attributing values of fibre strength, at each point $M$, through a Monte-Carlo process (§ 10.4.2). These values define therefore the inhomogenity of the failure strengths through a scalar field $\mathcal{P}^R$ for which the value at point $M$ is $\mathcal{P}^R(M)$. It is convenient to note that $\mathcal{P}^R(M)$ is a field defined at the macroscopic scale: it is therefore clearly linked with the field $\sigma_R$ of the fibre failure strengths (§ 10.4.2), but it is not in any case a priori directly the same field.

The initial state of the structure being clearly defined it is now possible to consider it under load and launch the multi-scale simulation. The load on the structure increases from zero and increases until it reaches a value at time $t$, noted $\mathcal{S}(t)$. Within the structure at the macroscopic scale there are stresses and strains imposed by the load, represented by the fields $\sigma(M, t)$ and $\varepsilon(M, t)$.

Due to the existence of these imposed macroscopic fields there exists within RVE($M$) the states of stress $s(P,t)$ and strain $e(P,t)$ at time $t$ at the point $P$ at the microscopic scale. The calculation of the microscopic fields $s(P,t)$ and $e(P,t)$ induced by the macroscopic fields $\sigma(M,t)$ and $\varepsilon(M,t)$ is called the localisation step in the multi-scale process. The point $P$ covers the geometry of the RVE($M$) and notably the fibres and matrix. Noting $s^{F_i}$ the restriction of the $s$ field for the case where $P$ is in the fibre $F_i$, and $s^m$ the restriction of $s$ where $P$ is in the matrix. Thus for the specific case considered (the failure of fibre $F_i$ at a point on its longitudinal axis), it is the average of the longitudinal component $s_{11}^{F_i}(P,t)$ over the cross section $S_{F_i}(y_1)$ situated on the abscissa $y_1$ which is of interest:

$$\langle s_{11}^{F_i}(P,t)\rangle(y_1,t) = \frac{1}{S_{F_i}(y_1)} \int_{S_{F_i}(y_1)} s_{11}^{F_i}(y_1,y_2,y_3,t)dy_2dy_3 \qquad (10.3)$$

Because of the stochastic nature of failure of fibre $F_i$ along its axis (§ 10.4.2), noted as $s_R^{F_i}(y_1)$, some fibres in the population are likely to break if the following condition is met:

$$\langle s_{11}^{F_i}(P,t)\rangle(y_1,t) \geq s_R^{F_i}(y_1)? \qquad (10.4)$$

It should be noted that the values $s_R^{F_i}(y_1)$ applied at the microscopic scale at points $P$ of the RVE($M$), are values which come from $\mathcal{P}^R(M)$ constructed by the associated Monte-Carlo process.

Thus at point $M$ at the macroscopic scale at instant $t$, a damaged material replaces the initially undamaged material. The value in $M$, at $t$, of the scalar field $\mathcal{D}$ becomes $\mathcal{D}(M,t)$. This value obviously depends on the number of broken fibres in RVE($M$) at time $t$. As for the initial undamaged state the behavioural properties $\mathcal{P}_i^C(M,t)$ ($i = 1,...,p$) are evaluated by a homogenisation step.

The time and loading continues to increase and the process described above is repeated until the structure fails. It is therefore necessary to clearly define the failure state of the structure which marks the end of the multi-scale process.

## Stopping the process: failure of the structure

Generally a loading curve is defined with an objective of finding a quantity $Q$ representative of the failure of a structure and more precisely, so as to exhibit a quantity $Q$ for which the rate function of time or in another quantity becomes infinite.

For the case of monotonic loading until failure, a loading curve of a structure can be defined as a curve giving $F$, a quantity characteristic of the loading of the structure, versus $U$, a quantity characteristic of the response of the structure. This curve can be experimentally or numerically (using Fracture or Damage Mechanics) obtained. Depending on the considered case, the definition of the failure of the structure will be specific to each case and will not be the same. The (physical or experimental) failure of a structure is defined as the breakage into two parts of the structure (Figure 10.6). The failure point of a structure is the (experimental instability) point $B_{INST\_EXP}$ on the (experimental) loading curve where the failure appears. The failure load of a structure is defined as the value $F_{INST\_EXP}$ of $F$ at the failure point. These concepts are very clear, experimentally, as well as in the framework of Fracture Mechanics, because in these cases, the damage is clearly identified with the appearance of discrete surfaces where the displacement becomes discontinuous, cutting the structure into two parts.

In the case of Damage Mechanics, it is difficult to apply these definitions because in this framework, damage never appears in a discrete way: the medium making up the structure is never physically cut. The concept of "the failure of a structure is defined as the breakage into two parts of the structure", but in the case of Damage Mechanics this should be "the failure of a structure is defined as the existence of a path-connected domain crossing the structure for which the medium

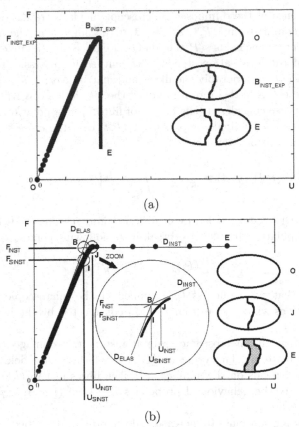

(a)

(b)

Figure 10.6: Definition of the failure point of a structure. (a) Typical experimental loading curve of a structure until failure, highlighting the experimental instability point $B_{INST\_EXP}$ defining the break into two parts of the structure. This is also a typical simulated curve of a structure until failure, made in the Fracture Mechanics framework. In both cases, the failure of the structure is induced by the breakage into two parts of the structure, showing discontinuities of displacement in a created surface. (b) Typical simulated loading curve of a structure in the framework of the Damage Mechanics, highlighting the numerical instability starting at the Start of INStability Point $I$ and assumed to finish at $J$ (Failure Point or Instability Point) close to the real breakage into two parts of the structure.

is totally degraded": the point $J$ marks this state on the simulated loading curve (Figure 10.6, point $J$), and point $E$ illustrates the totally degraded but not cut material (Figure 10.6, point $E$, grey area). The point $J$ is then denoted as the Failure Point or Instability Point of the structure. This point is also characteristic of the fact that around it, the simulated loading curve shows the beginning of a plateau indicating that the rate of $U$ versus $F$ becomes infinite.

To find an existing path-connected domain where the material is totally degraded with, as an objective, the identification of the point $J$ is not easy. An easier, clear and reproducible method method giving this point (or a very close one) is necessary. This is explained in the following. The failure (instability) load $F_{INST}$ of the structure is defined as the value of $F$ at the intersection (Figure 10.6, point $B$) of the tangent of the instability plateau $D_{INST}$ and the tangent $D_{ELAS}$ at the origin of the loading curve. The characteristic displacement corresponding to $F_{INST}$ on the simulated loading curve is denoted as $U_{INST}$. The characteristic displacement corresponding to $F_{INST}$ on the tangent at the origin of the simulated loading curve is denoted as $U_{SINST}$. The value of $F$ corresponding to $U_{SINST}$ on the simulated loading curve is denoted as $F_{SINST}$.

The point $I$ is defined by the coordinates $(U_{SINST}, F_{SINST})$. The point $J$ is defined by the coordinates $(U_{INST}, F_{INST})$.

In the case of the composite considered here (carbon fibre/resin), it is well known that the loading curve of structures made with such materials have the form shown in Figure 10.6: nothing can be significantly seen which indicates that the break will appear. This is a so-called sudden-death failure. In other terms, the point $B_{INST\_EXP}$ is not readily observable experimentally because of the difficulty in controlling the failure process of a unidirectional carbon fibre composite. For this reason the point $I$ denoted as the Start of the INSTability point coming from the simulated loading curve more realistically marks the end of the experimental loading curve and of the failure of the structure.

In the case of monotonic loading followed by sustained loading, a loading curve can also be defined but in another way compared to the monotonic loading until failure. The concept of failure is evidently the same, but the quantity to consider should in general be explicitly the time variable.

Exactly analogous cases as described above but specifically for a pressure vessels are shown in Figure 10.7.

### 10.6.3  Specificity of a calculation for which the behaviour of the composite is anisotropic. A few general reminders: anisotropic linear elastic behaviour in an anistropic local framework

For the case of a continuum consisting of isotropic linearly elastic material, even if inhomogeneous, the data for the point $M$ consist of two coefficients (Young's modulus and Poisson's ratio) which, in a continuous framework, are sufficient for a complete definition of the behaviour and Finite Element calculations of the structure. These two coefficients allow the construction, at point $M$, of the 4th-order tensor linear elastic behaviour (rigidity or flexibility), noted as $c(M)$ for which the expression is identical in all frameworks obtained within an orthogonal transformation: the 4th-order tensor of the isotropic linear elastic behaviour of a isotropic linear elastic material is invariant for changes of orthonormal bases. The dependence with respect to $M$ in $c(M)$ indicates that the continuum consists of inhomogeneous material.

In the case considered here the material making up the continuum is inhomogeneous at the macroscopic scale as it is made up of fibre volume fractions which differ at each point, $\mathcal{V}(M) = V_f(M)$. Thus the 4th-order tensor of the linear elastic behaviour (rigidity or flexibility) of the material including the dependence of the fibre volume fraction can be noted as $c(\mathcal{V}(M))$. However the material is also anisotropic. In this case it is no longer defined uniquely by two characteristics as for the isotropic case but in the more general anisotropic case it is defined by all its twenty one independent components. However knowledge of these components is not sufficient to completely define the anisotropic linear elastic behaviour: it is also necessary to define the framework used so as to complete the identification of these components. This framework is called the anisotropic framework of the material. It can be noted as $R_{loc} = (O_{loc}, b_{loc})$, $O_{loc}$ its origin and $b_{loc} = (\vec{x_1}^{loc}, \vec{x_2}^{loc}, \vec{x_3}^{loc})$ its base. In the cases considered here, the geometrical organisation of the constituents within each RVE for each fibre volume fraction are identical (Figure 10.5): it is therefore possible to take the same anisotropic framework for all the RVE. Here we take $R_{loc} = R_y$, the framework at the microscopic scale. Thus the combined data of $c(\mathcal{V}(M))$ and $R_{loc}$ define the linear elastic behaviour of the anisotropic material making up the continuum of which the structure is made, describing its rigidity or its flexibility in a 4th-order tensor noted as $c^{loc}(\mathcal{V}(M))$ ($C^{loc}(\mathcal{V}(M))$ for the rigidity and $S^{loc}(\mathcal{V}(M))$ for the flexibility). However, $c^{loc}(\mathcal{V}(M))$ is not sufficient for the Finite Element calculation to be carried out. This is because the anisotropy framework has no reason to coincide with the macroscopic framework $R$ in which the calculation is carried out (in the sense of a framework of projection) because of the arrangement of the fibres in the structure at a point $M$.

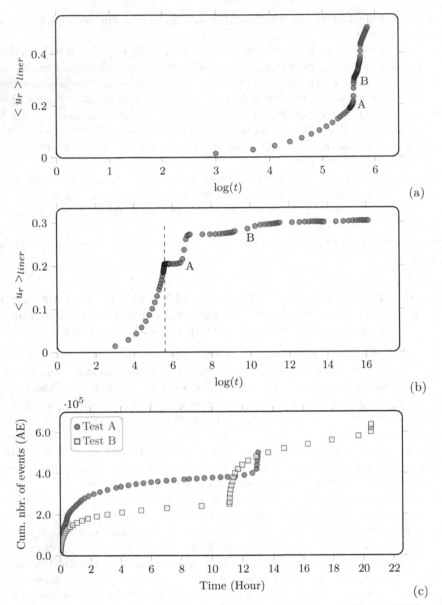

Figure 10.7: Typical loading curves for Monotonic Increasing Pressure Test (MIPT) and Monotonic Increasing then Sustained Pressure Test (MISPT) for a pressure vessel: averaged, over the liner, of the radial displacement $u_r$ as a function of the time, i.e., $< u_r >_{liner}$ versus $t$ or $log(t)$. (a) Monotonic Increasing Pressure Test until failure. The loading rate is 1 MPa/s to avoid the effects of the viscosity of the matrix of the composite. The Point $A$ defines the first instability point of the loading curve. The Point $B$ is another instability point. From a security point of view, the Point $A$ is taken as the failure point of the vessel. (b) Monotonic Increasing then Sustained Pressure Tests until failure. In the monotonic part of the loading (time before the vertical bar), the loading rate is 1 MPa/s to avoid the effects of the viscosity of the matrix of the composite. Then (time after the vertical bar), the sustained pressure $P_{MISPT}$ is applied for 4 months ($10^7$ seconds). The Point $A$ defines the first instability point of the loading curve. The Point $B$ is another instability point, which does not correspond to the total destruction of the vessel. From a security point of view, the Point $A$ is taken as the failure point of the vessel. (c) The curve at (b) is a curve obtained by simulation. But, this is confirmed by experimental test.

In mathematical terms this means that the components of $c^{loc}(\mathcal{V}(M))$ in $R_{loc}$ have no reason to be equal to the components of $c^{loc}(M)$ in $R$ at point $M$, as in general, $R_{loc} \neq R$. Also, in addition to $c^{loc}(\mathcal{V}(M))$, the 3 angles of Euler have to be given and then allowed, to access to the local positioning of $R_{loc}$ at each point $M$ of the structure, with respect to the $R$ framework. Symbolically it can be noted that $R_{loc}(M)$ represents all these data. Finally, it becomes possible to calculate the components $c^{loc}(\mathcal{V}(M))$ in the framework of projection where the calculation is made, that is $R$. This calculation takes place knowing $R_{loc}(M)$ and the usual formulae for the change of base for a 4th-order tensor. We can then write $c^{ref}(M, \mathcal{V}(M))$ the 4th-order tensor (rigidity or flexibility) of the anisotropic linear elastic behaviour in $M$, expressed in the $R$ framework: $C^{ref}(M, \mathcal{V}(M))$ for the rigidity and $S^{ref}(M, \mathcal{V}(M))$ for the flexibility.

Up to this point, we have omitted that the above reasoning does take into account the damage that can appear and be different at each $M$ point and modifies the components of the previous 4th-order tensors. To take account of this we can write: $c^{loc}(\mathcal{V}(M), \mathcal{D}(M, t))$, $C^{loc}(\mathcal{V}(M), \mathcal{D}(M, t))$, $S^{loc}(\mathcal{V}(M), \mathcal{D}(M, t))$, $c^{ref}(\mathcal{V}(M), \mathcal{D}(M, t))$, $C^{ref}(\mathcal{V}(M), \mathcal{D}(M, t))$ and $S^{ref}(\mathcal{V}(M), \mathcal{D}(M, t))$.

## 10.7   A first approach to resolve the multi-scale process by a $FE^2$ method and orders of magnitude of length of time of the calculations. Justification of the simplified $FE^2$ approach

### 10.7.1   Details of the steps in the $FE^2$ multi-scale process

The intellectual steps for carrying out numerical calculations on real composite structures will be outlined in this section. At this stage it must be understood that the great number of very fine fibres making up the structure compared to the characteristic length of an industrial composite structure poses great difficulties in doing so in a reasonable time. In later sections the means to shorten calculation times will be explained but here will be outlined necessary steps to be tackled even if this stage they cannot give a final solution for the problem.

The Finite Element technique is the numerical approach used for solving the calculations. The first input data is the discretisation of the structure. In this way the number of degrees of freedom $N_{ddl}$, the number of elements $N_{elem}$ and the number of Gauss Points $N_{Gauss}$ are known. The Gauss Points are the essential points for the calculation: they play the role of the $M$ points in the model. They are denoted here as $M_{Gauss}$. The number of degrees of freedom defines the size of the linear system to be solved during the iterations for convergence at each step of the calculation at the macroscopic level. The number of Gauss Points defines the number of times that the localisation step and that of homogenisation for each iteration of each increment of the calculation have to be carried out at the macroscopic scale.

The different steps in the calculation explained (§ 10.6.2) are as follows:

- Step 0 – Execution of Monte-Carlo processes which affect the values (these will be discussed later and their number defined) of strengths of fibres at each Gauss Point and also the local fibre volume fraction at each Gauss Point;

- Step I – $I$-th time increment. The change in load or not on the structure (the latter in the case of a steady load for example);

- Step RMi – Resolution to convergence of the $I$-th time increment at the macroscopic scale so as to obtain the macroscopic stress and strain fields at each of the $N_{Gauss}$ Gauss Points. So as to arrive at convergence, $N_{RMi}$ iterations are necessary. At each of these iterations at each Gauss Point $M_{Gauss}$:

    · Step LM$_{Gauss}$ – Localisation step - This is a Finite Element calculation on the

RVE characteristics of the Gauss Point which allow the stress and strain fields to be known in all the RVE which then permits an increase in the number of fibres which break (Eq. 10.4);

· Step $\text{HM}_{Gauss}$ – Homogenisation step - This is carried out using Finite Elements to resolve six elementary problems (loading of the RVE with the six usual symmetrical 2-tensor base) on the RVE characteristic of the Gauss Point, followed by averaging on the six solutions. This step gives $C^{loc}(\mathcal{V}(M), \mathcal{D}(M,t))$, $S^{loc}(\mathcal{V}(M), \mathcal{D}(M,t))$ then $C^{ref}(\mathcal{V}(M), \mathcal{D}(M,t))$, $S^{ref}(\mathcal{V}(M), \mathcal{D}(M,t))$, used for the calculation of the macroscopic scale;

- Step Failure – The condition of the structure is examined to determine if failure has happened at the end of the $I$-th step in the calculation (when convergence has occurred). If failure has occurred the calculation is stopped but if not the time-loading is increased (Step I).

## 10.7.2 A simple example allowing the calculation time to be evaluated showing that the technique cannot give a solution

Consider a pressure vessel which is made by filament winding carbon fibres over a mandrel which becomes the liner. Locally the fibres are arranged in a unidirectional manner and are subjected only to tensile loads when the vessel is pressurised as they are automatically placed on geodesic paths during manufacture. Consider such a pressure vessel for which its internal length is 700 mm, its internal diameter is 100 mm. It is supposed that the wall thickness of the composite having a fibre volume fraction of 64 % is 3.2 mm. As above it is taken that the composite assures the mechanical strengths of the pressure vessel and that the fibres can fail in the same manner as in a unidirectional composite.

If only the cylindrical part of the pressure vessel, of length 500 mm, is considered with a discretisation at the macroscopic scale with elements having approximately the dimensions of eight RVE (0.1 mm × 0.1 mm × 8.0 mm, § 10.5), this part of the composite structure contains about 400000 hexahedral elements each having eight nodes and eight Gauss Points giving 800000 degrees of freedom and 3200000 Gauss Points at the macroscopic scale. It can be estimated then that iteration in the Finite Element calculation at the macroscopic scale involving the inversion of a linear system containing 800000 unknowns would last approximately 60 seconds with a standard configuration computer. Furthermore, so as to carry out the localisation and homogenisation (once more with the Finite Element technique, which explains the name $FE^2$), the multi-scale process needs the dicretisation of the RVE at the microscopic scale. For a sufficiently fine dicretisation to be carried out so as to model fibre failure it would involve 250000 degrees of freedom. The steps of localisation and homogenisation require, as mentioned above, the resolution of 1+6 problems, which by the Finite Element technique means seven inversions of a linear system with 250000 unknowns and of the average operations: that is estimated to take ten seconds. These operations at the microscopic scale must be carried out as many times as the number of Gauss Points, which means that one iteration in one increment of the multi-scale calculation takes 32 million seconds (rather more than one year). The calculation time at the macroscopic scale ($\approx$ 60 seconds) is negligible compared to the calculation which is necessary at the microscopic scale ($\approx$ 32000000 seconds). Knowing that a single increment of the calculation made close to the failure point of the structure can require several tens of iterations to achieve convergence it is easy to understand that the complete simulation of the failure of a pressure vessel by this approach would be inaccessible to an industrial design team.

It is therefore necessary to find another solution. To achieve this we will take an analytical approach for the steps at the microscopic scale. The precise details of this approach are given in the following section. However the main outlines are as follows:

- investigate, by Finite Element analysis, a large number of microscopic configurations for the RVE which are possible to encounter during the calculation;
- store in a data base all the possible cases considered and their results.

Then, during the calculation process which conserves the resolution of the Finite Element analysis on the macroscopic scale, at each Gauss Point, the configuration of the RVE will be identified and its state found from the data base. This approach will be called the $FE^2$ simplified method. It will be shown that with this technique the calculation time at the microscopic scale will be drastically reduced ($\approx$ several seconds instead of 3200000 seconds) and that it becomes negligible when compared to the macroscopic step ($\approx$ 60 seconds).

Using this approach a provisional study of the pressure vessel becomes possible. However the construction of the data base is long as will now be demonstrated.

## 10.8 Creation of the simplified multi-scale $FE^2$ process

### 10.8.1 Resolution of the homogenisation step by an analytical approach: behaviour at the macroscopic scale, hypotheses

The first hypothesis is that the principal process governing failure of the structure is the breakage of the reinforcing continuous fibres. As a consequence of this supposition the non-linearity of the macroscopic behaviour of a unidirectional composite which occurs when the material is loaded perpendicularly to the fibres or in shear (due to intralaminar cracking or viscoelastic behaviour) is not considered. The second hypothesis concerns the behaviour parallel to the fibre direction. Based on experimental observations (§ 10.4.4), and in contrast to the microscopic scale, it is considered that any non-linearity parallel to the fibres, due to the viscoelasticity of the matrix, can be ignored. It is accepted that the non-linearity must exist but that it is negligible compared to the effects due to the failure of the fibres.

Finally, the only non-linearity visible at the macroscopic scale is that in the axis of the fibres and is due to their failure. This effect is directly due to the local density of broken fibres described with $\mathcal{D}(M,t)$, at the point $M$ at time $t$. Thus, finally, in the local anisotropic framework of the material, the dependence of the behaviour on damage is revealed by writing the 4th-order tensor of the linear elastic behaviour: $C^{loc}(\mathcal{V}(M), \mathcal{D}(M,t))$. As has already been shown the simulation of this behaviour (resolution of problems of homogenisation) is made by a numerical route, in particular by Finite Element analysis, which can be long and costly. However, in the case considered only the component $C^{loc}_{1111}$ from $C^{loc}(\mathcal{V}(M), \mathcal{D}(M,t))$ is affected by fibre failure. It can therefore be shown that due to the large differences in properties between the fibres and matrix a good approximation for the value of the component $C^{loc}_{1111}(\mathcal{V}(M), \mathcal{D}(M,t))$ is given by the classic law of mixtures analysis which gives:

$$
\begin{aligned}
C^{loc}_{1111}(\mathcal{V}(M), \mathcal{D}(M,t)) &= V_f(M) C^f_{1111}(1 - \mathcal{D}(M,t)) \\
&+ (1 - V_f(M)) C^m_{1111}
\end{aligned}
\tag{10.5}
$$

with:

- $\mathcal{D}(M,t) = \frac{N_R(M,t)}{N_T}$ where $N_R(M,t)$ designates the number of broken fibres in the RVE RVE($M$) at $t$ and $N_T = 32$ the total number of fibres which is identical for all the RVE;
- $C^f_{1111}$ and $C^m_{1111}$ designates the component along the vector $\vec{x_1}^{loc}$ in the $R_{loc}$ framework of the 4th-order tensor of rigidity respectively of the fibres and the resin, both seen as continuous phases at the microscopic scale. In reality as these constituents are considered isotropic for this calculation this precision is not necessary. Nevertheless it

is accepted that carbon fibres are anisotropic but are transversely isotropic which is only what matters in this simulation but could be an issue in a further development;

- $V_f(M)$ designates the local fibre volume fraction in the RVE($M$) and is time independent.

Finally, at the macroscopic scale and in the $R$ framework, the anisotropic and inhomogeneous linear elastic behaviour of the material, which can be damaged so as to become non-linear, obtained from $C^{loc}(\mathcal{V}(M), \mathcal{D}(M,t))$ and $R_{loc}(M)$, is written as:

$$\sigma(M,t) = C^{ref}(M, \mathcal{V}(M), \mathcal{D}(M,t))\varepsilon(M,t)$$

$$\Longleftrightarrow$$

$$\varepsilon(M,t) = S^{ref}(M, \mathcal{V}(M), \mathcal{D}(M,t))\sigma(M,t)$$

More precisely, the access to $C^{ref}(M, \mathcal{V}(M), \mathcal{D}(M,t))$ takes place in two steps:

- to begin with the resolution in the $R_{loc}$ framework, of the problems of homogenisation concerning RVE($M$) degraded by the failure of fibres at point $M$ at $t$, which gives $C^{loc}(\mathcal{V}(M), \mathcal{D}(M,t))$. Here it is the analytical formula (Eq. 10.5) which is used. The use of this formula is much less onerous than using Finite Element analysis and allows a considerable reduction in calculation time;

- then by knowing $R_{loc}(M)$ and the usual formulae for the changes of bases for a 4th-order tensor, which give $C^{ref}(M, \mathcal{V}(M), \mathcal{D}(M,t))$.

The choice above is that it is the failure of the fibres which is by far the principal damage process. It is the only phenomenon to be considered and its impact described by $C^{loc}_{1111}(\mathcal{V}(M), \mathcal{D}(M,t))$, justifies totally that only the axial stress supported by the fibres $F_i$, $s^{F_i}_{11}(P,t)$, is finally important in the failure process of the structure. It is therefore the reason why our interest will be concentrated on it at the microscopic level and notably on the increase in load on the remaining intact neighbouring fibres which can break when the matrix relaxes. So as to reinforce the importance of this effect it should be remembered that it is this stress averaged on the cross-section $S_{F_i}(y_1)$ placed on the abscissas $y_1$, $\langle s^{F_i}_{11}(P,t)\rangle(y_1,t)$, which causes, or not, the failure of the fibre $F_i$.

## 10.8.2  Preparation for the resolution of the localisation step by an analytic process. Modelling of the phenomenon of fibre failure at the microscopic scale

Following from what has been described above it becomes clear that the phenomenon which must be understood is the transfer of axial loads between neighbouring fibres within a RVE representing a given fibre volume fraction:

- when fibres break;
- when the failure of fibres induces debonding from the matrix at the point of the break;
- when, in addition to the previous phenomena the matrix is considered to be viscoelastic and so leads, as a function of time, to a transfer of load from the broken fibre to its neighbours an increase in load which evolves with time.

### Hypotheses concerning the failure of fibres at the microscopic scale. Notions of i-plets

The work of Baxevanakis (1994) in 2 dimensions and Rojek (2020) in 3 dimensions showed that the state of failure of the RVE, on average, had the following characteristics:

- the fibres are broken once in the RVE;
- the breaks are considered to be in the same plane within the RVE.

For the sake of this simulation the fibres are considered to be organised as a square array (§ 10.5). In the same way it will be supposed that the fibre breaks respect a regular periodicity in the geometry of the RVE. In this way it is assumed that the passage from an intact (no broken fibres in the RVE) described by the RVE noted as $CS32$ (Figure 10.8), to a material totally degraded in which all the fibres of the RVE are broken in the same plane, noted as the RVE described as $C1$ (Figure 10.8) goes through five distinct states of damage in which the RVE contains 1, 2, 4, 8 and 16 broken fibres. The RVE representing the different states of damage are respectively $C32, C16, C8, C4, C2, C1$ (Figure 10.8) and define i-plets representing the numbers of broken fibres ($i$ broken fibres in a RVE): $C32$ defines 1-plet (1 broken fibre of the original 32), $C16$ defines a 2-plet (2 broken fibres of the 32), $C8$ defines a 4-plet (4 broken fibres of the 32), $C4$ defines a 8-plet (8 broken fibres of the 32), $C2$ defines a 16-plet (16 broken fibres of the 32), $C1$ defines a 32-plet (all 32 fibres are broken). The RVE $CS32$ defines therefore a 0-plet which is characteristic of the undamaged state of the composite.

**The load transfer coefficient**

The longitudinal coefficient of load transfer, noted as $k_r$, is defined for:

- a fibre volume fraction $V_f$;
- a damage state represented by the RVE $C$ ($C = CS32$, $C32$, $C16$, $C8$, $C4$, $C2$, $C1$);
- a fibre $F_i$ ($i = 1, \ldots, 32$);
- a debonding length $d$;
- the time $t$;
- a value $Y_1$ of the coordinate $y_1$ along the fibre axis measured from the plane of fibre failure, ($y_1 = 0$),

by:

$$k_r\left(V_f, C, F_i, d, t, Y_1\right) =$$
$$\frac{\int_{Y_{1j}}^{Y_{1j+1}} \left(\int_{S_{F_i}(y_1)} s_{11}^{F_i}\left(V_f, C, d | y_1, y_2, y_3, t\right) dy_2 dy_3\right) dy_1}{\int_{Y_{1j}}^{Y_{1j+1}} \left(\int_{S_{F_i}(y_1)} s_{11}^{F_i}\left(V_f, CS32, d = 0 | y_1, y_2, y_3, t = 0\right) dy_2 dy_3\right) dy_1} \tag{10.6}$$

where:

- $y_2$, $y_3$ designates the coordinates identifying the cross-section of the fibre;
- $Y_{1j}$ and $Y_{1j+1}$ are the axes of the cross-sections between which are calculated the coefficient $k_r$;
- $Y_1 = \frac{Y_{1j+1}+Y_{1j}}{2}$;
- $S_{F_i}(y_1)$ designates the cross-section of the fibre $F_i$ considered, in $y_1$. Here, it should be recalled that given the hypothesis (§ 10.5), $S_{F_i}(y_1) = S_F$;
- $s_{11}^{F_i}\left(V_f, C, F_i, d | y_1, y_2, y_3, t\right)$ is the axial stress $s_{11}^{F_i}(y_1, y_2, y_3, t)$ in the fibre $F_i$ for the case where the fibre volume fraction is equal to $V_f$, of the damage state RVE $C$ for a debonding length of $d$.

It is considered that the relationships between failure, debonding and relaxation of the matrix are of second order in importance. Also so as to separate their effects on the load transfer we

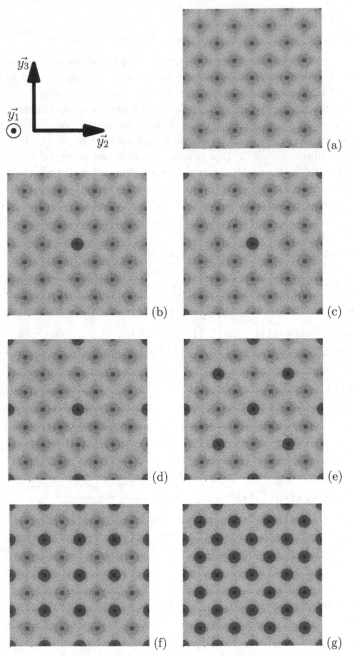

Figure 10.8: Representative Volume Element of the material damage states and corresponding i-plets. Broken fibres are in red colour. (a) RVE $CS32$ / 0-plet. (b) RVE $C32$ / 1-plet. (c) RVE $C16$ / 2-plet. (d) RVE $C8$ / 4-plet. (e) RVE $C4$ / 8-plet. (f) RVE $C2$ / 16-plet. (g) RVE $C1$ / 32-plet.

write:

$$k_r\left(V_f, C, F_i, d, t, Y_1\right) = 1+$$
$$K_r\left(V_f, C, F_i, d = 0, t = 0, Y_1\right) +$$
$$K_d\left(V_f, C, F_i, d, t = 0, Y_1\right) +$$
$$K_v\left(V_f, C, F_i, d = 0, t, Y_1\right)$$

$$(10.7)$$

and the following are calculated separately:

- $K_r\,(V_f,C,F_i,d=0,t=0,Y_1)$ which is the part of the load transfer exclusively due to the fibre breaks;
- $K_d\,(V_f,C,F_i,d,t=0,Y_1)$ which is the part due to the load transfer associated with the length of debonding. This length varies between 0 and 35 $\mu$m from the plane of the fibre break. The value of 35 $\mu$m has been determined experimentally and results in a maximum of locally transferred load to neighbouring intact fibres;
- $K_v\,(V_f,C,F_i,d=0,t,Y_1)$ is due to the load transfer due to the relaxation of the matrix.

Because of the positions of the broken fibres which define the different damage states it is only necessary to carry out the calculation on only one quarter of the RVE (Figure 10.9).

**The most unfavourable value of the load transfer coefficient. Construction of a data base for the localisation step**

For safety considerations a conservative evaluation of failure is considered by using the most unfavourable value of the load transfer coefficient. This means that the maximum values of the above functions will be considered, so that:

$$
\begin{aligned}
k_r^{MAX}\,(V_f,C,t) = 1 + \\
K_r^{MAX}\,(V_f,C) + \\
K_d^{MAX}\,(V_f,C) + \\
K_v^{MAX}\,(V_f,C,t)
\end{aligned}
\tag{10.8}
$$

with:

$$
\begin{cases}
K_r^{MAX}\,(V_f,C) = \max_{F_i,Y_1}\ K_r\,(V_f,C,F_i,d=0,t=0,Y_1) \\[2mm]
K_d^{MAX}\,(V_f,C) = \max_{F_i,Y_1,d}\ K_d\,(V_f,C,F_i,d,t=0,Y_1) \\[2mm]
K_v^{MAX}\,(V_f,C,t) = \max_{F_i,Y_1}\ K_v\,(V_f,C,F_i,d=0,t,Y_1)
\end{cases}
\tag{10.9}
$$

A smoothing is carried out of the calculated points obtained above by simple polynomials. During the multi-scale process the value $V_f$ and $C$ lead directly to the value of the maximum transferred load transfer in an analytical form and not as a Finite Element result.

**Evolution of the fibre break phenomenon at the microscopic scale during the simplified $FE^2$ multi-scale process**

The evolution of the fibre breaks is governed, in the most general case, by the condition Equation (10.4) (§ 10.6.2). However, taking into account the preceding section (§ 10.8.2), at the level of the RVE in the simplified process $FE^2$ the evolution of the process can be evaluated as follows:

- at each point $M_{Gauss}$, for a fibre volume fraction of $V_f$, at each RVE with characteristics of the damage states which are likely to be met, $C$ ($C = CS32, C32, C16, C8, C4, C2$) is associated a failure value noted as $s_R^F(V_f,C)$ from the Monte-Carlo process given in the step 0 (§ 10.7.1) which assigns six values to each Gauss Point;
- if the value is known for $M_{Gauss}$:
  - the loading (which is to say the macroscopic stress $M$ applied to the RVE RVE($M_{Gauss}$);
  - the local fibre volume fraction $V_f$ (Step 0, § 10.7.1);

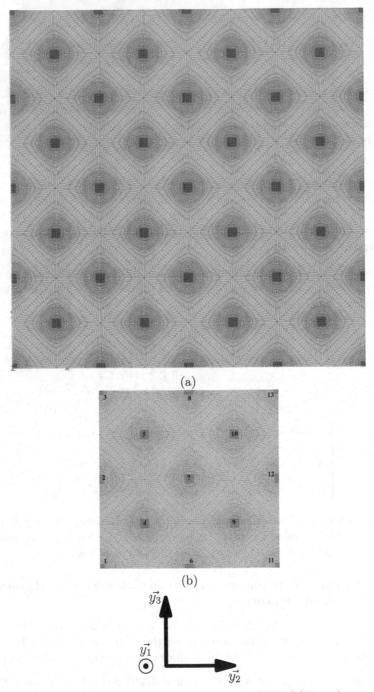

Figure 10.9: (a) RVE $CS32$. (b) Upper right quarter of the RVE $CS32$ and numerotation of the fibres for the calculation of the load transfer.

    · the actual damage state of RVE($M_{Gauss}$) characterised by RVE $C$;

   · the time $t$;

the data base $k_r^{MAX}(V_f, C, t)$ supplies the value of the maximum stress in the fibres within the RVE ($M_{Gauss}$), noted as $s^F_{MAX}(V_f, C, t)$;

- to make evolve the process of fibre breakage in $M_{Gauss}$, the following condition has to be evaluated:

$$s^F_{MAX}(V_f, C, t) \geq s^F_R(V_f, C)? \qquad (10.10)$$

If the condition is not verified nothing happens. If it is verified the fibre breaks at $M_{Gauss}$ evolve:

- if $C = CS32$, so one fibre breaks and the state of the RVE $CS32$ evolves to $C32$;
- if $C = C32$, then one more fibre breaks and the state of the RVE $C32$ evolves to $C16$;
- if $C = C16$, then two further fibres break and the state of damage of the RVE $C16$ evolves to $C8$;
- if $C = 8$, then four more fibres break and the state of damage of the RVE $C8$ evolves to $C4$;
- if $C = 4$, then eight more fibres break and the state of damage of the RVE $C4$ evolves to $C2$;
- if $C = 2$, then sixteen more fibres break and the state of damage of the RVE evolves to $C1$.

## 10.9 Illustration of the approach to the simulation: solution of the problem of calculating the behaviour of the composite structure until failure

### 10.9.1 Outline of the simulation

All the material systems studied are being displaced in the physical space $\varepsilon^3$ with respect to the framework $R = (O, b)$ (macroscopic scale), supposed to be Galilean, for which the base is $b = (\vec{x}_1, \vec{x}_2, \vec{x}_3)$ and $O$ the origin. They are all, or in part, composed of unidirectional composite material composed of fibres and resin (carbon fibres and epoxy resin to be precise unless otherwise stated) of which can be noted $b_{loc} = (\vec{x}_1^{\ loc}, \vec{x}_2^{\ loc}, \vec{x}_3^{\ loc})$ the base of the local anisotropy framework $R_{loc} = (O_{loc}, b_{loc})$ for which the vector $\vec{x}_1^{\ loc}$ is aligned with the fibre axis (local anisotropy framework, § 10.6.3). For the parts of the structure made of composite material, $R_{loc}(M)$ designates the position of $R_{loc}$ at the point $M$, knowing, for example, the Euler angles at the point $M$ which allow to bring $R_{loc} = R$ (by definition of Euler's angles) on $R_{loc}(M)$ (§ 10.6.3). The initial state of the considered structure could be the following: a local fibre volume fraction presenting a random characteristic but supposed constant as a function of time, but in the following applications the random fibre volume fraction will not be considered Then the initial state of the considered structure is: at all points in the structure the volume fraction is considered identical and constant during time and any broken fibres are present.

We shall consider the following hypotheses:

- small displacements;
- at all times the system is supposed to be in a quasi-static equilibrium;
- Gravity effects are neglected;
- it is supposed that there are negligible variations in temperature due to dissipation processes, particularly the failure of fibres. The temperature of the system is therefore uniform and constant with time;
- the behaviour of the material considered is, for the given damage state, anisotropic linear elastic.

### 10.9.2    Common introduction to all the calculations

To begin the calculation, whatever it could be, a Monte-Carlo process attributes six failure values at each Gauss Point of the structure. These values are necessary to enable the kinetics of fibre failure to be included in the simulation (§ 10.8.2). Another Monte-Carlo process could attribute a local fibre volume fraction to each Gauss Point of the structure: but it is not the case here.

### 10.9.3    Formulation of the most general problem. Resolution at the macroscopic scale

#### Statement of the problem

The material system $S$ studied is being displaced and coincides in time with the domain $D = \Omega \cup \partial\Omega$. $D$ is a continuous path-connected bounded and closed set in $\varepsilon^3$, $\Omega$ designates its interior and $\partial\Omega$ its boundary which is considered to be regular. More precisely, $D$ coincides with $D(t) = \Omega(t) \cup \partial\Omega(t)$ at time $t$ and with $D_0 = D(t_0) = \Omega(t_0) \cup \partial\Omega(t_0)$ at time $t_0$. At time $t$, the outer unit vector in a point $M$ of $\partial\Omega$ is noted as $\vec{n}(M,t)$.

The evolution of the domain is studied between times $t_0$ and $t_{max}$.

The loadings applied to the domain can be: remote actions acting on $\Omega(t)$ and induced by a volume density of force denoted $\rho(M)\vec{f}(M,t)$ (induced by Gravity for example) and contact actions on $\partial\Omega(t)$. Here the hypotheses assume that the effect of gravity is negligible with respect to other loads. Thus: $\rho(M)\vec{f}(M,t) = \vec{0}$.

To construct a general framework for the contact actions it is necessary to divide the frontier of the domain into three distinct parts $\partial\Omega_U$, $\partial\Omega_F$ and $\partial\Omega_{FU}$. It is supposed that this division is independent of time and that it forms a partition of $\partial\Omega(t)$, which is to say that summing the different parts reforms the totality of $\partial\Omega(t)$ and that their intersection two by two is equal to the empty set:

$$\partial\Omega(t) = \partial\Omega_U \cup \partial\Omega_F \cup \partial\Omega_{FU} \qquad \begin{cases} \partial\Omega_U \cap \partial\Omega_F = \emptyset \\ \partial\Omega_F \cap \partial\Omega_{FU} = \emptyset \\ \partial\Omega_U \cap \partial\Omega_{FU} = \emptyset \end{cases}$$

One or two of these parts can be equal to the empty set. But, when a part is not equal to the empty set, it is imposed that its measurement (its area) is strictly positive. Starting from this division, we define the following general boundary conditions:

- on $\partial\Omega_F$, the surface density of force $\vec{F}(M,t)$ is given;
- on $\partial\Omega_U$, the displacement field $\vec{U}(M,t)$ is given;
- on $\partial\Omega_{FU}$, a local vector of information, for which one component can be a component of a surface density of force or a component of a displacement field, is given. For instance:

$$\begin{pmatrix} F_1'(M,t) \\ U_2'(M,t) \\ F_3'(M,t) \end{pmatrix}$$

where $F_1'(M,t)$ and $F_3'(M,t)$ are the first and third components of a surface density of force $\vec{F}'(M,t)$ and $U_2'(M,t)$ the second component of a displacement field $\vec{U}'(M,t)$.

These boundary conditions are those of a so-called standard problem. Associated with linear elastic behaviour (as considered here, for a given damage state) they are those of a standard problem of linear elasticity (given in the following section). In this case, theorems exist that give the existence and uniqueness of the solutions of this problem.

**The equations of the problem**

The problem which has to be solved consists of finding, at each $M$ point of the domain the following unknowns:

- the displacement vector $\vec{u}(M,t)$;
- the stress tensor $\sigma(M,t)$;
- the damage variable $\mathcal{D}(M,t)$ relative to the number of broken fibres in $M$ at $t$, $N_R(M,t)$,

verifying:

- the local equation of equilibrium (including the hypotheses given):

$$\overrightarrow{div}\,\sigma(M,t) = \vec{0} \;\; \forall M \in \Omega(t)$$

- the behaviour law

  . state law: $\sigma(M,t) = C^{ref}(M,\mathcal{D}(M,t))\varepsilon(\vec{u}(M,t)) \iff \varepsilon(\vec{u}(M,t)) = S^{ref}(M,\mathcal{D}(M,t))\sigma(M,t)$;

  . evolution law of the fibre failure at point $M$: knowing $\sigma(M,t)$, $V_f(M)$ and $\mathcal{D}(M,t)$ (therefore the RVE characteristic $C$) and $t \Rightarrow$ analytical localisation step in the simplified multi-scale calculation and fibre break evolution (§ 10.8.2)

    * consultation of the database (§ 10.8.2) giving $k_r^{MAX}(V_f,C,t) \Rightarrow$ access to the maximum axial stress $s_{MAX}^F(V_f,C,t)$ in the fibres of RVE $C$
    * knowing the value of the failure stress of fibres (Monte-Carlo process at the beginning of the calculation) likely to cause further fibre breaks, $s_R^F(V_f,C)$
    * examine the condition $s_{MAX}^F(V_f,C,t) \geq s_R^F(V_f,C)$ (§ 10.8.2) ?
    * if the condition is verified the damage state will increase. Then the mechanical properties of the new state of the material can be calculated at $M$ ( $\Rightarrow$ analytical homogenisation step of the simplified multi-scale process, § 10.8.1)
    * if the condition is not verified the damage state remains the same

- boundary conditions:

$$\begin{cases} \text{on } \partial\Omega_F \;:\; \sigma(M \in \partial\Omega_F,t) \times \vec{n}(M \in \partial\Omega_F,t) = \vec{F}(M,t) \\[2mm] \text{on } \partial\Omega_U \;:\; \vec{u}(M \in \partial\Omega_U F,t) = \vec{U}(M,t) \\[2mm] \text{on } \partial\Omega_{FU} \;:\; \begin{cases} \{\sigma(M \in \partial\Omega_{FU},t) \times \vec{n}(M \in \partial\Omega_{FU},t)\}_1 = F_1'(M,t) \\ \{\vec{u}(M \in \partial\Omega_{FU},t)\}_2 = U_2'(M,t) \\ \{\sigma(M \in \partial\Omega_{FU},t) \times \vec{n}(M \in \partial\Omega_{FU},t)\}_3 = F_3'(M,t) \end{cases} \end{cases}$$

## 10.10 The first structure considered. A laboratory structure: a parallelepiped specimen

### 10.10.1 Description of the structure and loading conditions

The system studied $S$ is a parallelepiped specimen (Figure 10.10). The specimen considered is taken to be a composite plate delimited by the following surfaces:

- $S_{-a}$ situated at $x_1 = -a$ and $S_{+a}$ situated at $x_1 = +a$;
- $S_{-b}$ situated at $x_2 = -b$ and $S_{+b}$ situated at $x_2 = +b$;

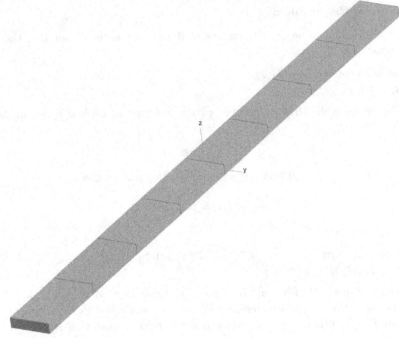

Figure 10.10: Simple parallelepiped specimen.

- $S_{-c}$ situated at $x_3 = -c$ and $S_{+c}$ situated at $x_3 = +c$.

The long axis of the specimen is described by the vector $\vec{x_1}$. The width is oriented by the vector $\vec{x_2}$ and the thickness by the vector $\vec{x_3}$. The plane of the specimen is defined by the vectors $\vec{x_1}$ and $\vec{x_2}$. The cross section of the specimen is therefore defined by the vectors $\vec{x_2}$ and $\vec{x_3}$. For the simulation the dimensions are $2a = 40$ mm, $2b = 4$ mm, $2c = 1$ mm. The specimen is made up of a unidirectional composite at $0°$, which is to say that at every point of the structure $R_{loc}(M) = R$.

The boundary conditions are taken to be the following (with $F_s(t) \geq 0$):

- on $S_{-a}$, a uniform surface density of force is applied $\vec{F}_{-a}(M, t) = -F_s(t)\vec{x_1}$;
- on $S_{+a}$, a uniform surface density of force is applied $\vec{F}_{+a}(M, t) = +F_s(t)\vec{x_1}$;
- the other surfaces were free of force;
- giving $\partial\Omega_F = \partial\Omega(t)$, $\partial\Omega_U = \emptyset$, $\partial\Omega_{FU} = \emptyset$.

Two types of loadings are considered:

- monotonic loadings increasing until failure, called $ML$ (Monotonic Loading): the effect of loading rate on failure strength of the structure is studied. Two loading rates are considered: 0.02 MPa/s and 20 MPa/s and are called $ML002$ and $ML20$. Only one Monte-Carlo run, denoted as $MCR001$, has been considered here for each calculation. To clearly see the rate effect the same Monte-Carlo run has been used for these two calculations. The value of the load at the SINST Point for each calculation, for this Monte-Carlo run $MCR001$, is noted respectively $F_{SINST\_002}^{MCR001}$ and $F_{SINST\_20}^{MCR001}$;
- increasing monotonic loadings which are stopped before failure of the composite and held steady until delayed failure occurs, called here $SL$ (Sustained Loading). Here, the load is increased until $0.90 \times F_{SINST\_20}^{MCR001}$ at 20 MPa/s then sustained. The point noted as $SSL$ Point (Start of the Sustained Loading) defines the load from which the loading

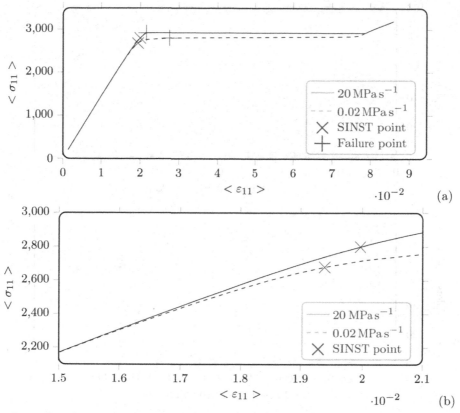

Figure 10.11: Simple specimen: Monotonic loading increasing to failure. Comparison of loading curves showing the influence of speed of loading. (a) Load curves (b) Zoom.

is maintained constant. Here also only one Monte-Carlo run has been considered, the same $MCR001$ as for the previous case. The objective here is to point out that the clustering effect is completely different in the $ML$ and $SL$ cases.

For the $ML$ cases, the loading curve (§ 10.6.2) is defined by $F = < \sigma_{11}(t) = F_s(t) >$ function of $U = < \varepsilon_{11}(t) >$ (here this definition is well adapted but it was possible that it was not the case).

## 10.10.2   Results of the calculations

The results and analysis of the following curves:

- the loading curves: Figure 10.11 for $ML002$ and $ML20$ cases, Figure 10.18 for $SL$ case;
- the populations of the i-plets:
    - 0-plets and 32-plets: Figure 10.12 for $ML002$ case, Figure 10.15 for $ML20$, Figure 10.19 for $SL$ case;
    - 2-plets and 4-plets: Figure 10.13 for $ML002$ case, Figure 10.16 for $ML20$, Figure 10.20 for $SL$ case;
    - 8-plets and 16-plets: Figure 10.14 for $ML002$ case, Figure 10.17 for $ML20$, Figure 10.21 for $SL$ case.

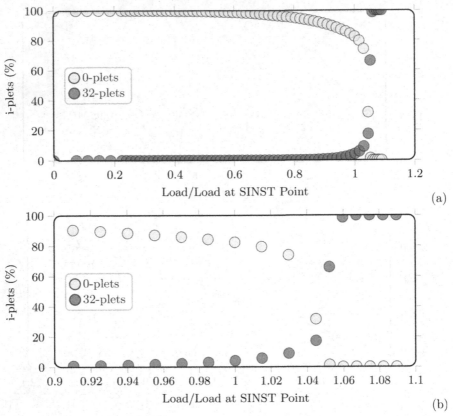

(a)

(b)

Figure 10.12: Simple specimen: Monotonic increasing load until failure. Speed of loading: 0.02 MPa/s. Loading curve. (a) Evolution of the populations of 0-plets and 32-plets. (b) Zoom.

Figure 10.11 shows clearly the effect of speed of loading. The reduced rate of loading allows the visco-elasticity of the matrix to increase damage so that the critical point in loading is reduced, resulting in a decrease in breaking load. There is no attempt here to include an affect which is also induced by the viscoelastic behaviour of the matrix which is to allow straightening of the fibres under the loads which they are supporting. This is known to increase the strength of carbon fibre composites and can lead to an increase in strength of a structure after prolonged steady loading or cyclic loading. The effects of increasing fibre breaks and straightening of the fibres are in competition but whilst the latter effect is expected to slow and cease the former progressively increases damage. Equally there is no attempt to include other mechanisms seen with other reinforcements such as glass fibres which suffer from stress corrosion or creep seen with aramid fibres. There are no reasons why these effects could not be added to the simulation if required. The plateaux shown in the curves are a consequence of the type of calculation carried out and should not be considered to represent what happens in reality although they are analogous to those described in the most general model of composite behaviour described by Aveston et al. (1971) and Aveston and Kelly (1973).

The effects of loading a composite plate are illustrated in Figure 10.12 to Figure 10.14 and are increasing numbers of fibre breaks leading to the formation of groups of breaks. These are described as i-plets where *i* indicates the number of breaks in a cluster. Initial damage in the composite is similar to that seen in a fibre bundle. That is to say that individual fibres break at their weakest points and these points of failure are both unrelated to each other and are

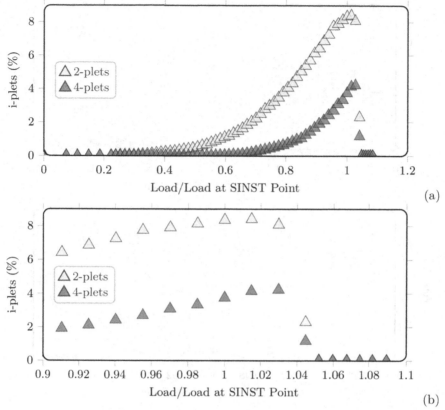

Figure 10.13: Simple specimen: Monotonic increasing load until failure. Speed of loading: 0.02 MPa/s. Loading curve. (a) Evolution of the populations of 2-plets and 4-plets. (b) Zoom.

randomly arranged in the composite. Nevertheless chance will play a part in this process so that neighbouring fibres could break in the same plane and this may even be encouraged by contacts between fibres. Increasing numbers of small groups of breaks lead to a concentration of local stresses with multiple associated breaks occurring producing an increasingly unstable situation, although no overall change in composite behaviour will be observed. This is because the breaks are isolated in the composite by shearing of the matrix around the breaks and have an insignificant effect on the overall behaviour. Figure 10.12 shows the most dramatic situation for loading at 0.02 MPa/s where all the fibres in the RVE break. This is the 32-plet case. It can be seen that this situation occurs just before complete failure and clearly is its cause. Significantly the changes in the numbers of 0-plets show that even just before the point of failure about 75% of all the fibres in the composite are unbroken. Figure 10.13 shows how early damage at 0.02 MPa/s creates 2-plets and 4-plets with the former being the most numerous. Small groups of breaks can evolve into larger groups so that Figure 10.14 shows how as 8-plets and 16-plets develop at a loading speed of 0.02 MPa/s as failure is approached. The model allows the effects of loading speed to be examined. Figure 10.15 shows results at a loading speed of 20 MPa/s and shows that as the higher speed failure occurs more rapidly when 32-plets develop. Again most fibres in the composite are undamaged just before this sudden death failure of the composite. Figure 10.16 and Figure 10.17 show similar trends of increasing numbers of 2, 4, 8 and 16-plets as at the lower loading speed emphasising the critical damage level of 32-plets.

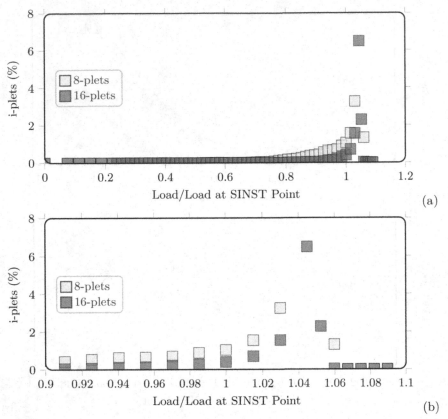

Figure 10.14: Simple specimen: Monotonic increasing load until failure. Speed of loading: 0.02 MPa/s. Loading curve. (a) Evolution of the populations of 8-plets and 16-plets. (b) Zoom.

### 10.10.3  Implications for complex composite structures of the above model supported by experimental results

The model explains how it is the viscoelastic nature of the matrix which determines damage accumulation in the form of fibre breaks when the composite is subjected to stress, particularly over long times. Figure 10.18 shows an experimental curve obtained with a unidirectional carbon fibre composite loaded to very near its expected tensile breaking load. Damage has been monitored using acoustic emission and the curve of damage as a function of time is exactly similar to that obtained at lower loads and importantly, as will be further explained in the next section, to that obtained when a carbon fibre pressure vessel is held at a constant pressure or cycled up to a given pressure. The advantage of loading to near the failure load was that results could be obtained in just a few hours and as will be explained loading at lower loads leads to much greater scatter in lifetimes making experiments costly in time and money if applied to pressure vessels. Figure 10.19 shows that during the steady loading damage increases at a decreasing rate but does not stop. Eventually the curve of damage against time increases rapidly and failure occurs. Figure 10.19 shows that the model described above based on the physical processes observed in composite failure describes well the behaviour. Modelling of damage accumulation in a composite structure such as a carbon fibre composite pressure vessel allows the kinetics of damage accumulation to be determined for very long periods of test. If a structure is subjected to a constant load, necessarily lower than its breaking load as determined by monotonic loading, it will be able to sustain greater damage without breaking than that at a higher load. Figure 10.20 shows this

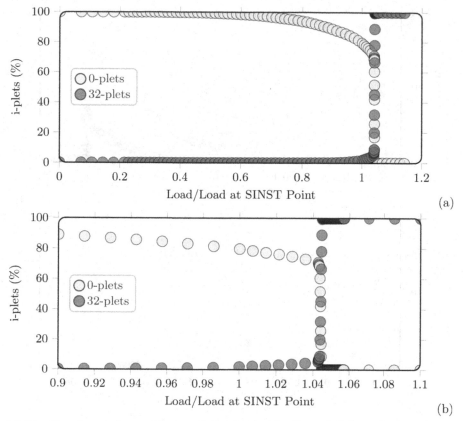

Figure 10.15:  Simple specimen: Monotonic increasing load until failure. Speed of loading: 20 MPa/s. Loading curve. Evolutiion of 0-plets and 32-plets. (b) Zoom.

is the case as when 32-plet clusters develop under a steady load, the composite can continue to function without rapidly breaking. Figure 10.20 and Figure 10.21 show how the i-plets develop under sustained constant loading. If the load is high enough or the test time long enough failure could occur without warning.

## 10.11   The second structure considered. An industrial structure consisting of a filament wound continuous carbon fibre vessel

The effects of long-term loading on advanced composite structures is now a major interest as they are now being used for major critical civil applications. This is nowhere more evident than for the use of carbon fibre composite pressure vessels which will enable the forthcoming hydrogen economy to be realised. Hydrogen will be and is at present being stored in these vessels, typically at a pressure of 70 MPa that is to say 700 atmospheres. This is a very high pressure and certainly the smallest risk of failure has to be avoided. The problem is that pressure vessels are subjected to proof testing which have been developed over many years for steel pressure vessels but not for composites. Typically a proof test consists of over-pressurising a pressure vessel by 50% above the highest designed in-service pressure. If the pressure vessel survives the test it is deemed to be able to continue in service. A metal fails by crack propagation so that the test is appropriate.

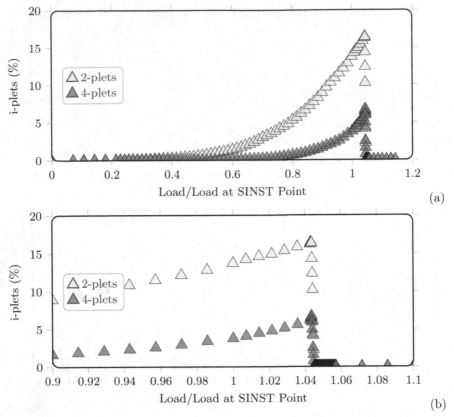

Figure 10.16: Simple specimen: Monotonic increasing load until failure. Speed of loading: 20 MPa/s. Loading curve. Evolution of 2-plets and 4-plets. (b) Zoom.

The stress field around the tip of any incipient crack will be increased by such a proof test which will induce plastic deformation at the crack tip in the metal. At the lower in-service loads this region of plastic deformation will hinder crack propagation and render the structure safe. No such mechanism exists in a composite structure and the above model shows that increasing the stress on the composite, above the in-service level, will induce increasing numbers of failures of the reinforcing elastic carbon fibres. Such a proof test results in the structure being closer to failure than before the test and says nothing about the residual lifetime of the composite structure. Composite pressure vessels today are over designed so as to avoid failure and with very few exceptions this has been successful in avoiding accidents however it is clearly necessary to design structures such as composite pressure vessels based on a quantifiable knowledge of damage kinetics for use over periods of decades. One significant advantage would be economic as by far the greatest part of the costs of producing carbon fibre pressure vessels is the purchasing of the fibres. The above model allows such a quantifiable understanding of the damage kinetics when a carbon fibre composite pressure vessel is either pressurised monotonically to burst or is subjected to use under high pressure for prolonged periods of perhaps decades. Of course the above analysis considers only the intrinsic nature of damage in a composite structure and not effects such as human error during manufacture or external damage which might occur but for the first time it allows damage to be quantified within such an advanced composite structure.

The following section applies the lessons learned from the preceding calculation of loading a composite plate to a filament wound pressure vessel for which both the effects of monotonic

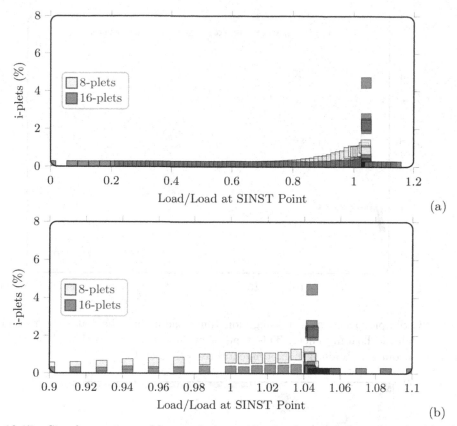

Figure 10.17: Simple specimen: Monotonic increasing load until failure. Speed of loading: 20 MPa/s. Loading curve. Evolutiion of 8-plets and 16-plets. (b) Zoom.

pressurisation to failure and constant pressure held over long periods will be examined.

The material system to be studied $S$ is a small but representative pressure vessel consisting of a central cylindrical part $C$ of length $L$ enclosed at each end by two hemispherical domes $H_-$ and $H_+$ with centres $O_-$ and $O_+$ (Figure 10.22). In the observation framework $R$, the main axis of the pressure vessel is orientated by the vector $\vec{x_3}$ and $\overrightarrow{OO_-} = -\frac{L}{2}\vec{x_3}$, $\overrightarrow{OO_+} = \frac{L}{2}\vec{x_3}$. It can be noted respectively that $S_{int}$ and $S_{ext}$ define the internal and external surfaces of the vessel. They are characterised by the radii $r_{int}$ and $r_{ext}$ and the total wall thickness of the pressure vessel is $e$. The pressure vessel consists of a liner of thickness $e_l$ and a composite wrapping produced by filament winding. The stratified composite wrapping covering the liner is described by the winding angles as $(90°, +20°, -20°)$ (from the outer to the inner of the vessel) such that:

- in the cylindrical part,

$$b_{loc}(M) = (\vec{x_1}^{loc}(M), \vec{x_2}^{loc}(M), \vec{x_3}^{loc}(M))$$
$$= (\vec{e_\varphi}(M), \vec{x_3}(M), \vec{e_l}(M))$$

where the vectors $\vec{e_l}(M)$, $\vec{e_\varphi}(M)$, $\vec{x_3}(M)$ are those of the usual local framework at $M$ defined by cylindrical coordinates $(l, \varphi, x_3)$;

- in the hemispherical domes,

$$b_{loc}(M) = (\vec{x_1}^{loc}(M), \vec{x_2}^{loc}(M), \vec{x_3}^{loc}(M))$$
$$= (\vec{e_\theta}(M), \vec{e_\varphi}(M), \vec{e_r}(M))$$

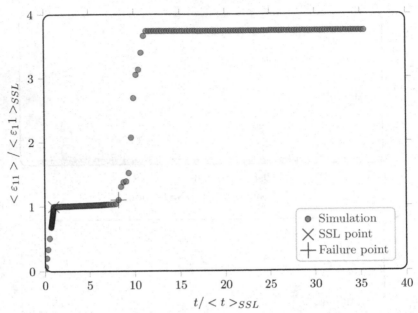

Figure 10.18: Simple specimen: Increasing monotonic loading until held at a constant load until its eventual failure. Loading curve. This type of evolution can be explained by the acoustic emission hits evolution obtained in experimental tests (Figure 10.7).

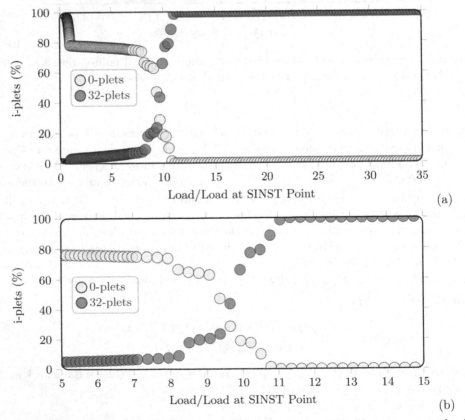

Figure 10.19: Simple specimen: Increasing monotonic loading until held at a constant load until its eventual failure. (a) Evolution of the populations of 0-plets and 32-plets. (b) Zoom.

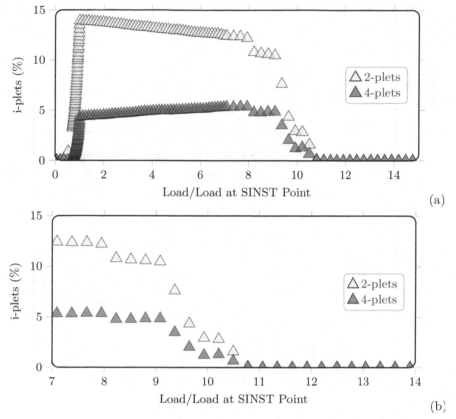

Figure 10.20: Simple specimen: Increasing monotonic loading until held at a constant load until its eventual failure. (a) Evolution of the populations of 2-plets and 4-plets. (b) Zoom.

where the vectors $\vec{e_r}(M)$, $\vec{e_\theta}(M)$, $\vec{e_\varphi}(M)$ are those of the usual local framework at $M$ defined by spherical coordinates $(r, \theta, \varphi)$ of which the centres are respectively $O_-$ and $O_+$.

The dimensions chosen are: $L = 10$ mm, $r_{int} = 7$ mm , $r_{ext} = 8.20$ mm, $e = 1.20$ mm, $e_l = 0.2$ mm. The thickness of the composite is therefore 1 mm distributed in the following manner: first ply 1 at $90°$ with a thickness of 0.4 mm, second ply 2 at $+20°$ with a thickness of 0.3 mm, third ply 3 at $-20°$ with a thickness of 0.3 mm. The liner is considered to be made of aluminium.

The following boundary conditions are used:

- a uniform pressure $p_{int}$ having a value $P(t) \geq 0$ applied to the surface $S_{int}$ giving a surface density of force $\vec{F}_{int}(M,t) = -P(t)\vec{n}(M,t)$;

- a uniform zero pressure $p_{ext}$ was applied to the surface $S_{ext}$ giving a surface density of force $\vec{F}_{ext}(M,t) = \vec{0}$ (surface free of force);

- giving $\partial\Omega_F = \partial\Omega(t)$, $\partial\Omega_U = \emptyset$, $\partial\Omega_{FU} = \emptyset$.

Two types of loadings are considered:

- monotonic loadings increasing until failure, called $ML$ (Monotonic Loading). In the first case the effect of loading speed on failure strength of the structure has been studied. Two loading rates are considered: 0.1 MPa/s and 100 MPa/s and are called $ML01$

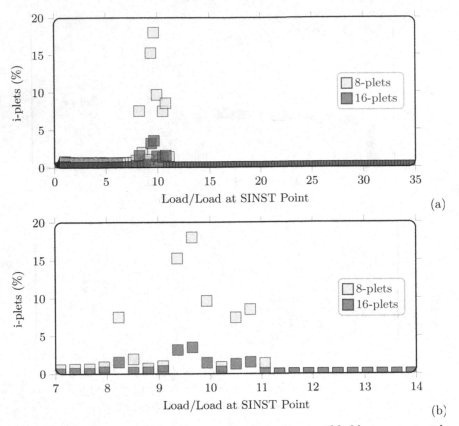

Figure 10.21: Simple specimen: Increasing monotonic loading until held at a constant load until its eventual failure. (a) Evolution of the populations of 8-plets and 16-plets. (b) Zoom.

and $ML100$. Only one Monte-Carlo run, denoted as $MCR001$, has been considered here for each calculation. To compare clearly the effect of the speed effect the same Monte-Carlo run has been used for these two calculations. The value of the load at the $SINST$ Point for each calculation, for this Monte-Carlo run $MCR001$, is noted respectively $F_{SINST\_01}^{MCR001}$ and $F_{SINST\_100}^{MCR001}$;

- increasing monotonic loadings which are stopped before failure of the composite and held steady until delayed failure occurs, called here $SL$ (Sustained Loading). Here, the load is increased until $X$ $F_{SINST\_100}^{MCR001}$ at 100 MPa/s then sustained ($X = 0.907, 0.926, 0.944, 0.963, 0.981, 1.000$). The point noted as $SSL$ Point (Start of the Sustained Loading) defines the load at which the loading is maintained constant. Here for each value of the sustained loading, 20 Monte-Carlo runs have been considered (the same 20 Monte-Carlo runs for each load). For each of the load and each Monte-Carlo run, the time-to-failure has been determined. This calculation give the statistical aspect of the burst pressure of the vessels.

For the $ML$ cases, the loading curve (§ 10.6.2) is defined by $F = P(t)$ function of $U = < u_1(M,t) >$ where $M$ are the points situated in the liner at the median cross-section of the vessels.

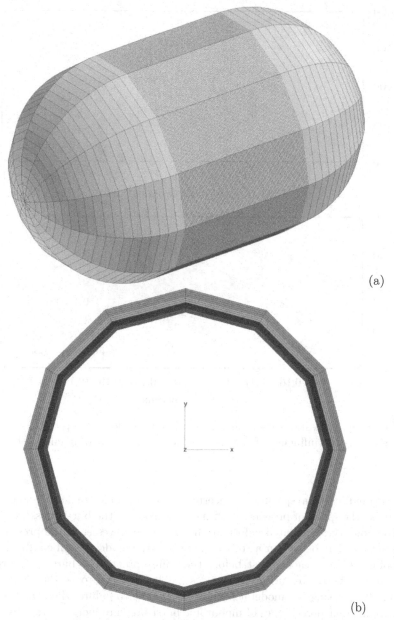

(a)

(b)

Figure 10.22: Simple pressure vessel. (a) Mesh of the pressure vessel. (b) Mesh of the cross section of the pressure vessel (green: ply 1, 90° / yellow: ply 2, +20° / blue: ply 3, −20° / red: liner).

## 10.11.1   Results of the calculations

The results and analysis of the following curves:

- the loading curves: Figure 10.23 for $ML01$ and $ML100$ cases;
- statistical analysis of the failure under sustained loading: Figure 10.24.

The conclusions that can be made in the case of the pressure vessels are exactly the same as in

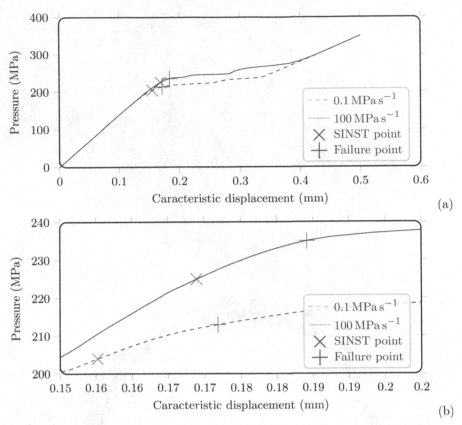

Figure 10.23: Simple pressure vessel : increasing monotonic loading up to burst. Comparison of load curves showing the influence of the speed of loading. (a) Loading curve. (b) Zoom.

the case of the unidirectional specimen, concerning, the effect of loading rate and clustering effect. In particular, as the speed of pressurisation increases so does the burst pressure (Figure 10.23). Figure 10.23 shows the calculated deformation of a pressure vessel as the pressure is increased and reveals very similar behaviour as to that shown in the model described above with a failure initiation point *SINST* reached just before the failure point. The burst pressure is shown to be lower with a reduced pressurisation rate, as has been shown to be the case experimentally (Chou et al., 2013). Using the model it is shown that time to failure when the pressure vessel is pressurised to different percentages of monotonic burst pressure increases rapidly as the applied pressure is reduced, as shown in Figure 10.24. The scatter in lifetimes can be seen to increase as the mean lifetime increases with decreasing pressures. Using these results it can be seen that the curve is asymptotic to a threshold below which failure would not occur. This value is for a pressure level of approximately 80%. This is likely to be a conservative value given the assumptions made in the model and gives a value for an intrinsic safety value of 1.25 meaning that the intrinsic in-service pressure could be as high as 80% burst pressure. However, as emphasised above, this is based on the intrinsic properties of the composites and of its components (fibres and matrix). It does not take account of manufacturing defects or accidental loadings such as impacts.

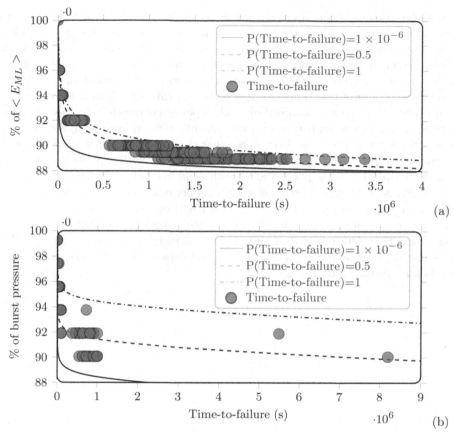

Figure 10.24: Scatter of the time-to-failure. (a) Specimen. (b) Pressure vessel.

## 10.12 Conclusions

The objective of this chapter has been to explore the mechanisms of failure in composite materials. Although the discussions are clearly relevant to all composites, it is the behaviour of carbon fibre reinforced matrices which has preoccupied most of the thinking described. There are several reasons for this: carbon fibres represent the fibre reinforcements with the highest performance; they are elastic and seemingly there are no intrinsic mechanisms which cause their delayed failure, at least at room temperature. They are also the materials which are used for the most demanding of applications. The emphasis has been on the most widely used arrangements of fibre reinforcements in which they play the most important role in supporting the loads applied to a composite structure. In the simplest structure, that of a unidirectional composite with a typical fibre volume fraction of 60%, the fibres support approximately 99% of the load in the fibre direction. It has been shown experimentally however that the properties of such a composite under load, unlike the individual fibres, are time dependent due to the viscoelastic behaviour of the matrix. This clearly has implications for their long-term use. As the fibres support the greatest part of the load their failure represents the most damaging process which can occur in a composite structure. A model has been presented which is based on physical processes including fibre failure and load transfer through the matrix to induce increased loads in intact fibres neighbouring fibre breaks which can lead to progressive fibre failure. The multi-scale model presented considers failure at the level of the constituents, the fibres the matrix and their interfaces and it is shown how using this information can simulate the behaviour of the macroscopic composite structure. Given the numbers of fibres making up a composite structure it is necessary to reduce computing times to reasonable levels. To do this the most damaging cases are considered so as to obtain a conservative evaluation of its macroscopic behaviour. This is shown to reflect accurately the tensile behaviour of both unidirectional composites loaded in the fibre direction and also filament wound pressure vessels under load. The viscoelastic nature of the matrix introduces time into the behaviour of the composite structures and this is accurately shown to correspond closely to experimental results obtained on both types of composite structure. The model has been applied to filament wound structures with the objective of examining the time dependent behaviour of pressure vessels. It has been shown that the burst pressure of a carbon fibre pressure vessel decreases as pressurisation rate is lowered which agrees with experimentally observations. Failure under steady loads has been studied both experimentally and theoretically in order to examine the effects of using pressure vessels over long periods of time. This approach has enabled safety factors to be quantified for these structures based on the real intrinsic damage processes which occur in composites. In the case of carbon fibre composite pressure vessels the minimum safety factor has been determined to be around 1.25, although it should be emphasised that this is based on the intrinsic properties of the material and does not take into account manufacturing errors and damage such as by impact.

# Revision exercises

**Exercise 10.1**

Why is a multiscale process particularly well adapted to calculate a pressure vessel's burst pressure?

**Exercise 10.2**

Even if a pure $FE^2$ finite element process is based on a multi-scale process, why is this method not the best solution, with such a structure made with unidirectional material, to calculate the burst failure pressure?

**Exercise 10.3**

Does increasing speed loading increase or decrease the burst pressure of such a vessel? Why?

**Exercise 10.4**

The calculations show considerable scatter for the time-to-failure at burst for a pressure vessel submitted to sustained loading. Why? What type of phenomenon is the controlling mechanism? Describe other physical phenomena that can also induce scatter on the burst pressure?

**Exercise 10.5**

The scatter appearing on the time-to-failure of burst pressure of a vessel is numerically obtained with many similar finite element calculations. Which data are different from one calculation to another?

**Exercise 10.6**

At a state close to failure (of specimens or pressure vessels), is the pourcentage (or the density) of fibre breaks high or low?

**Exercise 10.7**

Explain the fibre break clustering effect for a specimen monotonically loaded until failure? What type of i-plets advertises the proximity of the failure state?

**Exercise 10.8**

How under, sustained loading, which physical phenomenon can explain that fibre breaks continue to appear? Explain this phenomenon.

**Exercise 10.9**

What is the order of magnitude of the size of the RVE of a unidirectional composite? Approximately how many fibres does it contain?

**Exercise 10.10**

Is Damage Mechanics able to describe the experimental failure? Then, how is the failure point of a structure obtained?

# Environmental ageing

Water is everywhere and most composites will inevitably absorb water, from the air, which holds water and this is measured as relative humidity, or from being in direct contact with liquid water by being immersed in it, or as a composite pipe, by transporting it. Although composites may be in contact with many forms of corrosive liquids, it is water which presents an all pervading concern as absorbed water has the potential of breaking the interfacial bonds between the fibres and the matrix on which the composite depends for its mechanical properties.

Fibre reinforced resins are often used, not only for their good mechanical properties and light weight, as well as their ability to be made into useful and sometimes complicated forms, but also because of their chemical inertness. Tanks for the storage and transport of corrosive chemicals are often made of composites. The preferred choice of materials in environments for which even stainless steel would be quickly corroded, such as in deep seated oil drilling, is also moving towards fibre reinforced resins. Nevertheless composites do have an Achilles heel which is the fibre matrix bond. The adhesion between fibre and matrix is often controlled by rather weak atomic bonds which can be broken by the arrival of water molecules in the vicinity of the fibre matrix interface. Dramatic falls in composite properties can be expected if the bonds are broken in this way, even if the fibres and the bulk matrix material are not modified. It should be noted that drying the composite often seems to allow some of the initial properties to be regained. However this is not the case. Drying removes the water and allows the matrix to shrink back onto the fibres allowing a physical, rather than a chemical, interface to be created which permits loads to be transferred between the fibres and the matrix. This regain in properties is rapidly lost when the composite experiences humid conditions and the interfacial chemical bonds are irreparably lost. However in some cases either the reinforcements or the matrix, or both, can also be deteriorated by water. Although composites may be exposed to many different types of environments, some of which may cause degradation, it is most often water absorption which can be the most damaging and which must be understood.

The fibre-matrix interfaces in resin matrix composites provide preferential channels for the ingress of outside agents which can modify the interactions between the fibres and the matrix. This means that water absorption can proceed much more quickly in the direction parallel to the fibres than in other directions so that we shall see that not only are mechanical properties of composites dependent on fibre arrangements but also the uptake of water and therefore the rate of ageing of the composites. The characteristics of the reinforcements and of the matrix materials can also be altered by the absorption of agents or by the temperature at which the materials are used.

Such deterioration is usually irreversible and strikes at the basic mechanism of fibre reinforcement as it removes the ability of the matrix to transfer load to the fibres and can reduce the performance of the component parts making up the composite. This is made worse as the water at the interface acts as a lubricant. Applications for which the composite must be immersed in water obviously raise the question of its effects but it is also in the air, in the form of humidity that problems can arise. Resin matrix composites will absorb water to a lesser or greater extent and although measures such as painting can hinder its penetration, eventually the composite will either come to an equilibrium with its environment or will be degraded by it. It is therefore necessary to understand and to be able to calculate the rate of diffusion of water in these materials as a function of water concentration and temperature, as the process is thermoactivated. Water also has the effect of a lubricant at the molecular level on the bulk structure of the resin which is revealed most strikingly by a fall in glass transition temperature, $T_g$, as absorption proceeds. In some resin systems this can lead to dramatic falls in $T_g$ so that at the beginning of exposure the composite may be well below the $T_g$ but as water is absorbed the $T_g$ falls below the ambient temperature and the physical behaviour of the matrix is as a consequence greatly changed. This effect can be totally or partially reversible. Water can also provoke the hydrolysis of the resin leading to a deterioration of the matrix and composite. In the case of thermoplastic matrix composites this can lead to an increase in crystallinity and progressive embrittlement of the matrix. In some high temperature resin systems, oxidation of the polymer matrix can also be a source of degradation.

Without loss of generality, this chapter is built around a particularly illustrative example of the various environmental ageing problems raised above. For several decades, polymeric materials have been increasingly used in many industrial fields facing a humid environment and amongst others, water distribution equipment Laiarinandrasana et al. (2011) is a sector for which environmental ageing must be considered in the foreground. Indeed, ease of implementation, low production costs and adapted mechanical properties are assets that now allow polymers to replace traditional metallic materials such as copper alloys. For domestic "hot" (hot is considered here to be up to 70 °C) water supply and structural components under water pressure, some of high-performance polyamides such as reinforced polyphthalamides (PPA) have proved to be efficient Mazé (2012). The sustainability of hydraulic equipment is nevertheless a crucial issue for the industry. As already indicated, most materials evolve by interaction with their hydro-thermal environment Bunsell and Renard (2005) which implies a simultaneous exposure to relative humidity or water immersion and to elevated temperatures. As an example, most of the results reported in this chapter relate to a PPA matrix reinforced with short glass fibres but the trends observed would apply equally well to other classes of composite materials.

Before going into the details and the hydro-thermal effects on this material, let's take a look at the fundamentals of transport phenomena.

## 11.1   Some basics about water diffusion

Diffusion laws have generally been applied to the modelling of simple water uptake and the simple Fick's law can be applied to a large number of composite materials subjected to humid environments. In this model the water is considered to remain in a single free phase driven to penetrate the resin by the water concentration gradient. Other studies have indicated, however, that non-Fickian processes do occur which can complicate the understanding of the role of the water and which may lead to irreversible changes in mechanical properties. Water penetration can cause swelling and plastification of the resin, and ingress can occur by capillary action along the fibre-matrix interface. Composite structures consisting of several layers of differing types of fibre lay-up, often with a thick resin coat or gel coat, can suffer from water being trapped in a particular layer or at the interface between layers. This effect may be irreversible because of osmosis, and blistering may result. This is particularly a problem with glass fibre reinforced boats for which considerable efforts are made, by a judicious choice of resin materials and manufacturing techniques, to avoid this type of damage. In addition to changes in mechanical properties, modifications to other physical characteristics may be observed, such as changes in dielectric properties.

## 11.1.1 Fick's law and single phase diffusion

Fick's law is applied to simple single free-phase diffusion of water into a material. In 1855, based on the experiments of T. GRAHAM Graham (1850), A. FICK was the first to propose a conceptualisation of molecular diffusion process by studying salt movement in liquids Fick (1855b,a). He drew considerable inspiration from the analytic theory of heat conduction published by J. FOURIER thirty-three years earlier Fourier (1822). Diffusion is the phenomenon governing the transport of material just as conduction is the mechanism of heat transfer. In the case of molecular diffusion, molecules are transported from a higher concentration region to one of lower concentration; it is a spontaneous phenomenon. For a transient state diffusion ($t$ denoting time), to know the rate at which the concentration $C$ is changing at any given point in space, we have to use Fick's second law Equation (11.1). Where $\underline{\nabla}$ denotes the divergence operator, $\underline{J}$ the "diffusive flux vector field" and $D$, the "diffusion coefficient" or "diffusivity" with dimensions of [length$^2$.time$^{-1}$]. The diffusion coefficient can be seen to be analogous to the "thermal diffusivity" of the heat conduction equation, i.e. the thermal conductivity divided by density and specific heat capacity at constant pressure.

$$-\frac{\partial}{\partial t}\int_\Omega C\,\mathrm{d}\Omega = \int_\Omega \underline{\nabla}.\underline{J}\,\mathrm{d}\Omega \quad\Rightarrow\quad \frac{\partial C}{\partial t} = -\underline{\nabla}.\underline{J} = \underline{\nabla}.\left(D\,\underline{\nabla}C\right) \tag{11.1}$$

Equation (11.1) represents a molecular transport mechanism which tends to homogenise the intensive quantity (concentration) even in the absence of macroscopic motion of the medium. $D$ appears here as a scalar coefficient but in general, water can be absorbed in different directions which may have different absorption coefficients and it is sometimes necessary to use the diffusivity in a tensorial form, i.e. $\underline{D}$. Considering a Cartesian coordinate system such that the diffusivity tensor can be expressed in diagonal form, Equation (11.1) reduces to Equation (11.2), $x$, $y$ and $z$ indicating the space variables.

$$\frac{\partial C}{\partial t} = \frac{\partial}{\partial x}\left(D_x\frac{\partial C}{\partial x}\right) + \frac{\partial}{\partial y}\left(D_y\frac{\partial C}{\partial y}\right) + \frac{\partial}{\partial z}\left(D_z\frac{\partial C}{\partial z}\right) \tag{11.2}$$

Most composite structures are however basically two dimensional as they are made of layers of fibres, often in the form of unidirectional lamina or woven cloth. This means that for many composite structures the thickness of the material is much less than the lateral dimensions so that the problem can be reduced to the penetration of water at right angles to the plane of the fibres. This can cause difficulties in interpreting laboratory specimens subjected to ageing experiments in a humid environment as such specimens are often of small size so that the surface areas of the edges can be a significant fraction of the total surface area of the material. This is exacerbated by the diffusion coefficient being greater in the direction of the fibres. We have all seen examples of the effects of water uptake in the fibre direction leading to accelerated deterioration of the material. Wooden posts placed in the ground rot from the bottom up because of the diffusion of water up the wood along the grain, or fibre, direction. However most artificial composite structures are big enough for the effects of water uptake in the fibre direction can be neglected and the interpretation of laboratory results can be misleading. Great care should be taken with samples, but if only one direction of diffusion can be considered; for example in the case of a thin plate where the lateral diffusion can be neglected compared to that in the thickness, Equation (11.2) simplifes further and becomes Equation (11.3).

$$\frac{\partial C}{\partial t} = \frac{\partial}{\partial x}\left(D\,\frac{\partial C}{\partial x}\right) \tag{11.3}$$

In Equation (11.3), $D$ is implicitly considered to be dependent on the values of $x$ but considering it constant leads to Equation (11.4), usually considered as a first approach for thin plates.

$$\frac{\partial C}{\partial t} = D\,\frac{\partial^2 C}{\partial x^2} \tag{11.4}$$

### 11.1.2   Water sorption, experimental observations

The water uptake in a material can be characterised by two physical quantities: the "diffusion coefficient" related to the kinetics of absorption and the "saturated concentration" related to the maximum amount of water that the material can absorb. These two parameters can be determined with thin plates and sorption curves connecting the exposure time to the hydro-thermal environment. In practice it is the variation in weight as the composite plate absorbs water which is measured.

As already indicated, to produce the results presented in this chapter, a short glass fibre reinforced PPA has been considered. The PPA considered was a block copolymer PA6T-PA6I for which all the adipic entering the composition of PA66 had been replaced by terephthalic and isophthalic acids. This copolymer was then reinforced by short glass fibres (50 % in mass). The samples were one millimetre thick injection moulded plates, with an area of 100 mm×100 mm, allowing a uni-dimensional diffusion through the thickness to be considered, thus eliminating the difficulties associated with the injection induced anisotropy Mazé (2012); Boukhoulda et al. (2006). It is worth mentioning that as long as the diffusion was kept uni-dimensional, experiments with greater thicknesses, although longer, allowed the standardisation of the sorption curves in relation to the thickness to be verified. This has been verified for the material considered up to 4 mm Mazé (2012) and it is very useful for modeling purposes.

Before ageing, all samples were dried over silica gel in an oven at 40 °C and stayed at this temperature until stabilisation of the weight. In this way the reference dry level for measuring the water uptake could be identified. Mass gains as a result of water uptake, were then recorded by removing the sample from its ageing environment and by weighing it periodically on a precision balance. During the sorption experiment, the evolution of the water concentration $C$ in the composite could be followed as a function of time $t$ by evaluating the ratio between the mass uptake $\Delta M = M(t) - M_0$ and the initial reference mass of the specimen $M_0$. It was then possible to know the sorption kinetics and if an equilibrium occurred, determine the saturated concentration value $C_\infty$. The saturated concentration value is clearly related to the dry reference level presented above.

Sorption kinetics are presented in Figure 11.1 for a multitude of hydro-thermal ageing conditions. Experimental data points correspond to the average of three samples with a corrected standard deviation of less than 3 % for the largest variability.

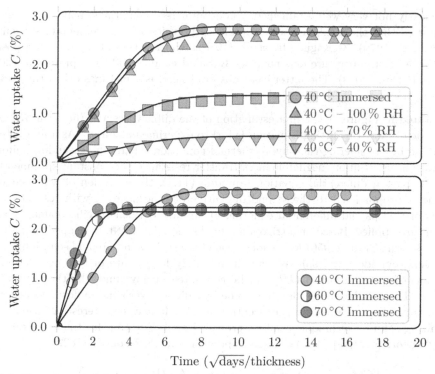

Figure 11.1: Experimental data points (symbols) obtained from the average value of three samples (short glass fibre reinforced PPA) for each hydro-thermal ageing condition: isothermal (top) and immersed (bottom) conditions. The fitted (solid) curves correspond to the analytical identification of the diffusion parameters $D$ and $C_\infty$ using Equation (11.4).

These curves are very informative and several observations can be made on them:

- Below 70 °C and whatever moisture conditions studied, no significant mass loss of the material was observed.

- According to Fick's equation, for a plate of finite thickness, water will initially diffuse and create a weight increase as a linear function of the square root of time (see Equation (11.7)). Water uptake then tends to saturation, clearly visible on Figure 11.1.

- Considering the results obtained for isothermal conditions (40 °C) but with varied humidities or water immersion (Figure 11.1, top), it seems at first sight that only the saturated concentration is affected by the water environmental content. The higher the water concentration in the surrounding environment, the more the saturated concentration is high. This is a classical result often observed for many polymer matrix composites. Nevertheless, as will be discussed in the next section, for the material under consideration, diffusivity also increases with the water content. This will be revealed by a normalisation of the curves.

- Considering now the case of water immersion at different temperatures (Figure 11.1, bottom), it can be clearly seen that the diffusion kinetics were thermally activated, i.e. the higher the temperature, the larger is the diffusivity and the faster is the diffusion. This result is expected and observed for most composite materials.

- Considering again these water immersion results (Figure 11.1, bottom), it is obvious that the saturated concentration dropped as the temperature increased. From a scientific point of view, the temperature dependence of the equilibrium moisture content is clearly not very well established. Under immersed conditions, many authors have reported that the maximum water uptake is insensitive to the temperature Shen and Springer (1976); McKague, Jr. et al. (1976), whilst others observed either positive or negative temperature dependencies El-Sa'ad et al. (1990); Chaplin et al. (2000); Karad et al. (2002). The latter case, observed here, is sometimes called the "reverse thermal effect".

As plotted in Figure 11.1, a first estimation of the diffusivities and the saturated concentrations can be obtained by fitting Equation (11.4) to experimental curves. As a first attempt, this can be done by considering each hydro-thermal condition separately, assuming uniform initial concentration distribution, negligible mass transfer resistance and constant specimen thickness.

For the partial differential equation Equation (11.4), the evolution of water concentration with time is proportional to the second derivative of concentration with respect to distance. Solutions of this partial differential equation are obtained when specific boundary and initial conditions are applied. Based on a trigonometrical series, J. CRANK developed simplified mathematical solutions Crank (1956) for various geometries such as infinite and semi-infinite medium, plane sheets, cylinders and spheres. For an infinitely large plate of thickness $h$ (membrane), the amount of water absorbed $C(t)$ can be computed at any time $t$ by the expression Equation (11.5). The membrane is considered to be initially at a zero and uniform concentration and its surfaces are kept at a constant concentration $C_\infty$. It may be interesting to point out that the following solution technique was first proposed by J. FOURIER for the heat equation in his treatise "Théorie analytique de la chaleur", published in 1822 Fourier (1822).

$$\frac{C(t)}{C_\infty} = 1 - \frac{8}{\pi^2} \sum_{n=0}^{\infty} \frac{1}{(2n+1)^2} \exp\left(\frac{-D\pi^2(2n+1)^2}{h^2}t\right) \tag{11.5}$$

The two parameters of a Fickean diffusion can be clearly seen to appear in Equation (11.5): $D$, which is here independent of time, of space and of the concentration of the diffusing species, and $C_\infty$ which is the water uptake at saturation. For a long-term absorption, when $t > \tau/15$ (with $\tau = h^2/D$, the characteristic diffusion time), we can approximate the Fickean kinetic behaviour described in Equation (11.5) by the relation Equation (11.6). This simplified expression allows a first estimation of the diffusivities and saturated concentrations to be obtained. As indicated above, fitted curves are shown in Figure 11.1.

$$\frac{C(t)}{C_\infty} \approx 1 - \frac{8}{\pi^2} \exp\left(\frac{-D\pi^2}{h^2}t\right) \tag{11.6}$$

For short-term absorption, i.e. when $t < \tau/15$, Equation (11.5) can be further simplified leading to Equation (11.7). The initial linear relationship, of $C(t)/C_\infty$ as a function of the square root of time is to be noted.

$$\frac{C(t)}{C_\infty} \approx \frac{4}{h}\sqrt{\frac{Dt}{\pi}} \tag{11.7}$$

It is considered that the first molecular layer of the specimen reaches equilibrium with the environment instantly so that the water concentration at this point is always equal to the equilibrium concentration $C_\infty$. A composite, or other material, which absorbs water and which is subjected to varying degrees of humidity in the environment will experience a variation of water content at the surface and the centre of the material may only begin to experience water diffusion after a delay which can be very long depending on the thickness.

If we consider a large, initially dry, composite plate of thickness $h$, placed in constant humid conditions at a constant temperature, water will diffuse through both surfaces in the direction, normal to the surfaces, which we shall designate as the $x$ direction, with the origin being at the centre of the plate. The surfaces are considered to reach saturation instantaneously but the

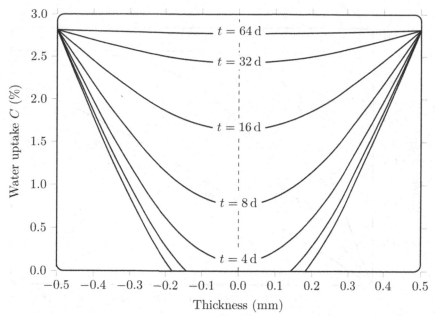

Figure 11.2: Water concentration profiles as diffusion proceeds through both faces of a large composite specimen of thickness $h = 1$ mm, for increasing lengths of time (from 1 day to 64 days, when the saturated level is reached). The material data are those of Figure 11.1, 40 °C immersed condition. Although the thickness is small, the times for water to reach the centre of the composite and for the composite to become saturated are quite long, a little less than 4 days and a little more than 2 months respectively.

diffusion takes time so that, as Figure 11.2 shows, the water concentration will vary within the material as a function of time. The equation for the change of water concentration at a distance $x$, from the centre of the plate is given by Equation (11.8).

$$\frac{C(t)}{C_\infty} = 1 - \frac{4}{\pi} \sum_{n=0}^{\infty} \frac{(-1)^n}{2n+1} \cos\left(\frac{\pi(2n+1)}{h}x\right) \exp\left(\frac{-D\pi^2(2n+1)^2}{h^2}t\right) \qquad (11.8)$$

Figure 11.2 shows in graphic form the change of water concentration profile as a function of time, $t$, which is given by Equation (11.8). It can be seen that the water concentration is a maximum at $C_\infty$ at the surfaces from the beginning of the diffusion but that water has to diffuse to the centre and that this takes time. Saturation is reached when the water concentration profile is uniform across the thickness of the plate which in the case of thick specimens could take years. Figure 11.3 shows the results of calculations, using Eqs. (11.6) and (11.7), for the composite under consideration subjected to 40 °C immersed conditions. The coefficient of diffusion has been determined using the results presented in Figure 11.1 and its value is about $8 \times 10^{-8}$ mm$^2$ s$^{-1}$. It can be seen that for a thickness of 1 mm the composite is nearly completely saturated after two months immersion but that it would take nine months with a specimen twice as thick and almost three years for a specimen four times as thick.

The times for water to arrive at the centre of the composite specimens and also for them to become fully saturated are given in Figure 11.4.

It is only when the composite has reached saturation under the prevailing humidity conditions that a uniform water concentration across the specimen exists. This should be a warning for conducting experiments to reveal the effect of water uptake on composite behaviour as if the specimen has not reached saturation a complex diffusion state exists which most probably will give unreliable results.

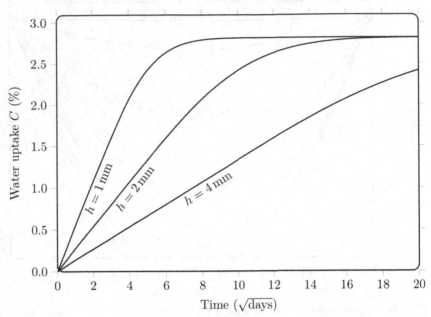

Figure 11.3: The initial increase in water concentration due to Fickian absorption of water by three composites plates (short glass fibre reinforced PPA subjected to 40 °C immersed condition) of different thicknesses. The time required to reach saturation is multiplied by a factor 4 by doubling the thickness.

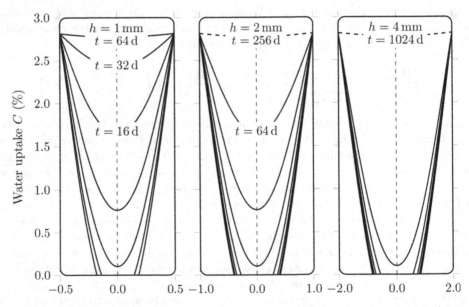

Figure 11.4: The change of water concentration in three similar composite plates (short glass fibre reinforced PPA subjected to 40 °C immersed condition) but of different thicknesses: 1 mm to 4 mm. To ease the comparison, the left profile is the one presented in Figure 11.2. The black curves correspond to the same exposure times whatever the thickness. For 2 mm and 4 mm, dashed curves correspond to the time to reach a quasi saturation.

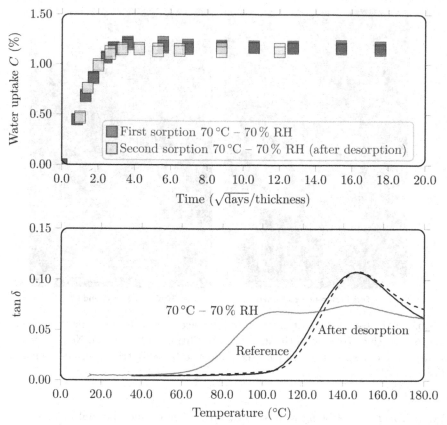

Figure 11.5: At the top, first and second water uptake measurements for three samples showing a small experimental scatter and supporting the hypothesis of reversible sorption effects. At the bottom, $\tan\delta$ curves obtained for unaged (reference), aged and rejuvenated (after desorption) samples. The good superposition of the "unaged" and "rejuvenated" curves is also a sign of the water sorption reversibility.

As has already been pointed out, no significant mass loss of the material under consideration can be observed in Figure 11.1. Additional analyses make it possible to study the reversibility of the sorption phenomenon. At the top of Figure 11.5, sorption data points are displayed for three samples under isothermal 70 °C and 70 % relative humidity conditions; the samples were subjected to a first sorption (dark symbols), desorption (drying at 70 °C not shown) and a further sorption (light symbols) for which the results are superimposed to the first obtained. Moreover, even if the $T_g$ was modified during sorption, it recovered after desorption (Figure 11.5, bottom).

From a microstructure point of view, no significant damage, i.e. fibre/matrix debonding, was observed in the saturated material at the microscopic level (SEM in Figure 11.6). For the material under consideration, no chemical ageing or microstructural irreversible changes seemed to have occurred. Nevertheless, effects of physicochemical changes are sometimes observed in resin matrix composites and this should be viewed with caution.

## 11.1.3 Physicochemical aspects of absorption

Most polymers absorb water to a greater or lesser degree so that, even in the absence of defects or preferential routes for water uptake such as are provided by poor fibre-matrix adhesion, resin-matrix composites will absorb water in humid environments. Water diffusion occurs in most

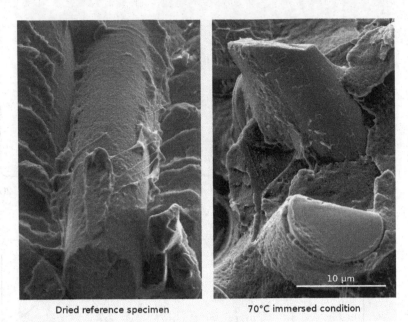

Dried reference specimen          70°C immersed condition

Figure 11.6: SEM observations of the fracture surface of a dried reference specimen (left) and after water saturation (70°C with the immersed condition on the right). No significant difference between dry and wet specimens can be seen and the fibre/matrix interface seems not to have been affected.

resins by hydrogen bonds being formed between the water and the molecular structure of the resin.

Epoxy resins are the most widely used resins in high-performance composite structures. They are mixed with a hardener, which can be of several types, to produce a crosslinked structure which provides many sites for hydrogen bonding for the water molecules such as hydroxyl groups, phenol groups, amine groups and sulfone groups. The hydrolysis of epoxy resins is in itself a cause of degradation and is not by any means limited to this class of matrix materials.

The use of an anhydride as the cross-linking agent, which is extremely common, produces a composite which, in a humid environment which is warmer than about 40 °C, degrades quite rapidly. Immersed in warm water this composite system shows an initial weight gain due to the absorbed water but this reaches a maximum after which the weight falls due to leaching of the resin system into the water. Saturation is not achieved and severe weight loss after drying is always observed. An anhydride is an acid with the water removed so that the combination of an anhydride, left over from the crosslinking process and warm water produces an acid which can quickly degrade the composite.

With a diamine hardener, it can been shown that the water uptake behaviour leads to an initial weight gain as a function of the square root of time as observed for the PPA in Figure 11.1. The composite reaches an equilibrium with its environment and saturates and the saturation level is a function of relative humidity and its independent of temperature. This behaviour indicates that the water is absorbed by the resin and is linked to it by hydrogen bonds. The saturated composite can reveal a somewhat increased tenacity due to a softening of the matrix which is accompanied by a fall in glass transition temperature.

A change to a dicyandiamide hardener produces, however, a significant modification in water-uptake kinetics: two plateaux in the sorption curves can be observed. The saturation is attained after considerably longer exposure than with the diamine-hardened resin but after equilibrium is reached both composites behave in a similar manner. This behaviour can no longer be handled by the single phase diffusion model. The Langmuir's model of diffusion is usually used. In applying this model it is often assumed that the diffusion coefficient remains independent of water concentration as in the Fickian model. However, the water is considered to be in two phases, one free to diffuse and the other trapped and so not free to move in the absorbing medium.

Polymer resins are not the only constituent to be affected by water. If carbon fibres remain unaffected, others, such as those of glass, can be damaged by prolonged exposure to water which arrives at the interface. Glasses are made up of silica in which are dispersed metallic oxides including those of the alkali metals. These latter non-silicate constituents represent micro-heterogeneities which are hydroscopic and hydrolysable. The absorption of water by glass is therefore characterised by the hydration of these oxides. The most common form of glass fibre is made of E-glass, which contains only small amounts of alkaline-metal oxides and so is resistant to damage by water. However, the presence of water at the glass-fibre surface may lower its surface energy and promote crack growth. Water trapped at the interface may also allow components of the resin to go into solution and if an acid environment is formed the fibre can be degraded.

Organic fibres such as the aramid family can absorb considerable quantities of water and although this does not lead to a marked deterioration of fibre properties the resulting swelling of the fibre may be a cause of composite degradation. It will certainly change the dielectric properties of the composite which in some applications, such as in radomes for which electromagnetic transparency is important, may present a problem .

The size, or coating, put onto many fibres to protect them from abrasion and to ensure bonding with the matrix also serves to protect them and the composite from damage from water absorption by eliminating sites at which the water can accumulate. However, this is not always adequate.

## 11.2 The hydro-thermal diffusivity dependency

From a mechanical point of view, the consequences of environmental ageing on the material are intimately linked to water concentration and temperature. It is also unlikely that the two faces of the material will be subjected to the same hydro-thermal conditions. It is therefore essential to identify the kinetics of the water uptake in order to establish a clear picture of the water concentration in the material, within the thickness, at any time during its use. Knowing the concentration gradient, it is then possible to determine the mechanical behaviour model of the material dependent on this local information. A lot of work in this direction is regularly published for composite structures. From an industrial point of view, however, the problem can be simplified by evaluating only the time required for the material to be saturated with water. The mechanical calculation is then based on a homogeneous concentration parameter over the entire composite material and constitutes the lower limit of its mechanical performance.

### 11.2.1 The thermal effect, an Arrhenius-type rate theory

Most reactions are faster when the temperature rises. This is the essence of the argument advanced by the Swedish scientist S. ARRHENIUS (1859–1927) for which he gained the Nobel Prize in Chemistry in 1903, by proposing an empirical law that reflects this phenomenon observed experimentally in many cases. He wrote his paper of 1889 by describing reaction rates of several published chemical reactions Arrhenius (1889). Although more rarely cited, Arrhenius' work was built on the earlier results of the Dutch scientist J.H. VAN'T HOFF (1852–1921) van't Hoff (1884), the first Nobel laureate in Chemistry in 1901.

Assuming that the "activation energy" $E_a$ does not depend on the temperature, which is only a reasonable assumption over a limited temperature range, the Arrhenius law for the diffusivity fits equation Equation (11.9) where $D_0$ represents the pre-exponential factor, $R$ denotes the ideal

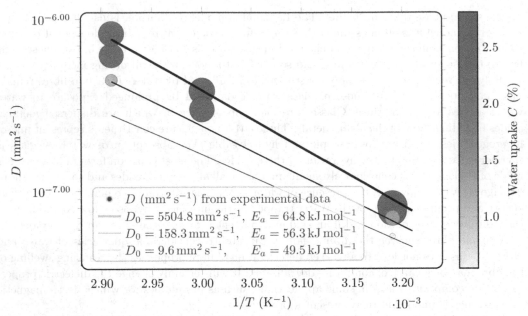

Figure 11.7: The diffusivity $D$ versus $1/T$: for similar concentrations, experimental points seems to be aligned. Bubble size and colour are associated with the equilibrium concentration and the diffusivity increases with the water environmental content.

gas constant ($R = 8.314\,\mathrm{J\,mol^{-1}\,K^{-1}}$) and $T$ the temperature in Kelvin. Depending on the resin used, $E_a$ is generally in the range $40\,\mathrm{kJ\,mol^{-1}}$ to $80\,\mathrm{kJ\,mol^{-1}}$ as will be seen for the short fibre reinforced PPA under consideration. It may be interesting to note that, from a historical point of view, it took almost thirty years and the work of S. DUSHMAN and I. LANGMUIR Dushman and Langmuir (1922) to consider that the temperature dependence of the diffusivity in solids obeys an Arrhenius law.

$$D = D_0 \exp\left(\frac{-E_a}{RT}\right) \tag{11.9}$$

By plotting, the coefficient of diffusivity, $D$ (obtained experimentally for various thermal conditions), as a function of $T$ (or rather $1/T$), it is possible to identify the values of $D_0$ and $E_a$ (Figure 11.7). Using a logarithmic scale, if the temperature dependence of the water uptake obeys an Arrhenius law, a straight line is obtained with $E_a$ as the directing coefficient and $D_0$ as the $y$-intercept.

Because of the alignment of experimental points obtained in Figure 11.7, we can postulate that, in the range of temperatures and concentrations considered, the water diffusion in the PPA reinforced composite studied followed an Arrhenius rate type relation. In Figure 11.7, a network of lines is used to schematically show this relation but also shows that the "dependency" is not only related to the temperature but also involves the water concentration in the material.

## 11.2.2   Modelling the diffusivity dependency

As pointed out previously for the short glass fibre reinforced PPA under consideration, if it seems at first sight that subjected to isothermal conditions only the saturated concentration is affected by the water environmental content (Figure 11.1, top), the kinetics of diffusion do also seem to be affected by the water concentration.

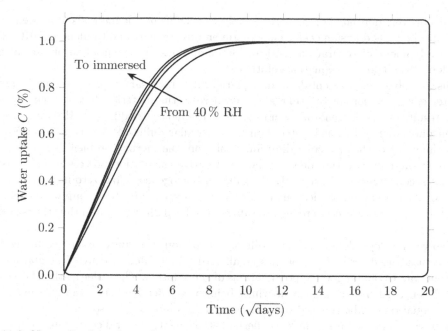

Figure 11.8: Normalised isothermal sorption kinetic simulations (Figure 11.1, top) highlighting the concentration dependency: the curves do not overlap.

Indeed, normalising and comparing the isothermal water uptake simulations shows a manifest variation between diffusivities (Figure 11.8)) since the curves do not overlap: diffusivity increases with the water content.

To take into account the above observations, the diffusivity can be described by a modified Arrhenius law for which both the pre-exponential factor and the activation energy depend on the water concentration. Two simple expressions for $D_0 : C \mapsto D_0(C)$ and $E_a : C \mapsto E_a(C)$ have been proposed in Joannès et al. (2014a). This additional degree of freedom can be justified by taking into consideration the effect of water on the polymer and in particular its affect on the glass transition temperature, which is closely linked to the activation energy. Indeed, $T_g$ is a major transition for many polymers as the material goes from a hard glassy to a rubbery state, as the temperature increases and occurs when large segments of the polymer start to move. Once the parameters of the model have been identified, it is possible to simulate non-symmetrical conditions of diffusion through the thickness of the composite. By following this approach, simulations should be limited to a temperature range from room temperature up to 70 °C (no irreversible phenomenon seems to occur). It is nevertheless possible to extend the lower bound by a few degrees, by extrapolation. Given this "double" dependence, Equation (11.4) to Equation (11.8) are no longer valid, Equation (11.3) still applies but another solution method is required. A rather simple alternative method is presented in the following section because it generally applies to differential equations based on Equation (11.3) but involving other physicochemical mechanisms.

## 11.2.3 Solving the concentration-dependent equation

For an arbitrary concentration dependency of the diffusion coefficient, it is usually not possible to give an analytical solution. An alternative Boltzmann-Matano Boltzmann (1894); Matano (1933); Philip (1960); Stenlund (2011) method permits the determination of a concentration-dependent diffusion coefficient from an experimental interdiffusion profile, which is nevertheless difficult to obtain. Consequently, the mathematics of concentration-dependent diffusion coefficients have almost exclusively depended on numerical analysis, even if specific methods have been developed to measure the mass diffusion coefficient experimentally Veilleux and Coulombe (2010). For most cases, development of accurate numerical approximations are essential to obtain quantitative evaluation of the transient equation solution.

Thus, to solve the transient diffusion equation, it is necessary to know the temperature and the water concentration at any time and anywhere in the material. It is of course possible to link the results of the thermal diffusion to those of the molecular diffusion. However, the thermal diffusion phenomenon is much faster than the molecular diffusion so it can be considered that at each time step of the molecular diffusion resolution, the thermal problem is in a steady state. Concerning the concentration dependency, an iterative calculation process must be used so as to impose a convergence criterion. By choosing an homogeneous and zero initial concentration, a first calculation can be carried out to distribute, in space and time, improved concentration values; these are then reused to refine the value of the local diffusivity and thus the concentration itself.

To solve the dependent coefficient diffusion equation, the finite difference method can be efficiently used as described in Joannès et al. (2014b). It offers a convenient and an easy-to-implement way to obtain good approximations of the solution. The Euler method is the simplest finite difference scheme to implement whilst being very effective. The concentration-dependent diffusion equation can be reduced to a matrix system, quite easy to solve.

The first step in the finite difference method is to construct a grid containing points on which we are interested in solving values Equation (11.3). To approximate this partial differential equation by a set of algebraic relations between the values of $C$ at points (grid nodes) $x_1, x_2, \cdots, x_i$, we need to replace the continuous derivatives with their finite difference approximations. Such approximations can be derived through the use of Taylor series.

The finite difference method gives an approximate solution for $C(x, t)$ at a finite set of $x$ and $t$. At this stage, we consider that discrete $x$ ($x_{i, i=1 \text{ to } N}$) and $t$ ($t_{m, m=1 \text{ to } P}$) are uniformly spaced. $C_i^m$ is the approximate numerical solution obtained by solving the finite difference equations. It might be as close as possible to $C(x_i, t_m)$ which is the continuous solution evaluated at the grid points. Let's choose the backward difference technique to approximate the time derivative on the left hand side of equation Equation (11.3) and a central difference for the right hand side. At a general node $x_i$, the partial time derivative term can be approximated using Equation (11.10) where $m$ is related to the current time step and $m - 1$ to the previous one.

$$\left. \frac{\partial C}{\partial t} \right|_{t_m, x_i} = \frac{C_i^m - C_i^{m-1}}{\Delta t} + \mathcal{O}(\Delta t) \tag{11.10}$$

The right hand side can be approximated using a central difference to estimate $\partial C / \partial x$ at the mid-points $x_{i+1/2}$ Equation (11.11) and $x_{i-1/2}$ Equation (11.12), followed by a central difference between those values to estimate the second derivative term Equation (11.13).

$$\left. \frac{\partial C}{\partial x} \right|_{t_m, x_{i-1/2}} = \frac{C_i^m - C_{i-1}^m}{\Delta x} + \mathcal{O}(\Delta x) \tag{11.11}$$

$$\left. \frac{\partial C}{\partial x} \right|_{t_m, x_{i+1/2}} = \frac{C_{i+1}^m - C_i^m}{\Delta x} + \mathcal{O}(\Delta x) \tag{11.12}$$

$$\frac{\partial}{\partial x}\left(D\frac{\partial C}{\partial x}\right)\bigg|_{t_m,x_i} = \frac{D_{i+1/2}^m \frac{\partial C}{\partial x}\big|_{t_m,x_{i+1/2}} - D_{i-1/2}^m \frac{\partial C}{\partial x}\big|_{t_m,x_{i-1/2}}}{\Delta x} + \mathcal{O}(\Delta x^2) \qquad (11.13)$$

$$= \frac{D_{i+1/2}^m\left(C_{i+1}^m - C_i^m\right) - D_{i-1/2}^m\left(C_i^m - C_{i-1}^m\right)}{\Delta x^2} + \mathcal{O}(\Delta x^2)$$

$$= \frac{D_{i-1/2}^m C_{i-1}^m - \left(D_{i+1/2}^m + D_{i-1/2}^m\right)C_i^m + D_{i+1/2}^m C_{i+1}^m}{\Delta x^2} + \mathcal{O}(\Delta x^2)$$

The values of $D$ at the mid-points $x_{i-1/2}$ and $x_{i+1/2}$ can be estimated by averaging $D^m$ respectively between $x_{i-1}$ to $x_i$ and $x_i$ to $x_{i+1}$ Equation (11.14).

$$D_{i-1/2}^m = \frac{D_{i-1}^m + D_i^m}{2} \quad \text{and} \quad D_{i+1/2}^m = \frac{D_i^m + D_{i+1}^m}{2} \qquad (11.14)$$

Thus by rearranging Equation (11.13), we obtain the right hand side approximation Equation (11.15).

$$\frac{\partial}{\partial x}\left(D\frac{\partial C}{\partial x}\right)\bigg|_{t_m,x_i} = \frac{D_i^m + D_{i-1}^m}{2\Delta x^2} C_{i-1}^m$$

$$- \frac{D_{i+1}^m + 2D_i^m + D_{i-1}^m}{2\Delta x^2} C_i^m$$

$$+ \frac{D_{i+1}^m + D_i^m}{2\Delta x^2} C_{i+1}^m + \mathcal{O}(\Delta x^2) \qquad (11.15)$$

By combining Equation (11.10) and Equation (11.15) and dropping the truncation error terms, we get the relation Equation (11.16) which is the backward time and centered space (BTCS) uniform discretisation of Equation (11.3).

$$\frac{1}{\Delta t} C_i^{m-1} = - \frac{D_i^m + D_{i-1}^m}{2\Delta x^2} C_{i-1}^m$$

$$+ \left(\frac{1}{\Delta t} + \frac{D_{i+1}^m + 2D_i^m + D_{i-1}^m}{2\Delta x^2}\right) C_i^m$$

$$- \frac{D_{i+1}^m + D_i^m}{2\Delta x^2} C_{i+1}^m \qquad (11.16)$$

At a given time $m$, Equation (11.16) is one equation in a system of equations for the values of $C$ at the internal nodes of the spatial mesh. To see the system of equations more clearly, we can rearrange the resulting equation by writing Equation (11.17).

$$d_i = a_i C_{i-1}^m + b_i C_i^m + c_i C_{i+1}^m \qquad \text{for} \qquad i = 2 \text{ to } N-1 \qquad (11.17)$$

$$\text{with} \quad \begin{cases} a_i = -\dfrac{D_i^m + D_{i-1}^m}{2\Delta x^2} \\[2mm] b_i = \dfrac{1}{\Delta t} + \dfrac{D_{i+1}^m + 2D_i^m + D_{i-1}^m}{2\Delta x^2} \\[2mm] c_i = -\dfrac{D_{i+1}^m + D_i^m}{2\Delta x^2} \\[2mm] d_i = \dfrac{1}{\Delta t} C_i^{m-1} \end{cases}$$

Since we have chosen to apply Dirichlet boundary conditions, $C_1^m = C(x_0, t_m) = C_0$ implies $b_1 = 1$, $c_1 = 0$ and $d_1 = C_0$. In the same manner, $C_N^m = C(x_h, t_m) = C_h$ implies $a_N = 0$, $b_N = 1$

and $d_N = C_h$; $h$ still denoting the thickness of the plate. The system of equations can be represented in a matrix form as Equation (11.18).

$$
\begin{bmatrix}
b_1 = 1 & c_1 = 0 & 0 & 0 & \cdots & 0 & 0 & 0 \\
a_2 & b_2 & c_2 & 0 & \cdots & 0 & 0 & 0 \\
0 & a_3 & b_3 & c_3 & \cdots & 0 & 0 & 0 \\
\vdots & \vdots & \vdots & \vdots & & \vdots & \vdots & \vdots \\
0 & 0 & 0 & 0 & \cdots & a_{N-1} & b_{N-1} & c_{N-1} \\
0 & 0 & 0 & 0 & \cdots & 0 & a_N = 0 & b_N = 1
\end{bmatrix}
\begin{bmatrix}
C_1^m \\
C_2^m \\
C_3^m \\
\vdots \\
C_{N-1}^m \\
C_N^m
\end{bmatrix}
=
\begin{bmatrix}
C_0 \\
d_2 \\
d_3 \\
\vdots \\
d_{N-1} \\
C_h
\end{bmatrix}
\quad (11.18)
$$

The BTCS scheme is implicit. As a result the system Equation (11.17) or Equation (11.18) must be solved at each time step and is unconditionally numerically stable Thomas (1995). This tridiagonal system can be solved by using the L. THOMAS algorithm Thomas (1949) which is a simplified form of Gaussian elimination without pivoting.

This first scheme can be improved to converge faster and improved versions can be found in Joannès et al. (2014b). This "tool" is a building block which can be combined and extended to other mechanisms. It is of course possible to combine it with the classical laminate theory for mechanical considerations or to add some chemical reactions such as hydrolysis; it can be carried out very quickly and is simpler than with closed commercial codes.

## 11.3    Climatic ingresses and hot water applications

In order to make the design of composite structures easier, mechanical simulations often assume that the material is completely "dry" or entirely "wet". In the latter case, the weakened properties of the material, encountered at the water saturation level, are considered. Nevertheless, as it has been seen in this chapter, water absorption to saturation can take many years and this modelling assumption, often combined with high temperatures, therefore represents the worst condition for a component. These severe conditions are not necessarily realistic. For some thick structures, water saturation will never be reached during operating life of the component and the temperature can fluctuate throughout its life.

While it is essential to guarantee the safety and reliability of structures, it is also necessary to have a better knowledge of what the material is actually experiencing so that potential mass savings can be achieved. It is this knowledge that should make it possible to modify the geometry of the parts accordingly and help to choose the ideal material.

### 11.3.1    Some applied examples with the short fibre reinforced PPA

The tool implemented in the previous section for solving the concentration dependency of the diffusion transient equation can be used to simulate realistic cyclic hydro-thermal conditions Joannès et al. (2014b). As an example, the water uptake of a hot water supply component, made from the short fibre reinforced PPA under consideration in this chapter, can be considered. For didactic reasons, a fictive 4 millimetres thick wall equipment is studied and it is supposed to be installed in Europe. This equipment is thus subjected to the weather ingress on the one side and to liquid water (50 °C in this example) on the other side. Daily hydro-thermal conditions have been obtained from the European Climate Assessment & Dataset project (http://eca.knmi.nl). Three different places have been chosen as an example: Amsterdam-Schiphol airport in Netherlands, Geneva airport in Switzerland (drier climate than Schiphol for equivalent temperatures) and Malaga airport in Spain (Mediterranean climate with dry hot summers and mild winters). Whatever the chosen location, under those hydro-thermal conditions and using the model presented above, more than four years are needed to reach the sorption equilibrium (starting from the reference dry level discussed in § 11.1.2). Nevertheless, differences in kinetics appear and we can wonder how the water uptake is affected by temperature and humidity and if these two parameters play similar roles. Due to the inertia of the sorption kinetics highlighted in the previous sections, we can also wonder if it is necessary to use such detailed and daily climate data to evaluate the time to reach saturation or if knowing average values are sufficient.

Figure 11.9 summarises the situation mentioned above. The first graph (Figure 11.9, top) shows the water uptake versus time for the three cities. Due to the inertia of the sorption kinetic, the results of a day by day simulation (solid lines) are very close to those obtained with annual average values (dashed lines); there is thus no need to get daily climate data to obtain good approximations of the water uptake under any hydro-thermal condition. The difference in relative humidity observed between Schiphol and Geneva does not seem to have any effect on the sorption kinetics. The water uptake curves are superimposed for those two places, even if the weather is different. A bigger difference appears with Malaga and we can notice a faster water uptake in the case of Spain, even if the climate is drier. The effect of the temperature "inside" the material is preponderant in relation to changes in external environmental conditions.

Just as easily, the simulation allows the water diffusion within the material to be determined, which is difficult to obtain experimentally. We can then plot the water profiles throughout the thickness and versus time. By comparing the water profiles obtained for Schiphol, Geneva and Malaga (Figure 11.10), we can actually see that the water uptake is faster in the case of Spain, although hydro-thermal conditions seem less severe. As aleady said, this reflects the key role of the temperature in the sorption kinetic. The temperature is of the first order for this composite and without temperature, high humidity will have little influence on the water uptake. Given the asymmetric hydro-thermal conditions, it is not in the centre of the plate that the concentration is the highest and this may be of importance given the orientation of the layers of fibres which vary in thickness.

Indeed, obtaining the information in Equation (11.9) and Equation (11.10) is not an end in itself. The previous development allows the water concentration at any point in the material to be known. By having evaluated the mechanical behaviour of the material for different water concentrations (at saturation), it is possible to attribute those properties locally to run more realistic mechanical simulations. These results can lead to modifications to the geometry of the part in order to better control the weak areas.

## 11.3.2  Withstanding hot temperatures

Families of resins able to withstand higher temperatures than epoxy resins have been developed and in general are less susceptible to water uptake. Prolonged exposure of these resins to hot water can, however, lead to irreversible changes. Such a resin is polystyrylpyridine (PSP) which is usually reinforced with carbon fibres and destined for aerospace applications. This resin can operate at 250 °C for 1000 h and can even withstand several hours at 400 °C. However, prolonged exposure to humid conditions, for example at 70 °C and 100 % RH, reveals complex behaviour. Initial water uptake appears to be Fickian. Infrared spectroscopy shows that the water is linked to the polar PSP functions by hydrogen bonding, with apparent saturation being reached at this stage. Longer exposures lead, however, to further water uptake and the creation of microcracks as well as the hydrolysis of ethylene double bonds left over after crosslinking of the polymer. The effects of water on other high temperature resins, such as PMR-15 and Avimid-N, which are rated to be used above 300 °C, when dry, are to produce a fall in $T_g$ of around 80 °C during hydrothermal cycling.

Degradation of this type of resin matrix can also occur during thermal cycling, particularly if water is present in microcracks. The carbon-fibre-reinforced PSP composites show increased water absorption after cycling from 70 °C to 150 °C due to the water being vaporised and forcing

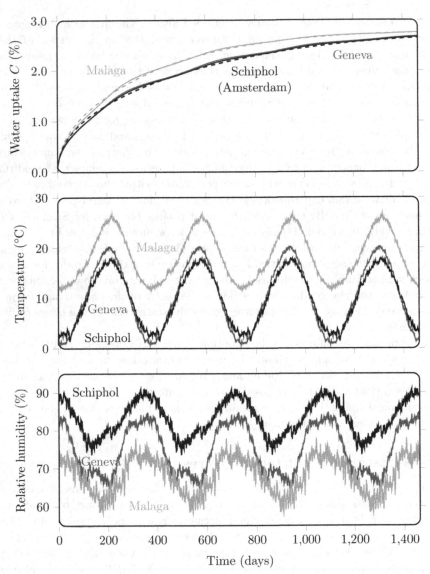

Figure 11.9: Top: comparison between a day by day sorption simulation and the water up-take obtained with temperature and relative humidity average values for three European cities. Centre and bottom: temperature and relative humidity obtained from the European Climate Assessment & Dataset project.

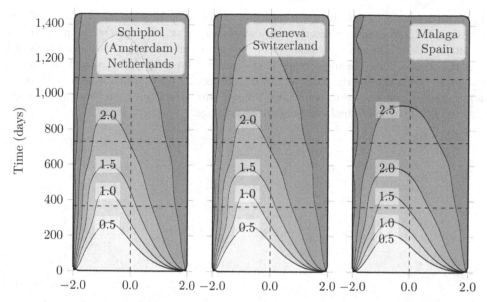

Figure 11.10: Water profiles (concentration percent) throughout the thickness of a four millimetre thick composite subjected to evolving environmental conditions on the left side and constant immersed temperature conditions on the opposite side for three European cities: Amsterdam-Schiphol (left), Geneva (center) and Malaga (right). The seasonal effect can be seen along the (left) vertical axis and the water uptake is faster in the case of Spain compared to the Netherlands and Switzerland. Horizontal dashed lines correspond to years.

the microcracks to further open. Cycling to 250 °C leads to a more marked fall in properties due to greater damage exacerbated by oxidation of the resin.

## 11.4   Conclusions

Environmental damage is a concern whether it is the rusting of steel, the oxidation of aluminium or water absorption by composites. Composites show themselves as being chemically resistant in many environments and the material of choice for many applications for which conventional materials would not be suitable. Water absorption is however often the principal concern when using composites. The properties of organic matrices are modified by water, which acts as a lubricant at the molecular level and this can be reversible, leading to a fall in $T_g$, or irreversible, due to hydrolysis, depending on the nature of the induced changes.

Water uptake kinetics are thermally activated. It nevertheless takes a long time for the water to penetrate the core of the material. On plates a few millimetres thick, this can take several years for water to reach the center of the plate. On thick composites, it is likely that the exposure time in service will not be sufficient to achieve the full saturation of the material. This means that a water concentration gradient exists through the thickness of a composite part and this should be considered when evaluating the mechanical performances of the structure. Significant mass gains can be obtained by considering the right assumptions. It should also be remembered that very humid environmental conditions do not mean that the material will take up water more quickly. It is the temperature that plays the main role and even in the Atacama desert, one of the world's driest place, the water contained in the air penetrates the composite materials due to the temperature. Under these conditions, it is clear that hot water is the most critical situation for composite materials.

Penetrating the material, the water infiltrates more easily between the fibres and the matrix. Debonding of the fibre-matrix interface is always damaging and irreversible in organic matrix composites. This is a possibility as the water molecules can break the hydrogen bonds adhering the matrix to the fibres. In most cases it is an understanding of the role of the interface or interphase between the matrix material and the reinforcing fibres which is seen as the key to the understanding of the long-term behaviour of composite structures even though, in some cases, the properties of the individual components can change.

# Revision exercises

### Exercise 11.1

What is the driving force for water diffusion into a material?

### Exercise 11.2

What are the parameters which need to be obtained so as to predict water absorption according to the Fickian model? What are the assumptions in this model?

### Exercise 11.3

If water absorption obeys Fick's law what does this mean if specimens of various thicknesses are considered? If the specimen thickness is doubled does this only double the time to reach saturation?

### Exercise 11.4

What are the effects on water absorption of varying the temperature and relative humidity of the environment on a composite which obeys Fick's law of water uptake?

### Exercise 11.5

In most cases, explain why it is important not to consider a fully saturated material to design an efficient and reliable equipment.

### Exercise 11.6

What effects can be expected to be seen on the properties of the matrix material? Which of these are reversible?

### Exercise 11.7

If a composite is found to have lost strength after absorbing water but regains most of this loss on drying, explain why this does not mean that the processes which occurred during water uptake are reversible.

### Exercise 11.8

Water absorption has a dramatic effect on Nylon but it is largely reversible. Explain what is observed as water is absorbed.

### Exercise 11.9

Explain how the choice of hardener can determine whether a composite is stable or not in a humid environment. Why are anhydride cured expoxies susceptible to degradation in the presence of warm water?

### Exercise 11.10

What are the mechanisms that can occur at the fibre-matrix interface when water penetrates the material?

# Bibliography

Ageyeva, T., I. Sibikin, and J. G. Kovács (2019, October). A review of thermoplastic resin transfer molding: process modeling and simulation. *Polymers 11*(10), 1555.

Alix, S., J. Goimard, C. Morvan, and C. Baley (2009). Influence of pectin structure on the mechanical properties of flax fibres: A comparison between linseed-winter variety (Oliver) and a fibre-spring variety of flax (Hermes). In H. A. Schols, R. G. F. Visser, and A. G. J. Voragen (Eds.), *Pectins and Pectinases*. Wageningen Academic Pub.

Allen, D. (2001). Homogenization principles and their application to continuum damage mechanics. *Composites Sciences and Technology 61*, 2223–2230.

Allen, D., C. Harris, and S. Groves (1987). A thermomechanical constitutive theory for elastic composites with distributed damage, Parts I and II. *International Journal of Solids and Structures 23*(9), 1301–1338.

Andrieux, S. (1983). *Un Modèle de Matériau Microfissuré, Applications Aux Roches et Aux Bétons*. Thèse, Université Paris VI, France.

Arrhenius, S. (1889). On the reaction velocity of the inversion of cane sugar by acids. *Zeitschrift fur Physikalische Chemie 4*(2), 226–248.

Ashby, M. (2016). *Material Selection in Mechanical Design (Fifth Edition)*, Butterworth-Heinemann.

Aussedat, E., A. Thionnet, and J. Renard (1995). Comportement en compression des composites par une définition du mode de sollicitation en mécanique de l'endommagement. *Comptes-rendus de l'Académie des Sciences de Paris, Série II 321*, 533–540.

Aveston, J., G. Cooper, and A. Kelly (1971). Single and multiple fracture, the properties of fibre composites. *Proc. Conf. National Physical Laboratories IPC Sci. and Tech Press Ltd., London*, 15–24.

Aveston, J. and A. Kelly (1973). Theory of multiple fracture of fibrous composites. *Journal of Material Sciences 8*, 352–362.

Azzi, V. and S. Tsai (1965). A theory of the yielding and plastic flaw of anisotropic metals. *Experimental Mechanics 5*, 283–288.

Bacon, R. and W. Shalamon (2005). *High Temperature Resistant Fibers from Organic Polymers*, (Ed.) Preston J., Interscience Publishers New York.

Baley, C., A. Le Duigou, C. Morvan, and A. Bourmaud (2018). Tensile properties of flax fibers. In A. R. Bunsell (Ed.), *Handbook of Properties of Textile and Technical Fibres (Second Edition)*, The Textile Institute Book Series, pp. 275–300. Elsevier-Woodhead Publishing.

Bansal, N. P. (2005). *Handbook of Ceramic Composites*. Springer US.

Batdorf, S. (1982a). Tensile strength of unidirectionally reinforced composites - 1. *Journal of Reinforced Plastics and Composites 1*, 153–163.

Batdorf, S. (1982b). Tensile strength of unidirectionally reinforced composites - 2. *Journal of Reinforced Plastics and Composites 1*, 165–175.

Baxevanakis, C. (1994). *Comportement Statistique à Rupture Des Composites Stratifiés.* Thèse, Ecole des Mines de Paris.

Berger, M. and D. Jeulin (2003). Statistical analysis of the failure stresses of ceramic fibres: Dependence of the Weibull parameters on the gauge length, diameter variation and fluctuation of defect density. *Journal of Material Science 38*, 2913–2923.

Betten, J. (1981). Damage tensors in continuum mechanics. *Journal de Mécanique Théorique et Appliquée 2*, 13–32.

Betten, J. (1986). Applications of tensor functions to the formulation of constitutive equations involving damage and initial anisotropy. *Engineering Fracture Mechanics 25*, 573–584.

Beyerlein, I., C. Zhou, and L. Schadler (1998). Time evolution of stress redistribution around multiple fiber breaks in a composite with viscous and viscoelastic matrices. *International Journal of Solids and Structures 35*, 3177–3211.

Bisanda, E. T. N. and M. P. Ansell (1991, January). The effect of silane treatment on the mechanical and physical properties of sisal-epoxy composites. *Composites Science and Technology 41*(2), 165–178.

Blassiau, S. (2005). *Modélisation Des Phénomènes Microstructuraux Au Sein d'un Composite Unidirectionnel Carbone/Époxy et Prédiction de Durée de Vie : Contrôle et Qualification de Réservoirs Bobinés.* Thèse, Ecole des Mines de Paris.

Blassiau, S., A. Thionnet, and A. Bunsell (2006a). Micromechanisms of load transfert in a unidirectional carbon-fibre epoxy composite due to fibre failures. Part 1: Micromechanisms and 3D analysis of load transfert, the elastic case. *Composite Structures 74*, 303–318.

Blassiau, S., A. Thionnet, and A. Bunsell (2006b). Micromechanisms of load transfert in a unidirectional carbon-fibre epoxy composite due to fibre failures. Part 2: Influence of viscoelastic and plastic matrices on the mechanism of load transfert. *Composite Structures 74*, 319–331.

Blassiau, S., A. Thionnet, and A. Bunsell (2008). Micromechanisms of load transfert in a unidirectional carbon-fibre epoxy composite due to fibre failures. Part 3: Multiscale reconstruction of composite behaviour. *Composite Structures 83*, 312–323.

Boisot, G., L. Laiarinandrasana, J. Besson, C. Fond, and G. Hochstetter (2011). Experimental investigations and modeling of volume change induced by void growth in polyamide 11. *International Journal of Solids and Structures 48*(19), 2642–2654.

Boltzmann, L. (1894). Zur Integration der Diffusionsgleichung bei variabeln Diffusionscoefficienten. *Annalen der Physik und Chemie 53*(2), 959–964.

Boukhoulda, B., E. Adda-Bedia, and K. Madani (2006). The effect of fiber orientation angle in composite materials on moisture absorption and material degradation after hygrothermal ageing. *Composite Structures 74*(4), 406–418.

Bourmaud, A., J. Beaugrand, D. U. Shah, V. Placet, and C. Baley (2018, August). Towards the design of high-performance plant fibre composites. *Progress in Materials Science 97*, 347–408.

Bunsell, A. and J. Renard (2005). *Fundamentals of Fibre Reinforced Composite Materials.* Taylor & Francis.

Bunsell, A. R. (2018). *Handbook of Properties of Textile and Technical Fibres.* The Textile Institute Book Series. Elsevier-Woodhead Publishing.

Bunsell, A. R. and M.-H. Berger (1999). *Fine Ceramic Fibers.* New York: Marcel Dekker. OCLC: 40762807.

Bunsell, A. R. and J. W. S. Hearle (1971, October). A mechanism of fatigue failure in nylon fibres. *Journal of Materials Science 6*(10), 1303–1311.

Camanho, P. P., P. Maimi, and C. G. Davila (2007). Prediction of the size effects in notched laminates using continuum damage mechanics. *Composites Sciences and Technology 67*, 2715–2727.

Camara, S., A. Bunsell, A., and D. Allen (2011). Determination of lifetime probabilities of carbon fibre composite plates and pressure vessels for hydrogen storage. *International Journal of Hydrogen Energy 36*, 6031–6038.

Cauvin, A. and R. Testa (1999). Damage mechanics : basic variables in continuum theory. *International Journal of Solids and Structures 36*, 747–761.

Chaboche, J. (1993). Development of continuum damage mechanics for elastic solids sustaining anisotropic and unilateral damage. *International Journal of Damage Mechanics 2*, 311–329.

Chaplin, A., I. Hamerton, H. Herman, A. Mudhar, and S. Shaw (2000). Studying water uptake effects in resins based on cyanate ester/bismaleimide blends. *Polymer 41*(11), 3945–3956.

Chou, H., A. Bunsell, G. Mair, and A. Thionnet (2013). Effect of the loading rate on ultimate strength of composites. Application: Pressure vessel slow burst test. *Composites Structures 104*, 144–153.

Chou, H., A. Thionnet, A. Mouritz, and A. Bunsell (2015). Stochastic factors controlling the failure of carbon/epoxy composites. *Journal of Materials Science 51*(1), 311–333.

Colomban, P. and V. Jauzein (2018). Silk: Fibers, films, and composites - types, processing, structure, and mechanics. In A. R. Bunsell (Ed.), *Handbook of Properties of Textile and Technical Fibres (Second Edition)*, The Textile Institute Book Series, pp. 137–183. Elsevier-Woodhead Publishing.

Cordebois, J. and F. Sidoroff (1979). *Damage Induced Elastic Anisotropy.* Colloque Euromech.

Cox, H. (1951). The elasticity and strength of paper and other fibrous materials. *British Journal of Applied Physics 12*, 72–79.

Crank, J. (1956). *The Mathematics of Diffusion.* Oxford University Press, London.

Crosky, A., C. Grant, D. Kelly, X. Legrand, and G. Pearce (2015, January). Fibre placement processes for composites manufacture. In P. Boisse (Ed.), *Advances in Composites Manufacturing and Process Design*, pp. 79–92. Woodhead Publishing.

Curtin, W. (2000). Dimensionality and size effects on the strength of fiber-reinforced composites. *Composites Sciences and Technology 60*, 543–551.

Curtin, W. and M. Ibnabdeljalil (1997). Strength and reliability of fiber-reinforced composites: Localized load sharing and associated size effects. *International Journal of Solids and structures 34*, 2649–2668.

Curtin, W. and N. Takeda (1998). Tensile strength of fiber-reinforced composites: I. model and effects of local fiber geometry. *Journal of Composite Materials 32*, 2042–2059.

Derksen, H. and G. Kemper (2002). Computational Invariant Theory: Encyclopedia of Mathematical Sciences. Springer - Verlag.

Dieudonné, J. and J. Carrel (1971). *Invariant Theory: Old and New*. Academic Press.

Dumontet, H. (1990). *Homogénéisation et effets de bords dans les matériaux composites*. Thèse d'état, Université Paris VI.

Dushman, S. and I. Langmuir (1922). The diffusion coefficient in solids and its temperature coefficient. *Physical Review 20*(1), 113.

Duvaut, G. and J. Lions (1972). *Les Inéquations En Mécanique et En Physique*. Dunod, Paris.

El-Sa'ad, L., M. Darby, and B. Yates (1990). Moisture absorption by epoxy resins: The reverse thermal effect. *Journal of Materials Science 25*(8), 3577–3582.

Eshelby, J. (1957). The determination of an elastic field of an ellipsoïdal inclusion and related problem. *Proceedings of The Royal Society of London, A 24/1226*, 376–396.

Feyel, F. (2003). A multilevel finite element method (FE2) to describe the response of highly non-linear structures using generalized continua. *Computer Methods in Applied Mechanics and Engineering 192*, 3233–3244.

Fick, A. (1855a). On liquid diffusion. *Philosophical Magazine and Journal of Science 10*(63), 31–39.

Fick, A. (1855b). Ueber diffusion. *Annalen der Physik und Chemie 170*(1), 59–86.

Fourier, J. (1822). *Théorie Analytique de la chaleur*. Firmin Didot, Paris.

Fuwa, M., A. Bunsell, and B. Harris (1976). An evaluation of AE techniques applied to carbon fibre composites. *Journal of physics D: Applied Physics 9*, 353–364.

Gorbatikh, L. and M. Kachanov (2000). A simple technique for constructing the full stress and displacement fields in elastic plates with multiple cracks. *Engineering Fracture Mechanics 66*, 51–63.

Goree, J. and R. Gross (1980). Stresses in a three-dimensional unidirectional composite containing broken fibers. *Engineering Fracture Mechanics 13*, 395–405.

Graham, T. (1850). On the diffusion of liquids. *Philosophical Magazine and Journal of Science 37*(251), 341–349.

GUM (2010). *JCGM 100: Evaluation of Measurement Data - Guide to the Expression of Uncertainty in Measurement*.

Hardy, B. L., M.-H. Moncel, C. Kerfant, M. Lebon, L. Bellot-Gurlet, and N. Mélard (2020, April). Direct evidence of Neanderthal fibre technology and its cognitive and behavioral implications. *Scientific Reports 10*(1), 4889.

Harlow, D. and S. Phoenix (1978a). The chain-of-bundles probability model for the strength of fibrous materials 1: Analysis and conjectures. *Journal of Composite Materials 12*, 195–213.

Harlow, D. and S. Phoenix (1978b). The chain-of-bundles probability model for the strength of fibrous materials 2: A numerical study of convergence. *Journal of Composite Materials 12*, 314–334.

Hashin, Z. (1985). Analysis of cracked laminates: a variational approach. *Mechanics of Materials 4*, 121–136.

Hearle, J. and J. Sparrow (1971, September). The fractography of cotton fibers. *Textile Research Journal 41*(9), 736–749.

Hearle, J. W. S., B. Lomas, and W. D. Cooke (1998, August). *Atlas of Fibre Fracture and Damage to Textiles, Second Edition* (Subsequent ed.). Boca Raton: CRC Press.

Hedgepeth, J. (1961). *Stress Concentrations in Filamentary Structures*. Rapport, NASA TND882, Langley research center.

Hedgepeth, J. and P. V. Dyke (1967). Local stress concentrations in imperfect filamentary composite materials. *Journal of Composite Materials 1*, 294–309.

Herrera Ramirez, J. M. and A. R. Bunsell (2005, March). Fracture initiation revealed by variations in the fatigue fracture morphologies of PA 66 and PET fibers. *Journal of Materials Science 40*(5), 1269–1272.

Hill, R. (1948). A theory of the yielding and plastic flaw of anisotropic metals. *Proceedings of the Royal Society (A) 193*, 281–297.

Hindersmann, A. (2019, November). Confusion about infusion: An overview of infusion processes. *Composites Part A: Applied Science and Manufacturing 126*, 105583.

Hoenig, A. (1979). Elastic moduli of non-randomly cracked body. *International Journal of Solids and Structures 15*, 137–154.

Hoffman, O. (1967). The brittle strength of orthotropic materials. *Journal of Composite Materials 1*, 200–206.

Humeau, C., P. Davies, P.-Y. LeGac, and F. Jacquemin (2018, December). Influence of water on the short and long term mechanical behaviour of polyamide 6 (nylon) fibres and yarns. *Multiscale and Multidisciplinary Modeling, Experiments and Design 1*(4), 317–327.

Islam, F. (2020). *Probabilistic Single Fibre Characterisation to Improve Stochastic Strength Modelling of Unidirectional Composites*. Thèse, Ecole des Mines de Paris.

Islam, F., S. Joannès, S. Bucknell, Y. Leray, A. R. Bunsell, and L. Laiarinandrasana (2020). Investigation of tensile strength and dimensional variation of T700 carbon fibres using an improved experimental setup. *Journal of Reinforced Plastics and Composites 39*(3-4).

Islam, F., S. Joannès, and L. Laiarinandrasana (2019, September). Evaluation of Critical Parameters in Tensile Strength Measurement of Single Fibres. *Journal of Composites Science 3*(3).

JEC (Ed.) (2017). *Overview of the Global Composites Market - at the Crossroads*. JEC Group.

Joannès, S., L. Mazé, and A. Bunsell (2014a). A concentration-dependent diffusion coefficient model for water sorption in composite. *Composite Structures 108*, 111–118.

Joannès, S., L. Mazé, and A. Bunsell (2014b). A simple method for modeling the concentration-dependent water sorption in reinforced polymeric materials. *Composites Part B: Engineering 57*, 219–227.

Joannès, S. and F. Islam (2019). Uncertainty in Fibre Strength Characterisation Due to Uncertainty in Measurement and Sampling Randomness. *27*(3), 165–184.

Jones, F. R. and N. T. Huff (2018). The structure and properties of glass fibers. In A. R. Bunsell (Ed.), *Handbook of Properties of Textile and Technical Fibres (Second Edition)*, The Textile Institute Book Series, pp. 757–803. Elsevier-Woodhead Publishing.

Kachanov, L. M. (1958). Mechanics of crack-microcrack interactions. *Isv. Akad. Nauk SSR Otd. Tekh. Nauk 8*, 26–31.

Kanatani, K. (1984). Distribution of directional data and fabric tensors. *International Journal of Engineering Science 22*, 149–164.

Karad, S., F. Jones, and D. Attwood (2002). Moisture absorption by cyanate ester modified epoxy resin matrices. Part II. The reverse thermal effect. *Polymer 43*(21), 5643––5649.

Kim, J., C. Kim, and D. Song (2003). Strength evaluation and failure analysis of unidirectional composites using monte carlo simulation. *Materials Science and Engineering A 340*, 33–40.

Ko, F. K. and L. Y. Wan (2018). Engineering properties of spider silk. In A. R. Bunsell (Ed.), *Handbook of Properties of Textile and Technical Fibres (Second Edition)*, The Textile Institute Book Series, pp. 185–220. Elsevier-Woodhead Publishing.

Kong, P. (1979). A Monte carlo study of the strength of unidirectional fiber-reinforced composites. *Journal of Composite Materials 13*, 311–327.

Krajcinovic, D. and G. Fonseka (1981). The continuous damage theory of brittle materials, part I: General theory. *Journal of Applied Mechanics 48*, 809–815.

Ku, H., H. Wang, N. Pattarachaiyakoop, and M. Trada (2011, June). A review on the tensile properties of natural fiber reinforced polymer composites. *Composites Part B: Engineering 42*(4), 856–873.

Ladeveze, P. and E. LeDantec (1992). Damage modeling of the elementary ply of laminated composites. *Composites Sciences and Technology 43*, 257–267.

Lagoudas, D., C. Hui, and S. Phoenix (1989). Time evolution of overstress profiles near broken fibers in a composite with a viscoelastic matrix. *International Journal of Solids and Structures 25*, 45–66.

Laiarinandrasana, L., E. Gaudichet, S. Oberti, and C. Devilliers (2011). Effects of aging on the creep behaviour and residual lifetime assessment of polyvinyl chloride (PVC) pipes. *International Journal of Pressure Vessels and Piping 88*(2–3), 99–108.

Landis, C., I. Beyerlein, and R. McMeeking (2000). Micromechanical simulation of the failure of fiber reinforced composites. *Journal of the Mechanics and Physics of Solids 48*, 621–648.

Landis, C. and R. McMeeking (1999). Stress concentrations in composites with interface sliding, matrix stiffness and uneven fiber spacing using shear lag theory. *International Journal of Solids and Structures 36*, 4333–4361.

Le Clerc, C., B. Monasse, and A. R. Bunsell (2007, November). Influence of temperature on fracture initiation in PET and PA66 fibres under cyclic loading. *Journal of Materials Science 42*(22), 9276–9283.

Leckie, F. and E. Onat (1981). *Tensorial Nature of Damage Measuring Internal Variables*. Springer.

Lee, J., C. Harris, and D. Allen (1989). Internal state variable approach for predicting stiffness reduction in fibrous laminated composites with matrix cracks. *Journal of Composite Materials 23*, 1273–1291.

Leguillon, D. and E. Sanchez-Palencia (1982). On the behaviour of a cracked elastic body with or without friction. *Journal de Mécanique Théorique et Appliquée (1) 2*, 195–209.

Lenain, J. C. and A. R. Bunsell (1979, February). The resistance to crack growth of asbestos cement. *Journal of Materials Science 14*(2), 321–332.

Léné, F. (1984). *Contribution à l'étude des matériaux composites et de leur endommagement*. Thèse d'état, Université Paris VI.

Lifschitz, J. and A. Rotem (1970). Time-dependent longitudinal strength of unidirectional fibrous composites. *Fibre Science and Technology 3*, 1–20.

Litewka, A. (1985). Effective material constants for orthotropically damaged elastic solids. *Arch. Mech. Sto. 37*, 631–642.

Luo, X. and N. Jin (2018). Fibers made by chemical vapor deposition. In A. R. Bunsell (Ed.), *Handbook of Properties of Textile and Technical Fibres (Second Edition)*, The Textile Institute Book Series, pp. 929–991. Elsevier-Woodhead Publishing.

Madsen, B., P. Hoffmeyer, A. B. Thomsen, and H. Lilholt (2007, October). Hemp yarn reinforced composites – I. Yarn characteristics. *Composites Part A: Applied Science and Manufacturing 38*(10), 2194–2203.

Mahesh, S. and S. Phoenix (2004). Lifetime distributions for unidirectional fibrous composites under creep-rupture loading. *International Journal of Fracture 127*, 303–360.

Manian, A. P., T. Pham, and T. Bechtold (2018). Regenerated cellulosic fibers. In A. R. Bunsell (Ed.), *Handbook of Properties of Textile and Technical Fibres (Second Edition)*, The Textile Institute Book Series, pp. 329–343. Elsevier-Woodhead Publishing.

Matano, C. (1933). On the relation between the diffusion-coefficients and concentrations of solid metals (the nickel-copper system. *Japanese Journal of Physics 8*(3), 109–113.

Matzenmiller, A., J. Lubliner, and R. Taylor (1995). A constitutive model for anisotropic damage in fiber-composites. *Mech. Mater. 20*, 125–152.

Mauge, C. and M. Kachanov (1994). Effective elastic properties of an anisotropic material with arbitrarily oriented interacting cracks. *Journal of the Mechanics and Physics of Solids 42*, 561–584.

Mazé, L. (2012). *Etude et modélisation du vieillissement hygrothermique de polyphthalamides renforcés fibres de verre courtes*. Ph. D. thesis, Mines-ParisTech, France.

McKague, Jr., E., J. Reynolds, and J. Halkias (1976). Moisture diffusion in fiber reinforced plastics. *Journal of Engineering Materials and Technology 98*(1), 92–95.

Menczel, J. D. and R. B. Prime (Eds.) (2009). *Thermal Analysis of Polymers: Fundamentals and Applications*. Wiley.

Militký, J., R. Mishra, and H. Jamshaid (2018). Basalt fibers. In A. R. Bunsell (Ed.), *Handbook of Properties of Textile and Technical Fibres (Second Edition)*, The Textile Institute Book Series, pp. 805–840. Elsevier-Woodhead Publishing.

Moreton, R., W. Watt, and W. Johnson (1967, February). Carbon Fibres of High Strength and High Breaking Strain. *Nature 213*(5077), 690.

Murakami, S. and N. Ohno (1981). A continuum theory of creep and creep damage. In *Creep in structures*, pp. 442–444. Springer-Verlag.

Muskhelishvili, N. (1953). *Some Basics Problems of the Mathematical Theory of Elasticity*. Noordhoff.

Nedele, M. and M. Wisnom (1994a). Stress concentration factors around a broken fibre in a unidirectional carbon fibre-reinforced epoxy. *Composites 25*, 549–557.

Nedele, M. and M. Wisnom (1994b). Three dimensional finite analysis of the stress concentration at a single fibre break. *Composites Science and Technology 51*, 517–524.

Ochiai, S., K. Schulte, and P. Peters (1991). Strain concentration for fibers and matrix in unidirectional composites. *Composites Science and Technology 41*, 237–256.

Okabe, T., M. Nishikawa, and N. Takeda (2008). Numerical modelling of progressive damage in fiber reinforced plastic cross-ply laminates. *Composites Sciences and Technology 68*, 2282–2289.

Osorio, L., E. Trujillo, F. Lens, J. Ivens, I. Verpoest, and A. V. Vuure (2018, June). In-depth study of the microstructure of bamboo fibres and their relation to the mechanical properties. *Journal of Reinforced Plastics and Composites 37*(17), 1099–1113.

Pegoretti, A. and M. Traina (2018). Liquid crystalline organic fibers and their mechanical behavior. In A. R. Bunsell (Ed.), *Handbook of Properties of Textile and Technical Fibres (Second Edition)*, The Textile Institute Book Series, pp. 621–697. Elsevier-Woodhead Publishing.

Philip, J. (1960). General method of exact solution of the concentration-dependent diffusion equation. *Australian Journal of Physics 13*(1), 1–12.

Phoenix, S. (1997). Statistical issues in the fracture of brittle matrix fibrous composites: Localized load-sharing and associated size effects. *International Journal of Solids and Structures 34*, 2649–2668.

Phoenix, S. and I. Beyerlein (2000). Statistical strength theory for fibrous composite materials. In A. Kelly and C. Zweben (Eds.), *Comprehensive Composite Materials. Pergamon-Elsevier Science*, 559–639.

Phoenix, S. and W. Newman (2009). Time-dependent fiber bundles with local load sharing. II. General Weibull fibers. *Physical Review 80*, 066115.

Purslow, D. (1981). Some fundamental aspects of composites fractography. *Composites 12*, 241–247.

Purslow, D. (1986). Matrix fractography of fibre-reinforced epoxy composites. *Composites 17*, 289–303.

Ramesh, M. (2018). Hemp, jute, banana, kenaf, ramie, sisal fibers. In A. R. Bunsell (Ed.), *Handbook of Properties of Textile and Technical Fibres (Second Edition)*, The Textile Institute Book Series, pp. 301–325. Elsevier-Woodhead Publishing.

Revest, N. (2011). *Comportement en fatigue des pièces épaisses en matériaux composites.* Thèse, Ecole des Mines de Paris.

Rojek, J. (2020). *Effect of voids in thick-walled composite pressure vessels: Experimental observations and numerical modelling.* Thèse, Ecole des Mines de Paris.

Rojek, J., C. Breite, Y. Swolfs, and L. Laiarinandrasana (2020, March). Void growth measurement and modelling in a thermosetting epoxy resin using SEM and tomography techniques. *Continuum Mechanics and Thermodynamics 32*(2), 471–488.

Rosen, B. (1964). Tensile failure of fibrous composites. *AIAA journal 2*, 1985–1991.

Rubinstein, M. and R. H. Colby (2003, June). *Polymer Physics.* Oxford, New York: Oxford University Press.

Sanchez-Hubert, J. and E. Sanchez-Palencia (1992). *Introduction aux méthodes asymptotiques et à l'homogénéisation.* Masson.

Scop, P. and A. Argon (1967). Statistical theory of strength of laminated composites. *Journal of Composite Materials 1*, 92–99.

Scop, P. and A. Argon (1969). Statistical theory of strength of laminated composites 2. *Journal of composite materials 3*, 30–44.

Scott, A., I. Sinclair, S. Spearing, A. Thionnet, and A. Bunsell (2012). Damage accumulation in a carbon/epoxy composite: Comparison between a multiscale model and computed tomography experimental results. *Composites Part A 43*, 1514–1522.

Selles, N., A. King, H. Proudhon, N. Saintier, and L. Laiarinandrasana (2018, November). Time dependent voiding mechanisms in polyamide 6 submitted to high stress triaxiality : experimental characterisation and finite element modelling. *Mechanics of Time-Dependent Materials 22*(3), 351–371.

Shen, C. and G. Springer (1976). Moisture absorption and desorption of composites materials. *Journal of Composite Materials 10*(1), 2–20.

Souza, F., D. Allen, and Y. Kim (2008). Multiscale model for predicting damage evolution in composites due to impact loading. *Composites Sciences and Technology 68*, 2624–2634.

Stenlund, H. (2011). Three methods for solution of concentration dependent diffusion coefficient. Technical report, Visilab Signal Technologies Oy. http://www.visilab.fi/bolmatano.pdf.

Summerscales, J., N. P. J. Dissanayake, A. S. Virk, and W. Hall (2010, October). A review of bast fibres and their composites. Part 1 – Fibres as reinforcements. *Composites Part A: Applied Science and Manufacturing 41*(10), 1329–1335.

Suquet, P. (1982). *Plasticité et homogénéisation.* Thèse d'état, Université Paris VI.

Talreja, R. (1985). *Fatigue of Composite Materials.* Technical University of Denmark.

Talreja, R. (1991). Continuum modelling of damage in ceramic matrix composites. *Mechanics of Materials 12*, 165–180.

Taya, M. and T. Chou (1981). On two kinds of ellipsoïdal inhomogeneities in a infinite elastic body: An application to a hybrid composite. *International Journal of Solids and Structures 17*, 553–563.

Thamae, T. and C. Baillie (2007, September). Influence of fibre extraction method, alkali and silane treatment on the interface of Agave americana waste HDPE composites as possible roof ceilings in Lesotho. *Composite Interfaces 14*, 821–836.

Thionnet, A. (1991). *Prévision d'endommagement sous chargement quasi-statiques et cycliques des structures composites stratifiées*. Thése de doctorat de l'Université Paris VI.

Thionnet, A. (2001). A model for the recovery of thermomechanical properties in strongly anisotropic damaged materials. *Journal of Composite Materials 35*, 731–750.

Thionnet, A. (2010). From fracture to damage mechanics: a behavior law for microcracked composites using the concept of crack opening mode. *Composites Structures 92*, 780–794.

Thionnet, A., L. Chambon, and J. Renard (2002). A theoretical and experimental study to point out the notion of loading mode in damage mechanics: Application to the identification and validation of a fatigue damage modelling for laminates composites. *International Journal of Solids and Structures 24*, 147–154.

Thionnet, A. and C. Martin (2006). A new constructive method using the theory of Invariants to obtain material behavior laws. *International Journal of Solids and Structures 43/2*, 325–345.

Thionnet, A., C. Martin, and S. Barradas (2003). *Mécanique et Comportements Des Milieux Continus, Tome 2 : Applications et Théorie Des Invariants*. Editions Ellipses, ISBN 2-7298-1807-3.

Thionnet, A. and J. Renard (1993). Meso-macro approach to transverse cracking in laminated composites using Talreja's model. *Composites Engineering 3*, 851–871.

Thionnet, A. and J. Renard (1998). Multi-scale analysis to determine fibre/matrix debonding criteria in SiC/Titanium composites with and without consideration of the manufacturing residual stresses. *Composites Science and Technology 58*, 945–955.

Thionnet, A. and J. Renard (1999). Modelling unilateral damage effect in strongly anisotropic materials by the introduction of loading mode in Damage Mechanics. *International Journal of Solids and Structures 36*, 4269–4287.

Thomas, J. (1995). *Numerical Partial Differential Equations: Finite Difference Methods*. Springer Verlag.

Thomas, L. (1949). Elliptic problems in linear difference equations over a network. Technical report, Watson Scientific Computing Laboratory Report, Columbia University, New York.

Trabelsi, W. (2013). *Approches multiéchelles d'expérimentation et de modélisation pour prédire la rupture d'un composite textile. Critère de classement des architectures tissées*. Thèse, Ecole des Mines de Paris.

Tsai, S. (1980). *Introduction to Composite Materials*. Technomic Publishing Company.

Tsai, S. and E. Wu (1971). A general theory of strength for anisotropic materials. *Journal of Composite Materials 5*, 58–80.

Vakulenko, A. and M. Kachanov (1971). Continuum theory of medium with cracks. *Mechanics of Solids 6*, 145–151.

Van Den Heuvel, P., S. Goutianos, R. Young, and T. Peijs (2004). Failure phenomena in fibre-reinforced composites Part 6: A finite element study of stress concentrations in unidirectional cfr epoxy composites. *Composites Science and Technology 64*, 645–656.

Van Den Heuvel, P., M. Wubbolts, R. Young, and T. Peijs (1998). Failure phenomena in two-dimensional multi-fibre model composites: 5. A finite element study. *Composites A 29*, 1121–1135.

van Oosterom, S., T. Allen, M. Battley, and S. Bickerton (2019, October). An objective comparison of common vacuum assisted resin infusion processes. *Composites Part A: Applied Science and Manufacturing 125*, 105528.

van Rijswijk, K. and H. E. N. Bersee (2007, March). Reactive processing of textile fiber-reinforced thermoplastic composites – An overview. *Composites Part A: Applied Science and Manufacturing 38*(3), 666–681.

van't Hoff, J. (1884). *Etudes de dynamique chimique*. Frederik Muller, Amsterdam.

Veilleux, J. and S. Coulombe (2010). Mass diffusion coefficient measurements at the microscale: Imaging a transient concentration profile using TIRF microscopy. *International Journal of Heat and Mass Transfer 53*(23–24), 5321–5329.

Vlasblom, M. (2018). The manufacture, properties, and applications of high-strength, high-modulus polyethylene fibers. In A. R. Bunsell (Ed.), *Handbook of Properties of Textile and Technical Fibres (Second Edition)*, The Textile Institute Book Series, pp. 699–755. Elsevier-Woodhead Publishing.

Voyiadjis, G. and P. Kattan (2007a). Evolution of fabric tensors in damage mechanics of solids with micro-cracks: Part 1 - theory and fundamental concepts. *Mechanics Research Communications 34*, 145–154.

Voyiadjis, G. and P. Kattan (2007b). Evolution of fabric tensors in damage mechanics of solids with micro-cracks: Part 2 - evolution of length and orientation of micro-cracks with an application to uniaxial tension. *Mechanics Research Communications 34*, 155–163.

Weibull, W. (1951). A Statistical Distribution Function of Wide Applicability. *Journal of Applied Mechanics 13*, 293–297.

Wisnom, M. and D. Green (1995). Tensile failure due to interaction between fibre breaks. *Composites 26*, 499–508.

Wollbrett-Blitz, J., S. Joannès, R. Bruant, C. Le Clerc, M. Romero De La Osa, A. R. Bunsell, and A. Marcellan (2016, February). Multiaxial mechanical behavior of aramid fibers and identification of skin/core structure from single fiber transverse compression testing. *Journal of Polymer Science Part B: Polymer Physics 54*(3), 374–384.

Yajima, S., J. Hayashi, M. Omori, and K. Okamura (1976, June). Development of a silicon carbide fibre with high tensile strength. *Nature 261*(5562), 683.

Zhang, L., X. Wang, J. Pei, and Y. Zhou (2020, June). Review of automated fibre placement and its prospects for advanced composites. *Journal of Materials Science 55*(17), 7121–7155.

Zweben, C. (1968). Tensile failure of fibers composites. *AIAA Journal 6*, 2325–2331.

# Index

Printed in the United States
By Bookmasters